Acrylamide in Food
ANALYSIS, CONTENT AND POTENTIAL HEALTH EFFECTS

Acrylamide in Food

ANALYSIS, CONTENT AND POTENTIAL HEALTH EFFECTS

VURAL GÖKMEN

Department of Food Engineering,
Hacettepe University, Ankara, Turkey

ELSEVIER

Amsterdam • Boston • Heidelberg • London
New York • Oxford • Paris • San Diego
San Francisco • Singapore • Sydney • Tokyo
Academic Press is an imprint of Elsevier

Academic Press is an imprint of Elsevier
125 London Wall, London EC2Y 5AS, UK
525 B Street, Suite 1800, San Diego, CA 92101-4495, USA
225 Wyman Street, Waltham, MA 02451, USA
The Boulevard, Langford Lane, Kidlington, Oxford OX5 1GB, UK

ISBN: 978-0-12-802832-2

British Library Cataloguing-in-Publication Data
A catalogue record for this book is available from the British Library

Library of Congress Cataloging-in-Publication Data
A catalog record for this book is available from the Library of Congress

For information on all Academic Press publications visit our website at http://store.elsevier.com/

Acquisition Editor: Nina Bandeira
Editorial Project Manager: Mariana Kuhl Leme
Production Project Managers: Karen East and Kirsty Halterman
Designer: Maria Inês Cruz

Typeset by TNQ Books and Journals
www.tnq.co.in

CONTENTS

LIST OF CONTRIBUTORS

Monica Anese
Department of Food Science, University of Udine, Udine, Italy

Bhawna Batra
Department of Biochemistry, M.D. University, Rohtak, Haryana, India

Edoardo Capuano
Wageningen University and Research Centre, Food Quality and Design Group, The Netherlands

Reinhold Carle
Hohenheim University, Institute of Food Science and Biotechnology, Chair Plant Foodstuff Technology and Analysis, Stuttgart, Germany; Biological Science Department, King Abdulaziz University, Jeddah, Saudi Arabia

Francisco J. Casado
Hohenheim University, Life Science Center, Stuttgart, Germany

Ayşegül ÇEBİ
Faculty of Health Sciences, Giresun University, Piraziz/Giresun, Turkey

Fang Chen
College of Food Science and Nutritional Engineering, National Engineering Research Centre for Fruits and Vegetables Processing, Key Laboratory of Fruits and Vegetables Processing, Ministry of Agriculture, China Agricultural University, Beijing, China

Zuzana Ciesarová
NPPC National Agricultural and Food Centre, Food Research Institute, Bratislava, Slovak Republic

Colin Crews
Fera Science Ltd., Sand Hutton, North Yorkshire, UK

Rosa M. Delgado
Instituto de la Grasa, Consejo Superior de Investigaciones Científicas, Seville, Spain

Piotr Dobrowolski
Department of Comparative Anatomy and Anthropology, Maria Curie-Sklodowska University, Lublin, Poland

Vincenzo Fogliano
Food Quality and Design group, Wageningen University & Research Centre, Wageningen, The Netherlands

Heinz Freisling
Dietary Exposure Assessment Group, International Agency for Research on Cancer (IARC-WHO), Lyon, France

Vural Gökmen
Department of Food Engineering, Hacettepe University, Beytepe, Ankara, Turkey

Eduardo J. Guerra-Hernández
Departamento de Nutrición y Bromatología, Universidad de Granada, Granada, Spain

Aytül Hamzalıoğlu
Department of Food Engineering, Hacettepe University, Beytepe, Ankara, Turkey

Francisco J. Hidalgo
Instituto de la Grasa, Consejo Superior de Investigaciones Científicas, Seville, Spain

Monika Hułas-Stasiak
Department of Comparative Anatomy and Anthropology, Maria Curie-Sklodowska University, Lublin, Poland

Sastry S. Jayanty
Department of Horticulture and Landscape Architecture, San Luis Valley Research, Colorado State University, CO, USA

Cheng Jin
Department of Food Science and Nutrition, College of Biosystems Engineering and Food Science, Zhejiang University Center, Hangzhou, PR China

Diganta Kalita
Department of Horticulture and Landscape Architecture, San Luis Valley Research, Colorado State University Center, CO, USA

Tolgahan Kocadağlı
Department of Food Engineering, Hacettepe University, Ankara, Turkey

Kristína Kukurová
NPPC VUP National Agricultural and Food Centre, Food Research Institute Bratislava, Priemyselná, Bratislava, Slovak Republic

Daotong Li
College of Food Science and Nutritional Engineering, National Engineering Research Centre for Fruits and Vegetables Processing, Key Laboratory of Fruits and Vegetables Processing, Ministry of Agriculture, China Agricultural University, Beijing, China

Jie Liu
College of Grain, Oil and Food, Henan University of Technology, Zhengzhou, China

Pieternel Luning
Department of Agrotechnology and Food Sciences, Wageningen University, Wageningen, The Netherlands

Marta Mesias
Institute of Food Science, Technology and Nutrition (ICTAN-CSIC), Madrid, Spain

Burçe Ataç Mogol
Department of Food Engineering, Hacettepe University, Beytepe, Ankara, Turkey

Hanna Mojska
Department of Food and Food Supplements, National Food and Nutrition Institute, Warsaw, Poland

Alfredo Montaño
Food Biotechnology Department, Instituto de la Grasa (CSIC), Seville, Spain

Francisco J. Morales
Institute of Food Science, Technology and Nutrition (ICTAN-CSIC), Madrid, Spain

Chandra S. Pundir
Department of Biochemistry, M.D. University, Rohtak, Haryana, India

Maimunah Sanny
Faculty of Food Science and Technology, Putra University, Serdang, Malaysia

Nadia Slimani
Dietary Exposure Assessment Group, International Agency for Research on Cancer (IARC-WHO), Lyon, France

Richard H. Stadler
Nestec Ltd, Corporate Quality Management, Vevey, Switzerland

Alfred Studer
Nestlé Research Centre, Vers-chez-les-Blanc, Switzerland

Yalçın Tepe
Biology Department, Faculty of Arts and Science, Giresun University, Giresun, Turkey

Ewa Tomaszewska
Department of Animal Biochemistry and Physiology, Faculty of Veterinary Medicine, University of Life Sciences, Lublin, Poland

Antonio D. Troise
Department of Agriculture and Food Science, University of Naples "Federico II", Naples, Italy, Food Quality and Design group, Wageningen University & Research Centre, Wageningen, The Netherlands

Pengpu Wang
College of Food Science and Nutritional Engineering, National Engineering Research Centre for Fruits and Vegetables Processing, Key Laboratory of Fruits and Vegetables Processing, Ministry of Agriculture, China Agricultural University, Beijing, China

Rosario Zamora
Instituto de la Grasa, Consejo Superior de Investigaciones Científicas, Seville, Spain

Ying Zhang
Department of Food Science and Nutrition, College of Biosystems Engineering and Food Science, Zhejiang University, Hangzhou, PR China; Zhejiang Key Laboratory for Agro-Food Processing, Zhejiang University, Hangzhou, PR China

Slađana Žilić
Department of Food Technology and Biochemistry, Maize Research Institute, Belgrade, Serbia

PREFACE

Human ancestors discovered the fire and the controlled use of it 1 million years ago. The fire gave the human the ability to get warmth and cook their foods. "How lucky that Earth has fire" says Richard Wrangham in his book named "Catching Fire: How Cooking Made Us Human." Cooking of foods, domestically or industrially, is an ancient practice that improves the organoleptic properties of the foods, nutrient bioavailability and microbiological safety. Today, cooking practices like baking, frying, and roasting are indispensable tools in human's daily life. Heat induces chemical change, and a plethora of new molecules are generated in foods, some of which have been claimed to impart positive health effects. On the other hand, certain compounds may also be formed that exhibit carcinogenic and in some cases mutagenic properties.

Over the past decades, researchers have gained increasing insight in understanding the presence, formation, and potential risk to public health posed by the compounds formed during domestic cooking and heat processing of different foods. The recent discovery of acrylamide in cooked carbohydrate-rich foods by Swedish researchers has raised much concern, and sparked intensive scientific studies into the occurrence, analysis, exposure, mechanisms of formation, possible measures of control, and toxicology of the compound.

Basically, acrylamide is formed when food is heated to high temperatures and is thus a naturally occurring compound. The foods rich in carbohydrate and asparagine have the highest potential to form acrylamide. It is found in all foods subjected to heating such as crisps, French fries, bread, bakery wares, breakfast cereals, coffee, chocolate, infant formula, and battered products. It is also found in nonheated foods such as dried prunes and black olives. This fact has shaken consumer confidence in the safety of these foods and highlighted that small–medium enterprises of the agrofood industry are poorly prepared to face such a problem. Reducing acrylamide levels in heated foods is still one of the major targets for agrofood industry. FoodDrinkEurope, in close cooperation with the national authorities and the European Commission, has developed the Acrylamide Toolbox that helps food manufacturers to identify the best ways to reduce acrylamide in their products taking into account the latest scientific and technological developments. In addition, the official journal of the European Union published new recommendations in 2013 regarding acrylamide levels in food, setting up new indicative values in the range of 100–1000 μg/kg for a wider range of food products.

Presence of acrylamide in heated foods has been considered an important food-related crisis by the authorities. The World Health Organization has declared that acrylamide represents a potential health risk as evidence shows that in some circumstances it can cause cancer or tissue damage. The caveat to this is the uncertainly at which levels it can induce pathology, the precise mechanism of its formation and the steps needed to reduce its concentration. Nevertheless, due to increasing awareness there has been in the past few years an increase in the published material on the formation and presence of acrylamide in different foods, the total amounts consumed in the diet, and its potential risk. This material has been somewhat sporadic, fragmentary, and hitherto unfocused.

There is thus a need to have a single source of material on acrylamide that enables those in one discipline to become familiar with the concepts and applications in other disciplines of food science. This is addressed in **Acrylamide in Food: Analysis, Content and Potential Health Effects**. There are reasons why this book is timely; the latest research findings on the mechanism of formation, occurrence in foods, dietary intake, risk assessment, effects on tissues and cancer, potential interactions and ways of reductions, and methods of analysis are all covered. Accordingly, this book should find importance as a textbook for students, both in graduate schools and at advanced undergraduate levels. It should also provide current researchers with jumping-off points to further their own work.

Vural Gökmen

INTRODUCTION: POTENTIAL SAFETY RISKS ASSOCIATED WITH THERMAL PROCESSING OF FOODS

Vural Gökmen
Department of Food Engineering, Hacettepe University, Ankara, Turkey

RISKS ASSOCIATED WITH THERMAL PROCESSING

Processing of foods is essential to improve microbiological safety, to increase nutritional quality, and to reduce the levels of potentially toxic compounds [1]. The thermal processing of foods, domestically or industrially, is an ancient practice that improves the organoleptic properties of the foods, their preservation, and microbiological safety [2].

Thermal processing is an important treatment for food preservation, especially in the manufacture of shelf-stable foods with specific nutritional properties. In addition, it is indispensable for determining the sensory properties, in particular color, texture, and flavor in fried, baked, and roasted products (Figure 1).

Heat induces chemical change and a plethora of new molecules are generated in foods, some of which have been claimed to impart positive health effects, for example, acting as health-promoting antioxidants and antimutagens [2]. On the other hand, it may lead to the formation of heat-induced toxic compounds, so-called **thermal process contaminants**, that exhibit carcinogenic and in some cases mutagenic properties, such as heterocyclic amines, acrylamide, furan, furfurals, and chloropropanols [3–6]. In particular, the heterocyclic aromatic amines and polycyclic aromatic hydrocarbons are classes of compounds that are formed under severe thermal conditions, and sources of the latter may also be occupational environment, tobacco smoke, and air pollution.

Thermally processed foods like bakery products, roasted coffee, snack foods, baby purees, and refined oils may contain above-mentioned compounds at varying amounts as a result of heat-induced conversions of certain compounds naturally available in food products (Figure 2). The final concentrations of thermal-process contaminants in foods depend on the concentration of their corresponding precursors as well as the severity of thermal-processing conditions.

CHEMISTRY OF THE MAILLARD REACTION

The chemical reactions that lead to the formation of thermal-process contaminants in foods are diverse. They mainly involve lipids, carbohydrates, and proteins with or without free amino acids as precursors. The formation of reactive intermediates from especially

Figure 1 Commonly consumed foods that are exposed to extreme temperatures during processing.

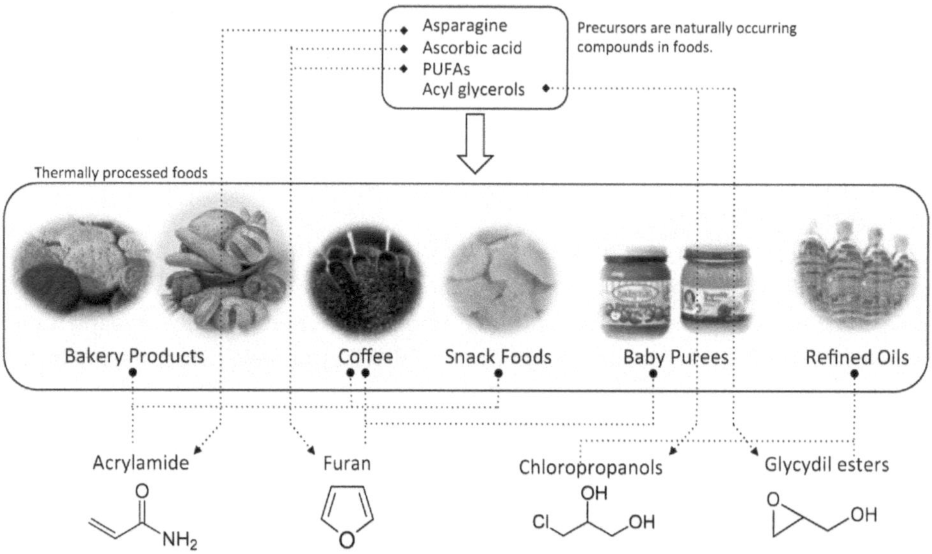

Figure 2 Examples of some naturally occurring compounds in foods leading to certain potentially harmful substances during thermal processing.

carbohydrates and lipids may play an important role. The reaction pathways are interrelated and strongly dependent on the size and composition of foods and processing conditions [1].

Among the reactions occurring in food during heating, the reaction between reducing carbohydrates and free amino groups of proteins or amino acids is the major source of neo-formed compounds. This reaction, so-called the Maillard reaction, causes the development of brown color in foods. The Maillard reaction gives unique color and flavor to food products. However, there are also certain undesired consequences of the Maillard reaction such as (1) the loss of nutrients like lysine, (2) generation of toxic compounds like acrylamide, and (3) formation of advanced glycation end products (AGEs) that are responsible for the development of certain age-related diseases. The Maillard reaction has three stages in which typical products are formed (Figure 3). Some of them like *N*-ε-fructosyllysine (FL) and carboxymethyllysine (CML) have been used to evaluate the degree of protein glycation in heated foods that indicate the loss of nutritive value [7–9]. Some other Maillard reaction products such as acrylamide and furfurals indicate the safety risks created in foods during thermal processing due to the potential adverse effects of these compounds on human health [1,10–12].

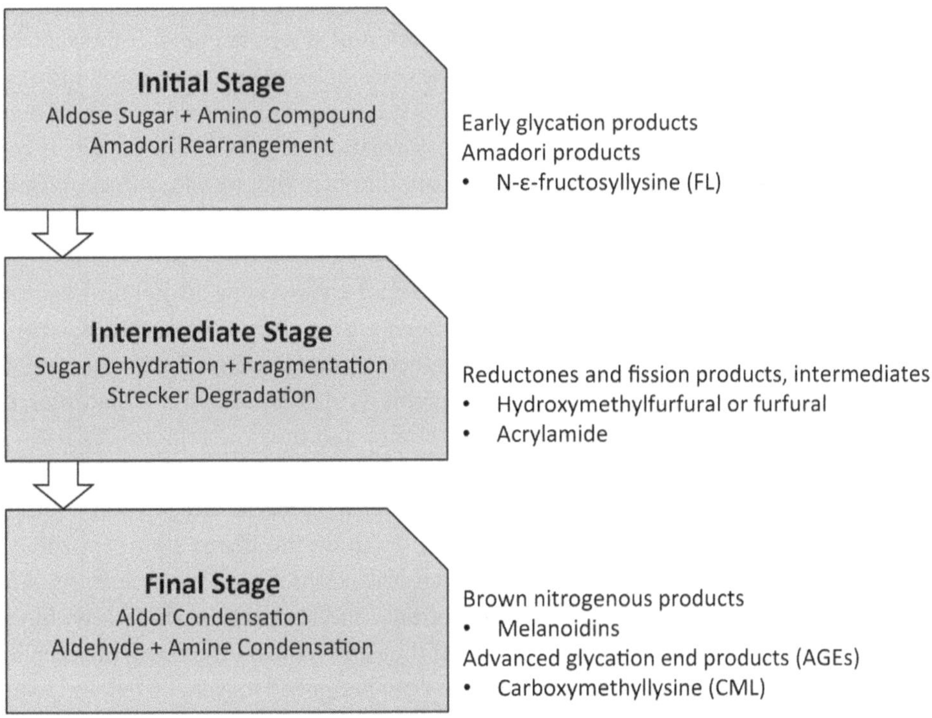

Figure 3 Typical products formed during the stages of Maillard reaction.

ACRYLAMIDE IN HEATED FOODS

In April 2002, a report by the Swedish Authorities on the presence of acrylamide (2-propenamide) in a wide range of fried and oven-cooked foods [4] attracted world-wide attention, due to the fact that acrylamide is classified as probably carcinogenic to humans (Group 2A) by the IARC [10]. Acrylamide is formed in food as a result of cooking practices, many of which have been used by mankind for many centuries. Initial surveys have shown that relatively high concentrations of acrylamide are found in high-carbohydrate foodstuffs such as potato chips, French fries, pan-fried potato products, and crisp bread [2]. The main pathway of acrylamide formation in foods is linked to the Maillard reaction and, in particular, the amino acid asparagine [13–15].

Presence of acrylamide in common heated foods is considered as an important food safety problem by international authorities. Acrylamide exposure varies depending upon the population's eating habits, and the way the foods are processed and prepared. Generally, fried potato products, bakery products, and roasted coffee appear as the most important food categories that contribute most to acrylamide exposure. The levels of acrylamide in heat-processed foods depend on certain recipe factors and thermal-processing conditions. The initial concentrations of precursors and processing temperatures are known to affect the rate of acrylamide formation in foods [16–20].

During baking or frying of foods, the evaporation of water, being an important barrier to internal energy increase, limits the amounts of acrylamide formed. However, greater energy input to the food at higher temperatures causes faster drying, and hence the temperature and moisture levels favor the formation of acrylamide to set in early during the process [19]. Time–temperature history differs in different locations of food as a result of simultaneous heat and mass transfer. Temperature rises much faster on the surface than the interior parts of food during heating under conventional conditions [21]. This makes the surface of foods more risky part from the viewpoint of chemical reactions leading to undesired toxic compounds. It has been reported that acrylamide formation takes place mainly at the surface and in near-surface regions, because during heating, the conditions in this part of foods become favourable [19]. This causes a large difference between the acrylamide concentrations of the surface and the core regions of food.

For optimization, it seems very useful to find a relationship between time and temperature to evaluate potential risks and benefits of the thermal process. Greenwood et al. [24] have first published a technique, which they used to study the destruction of thiamin in cured pork luncheon meat compared with microbial destruction. This evaluation allows determining the only acceptable combinations of time and temperature that fall within the area of "cooked sterile," all other combinations of time and temperature being unacceptable [25]. As we recently exemplified, similar evaluation can be applied to baked, fried, and roasted foods taking into account the formation of thermal-process contaminants and development of acceptable sensory properties [22]. In such case, the appropriate combinations of time and temperature would fall within the area "non-risky acceptable" [1,23].

CONCERNS OVER THERMAL PROCESSING OF FOODS

Thermal process induces several reactions in food products that may lead to impairment of nutritional quality of food, generation of toxic compounds such as acrylamide and furan, and formation of AGEs. Dietary intake of these exogenous compounds may cause their accumulation in bloodstream and in tissue proteins, then undergoing further reactions and several consequences in the body. Recent findings indicate that in vivo Maillard reaction processes are at least in part responsible for diabetic complications and the general aging process. The AGEs may contribute to the decline in tissue and organ function with age and are related chronic and degenerative diseases, such as diabetes, renal failure [26], atherosclerosis [27], Alzheimer's disease, and Parkinson's disease [28]. However, the mechanisms of these effects are largely unknown. Therefore, there is still a clear need for in-depth research not only to understand the in vivo effects of neo-formed compounds present in heated foods, but also to understand better the mechanisms and factors affecting their formation in foods.

REFERENCES

[1] Gökmen V. A perspective on the evaluation of safety risks in thermal processing of foods with an example for acrylamide formation in biscuits. Qual Assur Saf Crops Foods 2014;6(3):319–25.
[2] Studer A, Blank I, Stadler RH. Thermal processing contaminants in foodstuffs and potential strategies of control. Czech J Food Sci 2004;22:1–10.
[3] Skog KI, Johansson MAE, Jagerstad MI. Carcinogenic heterocyclic amines in model systems and cooked foods: a review on formation, occurrence and intake. Food Chem Toxicol 1998;36:879–96.
[4] Tareke E, Rydberg P, Karlsson S, Eriksson M, Törnqvist M. Analysis of acrylamide, a carcinogen formed in heated foodstuffs. J Agric Food Chem 2002;50:4998–5006.
[5] Crews C, Brereton P, Davies A. The effects of domestic cooking on the levels of 3-monochloropropanediol in foods. Food Addit Contam 2001;19:271–80.
[6] Maga JA. Furans in foods. CRC Crit Rev Food Sci Nutr 1979;11:355–400.
[7] Reddy VP, Beyaz A. Inhibitors of the Maillard reaction and AGE breakers as therapeutics for multiple diseases. Drug Discov Today 2006;11:646–54.
[8] Henle T. Protein-bound advanced glycation endproducts (AGEs) as bioactive amino acid derivatives in foods. Amino Acids 2005;29:313–22.
[9] Hodge JE. Dehydrated foods–chemistry of browning reactions in model systems. J Agric Food Chem 1953;1:928–43.
[10] Gökmen V, Serpen A, Açar ÖÇ, Morales FJ. Significance of furosine as heat-induced marker in cookies. J Cereal Sci 2008;48:843–7.
[11] Wellner A, Huettl C, Henle T. Formation of Maillard reaction products during heat treatment of carrots. J Agric Food Chem 2011;59:7992–8.
[12] The International Agency for Research on Cancer (IARC). Acrylamide. In: IARC Monographs on the Evaluation of the Carcinogenic Risk of Chemicals to Humans, vol. 60. 1994. p. 389–433. Lyon France.
[13] European Food Safety Authority (EFSA). Scientific opinion on the re-evaluation of caramel colours (E 150 a, b, c, d) as food additives. EFSA J 9(3):103. Available at: http://www.efsa.europa.eu/en/efsajournal/doc/2004.pdf; [accessed 06.01.14].
[14] Capuano E, Fogliano V. Acrylamide and 5-hydroxymethylfurfural (HMF): a review on metabolism, toxicity, occurrence in food and mitigation strategies. LWT-Food Sci Technol 2011;44:793–810.
[15] Mottram DS, Wedzicha BI, Dodson AT. Acrylamide is formed in the Maillard reaction. Nature 2002;419:448–9.

[16] Stadler RH, Blank I, Varga N, Robert F, Hau J, Guy PA, et al. Acrylamide from Maillard reaction products. Nature 2002;419:449–50.
[17] Yaylayan VA, Wnorowski A, Locas CP. Why asparagine needs carbohydrates to generate acrylamide. J Agric Food Chem 2003;51:1753–7.
[18] Becalski A, Lau BP-Y, Lewis D, Seaman S. Acrylamide in foods: occurrence, sources and modeling. J Agric Food Chem 2003;51:802–8.
[19] Rydberg P, Eriksson S, Tareke E, Karlsson P, Ehrenberg L, Törnqvist M. Investigations of factors that influence the acrylamide content of heated foodstuffs. J Agric Food Chem 2003;51:7012–8.
[20] Matthäus B, Haase NU, Vosmann K. Factors affecting the concentration of acrylamide during deep-fat frying of potatoes. Eur J Lipid Sci Technol 2004;106:793–801.
[21] Gökmen V, Palazoğlu TK, Şenyuva HZ. Relation between the acrylamide formation and time–temperature history of surface and core regions of French fries. J Food Eng 2006;77:972–6.
[22] Pedreschi F, Moyano P, Kaack K, Granby K. Color changes and acrylamide formation in fried potato slices. Food Addit Contam 2005;38:1–9.
[23] Açar ÖÇ, Gökmen V. Investigation of acrylamide formation on bakery products using a crust-like model. Mol Nutr Food Res 2009;53:1521–5.
[24] Greenwood DA, Kraybill HR, Feaster JF, Jackson JM. Vitamin retention in processed meats. Ind Eng Chem 1944;36:922.
[25] Açar ÖÇ, Gökmen V. A new approach to evaluate the risk arising from acrylamide formation in cookies during baking: total risk calculation. J Food Eng 2010;100:642–8.
[26] Šebeková K, Somoza V. Dietary advanced glycation endproducts (AGEs) and their health effects. Mol Nutr Food Res 2007;51:1079–84.
[27] Wang Z, Jiang Y, Liu N, Ren L, Zhu Y, An Y, et al. Advanced glycation end-product *N*-ε-carboxymethyl-lysine accelerates progression of atherosclerotic calcification in diabetes. Atherosclerosis 2012;221:387–96.
[28] Li J, Liu D, Sun L, Lu Y, Zhang Z. Advanced glycation end products and neurodegenerative diseases: mechanisms and perspective. J Neurol Sci 2012;317:1–5.

CHAPTER 1

Acrylamide Formation Mechanisms

Richard H. Stadler[1], Alfred Studer[2]
[1]Nestec Ltd, Corporate Quality Management, Vevey, Switzerland; [2]Nestlé Research Centre, Vers-chez-les-Blanc, Switzerland

INTRODUCTION

Since the discovery of acrylamide in certain types of food cooked at high temperatures in April 2002 by the Swedish National Food Authority [1,2], extensive work has been done to identify the molecular precursors and to elucidate the reaction mechanism(s) leading to acrylamide. The very early studies dealing with the formation of acrylamide in food identified the Maillard reaction as a key driver of acrylamide formation [3,4]. The Maillard reaction is the reaction between naturally present amino acids and reducing sugars (e.g., glucose or fructose) when foods are heated. It is responsible for the development of the desirable flavor and color in many cooked foods subjected to baking, frying, or roasting. Acrylamide is mainly formed from the amino acid asparagine and reducing sugars (glucose and fructose) both of which occur naturally in plant materials including, for example, cereal grains, vegetables (such as potatoes), cocoa, and coffee (for in-depth reviews see Refs [5–8]).

The first research papers on the formation of acrylamide had in common that they identified asparagine as the key amino acid that furnishes the acrylamide backbone [3,4,9], although the initial proposals of the chemical pathways described by the authors were divergent, providing only little evidence of the key intermediates. Stadler et al. [3] provided evidence of the involvement of the free amino acid asparagine through stable isotope dilution assays using ^{15}N-labeled asparagine, showing that >98% of the label is incorporated into acrylamide. The same paper showed that *N*-glycosides (*N*-(D-glucos-1-yl)-1-asparagine and *N*-(D-fructos-2-yl)-L-asparagine) are effective precursors of acrylamide, yielding >1.3 mmol acrylamide per mol of *N*-glycoside. Mass spectra of the pyrolysate of asparagine and glucose confirmed the presence of the corresponding *N*-glycoside [3].

The group of Mottram [4] proposed the involvement of the Strecker reaction in the formation of acrylamide, using 2,3-butanedione as the dicarbonyl source reacting with asparagine. Strecker aldehydes are formed via the oxidative deamination and decarboxylation of α-amino acids, driven by deoxysones. Model system studies showed that carbonyls furnish acrylamide under both wet (phosphate buffer) and dry conditions. The authors also proposed acrolein/acrylic acid as potential intermediates, albeit at lower yields due to the likely limitation of ammonia in the reaction [4].

Acrylamide in Food
http://dx.doi.org/10.1016/B978-0-12-802832-2.00001-2

Zyzak et al. [9] were the first to show that 3-aminopropionamide (3-APA), a biogenic amine formed during the Maillard reaction, may be one of the key precursors of acrylamide. The researchers also showed through experiments with 2-deoxy-glucose that carbonyls, but not dicarbonyls, are essential in the Maillard route to acrylamide: that is, Amadori rearrangement products are not needed. Further studies by other researchers [10] corroborated this early work on 3-APA, first in model systems and then in actual food samples such as cheese and cocoa [11,12].

MAJOR PATHWAYS OF ACRYLAMIDE FORMATION

Experiments based on binary mixtures of short carbon chain carbonyls and hydroxycarbonyls are useful to compare the efficacy of precursors and gain insight into the possible reaction mechanisms governing the formation of acrylamide. Several reports have shown that both dicarbonyls and hydroxycarbonyls react rapidly with asparagine to release acrylamide. These earlier studies provided useful information on the impact of temperature, moisture, pH, and other reactants on the pathway to acrylamide, conducted under both wet and dry conditions.

Aldo versus Keto Sugars

Binary model systems demonstrate that about equal amounts of acrylamide are formed when asparagine is added to fructose or glucose; significantly lower amounts of acrylamide are formed when sucrose is employed as reactant. The different reactivities of carbonyls have been reported by several authors [13,14]. However, the data of the reports are difficult to compare, due to the fact that heating conditions vary. However, there seems consensus in different reports that lower amounts of acrylamide are formed in glucose mixtures versus fructose. This may be due to the lower melting point of fructose (and hence higher mobility) and consequently faster interaction of the precursors to afford the early Maillard intermediates [14]. Alternatively, keto sugars such as fructose form the fructose Schiff intermediate that stabilizes the azomethyine ylide (H bonding), see also Figure 2. This extra stabilization can increase the rate of its formation and consequently acrylamide formation versus glucose systems. Yaylayan and Stadler [15] suggest that the reactivity is reversed in aprotic solvents, supporting the hydrogen-bonding theory.

Pyruvic acid (2-oxopropionic acid) and hydroxyacetone are highly efficient reactants that afford >6 mmol acrylamide/mol reactant, whereas mixtures containing glyoxal, methylglyoxal, or propanal yield about 2 mmol/mol [16,17]. Interestingly, pentosans and cellulose may also contribute to acrylamide formation when incubated together with asparagine; mixtures with pentosans providing even more acrylamide than mixtures with comparable concentrations of glyoxal.

Table 1 Comparison of the efficacy of different carbonyl reactants in the formation of acrylamide (mmol/mol reactant)

Reactant[a]	Acrylamide (mmol/mol)
2-Hydroxy-1-butanal	15.8
Hydroxyacetone	3.97
Glucose	2.22
Methylglyoxal	0.52
Glyoxal	0.38
Diacetyl	0.26
3-Hydroxy-propanamide	0.24
Glyoxylic acid	0.08
1-Butanal	0.01

[a]Samples heated at 180 °C for 5 min.
Adapted from Refs [13,18]

As depicted in Table 1, α-dicarbonyls are effective reactants that generate acrylamide. However, compounds such as glyoxylic acid without a dicarbonyl moiety may also lead to the formation of acrylamide in binary mixtures with asparagine.

Interestingly, hydroxyacetone afforded significantly more (ca. 10-fold) acrylamide in this dry model versus short carbon chain dicarbonyls. Similarly, only relatively low conversion rates were observed in incubations with the Strecker alcohol (3-hydroxy-propamanide). Such models are useful to compare relative reactivities under given conditions, but as observed by Blank et al. [18], may be highly variable even within a given series (>20% coefficient of variation). The highest yield of acrylamide was recorded for mixtures composed of 2-hydroxy-1-butanal. This compound tautomerizes rapidly to 1-hydroxy-2-ketone, demonstrating the importance of the α-hydroxy function in the formation of acrylamide via the Maillard reaction.

Intermediacy of 3-APA

Zyzak and coworkers were the first researchers to describe the intermediacy of 3-APA in the formation of acrylamide [9]. They proposed a carbonyl compound reacting with asparagine, to furnish the Schiff base that decarboxylates under dry conditions. The decarboxylated Schiff base then forms acrylamide via 3-APA (and subsequent deamination to acrylamide) or directly to release acrylamide via an elimination reaction (Figure 2).

Stable isotope-labeling experiments serve to unequivocally demonstrate the source of the label and hence involvement of a molecule (or parts of a molecule) in a reaction pathway. Extensive work conducted by Granvogl and Schieberle [11,12] provided concrete evidence of the intermediacy of 3-APA in the reaction to acrylamide. In fact, thermal decomposition of 3-APA without the addition of carbonyls affords acrylamide in a very high

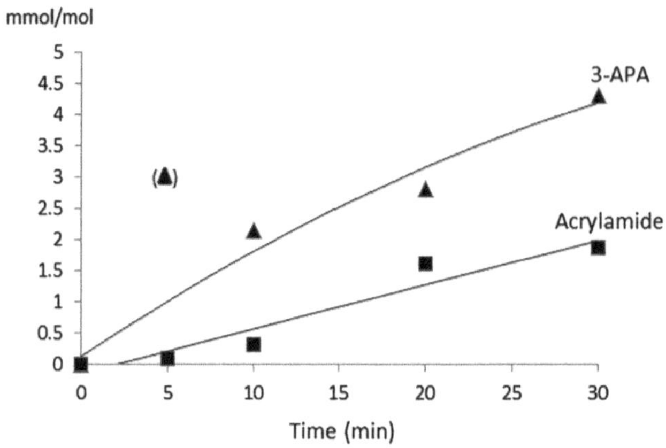

Figure 1 Time course of acrylamide and 3-APA formation (mmol/mol reactants) in a dry model system (170°C, closed vessels, reactants in silica gel containing 10% water). *Adapted from data presented by Ref. [17].*

yield, >60 mol %, under ideal reaction conditions [10]. Model experiments (dry, silica gel containing 10% water) with [$^{13}C_4{}^{15}N_2$]-asparagine mixed together with glucose and reacted over time afforded [$^{13}C_3{}^{15}N_1$]-acrylamide and [$^{13}C_3{}^{15}N_2$]-3-APA, following formation over time comparable to that of unlabeled asparagine. The amounts of 3-APA formed in their model was always lower as compared to acrylamide (ratio ca. 4:1, acrylamide:3-APA).

3-APA alone furnishes significant amounts of acrylamide via deamination. Incubation of 3-APA in a model system together with glucose in an aqueous environment (phosphate buffer 0.1 mol/L, pH 7, 170 °C) yielded 0.3 mol acrylamide/mol of 3-APA. Therefore, acrylamide is not only formed under "dry" conditions, but also in a heated aqueous system, provided that the reaction conditions/matrix are favorable and precursors are freely available.

Figure 1 compares the rate of formation of acrylamide and 3-APA, both following similar patterns but 3-APA is released more rapidly and very early in the reaction. The amount of 3-APA formed in this dry system clearly exceeds the yield of acrylamide.

The model system studies led to increased interest in the amounts of 3-APA in cooked or heated foods in general. A few publications and data have been published in the literature, and are summarized in Table 2. Some of the foods included in the study with highest amounts of 3-APA are popcorn and roasted cocoa.

Bagdonaite et al. [19] measured significantly higher amounts of 3-APA in potatoes (ca. 100-fold more than that reported by the group of Prof. Schieberle). This points to the activity of a specific decarboxylase enzyme in stored potatoes (no details were available on the potato storage temperature). The range of 3-APA in the same variety measured in three different farms was 9.5–14 mg/kg. However, the same authors did not detect 3-APA in roasted coffee.

Table 2 Amount of 3-APA measured in selected foods, compiled from different sources [11,19]

Food	3-APA (mg/kg)
Popcorn	1.88
Cocoa beans, fermented, roasted	1.8
Cocoa beans, unfermented, roasted	1.5
Gouda cheese, heated	1.32
Coffee surrogate	0.45
Wheat snack	0.28[a]
Potato, Likaria (fresh)	0.27
Potato, Likaria (warm stored)[b]	1.05
Potato, Nicola (fresh)	0.14
Potato, Saturna (stored ca. 20 weeks)	14
Cornflakes	0.17
Bread chips	0.15
Coffee (roast and ground)	0.12
Rusk	0.034

[a]Average of two independent measurements.
[b]12 days at 35 °C.

Intermediacy of the Glycoconjugates

The Maillard reaction commences with the formation of the Amadori reaction products, formed through the reaction of reducing sugars with amino acids. The *N*-glycosides can undergo Amadori rearrangement converting them to dicarbonyls (1-deoxysone or 3-deoxysone) via enolization and dehydration/hydrolysis reactions. These compounds can further decompose in the reaction cascade to generate color and flavor. However, under dry conditions or in low moisture systems *N*-glycosyl compounds mainly rearrange via the Schiff base. Early model experiments demonstrated that direct degradation of the *N*-glycosylamine of asparagine affords significantly higher yields of acrylamide under dry conditions than asparagine–sugar mixtures [13]. Notably, in relative terms, the yield of acrylamide in model systems from asparagine under favorable conditions is in the range of 0.1–1 mol%.

Other research groups confirmed the intermediacy of the decarboxylated Schiff base (imine 2, see Figure 2), which is relatively stable under low-moisture conditions, and the decarboxylated Amadori compound that effectively releases acrylamide and thus serves as direct precursors [9,14,20].

As shown in Figure 2, decarboxylation of the Schiff base may proceed via a cyclic intermediate or by a dipolar Schiff base betaine. Intramolecular cyclization of the Schiff base assists decarboxylation, forming the stable azomethine ylide (formation of the oxazolidin-5-one). Decarboxylation of the Schiff base may also proceed via the Schiff base betaine. Protonated ylides tend to undergo a 1,2-prototropic shift, to afford the decarboxylated Schiff base. Zyzak and coworkers [9] in their study reported evidence of the decarboxylated

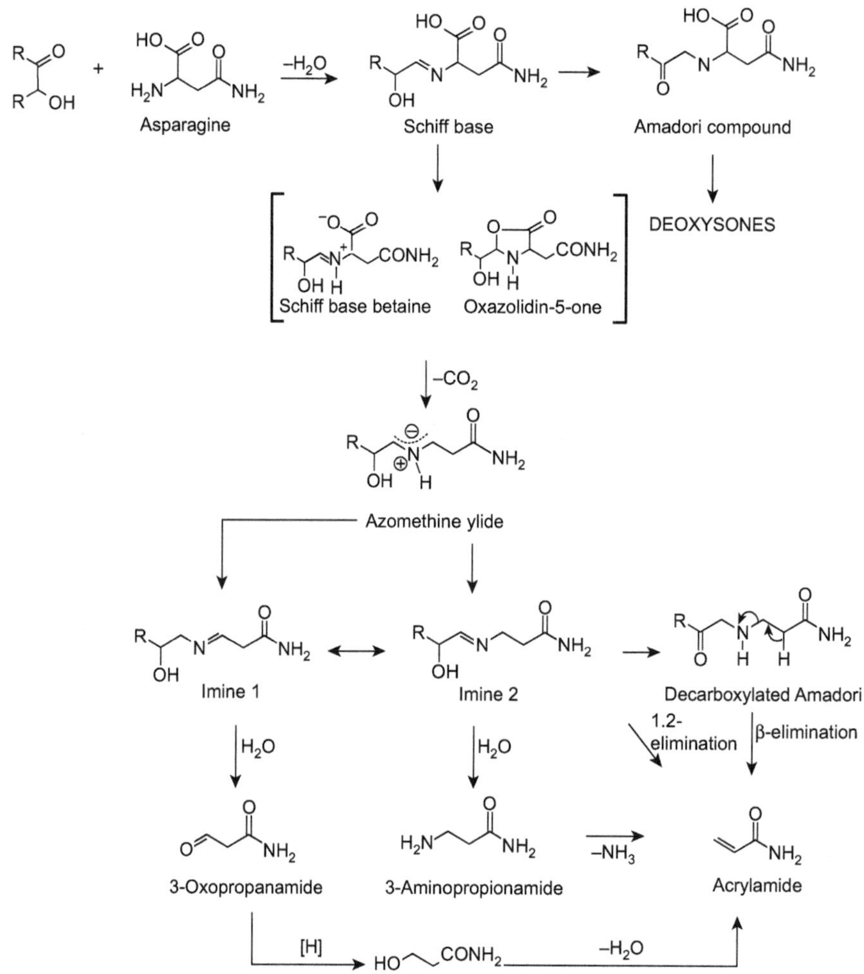

Figure 2 Mechanisms of acrylamide formation through the asparagine-reducing sugar (hydroxycarbonyl) route.

Schiff base of asparagine in a model system, through LC-MS measurements and detection of a signal at $m/z = 251$. The decarboxylated Schiff base after rearrangement leads to the corresponding Amadori product that upon β-elimination, initiated by the sugar moiety, directly furnishes acrylamide. The decarboxylated Schiff base (imine 2) releases 3-APA upon hydrolysis, or in an alternative direct route via 1,2-elimination acrylamide. The cleavage of the carbon–nitrogen bond by β-elimination to acrylamide (important is the availability of a β-proton) and the corresponding amino ketone is most likely the rate-limiting step in this reaction cascade as described by Stadler [13].

A feasible reaction pathway is the formation of 3-APA from the imine 2 via hydrolysis. The biogenic amine 3-APA then reacts further as described before (loss of ammonia) to procure acrylamide. Imine 1 in the reaction scheme leads to the well-known Strecker aldehyde (3-oxopropoanamide) that can react further with the corresponding alcohol and upon hydrolysis generate acrylamide. This sequence requires additional reduction/hydrolysis steps and most likely represents only a minor pathway in terms of its contribution to the overall acrylamide yield.

Evidence corroborating the formation of the cyclic 5-oxazolidinone in asparagine–sugar reactions has been provided using Fourier-transform infrared spectroscopy [20]. The same study demonstrated that the decarboxylated Schiff base generates acrylamide under both dry and wet conditions. In low-moisture model systems, the decarboxylated Schiff base afforded acrylamide in about fourfold higher amounts than 3-APA. With decreasing moisture in the system, the 5-oxazolidinone pathway becomes more dominant.

It is important to note that the starting carbonyl structure may impact the yields of acrylamide, especially the functional group in the beta position to the nitrogen [13,15]. A hydroxyl group favors the rearrangement of the decarboxylated Amadori compound, which may explain why hydroxyacetone (acetol) generates higher amounts of acrylamide (>4 mmol/mol) than methylglyoxal (<0.2 mmol/mol), using binary mixtures under low-moisture conditions (see previous section).

The reaction of asparagine with a dicarbonyl (Figure 3) can follow an analogous route, with the azomethine ylide after delocalization of the charge furnishing two possible imines [13]. The decarboxylated Schiff base releases 3-APA after hydrolysis, or the Strecker aldehyde, depending on the localization of the imine double bond. Further reaction of 3-APA (deamination) provides acrylamide. The Strecker aldehyde requires an additional reaction step; that is, reduction in the alcohol and subsequent hydrolysis of the alcohol to acrylamide, making this route far less likely when compared with the yields of acrylamide in models using the corresponding Strecker aldehyde and Strecker alcohol. The reaction of dicarbonyls with asparagine therefore most likely favors the 3-APA Maillard-driven pathway, considering the relatively high yields of acrylamide measured when reacting 3-APA under low-moisture conditions either alone or in combination with sugar. A minor route that is also feasible is the direct fragmentation by an elimination reaction of the imine to acrylamide. However, to date no experiments have been conducted to corroborate this route.

Which pathways finally govern or predominate in the formation of acrylamide is dependent on numerous factors, including pH, temperature (thermal input), reactants (and their mobility), the food matrix per se, and amount of water in the system. The generally low yield of acrylamide via the Maillard route is most likely attributable to the many different possible reaction pathways, many leading to other reactants and not acrylamide. In addition, low or minor amounts of key intermediates that from a structural perspective would favor higher yields, contributing to an overall low efficiency of chemical conversion of asparagine to acrylamide.

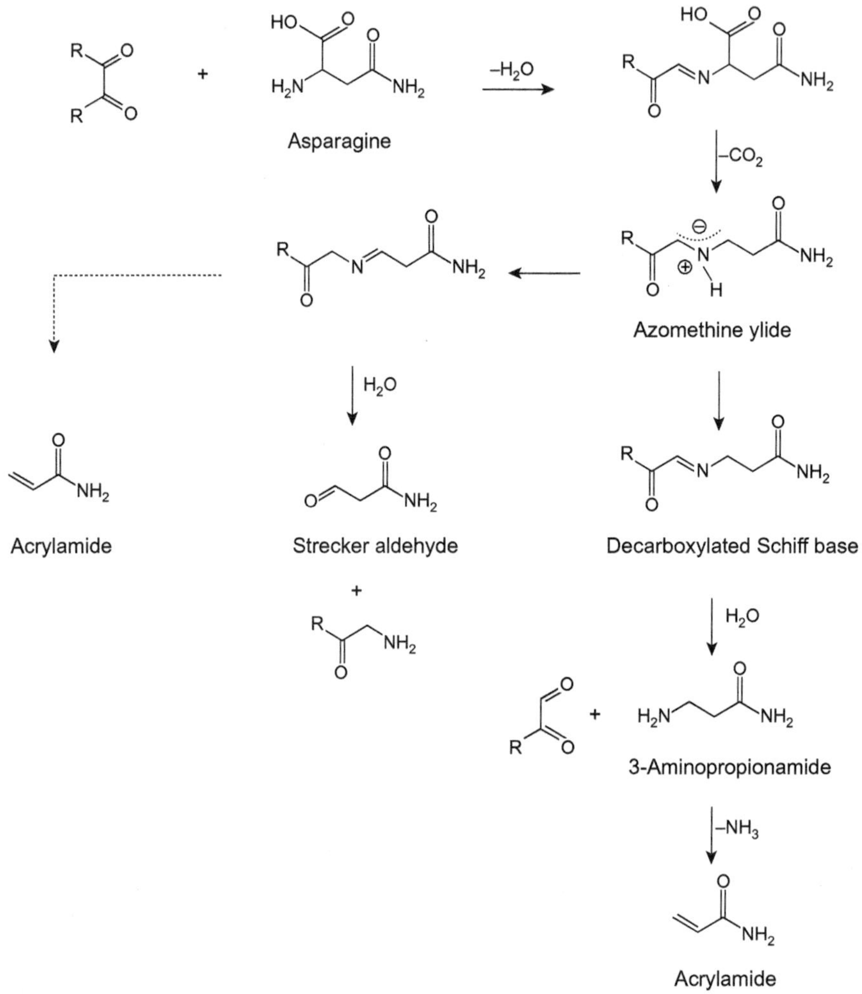

Figure 3 Mechanisms of acrylamide formation through the asparagine–dicarbonyl route.

MINOR PATHWAYS AND NON-MAILLARD ROUTES TO ACRYLAMIDE

Today it is well established that asparagine is the most effective precursor of acrylamide, when heated in the presence of reducing carbohydrates, α-dicarbonyl compounds, *n*-aldehydes, and 2-oxo acids. The pivotal role of asparagine has been demonstrated effectively through the addition of asparaginase, an enzyme that catalyzes the hydrolysis of the amide group of asparagine. Experiments in food models have shown that asparaginase can be highly efficient in lowering acrylamide, resulting in certain cases in reductions even above 95%.

Early during the investigations into acrylamide formation, several alternative proposals were raised that today in terms of importance to overall contribution can be considered marginal [15,21]. Acrolein (2-propenal), for example, is one candidate molecule.

This three-carbon aldehyde can be formed by different routes, including oxidative degradation of lipids, degradation of carbohydrates, amino acids and proteins, and through the Maillard reaction. For example, the amino acid methionine can furnish methional by the Strecker degradation, which in turn can fragment to acrolein.

Acrylamide formation has been studied particularly in edible oil/lipid systems [22]. Depending on frying conditions (oil temperature and quality grade), 5–250 mg acrolein can be formed per kg oil. Acrolein has a relatively low boiling point (51 °C) and will volatilize readily under high-temperature frying conditions. However, acrolein when oxidized can furnish acrylic acid, which can further react with ammonia via amino-dehydroxylation to afford acrylamide. In a typical food matrix, acrolein will react rapidly with other constituents rather than be oxidized to form acrylic acid.

Acrolein may also react with asparagine and under certain conditions lead to significant amounts of acrylamide. Lipid systems therefore may also be an indirect source of acrylamide, provided acrolein reacts further with either amino acids (asparagine) or is oxidized to the intermediate acrylic acid.

The acrylic acid route via the Maillard reaction was studied intensively by Stadler and coworkers [21]. The starting amino acid, aspartic acid (that provides the carbon backbone of acrylic acid, as does asparagine in the case of acrylamide), reacts analogously as described for acrylamide. The yields recorded in model systems were comparable to those reported for acrylamide, and reached up to 14 mmol acrylic acid per mol aspartate (heating mixtures for 5 min at 230 °C). However, acrylic acid requires an additional reaction step, that is, conversion of the acid moiety to the corresponding amide, and hence availability of ammonia. This is probably the rate-limiting step, as ammonia is volatile and not readily available. Cocktails of amino acids (those chosen that readily release ammonia under pyrolytic conditions) in model systems with acrylic acid have led to the formation of acrylamide at roughly 5% of the yield obtained by the asparagine route.

In particular, amino acids with a nitrogen in their side chain (asparagine, glutamine, lysine, and arginine) produced significant amounts of acrylamide in a model binary system composed of acrylic acid plus the amino acids (heated at 120 °C for 30 min), in the range 0.9–1.7 mg acrylamide per kg amino acid. Heating acrolein with the same amino acids did not lead to the formation of measurable acrylamide, suggesting that oxidation of acrolein is not a major route to acrylamide or its precursor acrylic acid [22].

Other amino acids and short carbon chain precursors have also been suggested to be involved in the formation of acrylamide (see Figure 4, adapted from Ref. [15]).

Acrylic acid can also be released directly from α-alanine or β-alanine via thermolytic deamination. Similarly, L-arginine can serve as a precursor of acrylic acid, albeit when heated at temperatures above 180 °C, and with yields within the same order of magnitude as registered for aspartic acid [15].

A further source of acrylamide besides asparagine and the Maillard pathway are proteins/peptides. Wheat gluten (peptides) may fragment under typical baking or roasting

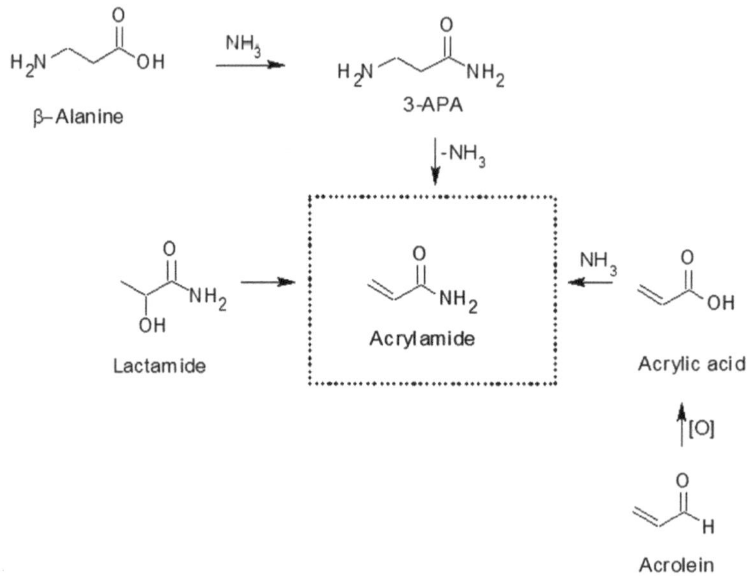

Figure 4 Other "marginal" pathways to acrylamide.

conditions and release acrylamide. A prerequisite for the release of acrylamide in this pyrolytic pathway is a β-proton in the amino acid adjacent to alanine [23]. For example, the sequence R_1–Gly–Ala–Phe–R_2 provides acrylamide and cinnamic acid amide, both detected by mass spectroscopy. The amount of alanine next to a β-proton varies depending on the type of protein, but reports estimate about 4.4% in gluten (on a mass basis). The fragmentation of peptides harboring alanine is therefore an alternative route to acrylamide. The same authors conducted an additional experiment using a food model (bread rolls), to which they added gluten at different concentrations. As expected, the amount of acrylamide increased with increasing gluten added to the dough (ca. 20% increase, depending on the amount of added gluten). However, the application of asparaginase to cereal-based foods results in a very significant reduction of acrylamide (ca. 70–80%), which shows that this alternative non-Maillard pathway can be considered as only marginal [15].

However, this is not the only study suggesting the involvement of peptides/other amino acids in the formation of acrylamide. Carnosine (*N*-β-alanyl-L-histidine) under pyrolytic conditions can release β-alanine followed by deamination and rearrangement to acrylamide, or via elimination of β-alanine amide and subsequent deamination to furnish acrylamide [15].

Granvogl et al. [10] propose an alternative pathway to 3-APA that avoid the requirement of a carbohydrate or carbonyl, that is, a non-Maillard pathway. They propose that 3-APA may be generated directly through the decarboxylation of asparagine, a plausible route in the presence of a suitable enzyme that accepts asparagine as a substrate and pyridoxal phosphate as a cofactor. In particular, histidine decarboxylase was effective in releasing 3-APA, albeit at a low conversion rate of around 0.1%.

As the reaction takes place under aqueous conditions, boiling or heating in, for example, a steamer may generate significant amounts of acrylamide if 3-APA is present, as the yield of acrylamide from 3-APA by far exceeds the yield from asparagine (comparable molar amounts and conditions). The enzymatic 3-APA pathway may explain the formation of acrylamide under aqueous conditions; for example, acrylamide formation in prune concentrates heated over a longer period of time and subject to drying at relatively low temperature (<100 °C). In fact, Becalski and his team [25] at Health Canada conducted interesting studies to better understand the low-temperature formation of acrylamide in prune-based food products and beverages. More than 0.2 mg/kg acrylamide can be formed in a prune juice heated for 24 h at 95 °C. This corroborates that acrylamide can be formed in an aqueous food system at temperatures well below 120 °C. Measurement of 3-APA or the decarboxylated Schiff base (see Figure 2) would provide a closer insight into the reaction pathways that govern the formation of acrylamide in such foods.

FACTORS IMPACTING THE FORMATION OF ACRYLAMIDE IN FOOD

Research in the field of acrylamide formation and mitigation over the past decade has revealed several key factors that may influence the formation of acrylamide in food [8,26]. In basic terms, the amount of precursors available for the reaction (asparagine, reducing sugars) is clearly determinant and in most cases success can be achieved through reduction of either asparagine or sugars in food materials [24]. Factors such as crop varieties, storage conditions, and seasonal weather conditions are known to have an effect on the levels of asparagine and consequently on acrylamide formed in potatoes when they are fried, baked, or roasted. Other agronomical factors such as the levels of sulfur in the soil and nitrogen fertilization have been found to have an influence on the levels of asparagine formed in cereals when they are cooked. Previous work on the asparagine content in selected European wheat varieties reveals at least a twofold range within the same crop year, and up to fivefold variation between varieties [5]. Selective breeding programs to produce new crop varieties (rye and wheat) that harbor naturally lower amounts of the free amino acid asparagine, would have a far greater impact on acrylamide reduction than any measures so far implemented. Unfortunately, the research gaps at agronomical level are not adequately addressed within the program of DG Research [27].

Another approach shown to work in many commercial applications is related to the processing conditions of the food (cooking temperature and time and moisture regime). Generally, the higher the cooking temperature, the greater the concentration of acrylamide in the food. However, it has been found that acrylamide is reduced at temperatures above 180 °C in some foods such as coffee and roasted/baked cereals. Over the past years, the food industry has studied the major factors impacting acrylamide formation, and has summarized the key measures applicable at a commercial scale in a "Toolbox" (see the next section).

MITIGATION, THE FOODDRINKEUROPE "ACRYLAMIDE TOOLBOX"

The European Food Safety Authority (EFSA), in its draft opinion issued in 2014, "confirms previous evaluations that, based on animal studies, acrylamide in food potentially increases the risk of developing cancer for consumers in all age groups" [28]. Acrylamide is a neurotoxicant in animals and humans, and a multiorgan carcinogen in both male and female rodent models. The Margin of Exposure (MoE) estimated for acrylamide is well below 10,000, indicating a concern from a public health perspective. Therefore, all efforts must be made to reduce exposure via the diet.

However, in this context it is important to understand that the elimination of acrylamide in foods is virtually impossible, as asparagine is an important amino acid in the nitrogen-storage cycle of plants [8]. A key challenge is to "decouple" the mechanistic route to acrylamide from the general Maillard pathway, retaining the desired attributes of cooked food with beneficial nutritional properties and health effects. The "Acrylamide Toolbox", established in 2006 by the Technical Process Contaminants Expert Group of the European Food & Drink Industries, summarizes the measures designed to achieve reduction of acrylamide in the pertinent food categories, using the fundamental knowledge attained in mechanistic studies [24].

In fact, the "Acrylamide Toolbox" brings together those methods of acrylamide reduction that are known to work in the production process (currently 33 commercially viable tools have been identified across the relevant food categories), and reviews other proposed methods that are at either the pilot plant or laboratory scale stages. The objective of the "Toolbox" is to provide practical tools that food manufacturers can evaluate to reduce acrylamide according to their particular situation. Using a selection of these tools food manufacturers have been able to reduce acrylamide levels in the range of 40–90% in certain selected foods. Where possible manufacturers have made changes to commercial products: for example, ammonium bicarbonate replacement by sodium bicarbonate or use of low fructose invert syrup. Recipe changes including ammonium bicarbonate replacement, reducing sugar modifications, and dilution of high-asparagine ingredients by more highly refined grains or other cereals, accounted for about two-thirds of the commercialized mitigation measures. In the case of the cereals group, most current efforts are focused on mitigation of acrylamide in whole grain and bran products, which are naturally high in asparagine; they continue to be monitored and are subject to ongoing development work.

Pretreatment with the commercially available enzyme asparaginase accounts for the majority of more recent measures in the Toolbox, and is more expected to be introduced in the 2015 revision cycle. This is despite limitations of use in some products and affordability for artisanal businesses. Some effective mitigation methods have been used only for a few specific products, or have not even been considered as potential approaches by some manufacturers, because they are in direct conflict with health targets. The replacement of wholemeal flour or bran with high extraction refined flour runs counter to the EU HEALTHGRAIN project that seeks to increase the intake of protective whole-grain components. Codex recognizes that when the use of different flours of varying asparagine content is being considered, manufacturers must balance the choice between the nutritional value and the minimization of acrylamide. Replacing ammonium bicarbonate with sodium salts is in direct conflict with WHO, and EU efforts to reduce hypertension and heart disease through sodium reduction in the diet, and targets set by the UK Food Standards Agency (typically replacement of ammonium bicarbonate by sodium bicarbonate could add up to 500 mg Na per 100 g product).

As technologies advance, new tools may be available or modifications to existing ones may be done. Food business operators are thus urged to make every reasonable effort to reduce acrylamide levels in final products and thereby reduce consumer exposure, following the ALARA (as low as reasonably achievable) approach. Where possible the commercial measures identified in the FDE Toolbox should be applied.

KEY FACTS

Risk assessment: Acrylamide is one of many different processing contaminants to which humans have been exposed since the mankind has begun to heat foods. Acrylamide is a neurotoxicant in animals and humans, and a multiorgan carcinogen in both male and female rodent models:

- Rats: Tumors in mammary gland, thyroid gland, and testes/epididymis (mesothelioma).
- Mice: Tumors in mammary gland, lung, skin, (fore)stomach, and Harderian gland.
- Reprotoxicant in rats/mice,
- Developmental toxicant in rats/mice.

The International Agency for Research on Cancer has classified acrylamide as a group 2A (probably carcinogenic to humans) based on sufficient animal evidence and limited human evidence. Intake of acrylamide through food is estimated to be about 0.5 μg/kg BW per day: 1000–10,000 times lower than doses in rodent studies.

Acrylamide is metabolized in the body to glycidamide by cytochrome P450 enzymes (CYP2E1). It is readily absorbed and distributed to all tissues. Both acrylamide and glycidamide are excreted in urine as mercapturic acids. Glycidamide is much more reactive than acrylamide with DNA and metabolism of acrylamide to glycidamide appears to be a prerequisite for the genotoxicity of acrylamide in vitro and in experimental animals.

EFSA, in its draft opinion issued in 2014, "confirms previous evaluations that, based on animal studies, acrylamide in food potentially increases the risk of developing cancer for consumers in all age groups" see: http://www.efsa.europa.eu/en/consultations closed/call/140701.pdf.

Risk management: There are no statutory maximum acrylamide levels set today for foodstuffs. However, in Europe the European Commission Recommendation (2013/647/EU) on investigations into the levels of acrylamide in food stipulates "indicative values" for acrylamide (see Mini Dictionary for definition of the term "indicative value"). See also: Commission Recommendation of 10.1.2011 on investigations into the levels of acrylamide in food (document C(2010) 9681 final, available at: http://ec.europa.eu/food/food/chemicalsafety/contaminants/recommendation_10012011_acrylamide_food_en.pdf).

Research: The European food and drink industry has contributed to important DG Research or national research programs, such as Heatox (http://heatox.org/), and Prom etheus(http://www.eusem.com/body/CS/EUproj/PROMETHEUS.htm) as well as a project to control asparagine in wheat. Other efforts concern cooking instructions to help consumers to reduce their exposure to acrylamide(www.goodfries.eu); which is available in 28 languages.

Acrylamide Pamphlets: The food industry sectors have worked closely with the European Commission and Member States to produce sector-specific pamphlets on ways to reduce acrylamide in food production. These pamphlets are aimed at small- and medium-sized enterprises and are available in 23 official languages of the European Union.

The pamphlets currently cover six food sectors:

- Biscuits
- Crackers and crisp breads
- Bread products
- Breakfast cereals
- Fried potato products (such as potato crisps and French fries)
- Infant food.

MINI DICTIONARY

Hazard A chemical agent in food with the potential to cause harm (relates to the inherent properties of the substance that make it capable of causing adverse effects).

Risk The probability of occurrence of an adverse health effect, weighted for its severity, that may result from the exposure under defined conditions to a chemical hazard in food.

Margin of Exposure (MoE) The carcinogenic effects of acrylamide follow a genotoxic mechanism, hence no health-based guidance values (such as the ADI/TDI) are available. In these cases, an MoE is calculated usually by dividing the (2.5%) lower confidence limit of the benchmark dose for a 10% increase in cancer incidence (BMDL10) by the exposure. If no BMDL10 is available, the no observed (adverse) effect level (NO(A)EL) is used. The MoE has been developed as a mechanism for prioritizing contaminants that require risk management measures and can provide some guidance on when exposure reductions may be considered adequate. EFSA has advised that contaminants with an MoE of >10,000 are not a priority for action. MoE summary:

- Dimensionless ratio.
- Does *not* denote a safe intake level or a numerical measure of risk but rather describes a level of concern.
- For genotoxic carcinogens, an MoE >10,000 proposed as of low concern from a public health perspective.
- The factor 10,000 is derived from a composite uncertainty factor of 100 and a default uncertainty factor of 100 for uncertainties associated with genotoxic carcinogens.

Genotoxic carcinogens Most genotoxic carcinogens are good electrophiles that interact directly with DNA through the formation of covalent bonds, resulting in DNA adducts. These adducts could result in DNA damage, and if not repaired result in mutations and may eventually cause tumors.

ALARA ("as low as reasonably achievable") In the context of acrylamide and other process contaminants, which are the result of naturally-occurring chemical reactions in heated foods and for which there are currently no levels that regulators have agreed upon as being "safe," ALARA means that the food manufacturer should make every reasonable effort (based upon current knowledge) to reduce levels in the final product and thereby reduce consumers exposure. ALARA could for example mean that a food business operator changes parts of a process or even a whole process if technologically feasible. The challenge with ALARA often is in defining what is "reasonably achievable."

Acrylamide "Indicative Values" The indicative values set for acrylamide are not intended as maximum limits but as a trigger to initiate investigations by enforcement authorities into food business operators' understanding of acrylamide and any action they have taken to mitigate its production. Exceedance of an indicative value does not trigger enforcement action.

The FoodDrinkEurope "Acrylamide Toolbox" The tools identified within the Toolbox are potential measures designed to limit acrylamide levels in final product through interventions at various stages of production (agronomical, recipe, processing, and final preparation). They are based upon scientific knowledge and practical application in specific circumstances. As technology develops, new and better tools for acrylamide reduction may become available. As part of a continuous process food business operators should review what tools are available on regular basis, and consider whether they can implement these tools into their processes or their product.

SUMMARY POINTS

- The main steps in the chemical reactions leading to acrylamide have been elucidated.
- Not only reducing sugars but also reactive carbonyl compounds are involved in the decarboxylation of asparagine via the early Maillard pathway.
- Several different chemical precursors may be involved, their role and importance depending on the reactants and the reaction conditions.
- Other "marginal" pathways have been identified, for example, peptides (in gluten), acrolein, or acrylic acid (all thermally driven).
- Some aspects of the reaction leading to acrylamide have so far not been clarified, such as the intermediates involved in the relatively low temperature formation in some fruits.
- Thermal input (heating temperature and heating time) is pivotal, as well as the end-moisture in the product.

- Any intervention steps to reduce acrylamide in food should also consider "risk–benefit" aspects, for example, impact on whole grain, salt, and other "beneficial" nutrients generated during the Maillard process.
- Funding through public–private partnerships is urgently needed to fuel the scientific programs targeted toward addressing the research gaps at agricultural level, that is, breeding programs for crops low in free asparagine.

REFERENCES

[1] Swedish National Food Administration. Information about acrylamide in food. April 24, 2002.
[2] Tareke E, Rydberg P, Karlsson P, Eriksson S, Törnqvist M. Analysis of acrylamide, a carcinogen formed in heated foodstuffs. J Agric Food Chem 2002;50:4998–5006.
[3] Stadler RH, Blank I, Varga N, Robert F, Hau J, Guy PA, et al. Acrylamide from Maillard reaction products. Nature 2002;419:449–50.
[4] Mottram DS, Wedzicha BL, Dodson AT. Acrylamide is formed in the Maillard reaction. Nature 2002;419:448–9.
[5] Taeymans D, Wood J, Ashby P, Blank I, Studer A, Stadler RH, et al. Review of acrylamide: an industry perspective on research, analysis, formation and control. Crit Rev Food Sci Nutr 2004;44:323–47.
[6] Taeymans D, Anderson A, Ashby P, Blank I, Gonde P, Van Eijck P, et al. Acrylamide: update on selected research activities conducted by the European food and drink industry. J AOAC Int 2005;88:234–41.
[7] Stadler RH, Scholz GS. Acrylamide: an update on current knowledge in analysis, levels in food, mechanisms of formation, and potential strategies of control. Nutr Rev 2004;62:449–67.
[8] Lineback DR, Coughlin JR, Stadler RH. Acrylamide in foods: a review of the science and future considerations. Annu Rev Food Sci Technol 2012;3:15–35.
[9] Zyzak DV, Sanders RA, Stojanovic M, Tallmadge D, Eberhart BL, Ewald DK, et al. Acrylamide formation mechanism in heated foods. J Agric Food Chem 2003;51:4782–7.
[10] Granvogl M, Jezussek M, Koehler P, Schieberle P. Quantitation of 3-aminopropionamide in potatoes-a minor but potent precursor in acrylamide formation. J Agric Food Chem 2004;52:4751–7.
[11] Granvogl M, Schieberle P. Thermally generated 3-aminopropionamide as a transient intermediate in the formation of acrylamide. J Agric Food Chem 2006;54:5933–8.
[12] Granvogl M, Schieberle P. Quantification of 3-aminopropionamide in cocoa, coffee and cereal products. Eur Food Res Technol 2007;225:857–63.
[13] Stadler RH, Blank I, Varga N, Robert F, Hau J, Riediker S, et al. In-depth mechanistic study on the formation of acrylamide and other vinylogous compounds by the Maillard reaction. J Agric Food Chem 2004;52:5550–5.
[14] Yaylayan VA, Locas CP, Wronowski A, O'Brien J. Mechanistic pathways of formation of acrylamide from different amino acids. In: Friedman M, Mottram DS, editors. Chemistry and safety of acrylamide in food. New York: Springer; 2005. p. 191–204.
[15] Yaylayan VA, Stadler RH. Acrylamide formation in food: a mechanistic perspective. J AOAC Int 2005;88:262–7.
[16] Schieberle P, Koehler P, Granvogl M. Chemical pathways of acrylamide formation: an update. In: Presentation at the acrylamide workshop Brussels, Belgium; 2006.
[17] Granvogl M, Schieberle P. Acrylamide and furan: mechanistics and modelling. In: Presented at the 2nd CIAA process contaminants mitigation workshop Brussels, Belgium; July 2010.
[18] Blank I, Robert F, Goldmann T, Pollien P, Varga N, Devaud S, et al. Mechanism of acrylamide formation: Maillard-induced transformations of asparagine. In: Friedman M, Mottram DS, editors. Chemistry and safety of acrylamide in food. New York: Springer; 2005. p. 171–89.
[19] Bagdonaite K, Viklund G, Skog K, Murkovic M. Analysis of 3-aminopropionamide: a potential precursor of acrylamide. J Biochem Biophys Methods 2006;69:215–21.
[20] Perez-Locas C, Yaylayan VA. Further insight into thermally and pH induced generation of acrylamide from glucose/asparagine model systems. J Agric Food Chem 2008;56:6069–74.

[21] Stadler RH, Verzegnassi L, Varga N, Grigorov M, Studer A, et al. Formation of vinylogous compounds in model Maillard reaction systems. Chem Res Toxicol 2003;16:1242–50.
[22] Ehling S, Hengel M, Shibamoto T. Formation of acrylamide from lipids. In: Friedman M, Mottram DS, editors. Chemistry and safety of acrylamide in food. New York: Springer; 2005. p. 223–34.
[23] Claus A, Weisz GM, Schieber A, Carle R. Pyrolytic acrylamide formation from purified wheat gluten and gluten supplemented wheat bread rolls. Mol Nutr Food Res 2006;50:87–93.
[24] Food Drink Europe Acrylamide Toolbox. http://www.fooddrinkeurope.eu/uploads/publications_documents/AcrylamideToolbox_2013.pdf; 2013 [accessed 17.04.15].
[25] Becalski A, Brady B, Feng S, Gauthier BR, Zhao T. Formation of acrylamide at temperatures lower than 100°C: the case of prunes and a model study. Food Addit Contam Part A 2011;28:726–30.
[26] Stadler RH. Acrylamide formation in different foods and potential strategies for reduction. Adv Exp Med Biol 2005;561:157–69.
[27] Stadler RH. Life sciences: strategic industries for Europe. In: The European files, http://www.europeanfiles.eu/; March 2015 – n°35 [accessed 19.04.15].
[28] EFSA draft opinion on acrylamide, http://www.efsa.europa.eu/en/consultationsclosed/call/140701.pdf [accessed 27.04.15].

CHAPTER 2

Challenges in Estimating Dietary Acrylamide Intake

Heinz Freisling, Nadia Slimani
Dietary Exposure Assessment Group, International Agency for Research on Cancer (IARC-WHO), Lyon, France

INTRODUCTION

Acrylamide, a known human neurotoxicant [1,2] and probable carcinogen [2,3], enters the food chain mainly as a by-product of heat-processed carbohydrate-containing foods such as not only French fries, potato chips, crackers, but also coffee, black olives, and others [1,2,4]. The presence of acrylamide in many different foods may represent a public health concern because virtually everyone is exposed to varying degrees depending on individual food preferences or dietary habits in a population [5]. Accurate and precise assessment of dietary acrylamide intake in different population groups is therefore essential for different areas of public health nutrition research. These areas include not only monitoring of intake/chemical exposure assessment for food safety aspects, but also for conducting nutritional epidemiological/health outcome research (Figure 1).

In public health nutrition research studies, dietary acrylamide intake is usually assessed through estimates based on self-reported dietary consumption of known acrylamide-containing foods, typically by means of short-term instruments such as 24-h dietary recalls (24-HR) or long-term instruments such as food frequency questionnaires (FFQ) [6,7].

Individual consumption of reported acrylamide-containing foods are then matched with mean occurrence values available in acrylamide occurrence databases and summed over the whole diet to estimate different summary statistics (e.g., mean, median, standard deviation, and percentile) for the population under study [8]:

$$I = \sum (W_1 C_1 + W_2 C_2 + W_3 C_3 + \cdots W_n C_n)$$

I = acrylamide intake; $W_{1,n}$ = weight of acrylamide-containing food 1, etc.; $C_{1,n}$ = acrylamide concentration in food 1, etc.

An alternative or complementary method of dietary acrylamide intake assessment employs biomarkers of acrylamide exposure. Currently available biomarkers include hemoglobin (Hb) adducts of acrylamide and its primary metabolite glycidamide, both measured in erythrocytes, the ratio of Hb adducts of glycidamide to Hb adducts of

Acrylamide in Food
http://dx.doi.org/10.1016/B978-0-12-802832-2.00002-4

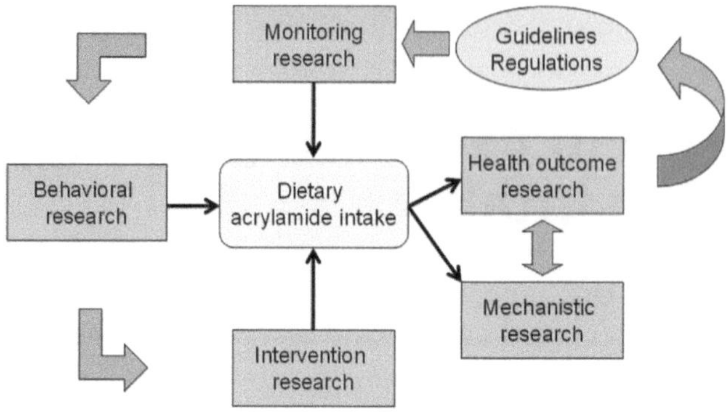

Figure 1 Areas of research, where dietary acrylamide intake data are needed.

acrylamide, and various acrylamide metabolites measured in urine [9]. Such biomarkers may then be used either as the main measure of exposure or as a reference method to evaluate self-reported dietary acrylamide intake [10,11].

All of these introduced methods to estimate dietary acrylamide intake are afflicted with specific shortcomings for a number of reasons, which will be discussed in more detail in the following sections.

METHODS TO ESTIMATE THE USUAL DIETARY ACRYLAMIDE INTAKE AND ITS SHORTCOMINGS

The dietary acrylamide exposure of interest is the "usual intake," which is the long-term average intake of acrylamide. This usual intake is a key concept in all areas of public health nutrition research studies including dietary monitoring and nutritional epidemiology and applies to any dietary exposure such as foods, nutrients, or other bioactive food compounds. Depending on the study objectives, the time frame of interest, which should be captured by the usual intake, can be one year, as it is the case for monitoring of dietary acrylamide exposure at population level, or much longer. In studies that aim to evaluate associations between dietary acrylamide intake and cancer risk, this time frame can also be decades because of the long latency period between exposure and developing first symptoms of cancer. For obvious reasons, the usual intake is not directly observable, but can be estimated from self-reported "actual" (or acute) intakes.

However, estimating dietary acrylamide intake is particularly difficult for a number of reasons. Exposure levels in even similar foods depend heavily on how foods are cooked (e.g., frying and roasting) and/or preserved (e.g., low moisture drying) and the extent of heat treatment (>120 °C) of foods that contain the amino acid asparagine and reducing sugars, such as fructose, glucose, or galactose. Acrylamide can also be found in foods that are not processed with high heat, but with low moisture drying; for example, black olives or prune juice. Both contain relatively high concentrations of asparagine and reducing sugars. Boiling or steaming of foods has not been reported to contribute to acrylamide formation [9,12,13]. Furthermore, it is difficult to take into account the potentially large variation in acrylamide concentrations within foods depending on the food composition (e.g., varying content of different types of reducing sugars), individual food preparation preferences at home (e.g., preferring dark roasted toast or else) or differences in food processing between different manufacturers of similar foods (e.g., degree of roasting of coffee), which means that brand or product information of acrylamide-containing foods is important [5,13]. Therefore, methods to provide unbiased estimates of acrylamide intake would be needed to collect such details about the acrylamide-containing foods consumed.

The main characteristics of common methods to estimate the usual dietary intake of acrylamide and related challenges are summarized in Table 1.

Short-Term Instruments

With short-term instruments such as 24-HR or food records, which are usually open-ended, it is theoretically possible to collect sufficient descriptive details about the food sources of acrylamide (e.g., cooking method, brand/product name information). In addition, 24-HR offer comparable estimations in multicenter studies because procedures such as administration and structure of interviews, food description and quantification, or quality controls can be standardized across centers [14].

However, in defining the level of detail of consumed foods that the study participants would be asked, the interview time and, thus, participant burden as well as the ability of the participant to provide all required information has to be considered. Thus, detail of food description and participant burden should be balanced. If more than one or two 24-HR administrations per participant are required—depending on the study aims and related sample size—costs also become an important factor because 24-HR are traditionally interviewer-administered.

For usual intake estimation, replicate 24-HR, or other short-term instruments (e.g., food records), need to be collected from each individual of a given population sample.

In dietary monitoring, two repeated nonconsecutive 24-HR per individual are sufficient to estimate the usual population intake and distribution; subsequently means or medians, percentiles, proportions above or below a specific cut-off can be calculated. However, a sufficiently large sample size of several thousand individuals covering equally all days of the week and all seasons of the year is needed. However, if the goal is to estimate only the mean intake of a population (e.g., for calibration purposes), then one 24-HR or food record is sufficient provided that all days of the week and months of the year are equally represented across a sufficiently large sample [6].

In diet–disease studies, however, more than two repeated 24-HR per individual are required to rank individuals according to their usual intake. Although it is currently not known how many short-term instrument replicates would be needed to reliably rank individuals according to their acrylamide intake, probably between 4 and 6 would suffice [15].

Table 1 Characteristics of methods to estimate dietary acrylamide intake in public health nutrition research studies

Method	Generic application	Challenges	Recommendation
Short-term instruments For example, 24-hour recalls	• Main method for estimating usual intake in dietary monitoring (repeated administration necessary) • Reference method for calibration and validation studies • Main method in modern large cohort studies to evaluate diet–disease relationships (at least four nonconsecutive administrations necessary)	Sufficient descriptive details about the food sources of acrylamide are required, which in theory can be collected with open-ended short-term instruments; however, a balance between details collected and interview duration and related participant burden and costs is needed	Collect sufficient detail on consumed acrylamide-containing foods including cooking method, extent of cooking, and brand name Collect replicates to capture sporadic consumption of some acrylamide-rich food items (e.g., crisps and fried potatoes) in order to decrease or remove within-person random error
Long-term instruments For example, food frequency questionnaires	• Main method in traditional large cohort or case–control studies to evaluate diet–disease relationships • Supplemental method (i.e., "food propensity questionnaire") for assessing usual intake of episodically consumed foods in dietary monitoring and in cohort studies using short-term instruments as main method	These instruments are generally not conceived to collect information about the processing and the cooking methods of all consumed foods; systematic and random errors can be substantial	Use nonquantitative long-term instruments to collect information on sporadically consumed acrylamide-containing foods, at least the propensity of consumption, supplementing short-term instruments
Biomarker measurements For example, hemoglobin adducts of acrylamide	• Main method for some dietary compounds in dietary monitoring (e.g., selenium and aflatoxin) and in (nested) case–control studies to evaluate diet–disease relationships • Reference method for calibration and validation studies • Complementary method with short- and long-term methods in modern cohort studies • Reference method to derive "empirical weights" to identify the most informative foods in a finite food list of long-term self-report instruments	Urinary acrylamide metabolites reflect exposure over recent days only; unless intake is constant over time, they are less informative for long-term intake Hemoglobin adducts of acrylamide and its primary metabolite glycidamide reflect exposure over 3–4 months, but is also related to nondietary exposure such as smoking; they also show large intra-individual variability	Collect replicates of biomarker measurements per individual to reduce intra-individual variability For validation studies, enroll nonsmokers to increase specificity for dietary exposure Use biomarkers to derive empirical weights Identify new and more specific biomarkers of dietary acrylamide exposure

Furthermore, since many acrylamide-containing foods are not regularly consumed (e.g., chips, crackers), the combined use of multiple 24-HR and FFQ may provide data superior to the use of either method alone [15]. No studies have been published to date that used repeated short-term instruments or a combination of multiple short-term instrument administrations and a long-term instrument. Until the advent of web-based tools to administer short- and long-term instruments, this approach was simply not feasible because of the high costs of interviewer-administered short-term instruments, and restricted to evaluation and calibration study designs with much smaller sample sizes.

Long-Term Instruments

In contrast to short-term instruments, a single administration of a long-term instrument, such as an FFQ or a diet history questionnaire, may aim to directly capture individual usual intakes at least to rank individuals according to their dietary acrylamide intake within a study population. Because of practical and feasibility aspects, this is still the most common approach in nutritional epidemiological studies; for example, to evaluate associations between dietary acrylamide intake and cancer risk [16–20]. In most cases, these long-term instruments were not designed for acrylamide studies.

FFQ typically consists of a limited number of generic or aggregated food items (e.g., crackers) and a frequency response section for participants to report how often each food was consumed. A few further details about quantity, preparation, or composition of a given food item may also be asked for. Depending on whether the objective is to measure intake of a few specific foods or dietary components or whether a comprehensive assessment of dietary intake is needed, the length of the food list may vary between 40 and 260 items. An important aspect is that an excessively long questionnaire can lead to fatigue for the participants, which can affect the completeness and accuracy of the reporting [7]. For these reasons, it is generally not recommended to collect too much detail about the consumed foods with FFQ. Moreover, most of the FFQ that are used in epidemiological studies were designed in the 1990s, when it was not yet known that acrylamide in food could become a public health issue. Thus, specific questions addressing acrylamide-formation in foods were not considered in the FFQ-design at that time, which may exaggerate systematic under- or overestimation of dietary acrylamide intake.

Biomarkers of Dietary Acrylamide Intake

Biomarkers of intake or exposure can be broadly defined as cellular, biochemical, or molecular consequences of the dietary intake of interest that can be measured in biological media (e.g., blood, urine) [10,11]. Compared to self-report instruments, biomarkers have the main advantages of being independent of potentially biased self-reported dietary intake and of potential errors or missing values in occurrence databases.

The main requirements of a valid biomarker for dietary acrylamide intake are summarized in Figure 2.

Currently available biomarkers of acrylamide intake—Hb adducts of acrylamide and of its primary metabolite glycidamide as well as urinary acrylamide metabolites [9]—only partially meet these requirements (see also Table 1).

Because the formation of glycidamide and its interaction with DNA may account for most of the mutagenic and other genotoxic effects of acrylamide, biomarkers of glycidamide formation are also of interest [9]. Although, strictly speaking, it is not a biomarker of exposure, but of internal dose, the ratio between Hb glycidamide- and Hb acrylamide-adducts is also often used [21,22].

Despite recent evidence that a dose–response relationship exists between dietary acrylamide exposures and Hb adduct levels [23], a number of correlation studies reported mixed results on the relationship between self-reported acrylamide intake and measured Hb adduct levels in humans. The correlation coefficients range from moderate [24–26] to trivial values [21,27,28]. One explanation is that nondietary factors are more strongly associated with Hb adduct variability (see also Box 1). For example, smokers have been shown

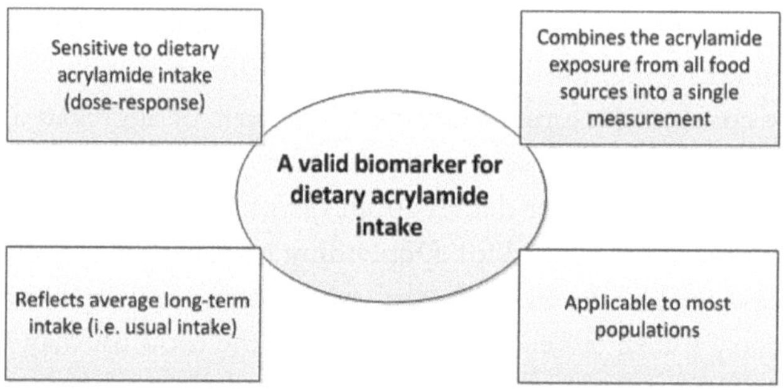

Figure 2 Main requirements for a valid biomarker of dietary acrylamide intake. *Adapted from [10,11].*

Box 1 Possible reasons for low correlations between dietary acrylamide intake and hemoglobin adducts of acrylamide are:

- Hemoglobin adducts of acrylamide reflect multiple sources of acrylamide exposure, not just the dietary exposure, and they also reflect individual differences in absorption and metabolism.
- Dietary assessment instruments aggregate multiple foods with varying acrylamide content into broad food groups, whereas acrylamide content could vary substantially even within single foods, with respect to cooking methods, cooking time, brand, and others aspects of food preparation practices, for which information is seldom available.
- Acrylamide occurrence databases often assemble data from different countries, because of a lack of country-specific analytic data on acrylamide content in foods, thus reducing the specificity of country-specific acrylamide estimates.
- Availability of only one replicate of 24-HR; the sporadic nature of consumption of some acrylamide-rich food items (crisps, fried potatoes, and so on) results in large day-to-day variability in 24-HR estimates.
- It is well accepted that dietary self-report instruments suffer from bias due to measurement errors, and thus do not provide accurate measures of intake in general.

to have almost three times higher acrylamide–Hb adduct concentrations than that of non-smokers [22]. While smoking is an important source of overall acrylamide exposure, increasing Hb adduct levels, lower Hb-glycidamide adducts and Hb-glycidamide- to Hb-acrylamide adduct ratios were observed with increasing alcohol intake, which competitively inhibits enzyme-mediated metabolism of acrylamide to glycidamide [29].

MEASUREMENT ERROR

Measurement error can be broadly defined as a deviation from the true value and is usually categorized into *random* and *systematic errors*. Both types of errors can occur at *individual level* (within-person) and *group/population level* (between-persons) [6,7]. No method is completely free of measurement errors including the biomarker measurements. However, typically the different sources and structure of measurement error are more complex and pronounced in self-report instruments (Figure 3).

Random Within-Person Errors

Random within-person error will not affect the mean dietary acrylamide intake of a population because these types of errors will cancel out. However, random within-person error contributes to, and thus inflates, the observed between-person variation of

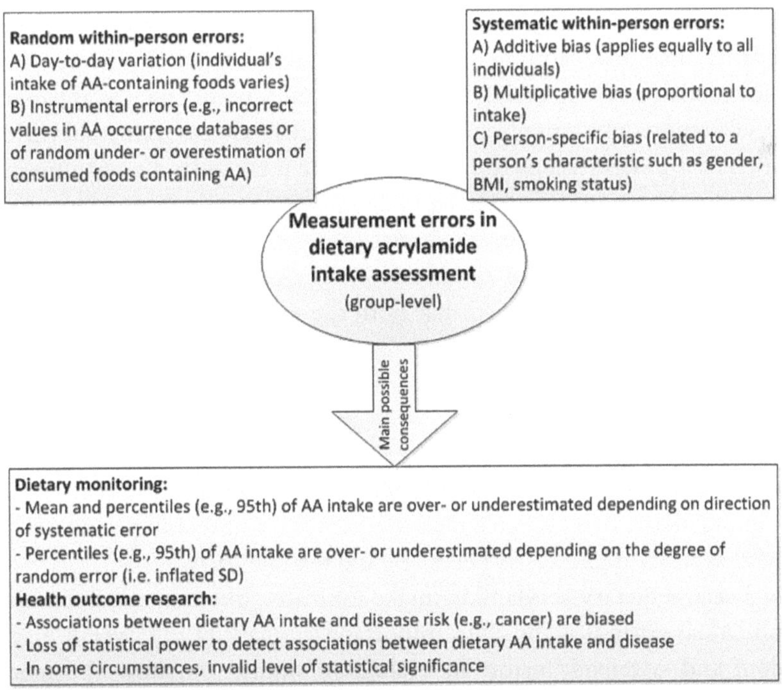

Figure 3 Summary of different types of measurement error in dietary acrylamide intake assessment using self-report instruments and the main possible consequences; AA, acrylamide; SD, standard deviation. *Adapted from [7,30].*

acrylamide intake. Therefore, the observed variation/distribution of acrylamide intake of a population is larger than the true variation. The main consequences of random between-person errors are that the 95th percentile dietary acrylamide intake estimate is overestimated, and furthermore, the strength of an association between dietary acrylamide intake and a health outcome is biased toward no effect (attenuated).

Systematic Within-Person Errors

In addition to the random within-person error, individuals may also systematically under- or overestimate their true food intake (consciously and subconsciously) and thus, their true acrylamide intake. This is referred to as systematic within-person error and is defined as the difference between the observed and the true, long-term average, acrylamide intake.

1. Systematic within-person errors can apply to all individuals equally; for example, caused by systematic errors in acrylamide occurrence databases, where the acrylamide content of a given food can be either too high or too low compared to its "true" intake.
2. Second, systematic errors can also be proportional to the level of individual intake; for example, individuals with higher intakes of acrylamide-containing foods, under-report relatively more than individuals with lower intakes report.
3. Third, systematic errors can also differ between individuals according to individual's specific characteristics such as age, sex, education, or other unmeasured characteristics (i.e., person-specific bias); for example, obese people tend to underestimate their food intake more than nonobese people do, which may lead to an underestimation of acrylamide intake depending on the body mass index (BMI) of a person. Freisling et al. showed that with increasing BMI, dietary underreporting was apparent on both dietary self-report instruments (i.e., 24-HR and FFQ) as evaluated with recovery biomarkers for protein—a proxy for total caloric intake—and potassium intake [31]. Although the degree of underreporting differed among the European countries, the BMI-effect on underreporting was similar [31].

In practice, all three forms of systematic within-person errors and random errors tend to be present simultaneously, so that both the population mean intake and its SD are measured with error.

Dealing with Measurement Errors

First, all possible and feasible measurement error reduction techniques should be applied at the stage of data collection. Information should be collected from each individual of a study sample with the best available measurement tools or techniques in a standardized way and with as many replicates as necessary and feasible [6,7].

Second, because dietary acrylamide intake estimates, like any measurement, can still contain substantial error, it is thus also important to quantify the overall magnitude of both random and systematic errors in evaluation studies and how the measurement errors affect the results of the main study [6].

Third, depending on the study objectives, it will often be necessary to correct for measurement errors using statistical approaches. With a technique referred to as linear regression calibration, random errors, in the form of within-person random error, and systematic errors, in the form of additive and multiplicative bias, can be at least partially corrected for or mitigated [7,30]. Calibration studies are needed to supply the best predictors of true usual intake [32]. Energy adjustment is another way to mitigate the effect of measurement errors [15,30,33]. A possible explanation is that energy intake contributes to the intake-related bias and part of it may be removed after correcting for energy intake. The same principle applies to other covariates such as age, gender, and BMI.

Finally, if the interest is in estimating dietary acrylamide intake distributions of usual intake of a population, then statistical techniques can be used to remove/reduce the within-person errors (day-to-day variation), leaving only the between-person variation. This is particularly needed in dietary monitoring, where short-term instruments such as 24-HR dietary recalls are the method of choice, in order to avoid overestimation of acrylamide intake levels at the higher end of the intake distribution (e.g., 95th percentile intake). Several methods to remove the within-person variation have been developed in the last years and there is also a wide range of software solutions available [34].

STANDARDIZED DIETARY ACRYLAMIDE INTAKE ACROSS COUNTRIES

Methodological differences in dietary acrylamide assessment across countries do not permit full comparison and interpretation of results, and it has been recommended to harmonize them better [9]. This is particularly important knowing that dietary acrylamide is a universal exposure affecting all dietary patterns and countries.

Comparison of Standardized Dietary Acrylamide Intake in Europe: Example from the EPIC Study

One of the very few examples of a reliable comparison of dietary acrylamide intake across countries is the study by Freisling et al., in which a harmonized occurrence database of acrylamide levels in foods has also been compiled [5]. This study used data from the European Prospective Investigation into Cancer and Nutrition (EPIC) calibration study, where standardized 24-HR were collected from almost 37,000 participants randomly selected in 27 centers of 10 European countries. Procedures such as administration and structure of interviews, food description and quantification, or quality controls were standardized across countries [32].

The harmonized acrylamide database was compiled using values from the EU monitoring database of acrylamide levels in food maintained by the Institute for Reference Materials and Measurements (IRMM). Food consumption data from the EPIC calibration study were matched with the foods in the acrylamide occurrence database. Whenever possible, the reported food with all its available descriptive details (e.g., on cooking) from the 24-HR was matched exactly with the corresponding food in the database. Often, an exact match was not possible, and then the reported food was matched to a similar item or to the mean of all items of the food group in the acrylamide database. For all countries, the same arithmetic mean of acrylamide content was assigned to the corresponding food [5]. One of the main challenges in the compilation and matching process were the large variations in acrylamide content within the same food groups potentially reflecting differences in food processing and/or food preparation across countries. For example, values for acrylamide content of "coffee, roasted" varied between 79 and 975 μg/kg (mean: 312 μg/kg) in the acrylamide database. Likewise, values for "potato chips" varied between 5 and 3428 μg/kg (mean: 294 μg/kg) [5].

The estimated country/center-specific mean intakes were about threefold higher in the centers of northern Europe as compared to those in southern European centers (Figure 4).

These results for the 24-HR are in line with an earlier study in EPIC, where higher levels of Hb adducts of acrylamide were found among nonsmokers from British and Dutch EPIC centers as compared to those from centers in southern Europe (e.g., Italy, Spain) [22], suggesting important differences in dietary intake. These results show the added value of collecting data with a common and standardized method in cross-country comparisons. Artificial differences in dietary acrylamide intake between countries that may be introduced by different instruments and data collection procedures can be minimized. This is highlighted in Figure 4, where dietary acrylamide intake estimates from the standardized 24-HR are compared with country-specific (validated) FFQ, where the latter shows substantial over- or underestimation relative to the 24-HR estimates.

The two major contributors to acrylamide intake in this study, as estimated with the standardized 24-HR, were bread and coffee. For example, in Denmark (where acrylamide intake

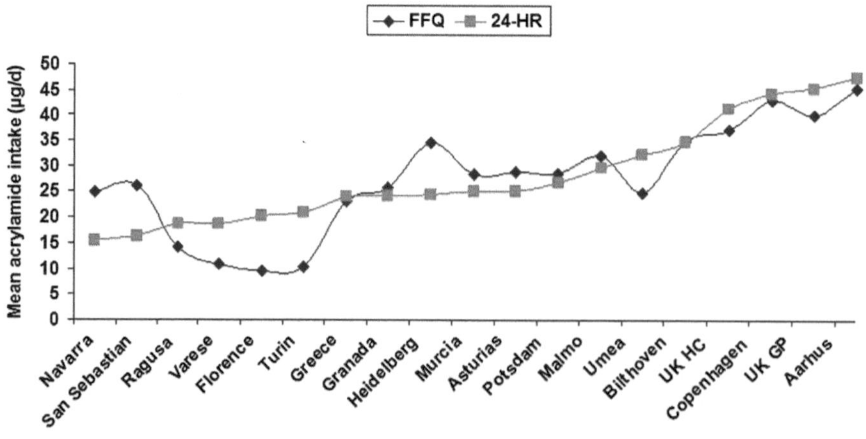

Figure 4 Dietary acrylamide intake in EPIC as estimated by 24-HR (GloboDiet: formerly EPIC-Soft) compared to FFQ, example: men. EPIC, European Prospective Investigation into Cancer and Nutrition study; 24-HR, 24-hour recall; FFQ, food-frequency questionnaire; UK HC, UK health conscious; UK GP, UK general population.

was among the highest) and Norway, these two food groups contributed to over 70% of the total estimated intake. This is similar to what has been found in other European studies [12].

One of the limitations of the study by Freisling et al. was that information relevant for dietary acrylamide intake was not collected during the 24-HR interview (e.g., extent of cooking such as frying or roasting), although information on the cooking method and brand name was collected [5].

Another limitation was that only a single 24-HR was collected per subject, thus, it was not possible to reliably calculate the population intake distribution (e.g., the 95th percentile). For food risk assessment purposes, different approaches are needed.

To better estimate the usual intake for such purposes usually two open-ended non-consecutive 24-HR are collected with sufficient details and ideally complemented with a food propensity questionnaire for episodically consumed foods [35].

Evaluation of Dietary Acrylamide Intake Using Biomarkers

Uncertainty remains with regard to how well acrylamide intake is measured with self-report instruments used in observational studies. Validation studies that were conducted so far reported rather inconsistent results regarding the relationship between self-report acrylamide intake and biomarkers of acrylamide exposure [21,25,27,28].

Within the EPIC study, Ferrari et al. compared estimates of dietary acrylamide intake based on 24-HR and FFQ to Hb adduct levels of acrylamide using data collected from 510 participants from nine countries [21]. After controlling for the most important confounders, individual level correlation coefficients between 24-HR and FFQ estimates and Hb adduct levels of acrylamide, and glycidamide, were overall low, both in smokers and never smokers, with correlation values consistently less than 0.20. Consistent with previous evidence on Hb adducts in the EPIC study [22], dietary acrylamide intake varied considerably across countries, particularly in FFQ, with 89% of the total acrylamide intake variability explained by the country. Variation was less apparent in 24-HR estimates overall and also by country (50% of the total acrylamide intake variability), possibly due to reduced measurement error variation across countries. This is also confirmed by the aggregate-level correlation coefficients ($n=14$) in never smokers using center-specific means of values from FFQ, 24-HR, and Hb adducts, which were stronger for the 24-HR estimates ($r=0.49$) as compared to FFQ estimates ($r=0.34$) [21].

A summary of possible explanations for the low correlation observed between dietary acrylamide intake and Hb adducts is provided in Box 1.

POTENTIAL NEW TOOLS TO ESTIMATE DIETARY ACRYLAMIDE INTAKE

One of the greatest challenges in public health nutrition research is obtaining unbiased estimates of usual intake using available self-report instruments. Traditionally, the most practical and affordable tools, which were deemed "sufficiently" valid for the purpose of a given study, were used to collect dietary intake data. However, these tools may not always be the most accurate ones. For example, it has been shown that traditional FFQ may contain large random and systematic measurement errors [33]. On the other hand, short-term instruments, such as 24-HR, are afflicted by high day-to-day variability, but less so with systematic error [15,33]. Thus, repeated administrations are necessary, which can be costly and challenging with regard to logistics.

Innovative Tools to Measure Usual Intake

Innovative tools (Box 2) may overcome some of the cost and feasibility issues, but not necessarily measurement errors, related to the data collection with short-term instruments by making use of new technologies. The main groups of new technologies described in the literature are: "personal digital assistant-," "mobile phone-," "interactive computer-," "web-," "camera- and tape-recorder-," and "scan- and sensor-based technologies." Their potential to improve, complement or replace short- or long-term self-report instruments has been evaluated recently [36].

Furthermore, innovative study designs may improve the validity (i.e., reducing measurement errors) of self-reported dietary intake data.

These new tools appear to be well suited to improve the usual dietary acrylamide intake estimation. The application of new technologies may allow to collect many replicates of a short-term instrument such as 24-HR in a cost-effective way. Moreover, additional details about acrylamide-containing foods can be queried for. Beside information on cooking method, which is already commonly queried in 24-HR, details about the extent of cooking and brand/product name could also be probed for. Secondly, the application of new methodologies also appears to be promising with respect to dietary acrylamide intake assessment. Combining the details of describing acrylamide-containing foods of short-term instruments with information from long-term instruments on frequency of consumption of episodically consumed acrylamide-containing foods (e.g., crisps) may substantially improve usual intake estimates. Adding information from a valid biomarker of acrylamide intake as an independent measure may pave the way for the future state-of-the-art dietary acrylamide intake assessment.

Studies to come will prove whether and to what extent innovative tools provide dietary acrylamide intake estimates that are superior to the currently applied methods.

Box 2 Innovative tools in dietary assessment

1. "New technologies" are making use of the internet and other communication technologies to collect, manage and exchange dietary data in a more efficient and cost-effective way.
2. "New methodologies" are strengthening complementarity through the combined application of different types of self-report instruments or self-report instruments with biomarkers.

Adapted from [36].

Empirical Prediction Equations

An alternative approach to the estimation of dietary acrylamide intake from matching self-reported acrylamide-containing food items with acrylamide occurrence databases is to use empirically derived weights or scores for each relevant food item. Empirical weights for acrylamide-containing food items are obtained by regressing each food, or food group, as the independent variable, on a reference measurement, ideally a valid biomarker for dietary acrylamide intake, as the dependent variable [7]. The food item-specific regression coefficients can then be multiplied with the individual frequency of consumption of these foods as reported on an FFQ. If sufficiently precise, these weights can then be used in the full sample of a cohort study or other studies with similar underlying dietary habits [7]. The idea is that exposure estimates based on scores derived from prediction equations, which weigh FFQ-food items by their relation to biomarker concentrations, may be more precise than estimates derived from matching FFQ-items to occurrence databases [7,37].

The advantage of such an approach is that an acrylamide occurrence database is not needed and thus, errors related to the compilation and matching procedures can be avoided. Many assumptions regarding food preparation, consumed portion size, and bioavailability can also be avoided. It may also reduce some of the errors inherent to self-reported dietary data. Empirical weights may be particularly useful for dietary compounds obtained from foods with highly variable bioavailability and assessment validity [37]. Dietary acrylamide may thus be a good candidate for such an approach.

However, the challenge is that a sufficiently large sample size (several hundreds of subjects) for the reference method is needed to calculate weights in order to get sufficiently precise food item weights [7]. Furthermore, unless a recovery biomarker is available as a reference method, absolute intakes cannot be estimated with empirical weights. This limits its use to studies evaluating diet–disease associations, where absolute intakes are not necessarily required.

As an example, in a subgroup of 800 nonsmoking postmenopausal women from the EPIC cohort, Obón-Santacana et al. (under review) investigated the most important determinants of Hb adducts of acrylamide and identified "salty biscuits, aperitif biscuits, and crackers," and "dry cakes and biscuits" explaining about 30% of the variability of Hb adduct levels of acrylamide. Whether these or similar empirical scores will predict cancer risk in relation to dietary acrylamide intake better than FFQ, has to be evaluated in future studies.

MAIN CHALLENGES IN ESTIMATING DIETARY ACRYLAMIDE INTAKE

The main challenges to estimate dietary acrylamide intake can be categorized into challenges pertaining to (1) dietary self-report instruments, (2) available biomarkers, and (3) acrylamide occurrence databases. Some of the shortcomings are specific to the acrylamide exposure, whereas others relate to the available methods in general.

1. Challenges pertaining to dietary self-report instruments:
   a. Sufficient descriptive details about the food sources of acrylamide are required (e.g., cooking method), which in theory can be collected with open-ended short-term instruments such as 24-HR; however, details about the consumed acrylamide-containing foods collected have to be balanced with interview duration and participant burden.
   b. Replicate administrations of short-term instruments are necessary to capture the sporadic consumption of some acrylamide-rich foods (e.g., crisps and fried potatoes) in order to decrease or remove within-person random error.
   c. FFQ were often not designed to collect sufficient details about the processing and cooking of all consumed acrylamide-containing foods; often, the food list of FFQ is limited and mainly contains generic or aggregated items (e.g., crackers); and most FFQ that are still currently in use, were designed in the 1990s, where acrylamide in food was not yet discovered and thus, not of concern.
   d. The presence of potentially large measurement errors, both random and systematic, is inherent in all dietary self-report instruments.
2. Challenges pertaining to biomarkers of dietary acrylamide intake:
   a. Low sensitivity of available biomarkers to dietary acrylamide intake; smoking status and alcohol intake are more strongly associated with biomarker variability.
   b. Biomarkers of dietary acrylamide exposure reflect a fairly recent period of exposure (from days to up to 4 months).
   c. Large intra-individual variability in biomarker levels in blood.
   d. Low validity in some subgroups of the population such as smokers or regular consumers of alcohol.
3. Challenges pertaining to acrylamide occurrence databases:
   a. Large variations in acrylamide content within the same food groups potentially reflecting differences in food processing or food preparation or laboratory analysis.
   b. Lack of representativeness of certain foods and of country-specific foods in the occurrence databases.
   c. Lack of analytical data on certain foods (e.g., brand name-specific foods).
   d. Limited number of samples available in some food groups.

POTENTIAL IMPROVEMENTS IN ESTIMATING DIETARY ACRYLAMIDE EXPOSURE

- Collect sufficient descriptive details about the food sources of acrylamide including cooking method, brand/product name information and information regarding the extent of cooking/roasting (e.g., potato chips).
- Collect replicate short-term estimates of food consumption from all study participants.

- Follow general recommendations for measurement error reduction such as standardization of measurements, choosing the best available methods and techniques.
- Develop new or validate further available biomarkers of dietary acrylamide intake.
- Evaluate the applicability of empirical prediction equations in studies investigating dietary acrylamide intake and associations with disease risk, notably cancer.
- Provide analytical data of acrylamide contents in foods accounting for differences in food preparation/processing and/or brand name information.
- Apply an integrated approach of dietary acrylamide intake assessment by combining repeated short-term instruments (e.g., 24-HR recalls) with long-term instruments (e.g., a food propensity questionnaire to collect information on episodically consumed acrylamide-containing foods) and biomarker measurements.
- Apply innovative technologies, for example, web-based self-administered 24-HR recalls, to collect up to six repeated 24-HR recalls asking sufficient details of acrylamide-containing foods.

KEY FACTS

- Currently available estimates of dietary acrylamide intake in humans show that infants, toddlers, and other children are the most exposed groups, where depending on the survey and age group, intake was between 0.5 and 1.9 μg/kg body weight per day and the 95th percentile was between 1.4 and 3.4 μg/kg body weight per day; dietary acrylamide intake of adolescents, adults, and elderly was estimated to be on average about 30% lower than in children [12].
- The top five acrylamide-containing foods contributing to dietary acrylamide intake in adults across Europe were (1) "bread, crisp bread, rusks," (2) "coffee," (3) "potato crisps and French fries," (4) "cakes/dry cakes," and (5) "salty biscuits" [5].
- A recent updated meta-analysis of 32 large prospective cohort studies on acrylamide intake and the risk of cancer of 14 different sites did not find associations for most cancers considered; however, a modest association for kidney cancer, and for endometrial and ovarian cancers in never smokers could not be excluded [38]. Similar results were reported in a systematic review of the literature of studies investigating association between dietary acrylamide intake and various cancers; however, they also concluded that exposure assessment has been inadequate leading to potential misclassification or underestimation of exposure and that studies with improved dietary acrylamide exposure assessment are encouraged [20].
- Measurement error in dietary self-report instruments used in epidemiological studies, usually FFQ instruments or similar, can affect diet–disease association estimates in several ways; the "classical" additive error in the dietary intake estimates, which applies equally to all participants in a study, attenuates the estimated relative risks (i.e., it brings the estimate closer to the value of "zero" risk); however, dietary measurement error also involves bias

proportional to the intake, which inflates (increases) the estimated relative risk, but the random-part of measurement error attenuates it; in combination, the random error usually prevails, leading to an overall attenuation of the relative risk estimates [30]. This means that a modestly increased, but still important, risk related to a dietary component (e.g., acrylamide) is difficult to detect in the presence of dietary measurement error.
- The EPIC Study is the largest prospective cohort worldwide; with more than 521,000 research participants followed for almost 15 years it is possible to study both common and rare diseases; major research fields include nutritional epidemiology using biomarker analysis, dietary questionnaire information in combination with a linear regression calibration component, where a single standardized 24-h recall was collected in more than 36,000 participants, as well as for genetic and lifestyle investigations [39].

MINI DICTIONARY

Dietary monitoring It is the repeated collection of quantitative data on diet and/or nutritional status to discern population (subgroup) differences and/or trends over time; often, the mean intake or distribution of main foods groups (e.g., fruits and vegetables) and/or essential nutrients is compared to recommended levels of consumption or dietary guidelines.

Nutritional epidemiology It seeks to investigate the causal relationship between the nutritional behavior of individuals and health outcome.

Usual intake It is the average long-term intake of specific foods or food groups, nutrients, or other dietary compounds present in the diet; it is not directly observable, but can be estimated from self-reported "actual" intakes; the usual intake of a population, which is the exposure of interest in dietary monitoring, can be described by the population mean intake and its distribution using, for example, two repeated nonconsecutive 24-h dietary recalls; the usual intake of an individual relative to other individuals of a population (i.e., ranking), which is the exposure of interest in nutritional epidemiology, is more difficult to estimate and requires between four and six repeated 24-h recalls.

24-hour (dietary) recalls It is an open-ended dietary assessment method asking the respondent in a structured way to recall and report all foods, beverages, and dietary supplements and their related quantities eaten during the preceding 24 h or on the previous day.

Linear regression calibration It is a statistical technique to partially correct measurement errors in dietary self-report, where the mean intake of the main (more error-prone, but less expensive) instrument is replaced by the predicted values from the reference (more accurate, but more expensive) instrument; for example, the mean dietary acrylamide intake estimated with an FFQ is replaced by the predicted values from a 24-h recall.

SUMMARY POINTS

In public health nutrition research studies, dietary acrylamide intake is usually assessed through estimates based on self-reported dietary consumption of known acrylamide-containing foods using short-term instruments (e.g., 24-HR dietary recalls), and long-term instruments (e.g., FFQ). Self-reported acrylamide-containing foods are then matched with acrylamide occurrence databases. Biomarkers of exposure (e.g., hemoglobin adducts of acrylamide) offer a complementary or alternative approach to assess dietary acrylamide intake. However, all methods are afflicted by a number of shortcomings. Challenges pertaining to dietary self-report instruments include that sufficient descriptive details about the food sources of acrylamide need to be collected (e.g., on cooking). Acrylamide occurrence databases often show substantial variation in acrylamide contents within the same food groups. Challenges pertaining to currently available biomarkers include their low sensitivity to dietary acrylamide intake. Potential improvements to estimate dietary acrylamide intake include the repeated collection of sufficient descriptive details about the food sources of acrylamide including cooking method, brand/product name information and information regarding the extent of cooking/roasting (e.g., of potato chips). Also, developing new or validating further available biomarkers of dietary acrylamide intake would be a valuable approach to improving dietary acrylamide intake assessment.

REFERENCES

[1] Parzefall W. Minireview on the toxicity of dietary acrylamide. Food Chem Toxicol April 2008;46(4): 1360–4.

[2] Tareke E, Rydberg P, Karlsson P, Eriksson S, Tornqvist M. Analysis of acrylamide, a carcinogen formed in heated foodstuffs. J Agric Food Chem August 14, 2002;50(17):4998–5006.

[3] Smith CJ, Perfetti TA, Rumple MA, Rodgman A, Doolittle DJ. "IARC group 2A Carcinogens" reported in cigarette mainstream smoke. Food Chem Toxicol April 2000;38(4):371–83.

[4] Dybing E, Sanner T. Risk assessment of acrylamide in foods. Toxicol Sci September 2003;75(1):7–15.

[5] Freisling H, Moskal A, Ferrari P, Nicolas G, Knaze V, Clavel-Chapelon F, et al. Dietary acrylamide intake of adults in the European Prospective Investigation into Cancer and Nutrition differs greatly according to geographical region. Eur J Nutr June 2013;52(4):1369–80.

[6] Slimani N, Freisling H, Illner AK, Huybrechts I. Methods to determine dietary intake. In: Lovegrove JA, Hodson L, Sharma S, Lanham-New S, editors. Nutrition research methodologies. 1st ed. West Sussex: Wiley; 2015. p. 48–70.

[7] Willett W. Nutritional epidemiology. 3rd ed. New York: Oxford University Press; 2013.

[8] Greenfield H, Southgate DAT. Food composition data – production, management and use. 2nd ed. Rome: FAO; 2003.

[9] Dybing E, Farmer PB, Andersen M, Fennell TR, Lalljie SP, Muller DJ, et al. Human exposure and internal dose assessments of acrylamide in food. Food Chem Toxicol March 2005;43(3):365–410.

[10] Jenab M, Slimani N, Bictash M, Ferrari P, Bingham SA. Biomarkers in nutritional epidemiology: applications, needs and new horizons. Hum Genet June 2009;125(5–6):507–25.

[11] Kuhnle GGC. Biomarkers of intake. In: Lovegrove JA, Hodson L, Sharma S, Lanham-New S, editors. Nutrition research methodologies. 1st ed. West Sussex: Wiley; 2015. p. 90–107.

[12] European Food Safety Authority. Results on acrylamide levels in food from monitoring years 2007–2009 and exposure assessment. EFSA J 2011;9(4):2133. [Ref Type: Generic].

[13] Hogervorst JG, Baars BJ, Schouten LJ, Konings EJ, Goldbohm RA, van den Brandt PA. The carcinogenicity of dietary acrylamide intake: a comparative discussion of epidemiological and experimental animal research. Crit Rev Toxicol July 2010;40(6):485–512.

[14] Slimani N, Casagrande C, Nicolas G, Freisling H, Huybrechts I, Ocke MC, et al. The standardized computerized 24-h dietary recall method EPIC-Soft adapted for pan-European dietary monitoring. Eur J Clin Nutr July 2011;65(Suppl. 1):S5–15.

[15] Carroll RJ, Midthune D, Subar AF, Shumakovich M, Freedman LS, Thompson FE, et al. Taking advantage of the strengths of 2 different dietary assessment instruments to improve intake estimates for nutritional epidemiology. Am J Epidemiol February 15, 2012;175(4):340–7.

[16] Hogervorst JG, Schouten LJ, Konings EJ, Goldbohm RA, van den Brandt PA. A prospective study of dietary acrylamide intake and the risk of endometrial, ovarian, and breast cancer. Cancer Epidemiol Biomarkers Prev November 2007;16(11):2304–13.
[17] Lujan-Barroso L, Gonzalez CA, Slimani N, Obon-Santacana M, Ferrari P, Freisling H, et al. Dietary intake of acrylamide and esophageal cancer risk in the European Prospective Investigation into Cancer and Nutrition cohort. Cancer Causes Control May 2014;25(5):639–46.
[18] Obon-Santacana M, Slimani N, Lujan-Barroso L, Travier N, Hallmans G, Freisling H, et al. Dietary intake of acrylamide and pancreatic cancer risk in the European Prospective Investigation into Cancer and Nutrition (EPIC) cohort. Ann Oncol October 2013;24(10):2645–51.
[19] Obon-Santacana M, Peeters PH, Freisling H, Dossus L, Clavel-Chapelon F, Baglietto L, et al. Dietary intake of acrylamide and epithelial ovarian cancer risk in the european prospective investigation into cancer and nutrition (EPIC) cohort. Cancer Epidemiol Biomarkers Prev January 2015;24(1):291–7.
[20] Virk-Baker MK, Nagy TR, Barnes S, Groopman J. Dietary acrylamide and human cancer: a systematic review of literature. Nutr Cancer 2014;66(5):774–90.
[21] Ferrari P, Freisling H, Duell EJ, Kaaks R, Lujan-Barroso L, Clavel-Chapelon F, et al. Challenges in estimating the validity of dietary acrylamide measurements. Eur J Nutr August 2013;52(5):1503–12.
[22] Vesper HW, Slimani N, Hallmans G, Tjonneland A, Agudo A, Benetou V, et al. Cross-sectional study on acrylamide hemoglobin adducts in subpopulations from the European Prospective Investigation into Cancer and Nutrition (EPIC) Study. J Agric Food Chem August 13, 2008;56(15):6046–53.
[23] Vikstrom AC, bramsson-Zetterberg L, Naruszewicz M, Athanassiadis I, Granath FN, Tornqvist MA. In vivo doses of acrylamide and glycidamide in humans after intake of acrylamide-rich food. Toxicol Sci January 2011;119(1):41–9.
[24] Wirfalt E, Paulsson B, Tornqvist M, Axmon A, Hagmar L. Associations between estimated acrylamide intakes, and hemoglobin AA adducts in a sample from the Malmo Diet and Cancer cohort. Eur J Clin Nutr March 14, 2007.
[25] Wilson KM, Vesper HW, Tocco P, Sampson L, Rosen J, Hellenas KE, et al. Validation of a food frequency questionnaire measurement of dietary acrylamide intake using hemoglobin adducts of acrylamide and glycidamide. Cancer Causes Control April 2009;20(3):269–78.
[26] Kutting B, Uter W, Drexler H. The association between self-reported acrylamide intake and hemoglobin adducts as biomarkers of exposure. Cancer Causes Control April 2008;19(3):273–81.
[27] Tran NL, Barraj LM, Murphy MM, Bi X. Dietary acrylamide exposure and hemoglobin adducts–National Health and Nutrition Examination Survey (2003–04). Food Chem Toxicol November 2010;48(11):3098–108.
[28] Bjellaas T, Olesen PT, Frandsen H, Haugen M, Stolen LH, Paulsen JE, et al. Comparison of estimated dietary intake of acrylamide with hemoglobin adducts of acrylamide and glycidamide. Toxicol Sci July 2007;98(1):110–7.
[29] Vikstrom AC, Wilson KM, Paulsson B, Athanassiadis I, Gronberg H, Adami HO, et al. Alcohol influence on acrylamide to glycidamide metabolism assessed with hemoglobin-adducts and questionnaire data. Food Chem Toxicol March 2010;48(3):820–4.
[30] Freedman LS, Schatzkin A, Midthune D, Kipnis V. Dealing with dietary measurement error in nutritional cohort studies. J Natl Cancer Inst July 20, 2011;103(14):1086–92.
[31] Freisling H, van Bakel MM, Biessy C, May AM, Byrnes G, Norat T, et al. Dietary reporting errors on 24 h recalls and dietary questionnaires are associated with BMI across six European countries as evaluated with recovery biomarkers for protein and potassium intake. Br J Nutr March 2012;107(6):910–20.
[32] Slimani N, Kaaks R, Ferrari P, Casagrande C, Clavel-Chapelon F, Lotze G, et al. European Prospective Investigation into Cancer and Nutrition (EPIC) calibration study: rationale, design and population characteristics. Public Health Nutr December 2002;5(6B):1125–45.
[33] Schatzkin A, Kipnis V, Carroll RJ, Midthune D, Subar AF, Bingham S, et al. A comparison of a food frequency questionnaire with a 24-hour recall for use in an epidemiological cohort study: results from the biomarker-based Observing Protein and Energy Nutrition (OPEN) study. Int J Epidemiol December 2003;32(6):1054–62.
[34] Souverein OW, Dekkers AL, Geelen A, Haubrock J, de Vries JH, Ocke MC, et al. Comparing four methods to estimate usual intake distributions. Eur J Clin Nutr July 2011;65(Suppl. 1):S92–101.

[35] Subar AF, Dodd KW, Guenther PM, Kipnis V, Midthune D, McDowell M, et al. The food propensity questionnaire: concept, development, and validation for use as a covariate in a model to estimate usual food intake. J Am Diet Assoc October 2006;106(10):1556–63.
[36] Illner AK, Freisling H, Boeing H, Huybrechts I, Crispim SP, Slimani N. Review and evaluation of innovative technologies for measuring diet in nutritional epidemiology. Int J Epidemiol August 2012;41(4):1187–203.
[37] Hendrickson SJ, Willett WC, Rosner BA, Eliassen AH. Food predictors of plasma carotenoids. Nutrients October 2013;5(10):4051–66.
[38] Pelucchi C, Bosetti C, Galeone C, La VC. Dietary acrylamide and cancer risk: an updated meta-analysis. Int J Cancer June 15, 2015;136(12):2912–22.
[39] Riboli E, Hunt KJ, Slimani N, Ferrari P, Norat T, Fahey M, et al. European Prospective Investigation into Cancer and Nutrition (EPIC): study populations and data collection. Public Health Nutr December 2002;5(6B):1113–24.

CHAPTER 3

Secular Trends in Food Acrylamide

Hanna Mojska
Department of Food and Food Supplements, National Food and Nutrition Institute, Warsaw, Poland

INTRODUCTION

Acrylamide is classified by the International Agency for Research on Cancer (IARC) as a compound "probably carcinogenic for humans" (Group 2A) [1]. In April 2002, the Swedish National Food Agency in collaboration with scientists from the Stockholm University [2] first reported the presence of high content of acrylamide in potato and cereal products processed at high temperatures, mainly by frying and roasting. The FAO/WHO Joint Expert Committee on Food Additives (JECFA) has recently concluded [3] that the estimated margins of exposure (MOE) to acrylamide from the diet are low, which indicates that dietary intake is a human health concern and that exposure assessment using determination of biomarkers of acrylamide in blood and urine were required to estimate the risk.

From the moment of publishing initial information [2] about the presence of acrylamide in thermally processed foods, intensive studies on the possibility of decreasing its content in foodstuffs were commenced around the world. The first guide that describes mitigation strategies of acrylamide formation in food has been already developed in 2005 by the Confederation of the European Food and Drink Industry (CIAA) [4]. CIAA (since June 2011 FoodDrinkEurope) issued a set of "tools" related to decreasing the level of acrylamide in food, the so-called "Toolbox," whose latest version was issued in 2013 [5]. In 2009, also the Codex Alimentarius Commission FAO/WHO developed the Code of Practice for the Reduction of Acrylamide in Food [6]. The aim of the developed guides is to support food manufacturers in taking up proper actions, both on the level of selecting raw materials and ingredients, and within the different stages of processing, in order to decrease the content of acrylamide in the final product.

Collection of data on acrylamide content in the highest possible number of various food products is crucial to evaluate the effectiveness of the tools described in the Toolbox. It would also provide the information helpful in risk assessment of dietary acrylamide intake and hence result in taking proper action within the scope of risk management.

Acrylamide in Food
http://dx.doi.org/10.1016/B978-0-12-802832-2.00003-6

The first database on acrylamide content in food products in the European Union (EU) was established by the European Commission's Directorate General Joint Research Centre (JRC) in close cooperation with CIAA. The EU database of acrylamide levels in food was kept between 2003 and 2006. Evaluation of results gathered in the database was published in 2007 [7].

Another general European database on acrylamide content in food was established by the European Food Safety Authority (EFSA) in 2007. The database contains analytical results, which are supplied by EU member states in accordance with Commission Recommendations on the monitoring of acrylamide levels in food (2007/331/EC and 2010/307/EC). EFSA reports published annually from 2009 to 2012 presented the results from European acrylamide monitoring done in subsequent years 2007 [8] and 2008 [9], and collective results from 2007 to 2009 [10] and 2007 to 2010 [11].

In 2011, JECFA also published assessment of results of acrylamide content in food, coming from 31 countries around the world [12].

It needs to be emphasized that food manufacturers collected their own data on acrylamide levels in food products, to assess whether the mitigation strategies identified within the Toolbox approach have been successfully implemented by the European food industry. Between 2002 and 2011, the European Snack Association (ESA) collected data on acrylamide levels in samples of potato crisps, which were supplied by European manufacturers. The results were published in 2013 [13]. There were also numerous results published on acrylamide levels in food in individual countries [14,15].

The results, collected within the course of several years from different databases, may be used for observing the secular trends in acrylamide levels for selected food categories, and they may serve to evaluate the effectiveness of actions taken up by the manufacturers in order to decrease the level of acrylamide in food. The aim of this chapter is to present tendencies of changes in acrylamide levels in food within the course of several years, on the basis of the review of analyses performed by various authors, and derived from available databases, and the own data.

The time trends in acrylamide content in various food categories were presented on the basis of analysis of results from two general European databases: JRC [7] and EFSA [11], a database of the food manufacturers' organization ESA concerns one food category [13], and the database of National Food and Nutrition Institute (NFNI) in Poland, containing results of acrylamide levels in several categories of food in a single country [17].

THE TIME TRENDS IN ACRYLAMIDE LEVELS FOR SELECTED FOOD CATEGORIES, CALCULATED FOR THE TIME PERIOD OF 2002–2006, ON THE BASIS OF "EUROPEAN UNION DATABASE OF ACRYLAMIDE LEVELS IN FOOD" [7]

The first general European database on acrylamide levels in foodstuffs was designed by the European Commission's Directorate General JRC in close cooperation with CIAA and opened for supply of results in April 2003. Both official food control and private laboratories were invited to submit analytical results of determination of acrylamide content in food products. In 2007, Wenzl and Anklam [7] published the evaluation of 9228 results on acrylamide levels in various products, collected in EU database between 2002 and 2006. The authors reviewed the results collected in the database and calculated the time trends in acrylamide content in several food categories. In order to apply only real and comparable results in the evaluation, there were some criteria developed, on the basis of which the results for further assessment were selected. These criteria first of all regarded to the quality and reliability of the analytical results and sensitivity of the analytical methods. Further evaluation excluded results, which the laboratories did not provide full information for, regarding, for example, limits of quantification (LOQ) and results of participation in proficiency testing (PT). Exclusion also pertained to the results, which were under a given LOQ, or the provided LOQ information was misleading, and the results, for which the provided LOQ and limits of detection (LOD) were the same. What is more, research results sent by the laboratories which obtained unsatisfactory results of participation in PT were excluded as well. Eventually, from the initial group of 9228 results, the evaluation was applied in case of 7239, which represented the following six categories of products: bakery products, biscuits (1068), potato crisps (839), gingerbread (1007), crisp bread (454), coffee (291), and cookies and biscuits for diabetics (402). A complex review of acrylamide content in the above mentioned food categories was performed together with comparison of the results between two 24-month long periods (May 2002–April 2004 vs May 2004–April 2006).

Figure 1 presents comparison of acrylamide content in the six food categories between two periods, when the samples of food were collected.

Wenzl and Anklam [7] found out a shift of distribution toward lower values for potato crisps (597 vs 391 μg/kg), crisp bread (257 vs 228 μg/kg), and cookies and biscuits for diabetics (290 vs 184 μg/kg). A minor increase of the median of acrylamide levels was observed in the group of bakery products, biscuits (131 vs 163 μg/kg) and a major one in relation to coffee (281 vs 432 μg/kg). In case of gingerbread, the median of acrylamide content was maintained on a similar level within the whole period. Attention is attracted by a high level of maximum determined levels of acrylamide in all of the above mentioned food categories, which between 2002 and 2004 was maintained on the level from 1112 μg/kg (coffee) to 7834 μg/kg (gingerbread) and from 1188 μg/kg (coffee) to 6141 μg/kg (gingerbread) between 2004 and 2006. This suggests a fact that in each category of products, regardless of the tendency of decrease or of increase, there were some samples of a considerably high content of acrylamide. The obtained results confirm a high level of variability of acrylamide content within separate product categories, observed in numerous studies [14–16].

While referring to representativeness of the results collected in the database, for product present in the European market, there is a necessity to draw attention to the fact that the collected results came mainly from one country, that is, Germany (77.6%). Only 10.3% of results were supplied by laboratories from other eight countries: Austria, Belgium, Finland,

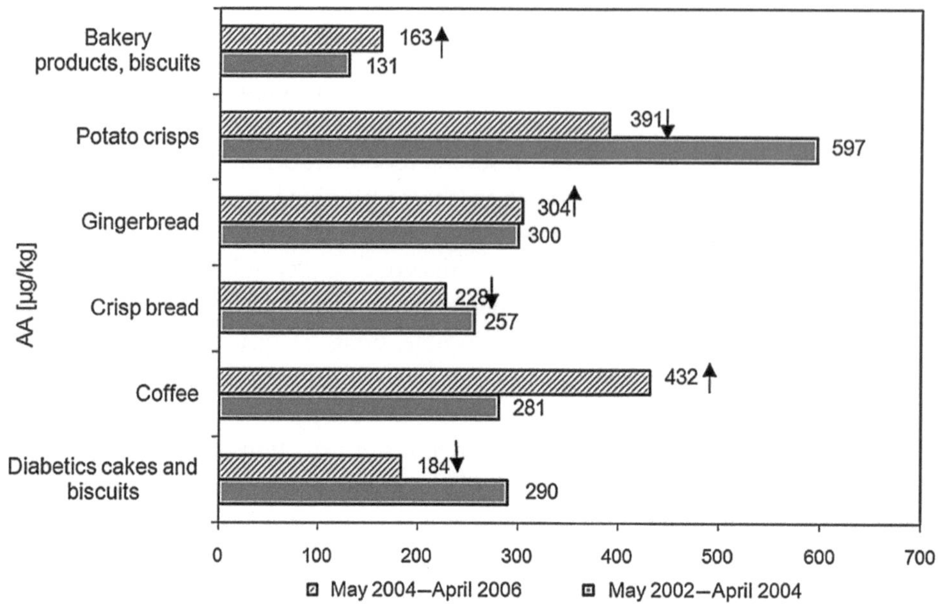

Figure 1 Comparison of a median of acrylamide levels in selected categories of food, between two periods of sampling. *Developed on the basis of Wentzl and Anklam [7].*

Greece, Ireland, Italy, the Netherlands, and Great Britain. With a general number of 9228 supplied results, it provides about 100 results from the listed eight countries on an average, if the distribution of results was even. Another 12.1% of results was provided by CIAA. However, the authors do not inform whether the countries, which the latter results came from, were identified. There is no information whether the food samples examined in terms of acrylamide levels were selected randomly or incidentally. Another problem, which is highlighted by the authors, is posed by differences in homogeneity of the food categories, for example, the category of bakery products, biscuits was highly differentiated (the results related to numerous products), while the categories like coffee or gingerbread were more homogeneous. In many cases the results sent to the EU database were not accompanied by information, among others on components of the researched products or technological processes of, for example, conditions of thermal processing. What is more, information about manufacturing and expiration dates was not always provided as well. As it is generally known, acrylamide levels change while storing the products. A crucial problem was also posed by various analytical methods used by laboratories and differences in confirmed LOQ and LOD for these methods.

To sum up, it needs to be emphasized that while analyzing the results collected between 2002 and 2006 in EU database, Wenzl and Anklam [7] concluded a lowering tendency of decreasing acrylamide contents in three categories of products: "potato crisps," "crisp bread," and "cookies and biscuits for diabetics." An increase of acrylamide levels was observed in "coffee" and "bakery products, biscuits." Experiences gained during the development of this first general European database served for preparation of principles ruling performance of monitoring research of acrylamide levels in food products in Europe in subsequent years.

"COMMON EUROPEAN TRENDS" AND *"MARGINAL EUROPEAN TRENDS"* OF CHANGES OF ACRYLAMIDE LEVELS IN FOOD BETWEEN 2007 AND 2010, CALCULATED ON THE BASIS OF THE DATABASE OF THE EFSA [11]

Since 2007, the systematic monitoring of European acrylamide levels in foodstuffs has been started under EU Commission Recommendation 2007/331/EC of 3 May 2007 [18]. According to the above-mentioned recommendation, Member States should carry out monitoring in the subsequent 3 years, 2007, 2008, and 2009, and annually provide the monitoring data to EFSA for evaluation. In 2010, the Commission issued another Recommendation on the monitoring of acrylamide levels in food (2010/307/EC) [19], with a revised food categorization. To provide collection of representative data, recommendations of the EU Commission describe in detail the principles of performing monitoring research and the form, for which the research results are to be submitted to EFSA.

EU Commission recommendation [18] lists 10 categories of food products, whose samples should be collected randomly within the territory of each EU member state. These are: French fries, sold as ready-to-eat; potato crisps; precooked French fries/potato products for home cooking; soft bread; breakfast cereals; biscuits, crackers, crisps bread, and similar products; coffee and coffee substitutes; baby foods; processed cereal-based food for infants and young children; other products, that is, potato, cereal, coffee, and cocoa products, together with infant food, which were not included in one of the previous categories, but they still reflect food habits of separate member states. In a subsequent Commission Recommendation from 2010 [19], the product categories were provided in more detail. All categories were complemented by two to five subcategories. Samples of the mentioned products, according to the quoted Commission Recommendations should be collected randomly and in order to provide their representativeness in relation to a batch size, which they are collected from, they should be sampled in line with Attachment B to the Commission Regulation 333/2007/EC [20].

For each member state, there has been a minimum number of samples agreed on, coming from each category, which should be collected and examined within each year of monitoring. Commission Recommendations also specified the demands in relation to the applied analytical methods, among other the requirements within the scope of methods validation, LOD and LOQ levels, correction of the result by the recovery, as well as the requirement to include information about the participation of the laboratory in a proper PT program and about the obtained results, in the report for EFSA.

In 2012, EFSA published a summary of monitoring research results from 2007 to 2010, together with presentation of the time trends of changes in the acrylamide levels in food products in Europe within those years [11].

Within 4 years of European monitoring (2007–2010), there were generally 13,162 results of acrylamide content in food supplied to EFSA, coming from 25 European countries. It needs to be emphasized that within this period only 16 states supplied the results from each year of monitoring research, the remaining nine countries provided the results two or three times within 4 years. A downward trend both in the number of states handing the results to EFSA (in 2007: 23 states and in 2010: 19 states) and in the number of provided results was observed. For instance, in 2010 only 57% on average of a minimum number of results for each category of products was supplied to EFSA.

After verification and reclassification of product categories in accordance with a detailed division recommended by the Commission, and after verification of all results in accordance with the accepted criteria [11], the evaluation was applied in 12,716 results.

Figure 2 presents the percentage share of results from separate product categories in a general number of results evaluated by EFSA.

A highly differentiated amount of the reported results in separate food categories may attract attention. The highest number of results reported to EFSA pertained to the group of "Biscuits, crackers, crisp bread and similar," and then "French fries, sold as ready-to-eat," and "Other foods." The least number of results was reported in the "Baby foods" category.

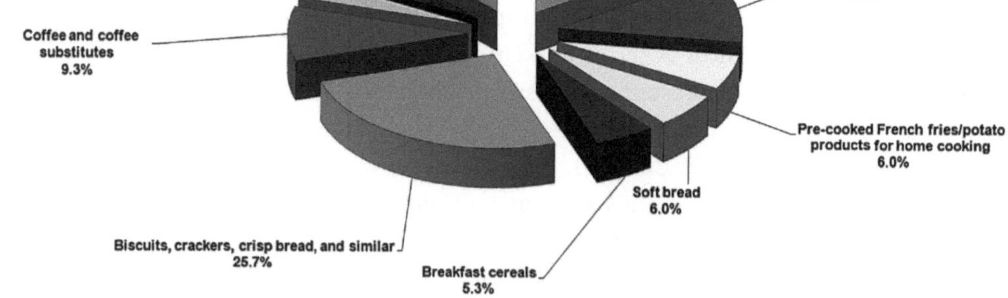

Figure 2 Percentage share of results from separate food products categories, in a general number of results from 2007 to 2010, evaluated by EFSA. *Developed on the basis of EFSA, 2012 [11].*

According to the quoted report, cut-off for left-censored (LC) data was set at the recommended LOQ values (30 and 50 μg/kg) [18,19]. Then there were three different scenarios applied in evaluation of analytical results sent to EFSA. The "lower bound" (LB) was obtained by assigning a value zero to all samples reported as being less than the respective left-censoring limit (LC), the "middle bound" (MB) was obtained by assigning half of the LC limit, and the "upper bound" (UB) was obtained by assigning the LC limit as the sample results. The EFSA report [11] presents the tendencies of changes in acrylamide levels in 10 food products' categories in Europe, between 2007 and 2010, and the applied analysis method. In short, the mixed effect model approach was applied to test whether a "*common European trend*" in acrylamide could be seen using log transformed MB and UB mean acrylamide content values. This "*common European trend*" was reported when the tendency in different countries headed in the same direction, and was shown to be statistically significant ($p<0.05$). A "*marginal European trend*" was also reported in the EFSA report. This term means that the group of evaluated states may incorporate a country, where the observed trend was opposite to the "*common European trend*." Time trends in acrylamide levels for different food categories were estimated with a linear model.

Figures 3–6 show changes in MB mean acrylamide levels of the 10 main food categories and of the eight subcategories, from 2007 to 2010, based on the results of the EFSA trend analysis [11].

A significant decrease ($p<0.05$) in MB mean acrylamide levels, evaluated in the EFSA report [11] as the "*common European trend*" was observed solely in one major food category: "Processed cereal-based foods for infants and young children" (a decrease by 49% between 2007 and 2010) and in two subcategories: "Biscuits and rusks for infants and young children" (a decrease by 43% between 2007 and 2010), and "Non-potato savory snacks" (a decrease by 35% between 2007 and 2010). In turn, a significant increase

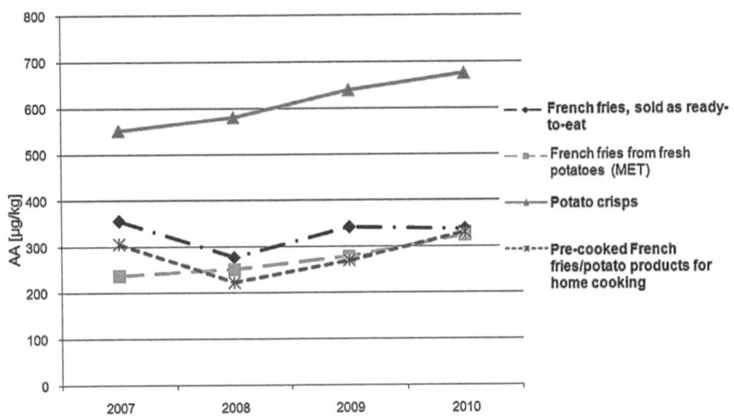

Figure 3 The time trends of changes in acrylamide levels for potato products. *Developed on the basis of EFSA, 2012 [11].*

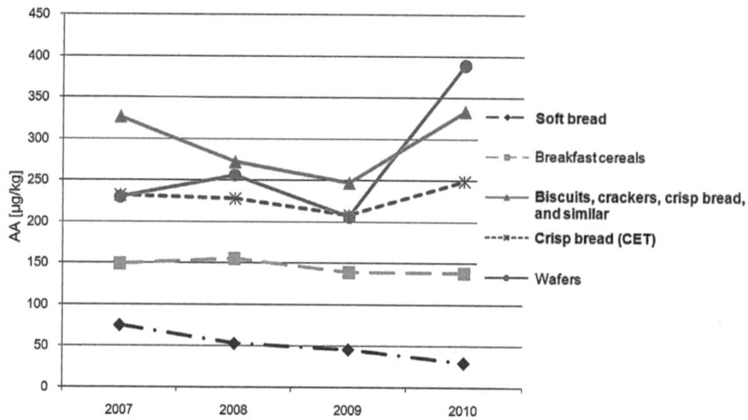

Figure 4 The time trends of changes in acrylamide levels for soft bread, biscuits, wafers, and breakfast cereals. *Developed on the basis of EFSA, 2012 [11].*

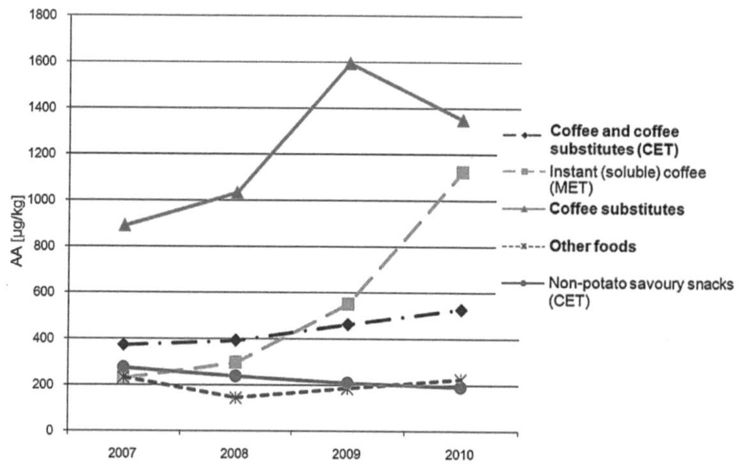

Figure 5 The time trends of changes in acrylamide levels for coffee and coffee substitutes, and group of other foods. *Developed on the basis of EFSA, 2012 [11].*

($p<0.05$) of acrylamide levels, evaluated as the "*common European trend*," was also observed in one major category: "Coffee and coffee substitutes" (an increase by 41% between 2007 and 2010) and in one subcategory called "Crisp bread" (an increase by 42% between 2007 and 2010). For the "Wafers" subcategory, the observed "*common European trend*" was only marginally significant ($p=0.08$). At the same time, changes in acrylamide levels, assessed in the EFSA report [11] as the "*marginal European trend*" were observed in three subcategories: a decrease in "Other processed cereal-based foods for infants and young children" (a decrease by 46% between 2007 and 2010), an increase in "French fries from fresh potato" (an increase by 39% between 2007 and 2010), and in "Instant (soluble) coffee" (an increase by 214% between 2007 and 2010).

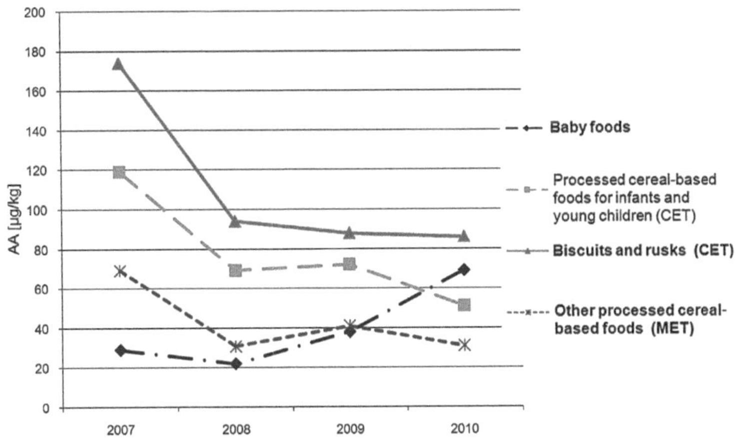

Figure 6 The time trends of changes in acrylamide levels for foods for infants and young children. *Developed on the basis of EFSA, 2012 [11].*

To sum up, it needs to be concluded that the analysis of tendencies of changes in acrylamide levels between 2007 and 2010 presented in the EFSA report [11], demonstrated that the significant "*common European trend*" in the decrease of acrylamide levels was observed only in one out of 10 examined food categories, and in two subcategories. Additionally, significant "*marginal European trend*" in the decrease of acrylamide levels was observed in only one category. It seems worrying that significant tendencies of increase of acrylamide levels were observed as well. This may point to lack or ineffectiveness of voluntary actions taken up by manufacturers on various stages of food production, in order to decrease acrylamide content in the final product. Simultaneously, EFSA in its report [11] has come to a conclusion that a four-year observational period seems too short to distinguish random fluctuations from real trends. This fact is related to a considerably high differentiation in acrylamide level in the same food categories, between both the countries and various manufacturers of a given food product's categories within a country. There are also significant differences between products from the same category or even products of the same brand, but from different manufacturing batches observed within separate food categories in a single country. Hence, it is of key importance, as suggested by EFSA conclusion, to examine an adequate number of samples from each food category each year. It is also significant to qualify the samples to proper food categories, to prepare them properly for analysis, and to document the applied analytical method.

There are also two aspects, which should be focused on in the context of monitoring research of acrylamide levels in food still performed in European countries. First of all, the highest number of results from the period of 2007 to 2010 came from only one country, Germany (12.5%), then from Great Britain, the Czech Republic, Belgium, and from Italy. Distribution of the results was not as uneven as in the case of the results presented by Wenzl and Anklam [7], where almost 80% of all results were provided by laboratories from

Germany. However, a significantly higher number of results from one country may influence the observed secular trends. Second of all, a decrease in the number of countries that supply the results to EFSA is disturbing, which may prove a decrease in interest of national authority bodies in the matter of acrylamide presence in food and its adverse effect for human health.

To provide protection of health of the European Community citizens, and in the light of lack of unambiguous results of decreasing the acrylamide content in food, the EU Commission issued in 2011 the Recommendation on investigations into the acrylamide levels in food [21], the latest version of which was published in 2013 [22]. The above-mentioned Recommendations provide the indicative values for acrylamide levels in separate categories of food products. It needs to be mentioned that these are not maximum allowable levels of acrylamide in food. In case the indicative value is exceeded in a certain product, risk analysis needs to be carried out. Moreover, the official food control employees should perform an investigation together with the manufacturer's representatives, to check whether critical control points were identified, and whether proper "tools"—proposed by FoodDrinkEurope Acrylamide Toolbox, intended to decrease the acrylamide levels in food—have been incorporated.

A DOWNWARD TREND IN ACRYLAMIDE LEVEL IN POTATO CRISPS IN EUROPE FROM 2002 TO 2011, BASED ON THE EUROPEAN SNACKS ASSOCIATION DATASET [13]

Results of acrylamide levels in one food category, examined within the subsequent 10 years, have been presented in the recently published work by Powers et al. [13].

The authors analyzed the most numerous, as for now, database encompassing 40,455 results of acrylamide levels in potato crisps. The results were supplied to the European Snacks Association between 2002 and 2011, by European manufacturers from 20 European countries. An asset of the analyzed collection of results was homogeneity because the samples for the research were in majority composed from crisps made from fresh, sliced potatoes. Furthermore, only those data for which acrylamide analyses had been carried out following authenticated analytical procedures using liquid chromatography with tandem mass spectrometry (LC-MS/MS) were included in the study. An analysis of variance was used to test the overall significance (F-test) of differences in acrylamide levels in crisps among subsequent years. Figure 7 shows the changes in the mean and 95% quartile (Q95) values for acrylamide in samples of potato crisps from 2002 to 2011, based on the analysis of the ESA dataset made by Powers et al. [13].

Analysis performed by Powers et al. [13] demonstrated a clear, statistically significant ($p<0.001$) influence on the mean acrylamide content in potato crisps, with a general tendency to lower this level, starting at 763 ± 91.1 μg/kg in 2002 to 358 ± 2.5 μg/kg in 2011. The authors also found out a generally significant ($p<0.05$) decrease in the acrylamide content in crisps by $53\% \pm 13.5\%$ between 2002 and 2011. A statistically significant decrease

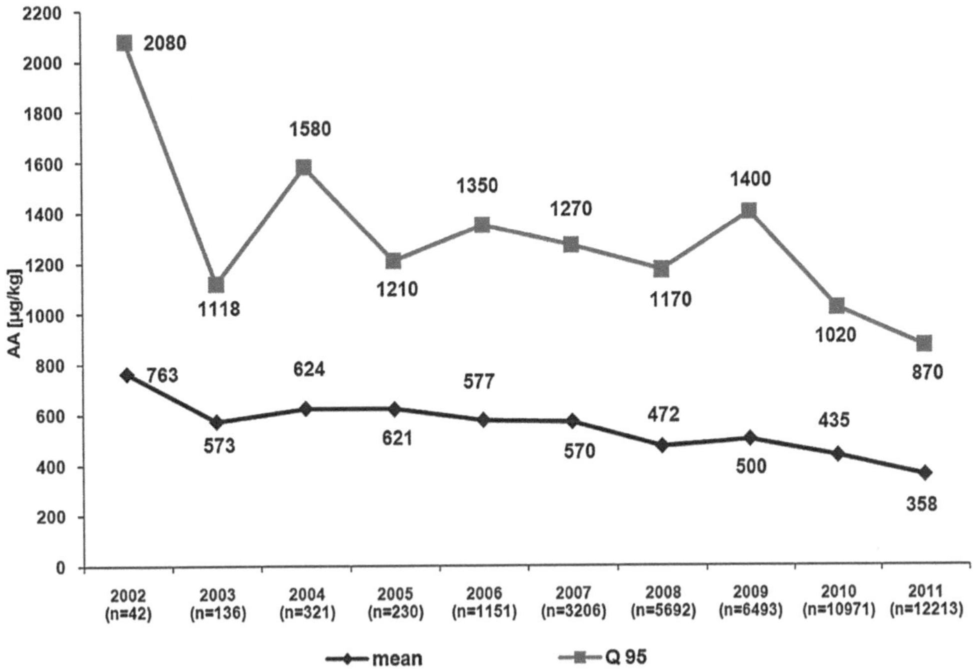

Figure 7 The mean and 95th percentile values for acrylamide in potato crisps samples from 2002 to 2011. *Developed on the basis of analyses by Powers et al. [13].*

in the acrylamide levels in crisps was found out between the results from the following subsequent years: 2005 and 2006, 2007 and 2008, 2009 and 2010, and 2010 and 2011. Similar observations were in the case of a decrease of the acrylamide levels in crisps, by 95th percentile. At the same time, attention is attracted to considerably high acrylamide levels found out in single samples, which fluctuated within the range from 1780 µg/kg in 2005 to 12000 µg/kg in 2010. It needs to be emphasized that Powers et al. [13] draw attention to the fact that the observed downward trend in acrylamide level in potato crisps in the whole of Europe does not necessarily need to be confirmed in separate countries. This is connected with influence of different factors in every country, among others those agronomic and technological, on acrylamide levels in the final product. The authors concluded the statistically significant tendency of decreasing the acrylamide levels in potato crisps between 2002 and 2011, which they demonstrated in the performed analysis, suggesting that "the *Toolbox* approach coordinated by FoodDrinkEurope has been effective."

To sum up, there are several crucial elements of Powers et al. [13] analysis of ESA dataset that needs to be taken into consideration. First of all, the long-standing observational period (10 years), which allows the elimination of influence of short-term changes on the acrylamide levels in potato crisps. Second of all, a very big ESA database with results, thanks to which the statistical analysis was performed on a collection of data composed on more than

40,000 units. What is more, the results collected in the database came from 20 European countries, which seem to assure representativeness of the data collection. A crucially important element is the matter of method applied to determine of acrylamide in potato crisps samples. All samples were analyzed with application of the same validated method (LC-MS/MS), which eliminates influence of errors arising from differences between methods of analyses. On the other hand, it needs to be borne in mind that the research was conducted in various laboratories, in at least 20, and there was no information pertaining to the LOD and LOQ for methods and results of laboratories participation in proper PT programs.

A shortcoming of the presented collection of data, mentioned by the Authors in their work, was also a changeable number of samples examined in subsequent years, from 42 in 2002 to 12,213 results in 2011. At the same time, there is a need to highlight an even increasing tendency in the number of results supplied each year to ESA, which exceeded 1000 per year already in 2006. An opposite situation, consisting in lowering the number of supplied results, took place in case of the previously discussed European monitoring results [11]. On the other hand there is also a need to attract attention to the fact that Powers et al. [13] did not provide any information on distribution of results for each country, in the general number of results sent from 20 countries. The previously presented EU database of acrylamide levels in food [7] suggests that majority of results (about 80%), supplied to this dataset, was from Germany. Also in case of the general European EFSA database, the biggest number of results came from Germany [11]. It remains unknown whether it was similar in case of ESA. There is also no information about the manner of sampling. Analytical research results were sent to the ESA database by the manufacturers, therefore the following information is crucial for evaluation of the results collection representativeness: Were the researched crisps samples collected randomly from the market within a given country? Did they come directly from the manufacturers—one or several manufacturers in a given state? If the samples came directly from the manufacturers, how were they collected from a product batch—from one batch or from several batches?

To sum up, it needs to be concluded that results of analysis of changes in acrylamide levels in crisps presented by Powers et al. [13], suggest a statistically significant tendency of decreasing the levels of acrylamide within 10 years, and they seem to confirm effectiveness of mitigation strategies used voluntarily by food manufacturers in relation to one category of potato products. At the same time, some concerns are triggered by a still very high content of acrylamide found out in single samples of potato crisps.

TENDENCIES OF CHANGES IN ACRYLAMIDE LEVELS IN CERTAIN CATEGORIES OF FOOD IN POLAND, BETWEEN 2004 AND 2010, ON THE BASIS OF NATIONAL FOOD AND NUTRITION INSTITUTE DATASET [17]

In 2004, monitoring research in acrylamide levels in food was performed for the first time in Poland. Tested material comprised 24 samples of potato crisps collected randomly in selected stores within the whole state. The monitoring research was carried on in subsequent years, and in 2006 it embraced another four product categories: French fries ready-to-eat; precooked French fries; breakfast cereals (cornflakes); bakery products, biscuits. Between 2007 and 2009, food samples were collected in accordance with the recommendation of the EU Commission [18]. Starting from 2004, the food samples were analyzed in the National Food and Nutrition Institute (NFNI) laboratory. Analysis of acrylamide levels in samples of products collected between 2004 and 2007 was performed through the gas chromatography method with mass spectrometry, and between 2008 and 2009 through the method of LC-MS/MS, as previously described [14,23,24]. Since 2011, determination of acrylamide in product samples, collected for monitoring research in Poland, is performed in two laboratories of official food control, with GC/MS method. The methods are validated and accredited and both the laboratories take regular part in appropriate PT programs, obtaining satisfactory results. The applied methods meet the requirements described in EU Commission recommendations [18,19]. Monitoring research results of acrylamide content in food, performed from 2004, are collected in the database of National Food and Nutrition Institute.

On the basis of the above-mentioned database, analysis of tendencies of changes in acrylamide levels was conducted in the following food categories: potato crisps (84), French fries ready-to-eat (59), precooked French fries (63), breakfast cereals (cornflakes, 73), bakery products, biscuits (71).

Evaluation of significance of differences in levels of acrylamide in the analyzed food categories was performed through the Kruskal–Wallis analysis of variance by ranks, assuming significance of differences on the level of $p < 0.05$.

Among five evaluated food categories, there was a decrease in the median of acrylamide levels found out between 2004, 2006, and 2012 in three categories: potato crisps (from 753 to 233 μg/kg), precooked French fries (from 1151 to 123 μg/kg), and breakfast cereals (from 327 to 87 μg/kg). In the category bakery products, biscuits, there was an increase of acrylamide levels in the studied period (from 111 to 290 μg/kg). A minor increase was also observed in case of French fries ready-to-eat (from 322 to 382 μg/kg). However, the significant ($p < 0.0001$) tendency of lowering the acrylamide levels was found out only in relation to precooked French fries.

Distribution of median, 25% and 75% quartiles and minimum and maximum values for acrylamide in samples of precooked French fries is shown in Figure 8.

We also found a significant ($p < 0.0001$) decrease in the content of acrylamide in potato crisps, however, only between 2010 and 2011 and the initial observation period (Figure 9).

The identified tendency of decreasing the level of acrylamide in precooked French fries as well as in potato crisps seems to point to effective application of mitigation strategy of acrylamide content by the manufacturers from the potato sector. However, in case of precooked French fries that have been fried in the laboratory before testing, most probably it is connected with a recommendation placed on labels of these products, suggesting to fry the

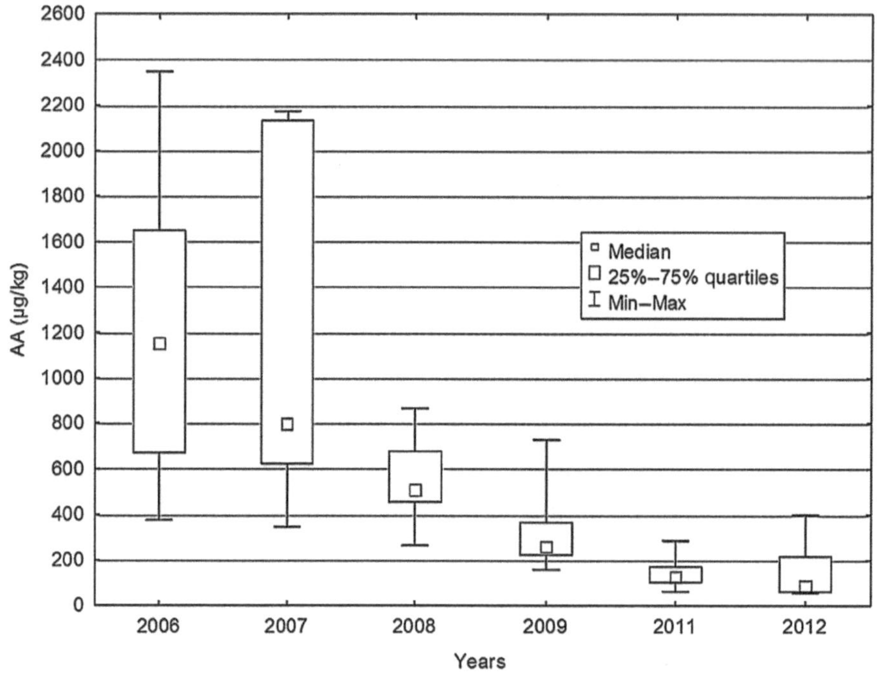

Figure 8 Distribution of median (Q 50), 25% (Q25), and 75% (Q75) quartiles for acrylamide in pre-cooked French fries samples, from 2002 to 2011.

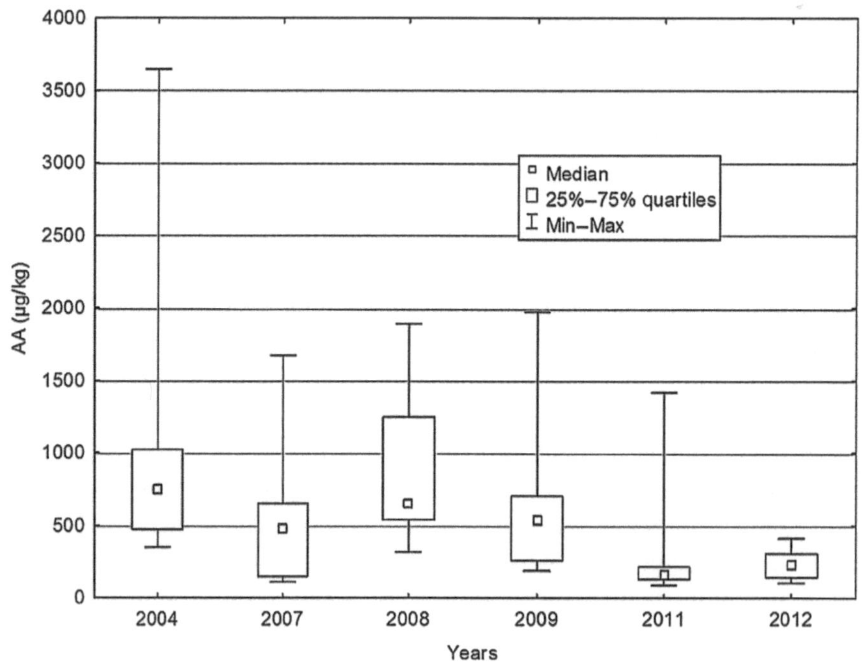

Figure 9 Distribution of median (Q50), 25% (Q25), and 75% (Q75) quartiles for acrylamide in potato crisps samples, from 2002 to 2011.

fries only until they become golden-colored. It is confirmed by lack of the tendency of lowering the level of acrylamide in the category called French fries ready-to-eat. This may prove that the mitigation strategy has not been fully incorporated in the stage of frying in restaurants and bars. It is also worth paying attention to the differentiation lowering within the course of time, related to the content of acrylamide inside categories of products. In the precooked French fries category, the minimal value of acrylamide in 2007 as 347 μg/kg and maximal as 2175 μg/kg, while in 2012 it was 55 and 402 μg/kg, respectively.

Advantages of the presented results include: (1) collecting samples for research in a random manner, within the whole country, by official food control employees, in accordance with sampling plans; (2) detailed information about the place and conditions of sampling, their raw material composition, manufacturer, conditions of thermal processing, etc.; (3) relatively even distribution of samples in each year of research; (4) analysis of acrylamide levels, between 2004 and 2009, only in one laboratory, by the same analyst, and between 2011 and 2012 by two laboratories of official food control; (5) applied methods: LC-MS/MS and GC/MS were validated and accredited, the analytical proficiency was confirmed in PT. A shortcoming is a relatively small number of the analyzed samples per year, in each product category, and the fact that they pertain only to one country.

To sum up, it needs to be emphasized that Poland commenced performance of the planned monitoring research in acrylamide levels in food in 2004, that is, 3 years before the EU Commission Recommendation (2007/331/EC) on the monitoring of acrylamide levels in food entered into force [18]. The observed significant tendency of lowering the acrylamide content in precooked French fries seems to confirm effectiveness of the mitigation strategy applied by manufacturers from the potato sector. On the other hand, the identified tendency of decreasing the acrylamide level only in one food category may point to the fact that the proposed "tools" for decreasing the acrylamide levels are ineffective or are not fully implemented by the manufacturers. The results obtained in monitoring research in Poland are confirmed in general European results published by Wenzl and Anklam [7] and EFSA report [11]. They show that despite the actions taken up by both authorities managing the risk and manufacturers, acrylamide is still present in food, and the presented tendencies of decreasing its levels relate to singular food categories. It also needs to be emphasized that the identified tendencies of decreasing the acrylamide level in all presented studies regarded various food categories, that is, there is no confirmation of the downward tendency for at least one food category.

CONCLUSIONS

Tendencies of changes in acrylamide levels in selected categories of food within the course of several years, based on four different available datasets were presented in this chapter. The evaluation of the distribution of acrylamide contents in food made by different authors, demonstrated that the time trends of changes in acrylamide levels are ambiguous. The downward trend of the acrylamide content in certain food categories was observed at the same period of time as the upward trend in the other food categories. It should be noted that in the two out of four analyzed databases the decrease in the content of acrylamide in potato crisps was observed, and in the other database the decrease of acrylamide level in precooked French fries was found. It seems to confirm effectiveness of mitigation strategies used voluntarily by food manufacturers from the potato sector. On the contrary to these results, the EFSA report demonstrates that significant "common European trend" of decreasing of acrylamide levels was observed only in "Processed cereal-based foods for infants and young children" category. It needs to be emphasized that the identified tendencies of decreasing level in all presented studies regarded various food categories. Presented results show that despite the actions taken by both authorities managing the risk and manufacturers, acrylamide is still present in food, and the presented tendencies of decreasing its levels relate to singular food categories. The identified tendency of changes in acrylamide level in food may point to the fact that the proposed "tools" for decreasing the level are ineffective or not fully implemented by the manufacturers.

In conclusion, there is a need to gather as much reliable data about acrylamide levels in food as possible and moreover special effort should be taken to reduce its content in food. Mitigation measures to avoid acrylamide in food should be a high priority for risk management.

KEY FACTS

Acrylamide is Classified as Probably Carcinogenic to Humans (Group 2A)

International Agency for Research on Cancer (IARC) has classified acrylamide already in 1994 as a compound "probably carcinogenic to humans" (Group 2A). This category is used when there is limited evidence of carcinogenicity in humans and sufficient evidence of carcinogenicity in experimental animals. IARC has established the following groups: Group1: Carcinogenic to humans; Group 2A: Probably carcinogenic to humans; Group 2B: Possibly carcinogenic to humans; Group 3: Not classifiable as to its carcinogenicity to humans; and Group 4: Probably not carcinogenic to humans.

The acrylamide still maintain the "probably carcinogenic to humans" status.

JECFA has Recommended the Use of MOE Approach for Risk Assessment of Acrylamide

The MOE values below 10,000 are considered of high human health concern. In 2011 JECFA recommended the use of MOE approach for risk assessment of acrylamide that is both genotoxic and carcinogenic. Two different values of $BMDL_{10}$ (lower limit on the benchmark dose for 10% response) were proposed: for induction of mammary tumors in rats—0.31 mg/kg bw/day and for Harderian gland tumors in mice—0.18 mg/kg bw/day. Taking into account the above JECFA concluded that the estimated MOE to acrylamide from the diet are low, which indicates a risk to human health.

Mitigation Strategy in Food Acrylamide

Tools to reduce acrylamide content in food were first described by CIAA (since June 2011 FoodDrinkEurope) in 2005 in the form of "Toolbox." The aim of the Toolbox is to provide national and local authorities, manufacturers and other relevant bodies, with brief descriptions of intervention steps which may prevent and reduce formation of acrylamide in specific manufacturing processes and products. Subsequent editions of Toolbox describe actual information of the various acrylamide reduction tools at the four stages of the production of food products: agronomy, recipe, processing, and final preparation. In the latest version of Toolbox (2013), the following 14 parameters were identified as being crucial in reduction of acrylamide in food: 1. Agronomy: Reducing sugars, Asparagine; 2. Recipe: Raising/Leavening agents, Other minor ingredients (e.g., glycine and divalent cations), pH, Rework, Dilution/Piece size, Fermentation; 3. Processing: Thermal Input and Moisture, Asparaginase, Pretreatment, Finished Product Color, Texture/Flavor; 4. Final Preparation: Consumer Guidance. Toolboxes are regularly updated based on new research results.

Europe-wide Monitoring on Acrylamide Levels in Food

First European monitoring of acrylamide levels in food was issued in 2007 according to EU Commission Recommendation (2007/331/EC), which were extended in 2010 (2010/307/EU). The purpose of monitoring studies was to collect reliable data on acrylamide levels in food across the Community in order to obtain unambiguous information about the levels of acrylamide in foods, which are known to provide acrylamide and foods that are eaten by specific vulnerable groups, such as infants and young children. Recommendation 2007/331/EC lists 10 categories of foods that should be taken each year for research in each Member State depending on the size and the number of residents in a given state. It was comprised in the scope between 4 and 24 units in each of 10 separated food products' categories. The monitoring data should be supplied in the form of a detailed report, identical for each member state. The new Recommendation includes a revised categorization of food products for monitoring purposes. Commission Recommendations also established specific criteria for the analytical methods to be used for the determination of acrylamide in food. Based on the results of monitoring supplied to EFSA annual reports are published.

Indicative Acrylamide Values in Food

The EU Commission Recommendation 2013/647/EU provide the indicative values for acrylamide levels in separate categories of food products. The lowest level of indicative acrylamide values was established in the following two categories of food: "Processed cereal-based foods for infants and young children" (excluded biscuits and rusks) and "Baby foods, other than processed cereal-based foods" (not containing prunes) at 50 μg/kg.

The highest one in "Coffee substitutes" was divided into "coffee substitutes mainly based on cereals" at 2000 μg/kg and "other coffee substitutes" at 4000 μg/kg. In other food categories acrylamide indicative values are in the range of 80 μg/kg ("wheat-based soft bread") to 1000 μg/kg ("Potato crisps," "Ginger bread"). Indicative values are not the maximum allowable levels of acrylamide in food. They are only intended to indicate the need for an investigation.

MINI DICTIONARY

Accreditation It is a process in which certification of competency, authority, or credibility is presented. Accreditation is a formal, third-party recognition of competence to perform specific tasks. Testing laboratories are accredited in accordance with requirements specified in EN ISO/IEC 17,025:2005.

Censoring It is a condition in which the value of a measurement is only partially known. *Censored values* are those reported as less than some value, greater than some value, or as an interval (between).

Left censored value It is one that is known only to be less than some value.

Limit of detection (LOD) It is expressed as the concentration or the quantity derived from the smallest measure that can be detected with reasonable certainty for a given analytical procedure.

Limit of quantification (LOQ) It is the minimum concentration that can be quantified at a specified level of precision or accuracy (or both).

Log transformation This consists of taking the log of each observation. If a measurement variable does not fit a normal distribution or has greatly different standard deviations in different groups, there is a need to try a data transformation. In some cases, transforming the data will make it fit the assumptions better. The log transformation can be used to make highly skewed distributions less skewed. This can be valuable both for making patterns in the data more interpretable and for helping to meet the assumptions of inferential statistics.

Margin of exposure (MOE) It is the ratio of the Benchmark Dose Lower Limit ($BMDL_{10}$) to the estimated human intake of compound. *Benchmark dose lower confidence* ($BMDL_{10}$) is the LB of a 95% confidence interval on the benchmark dose (BMD) corresponding to a 10% tumor incidence. JECFA has recommended the use of MOE approach for risk assessment of acrylamide that are both genotoxic and carcinogenic. The MOE values below 10,000 are considered of high human health concern.

Monitoring It means conducting a planned sequence of observations or measurements with a view to obtain an overview of the state of compliance with feed or food law, animal health and animal welfare rules (according to Regulation (EC) No 882/2004 of the European Parliament and of the Council of 29 April 2004 on official controls performed to ensure the verification of compliance with feed and food law, animal health and animal welfare rules).

Proficiency testing It means determination of laboratory testing performance by means of interlaboratory comparison (ISO/IEC Guide 43-1 Proficiency testing by interlaboratory comparisons). *Interlaboratory comparisons* means organization, performance, and evaluation of tests on the same or similar test items by two or more laboratories in accordance with predetermined conditions (ISO/IEC Guide 43-1 Proficiency testing by interlaboratory comparisons).

Risk assessment It means the identification, evaluation, and estimation of the levels of risks involved in a situation, their comparison against standards, and determination of an acceptable level of risk. According to FAO/WHO in food industry risk assessment consists of hazard identification, hazard characterization, exposure assessment, and risk characterization.

Risk management It refers to a coordinated set of activities and methods that is used to direct an organization and to control the many risks that can affect its ability to achieve objectives. According to the Introduction to ISO 31,000: 2009, the term risk management also refers to the architecture that is used to manage risk. This architecture includes risk management principles, a risk management framework, and a risk management process.

Secular trend It means changes over a long period of time, generally years or decades.

Validation It is the confirmation by examination and the provision of objective evidence that the particular requirements for specific intended use are fulfilled (EN ISO/IEC 17,025:2005 General requirements for the competence of testing and calibration laboratories).

SUMMARY POINTS

- The analysis of tendencies of changes in acrylamide levels in food based on the four different datasets demonstrates that there are no constant and firm trends across food groups.
- The downward trend of the acrylamide content in certain food categories was observed at the same period of time as the upward trend in the other food categories.
- Very high content of acrylamide is still found in single samples of food products.
- Tendencies of decreasing acrylamide levels in potato products, demonstrated by two different evaluations, seem to confirm effectiveness of mitigation strategies used voluntarily by food manufacturers only from the potato sector.
- It needs to be emphasized that the EFSA report demonstrates that significant "common European trend" of a decrease in acrylamide levels was observed only in the "Processed cereal-based foods for infants and young children" category.
- Presented results show that despite the actions taken by both authorities managing the risk and manufacturers, acrylamide is still present in food, and the presented tendencies of decreasing its levels relate to singular food categories.
- The identified tendency of changes in acrylamide level in food may point to the fact that the proposed "tools" for decreasing the level are ineffective or not fully implemented by the manufacturers.
- There is a need to gather as much reliable data about acrylamide levels in food as possible and moreover special effort should be taken to reduce its content in food.
- Mitigation measures to avoid acrylamide in food should be a high priority for risk management.

ABBREVIATIONS

CIAA Confederation of the European Food and Drink Industry (since June 2011 Food Drink Europe)
EFSA European Food Safety Authority
ESA European Snack Association
GC/MS Gas chromatography with mass spectrometry
IARC International Agency for Research on Cancer

JECFA FAO/WHO Joint Expert Committee on Food Additives
JRC European Commission's Directorate General Joint Research Centre
LC-MS/MS Liquid chromatography with tandem mass spectrometry
LOD Limit of detection
LOQ Limit of quantification
MOE Margins of exposure
NFNI National Food and Nutrition Institute
PT Proficiency testing

REFERENCES

[1] International Agency for Research on Cancer. Some industrial chemicals. Lyon, France: International Agency for Research on Cancer, 1994, http://www.iarc.fr/ENG/Databases/index.php.
[2] SNFA. Swedish National Food Administration. Information about acrylamide in food, http://www.slv.se/engdefault.asp.
[3] JECFA. Joint FAO/WHO Expert Committee on food additives. Evaluation of certain food additives and contaminants. In: 72nd report of the joint FAO/WHO expert committee on food additive. WHO technical report series, vol. 959. 2011.
[4] CIAA Technical Report. Acrylamide status report December 2004. Brussels January 2005. Available: http://www.ciaa.be. [accessed 23.09.05].
[5] Confederation of the food and drink industries in the EU (CIAA). 2014. Available: http://ec.europa.eu/food/food/chemicalsafety/contaminants/toolbox_acrylamide_201401_en.pdf. [accessed 28.11.14].
[6] Codex Committee on Contaminants in Food. Proposed Draft Code of Practice for the Reduction of Acrylamide in Food. Codex Alimentarius Commission FAO/WHO; January 2008.
[7] Wenzl T, Anklam E. European Union database of acrylamide levels in food: update and critical review of data collection. Food Addit Contam 2007;24(S1):5–12. Supplement 1.
[8] Results on the monitoring of acrylamide levels in food. EFSA Sci Rep 2009;285:1–26.
[9] Results on acrylamide levels in food from monitoring year 2008. Scientific Report of EFSA. EFSA J 2010;8(5):1599.
[10] Results on acrylamide levels in food from monitoring years 2007-2009 and exposure assessment. Scientific Report of EFSA. EFSA J 2011;9(4):2133.
[11] Update on acrylamide levels in food from monitoring years 2007-2010. Scientific Report of EFSA. EFSA J 2012;10(10):2938.
[12] WHO (Joint FAO/WHO Expert Committee on Food Additives). Evaluation of certain food contaminants. In: Seventy-second report of the joint FAO/WHO expert committee on food additives (Rome, 16–25 February 2010). WHO technical reports series 959. 2011.
[13] Powers SJ, Mottram DS, Curtis A, Halford NG. Acrylamide concentration in potato crisps in Europe from 2002 to 2011. Food Addit Contam Part A 2013;30(9):1493–500.
[14] Mojska H, Gielecińska I, Szponar L, Ołtarzewski M. Estimation of the dietary acrylamide exposure of the Polish population. Food Chem Toxicol 2010;48:2090–6.
[15] Sirot V, Hommet F, Tard A, Leblanc JC. Dietary acrylamide exposure of the French population: results of the second French Total Diet Study. Food Chem Toxicol 2012;50:889–94.
[16] DTU (National Food Institute, Technical University of Denmark). Chemical contaminants 2004–2011. Food monitoring 2004–2011. National Food Institute, Technical University of Denmark; 2013. Available at: www.food.dtu.dk.
[17] Mojska H, Gielecińska I, Stoś K, Jarosz M. Trends in acrylamide level in food over the last ten years in Poland. Ann Nutr Metab 2013;63(Suppl. 1). s. 1803.
[18] Commission recommendation of 3 May 2007 on the monitoring of acrylamide levels in food (2007/331/EC). Off J Eur Union L 123/33.
[19] Commission recommendation of 2 June 2010 on the monitoring of acrylamide levels in food (2010/307/EU). Off J Eur Union L 137/4.

[20] Commission Regulation (EC) No 333/2007 of 28 March 2007 laying down the methods of sampling and analysis for the official control of the levels of lead, cadmium, mercury, inorganic tin, 3-MCPD and benzo(a)pyrene in foodstuffs.
[21] Commission recommendation of 10.1.2011 on investigations into the levels of acrylamide in food (Text with EEA relevance). European Commission. Brussels, 10.1.2011 C(2010) 9681 final.
[22] Commission Recommendation of 8 November 2013 on investigations into the levels of acrylamide in food (Text with EEA relevance) (2013/647/EU). Off J Eur Union L 301/15.
[23] Mojska H, Gielecińska I, Stos K. Determination of acrylamide level in commercial baby foods and the assessment of infant dietary exposure. Food Chem Toxicol 2012;50(8):2722–8.
[24] Mojska H, Gielecińska I. Studies of acrylamide level in coffee and coffee substitutes: influence of raw material and manufacturing conditions. Rocz Panst Zakl Hig 2013;64(3):173–81.

SECTION A

Acrylamide, The Food Chain in the Context of Harm

CHAPTER 4

Acrylamide Intake, Its Effects on Tissues and Cancer

Ayşegül ÇEBİ
Faculty of Health Sciences, Giresun University, Piraziz/Giresun, Turkey

INTRODUCTION

Acrylamide is an industrial chemical that has been used since the middle of the 1950s in the manufacturing of polyacrylamide. It is not known to occur as a natural product. Acrylamide has been classified by International Agency of Research on Cancer (IARC) as a "2A Group" carcinogen [1]. A person can be exposed to acrylamide from diet, drinking water, smoking, secondhand smoke, occupational conditions and so on. Acrylamide is also used mainly in certain industrial processes, such as paper manufacturing, and drinking water and wastewater treatment. It is found in small amounts in some consumer products, such as caulk, food packaging, and some adhesives. Human exposure to acrylamide is widespread in the world. In the past few years, animal studies have shown that acrylamide was a multiorgan carcinogen in rodents [2,3]. In 2002, Swedish scientists first declared that acrylamide was formed in heated foods (frying, baking, and roasting of some foods) [4].

Acrylamide is metabolized to reactive glycidamide via epoxidation. Glycidamide is believed to have a role in the genotoxity of acrylamide by attaching itself to DNA. Acrylamide and its metabolite, glycidamide, react readily with a number of biomolecules including hemoglobin [5].

There have been many studies that claimed an association between acrylamide and increased risk of cancer in human. The potential carcinogenic effects of dietary acrylamide are controversial. After findings of acrylamide intake in daily food consumption many epidemiological studies have emerged investigating the association between acrylamide intake and cancer. Food frequency questionnaire (FFQ), a commonly used approach in the cohort studies, to measure the daily intake of acrylamide has been arguable. Generally, coffee, French fries, and fried potato products are the more commonly consumed food source of acrylamide in the world.

It was reported that acrylamide increased the incidence of benign and malignant tumors in some organs such as thyroid, adrenals, and tunica vaginalis. Hence, acrylamide induced lung and skin tumors, in a series of nonstandard carcinogenicity bioassays in mice [3,6].

Acrylamide in Food
http://dx.doi.org/10.1016/B978-0-12-802832-2.00004-8

A series of wide-ranging cohort studies have been carried out to lighten the carcinogenic effect of acrylamide on human in different countries including Sweden, the Netherlands, and the USA.

HUMAN EXPOSURE TO ACRYLAMIDE

Acrylamide has been considered as a probable carcinogen and classified as "2A Group" in 1994 by IARC [1]. People can be exposed to acrylamide from diet, drinking water, smoking, secondhand smoke, occupational conditions and so on. Acrylamide is used as a building block in making polyacrylamide and acrylamide copolymers. Therefore, it takes part in manufacturing of the water-soluble polymers which are used for paper manufacturing, sludge treatment, wastewater, soil stabilization, and so on [7]. Although these have a wide range of area of usage, its exposure to human is limited. Oral feeding, inhalation and dermal contact are the ways to intake of acrylamide. Factory and tunnel workers are at risk with respect to acrylamide intake by inhalation and skin contact. It can be absorbed through skin and mucous membranes. During the manufacture, the estimated exposure of acrylamide is 0.07 mg/kg/day. In previous animal studies, dermal application of radioactive-labeled acrylamide resulted in AA-val adduct levels, which were found to be lower than those observed following intraperitonal application in rats and mice [8]. Dermal absorption levels vary from about 3% to 100% in vitro and animal studies [7]. In vitro studies [8,9] showed that low doses (1.3 g/cm^2) of acrylamide (prepared using ethyl acetate) in creams and lotions had a 30% absorption, while higher doses (2000 g/cm^2) of acrylamide (also prepared in ethyl acetate) had a reduced absorption (20%) in pig. They concluded that the lower absorbed dose following the 2000 g/cm^2 application of acrylamide was attributed to a saturation in the penetration of acrylamide. Absorption of 20–30% was found in low doses (0.1 g/cm^2) of acrylamide in polyacrylamide gels in vitro [8,10]. Acrylamide is quickly absorbed by the skin and mucosa if inhaled. After taken by oral way, it is well absorbed and distributed to the tissues [11]. Acrylamide can be carcinogenic, after conversion to glycidamide which is DNA reactive [12].

In addition, acrylamide is found in filtered cigarette smoke. The levels of acrylamide have been declared to range between 1100 and 2340 ng per cigarette [13]. Indirect exposure of acrylamide via environment drinking water was presented by European Chemicals Bureau, a joint of European Union in 2002. The daily acrylamide dose through drinking water was calculated using three parameters. These are the maximum expected concentration of acrylamide in drinking water (0.125 μg/L), the daily intake of drinking water (2 L/day), and human average bodyweight. This equation is expressed as follows:

$$\text{Dose} = \frac{\text{Concentration}}{\text{Body weight}} \times \text{Intake}$$

This formula gives a dose of 0.0036 μg/kg bodyweight/day for a 70 kg average human [7].

Oral exposure of acrylamide is more effective to the body than the dermal exposure as the skin can act as a barrier. Firstly, acrylamide enters the body; after digestion and absorption, then excreted by urine. Its half life is 3.1–3.5 h in the body. In the liver, acrylamide comes over conjugation by glutathione (GSH) and form mercapturic acid to excrete. Additionally, epoxidation of acrylamide by cytochrome P450 2E1 (CYP2E1) enzyme results in the formation of glycidamide, which binds to DNA and has mutagenic effects. Glycidamide encounters further biotransformation through hydrolysis to form glyceramide or conjugation by GSH, resulting in the formation of mercapturic acid excreted in the urine [5]. Acrylamide displays biphasic elimination with an initial half life of about 5 h and a final half life of 8 days in tissues. In addition, it does not accumulate in the body.

DIETARY INTAKE OF ACRYLAMIDE

Acrylamide is one of the by-products of the Maillard reaction between reducing sugars (i.e., glucose and fructose) and free asparagine. Thus, it appears in starchy foods such as potato chips, crisps, and bread treated at temperatures above 120 °C. Cooking methods specify the formation of acrylamide. Baking, frying, deep-frying, overcooking, and microwaving produce high amounts of acrylamide, but not boiling foodstuffs.

Distribution of acrylamide levels in foods has been reported between 2007 and 2010 by European Food Safety Authority (EFSA) Scientific Committee. The coffee substitutes, potato crisps, and French fries have high acrylamide levels, while other processed cereal-based foods for infants and young children have low levels [14]. Food categories are shown in Table 1.

The basic sources of acrylamide in foodstuffs depend on the national/regional food habits. For example, in the USA, French fries and other potato products are highly consumed (the average daily acrylamide intake is 35%), while coffee and bread are consumed relatively less (coffee, 7% and toast and soft bread, 11%) [15].

Comparing dietary acrylamide intake across populations is challenging because of procedural variations in the estimation of the intake. It is known that food preparation methods change among different cultures. Using open-ended 24 h dietary recalls, an investigation including stratified analyses according to the habitual alcohol consumption, smoking status, physical activity, body mass index (BMI), and education was performed to evaluate the mean dietary acrylamide intake and its dietary sources in 27 centers in 10 countries in the EPIC study. The south and west of Europe take acrylamide mostly from bread, crisp bread, and rusks (range: 24–50%), whereas the north of Europe takes it from coffee (range 33–40%), followed by bread, crisp, and rusks (range: 20–35%). Among all Europe countries, these two food groups were settled at the first and the second order, whereas potatoes and its products (fries) and others such as cooking above 120 °C were

Table 1 Distribution of acrylamide levels[a] in foods in 2010

Food category	*n*	Median (µg/kg)	Mean (µg/kg)	P90 (µg/kg)	P95 (µg/kg)	Maximum (µg/kg)
French fries, sold as ready-to-eat	256	240	338 (336–339)	725	1024	2174
French fries from fresh potatoes	196	239	325 (325–326)	692	921	2174
French fries from potato dough	1	150	150	150	150	150
Unspecified French fries	59	240	382	1019	1377	1800
Potato crisps	242	450	675 (674–676)	1538	2080	4533
Potato crisps from fresh potatoes	173	543	758 (757–758)	1822	2193	4533
Potato crisps from potato dough	19	370	435	980	1000	1000
Unspecified potato crisps	50	313	481 (478–484)	890	1389	4039
Precooked French fries/potato products for home cooking	117	151	331 (329–333)	873	1159	3955
Fries baked in the oven (oven fries)[b]	28	410	690	1888	1991	3955
Deep-fried fries[c]	64	115	198 (195–201)	568	681	1155
Unspecified potato products for home cooking[d]	25	179	270	707	928	1295
Soft bread	150	18 (9–24)	30 (25–35)	63	94	425
Unspecified bread	0	–	–	–	–	–
Breakfast cereals	174	91 (91–100)	138 (132–144)	293	353	1290
Biscuits, crackers, crisp bread, and similar	462	129	333	833	1337	5849
Crackers	64	139	178	303	491	1062
Crisp bread	54	110	249 (248–250)	665	1443	1863
Wafers	37	225	389	880	1300	1300
Ginger bread	207	134	415 (414–415)	1187	1635	3191
Other biscuits, crackers, crisp bread, and similar	100	99	289 (288–290)	640	1061	5849
Coffee and coffee substitutes	151	242	527	1200	2000	8044
Roast coffee	103	200	256 (255–257)	462	641	1932
Instant (soluble) coffee	15	520	1123	2629	8044	8044
Coffee substitutes	24	870	1350 (1349–1350)	3300	3400	4200
Unspecified coffee	9	300	441	1800	1800	1800

Baby foods (excluding cereal based)	55	12 (0–18)	69 (64–74)	116	419	1107
Processed cereal-based foods for infants and young children	128	24(0–30)	51 (45–57)	144	175	578
Biscuits and rusks for infants and young children	46	57	86 (83–90)	175	250	470
Other processed cereal-based foods for infants and young children	82	13 (0–24)	31 (23–39)	60	130	578
Unspecified cereal-based foods for infants and young children	0	–	–	–	–	–
Other foods	336	82 (82–84)	225 (221–228)	612	811	3972
Muesli and porridge	14	56	80 (77–83)	104	420	420
Pastries and cakes	81	55	146 (144–148)	420	793	890
Nonpotato savoury snacks	80	115	192 (189–194)	389	618	1910
Other unspecified products	161	81 (81–82)	293 (289–298)	707	1330	3972

[a]Values indicate middle bound (MB) and ranges in brackets indicate lower bound (LB) and upper bound (UB) values. When LB and UB do not differ from the MB, only the MB value is presented.
[b]86% were reported to be analyzed as prepared, 14% analyzed as unprepared.
[c]73% were reported to be analyzed as prepared, 8% analyzed as unprepared, 19% type of preparations were unknown.
[d]56% were reported to be analyzed as prepared, 44% analyzed as unprepared [14].

third (range 4–26%). In the UK, potatoes were found to be higher acrylamide contributors (15%) [16]. Comparing acrylamide intake with life style factors, alcohol consumption was found to be associated with acrylamide intake in all population ($p = 0.002$ for men; $p = 0.03$ for women). Dietary acrylamide intake was found higher in women who consumed alcohol than abstainers. But, there were no consistent data revealing the association between acrylamide intake and alcohol drinking in men. It was found that acrylamide intake was related with smoking status in women ($p < 0.001$), and not in men ($p = 0.49$). Smokers and former smokers were more related with acrylamide intake than never smokers. Physical activity, BMI, and education were not related with dietary acrylamide intake both in women and in men [16].

RISK ASSESSMENT

Acrylamide has an electrophilic double bond, which can react with nucleophilic groups. Thus, it covalently interacts with cellular nucleophiles such as the sulfydryl groups in reduced glutathione and in proteins [12]. Different approaches are available for physiologically based pharmacokinetic (PBPK) modeling of acrylamide absorption, metabolism, and constitution for predicting human internal exposures to acrylamide and glycidamide to eliminate bias in risk assessment.

Acrylamide has a complex metabolism in vivo. After absorption, minimum four different urinary acrylamide metabolites have been identified in rats [17]. *N*-acetyl-*S*-(propionamide)-cysteine (APC) is a mercapturic acid that accounts for 48% of the dose. To analyze APC in urine, a biological monitoring method has been developed, which comprises acid hydrolysis of APC to produce *S*-carboxyethyl-cysteine [18]. Hemoglobin adducts of acrylamide are used to measure human exposure to electrophilic compounds over the first 4 months (human red blood cells' life span is 120 days). *N*-terminal valine of hemoglobin has the adduct formation, and it is used as a marker of in vivo exposure (internal dose) to acrylamide [19]. A modified Edman degradation is processed as an analytical procedure. Measurement of hemoglobin adducts of glycidamide is also used as a similar analytical approach. Hemoglobin adducts of acrylamide and glycidamide are found three to five times higher in smokers than in nonsmokers [20,21]. The average dietary acrylamide intake of 0.8 μg/kg/day has been estimated in nonsmoker US population [21]. In Vesper's study, the levels of hemoglobin adducts, that is, HbAA and HbGA have expanded respectively between 3–910 and 4–756 pmol/g hemoglobin. Mexican-Americans have the highest level of hemoglobin adducts, whereas non-Hispanic blacks have the lowest in the US population [21]. However, in children and adolescents, a higher dietary acrylamide intake has been found than the adults because of the elevated consumption of acrylamide rich foods such as potato chips, French fries, and the lower body weight [12]. After 2003, some mitigation strategies have been reported for the food types with high acrylamide levels. Although there are preventive studies, generally in all countries, they have not been considered an effective protection from acrylamide exposure but only in some population subgroups [22,23].

There are some studies that has shown a relationship between the dietary intake of acrylamide and the increased risk of cancer in human. But, this relationship is considered inconsistent. A meta-analysis of some epidemiological studies has indicated that dietary acrylamide is not associated with the risk of most common cancers [24].

The Joint FAO/WHO Expert Committee on Food Additives (JECFA) performed the margin of exposure (MoE) as an alternative approach for the health risk assessment of acrylamide. MoE is based on the benchmark dose, derived from dose–response modeling, and a qualitative description for a possible prioritization of risks. According to the JECFA, dietary exposure of acrylamide is calculated to be 0.13 μg/kg bw/day and the value of MoE is 1385 for average, 0.69 μg/kg bw/day, value of MoE is 261 for high consumer. EFSA Scientific Committee has proposed that there is a health concern among the local population because of the relatively low MoE values. They have suggested this approach for compounds that have genotoxic and carcinogenic features like acrylamide [25].

The 64th meeting of the JECFA noted that in 2005 the average acrylamide intake of the general population was 0.001 mg/kg bw/day and the BMDL (benchmark dose lower limits for a 10% extra risk in tumors) of 0.30 mg/kg bw/day for induction of mammary tumors in rats (MoE: 300), whereas high acrylamide consumers have the value of 0.004 mg/kg bw/day (MoE: 75) [26]. In the 72nd meeting of JECFA, No changes have been observed in the average and high dietary acrylamide exposure values and this remained the same in 2010. The JECFA has declared that the $BMDL_{10}$ has been 0.18 mg/kg bw/day, and the MoE values of the average and high exposures, respectively, have been 180 and 45 for the induction of Harderian gland tumors in mice. The $BMDL_{10}$ has been 0.31 mg/kg bw/day, and the MoE values have been 310 and 78, respectively for the mammary tumors in rats. MoE values state human health concern. Acrylamide was considered as both genotoxic and carcinogenic by FAO/WHO in 2010 [27].

TOXIC EFFECTS OF ACRYLAMIDE IN TISSUES

Previous studies have shown the toxicological effects of acrylamide in tissues. In a study, the hazardous effects of a single oral dose of 150–100 mg/kg acrylamide were histopathologically examined on the testis in pre-pubertal and adult mice. Acute testicular damage and necrosis of the late elongated spermatids were seen one day after administration of acrylamide in the prepubertal mice. The round spermatids had been destroyed in stages I–II and IV–VIII one day after administration of acrylamide in adult mice [28]. In a study, acrylamide was administered to the rats at doses of 0, 5, 15, 30, 45, and 60 mg/kg/day for 5 consecutive days by oral gavage, and morphological changes were shown in the testicular histology. The deterioration was observed in the germ cells of the seminiferous epithelium of rats administered with acrylamide at a dose of 60 mg/kg/day. Furthermore, a wide range of multinucleated giant cells with sloughed seminiferous epithelium were seen in the lumen of the seminiferous tubule. Decreased sperm concentration and morphological defects of sperm in cauda epididymis of the acrylamide-treated group were seen in a dose-dependent manner [29].

Numerous experiments related with acrylamide were done within National Toxicology Program (NTP) and were reported in 2002. Administration of 0.14, 0.35, 0.70, 1.41, 3.52, and 7.03 mM acrylamide in the drinking water corresponds to the consumed amount of about 1.4, 3.8, 7.8, 15.4, 37.4, and 67.6 mg/kg bw/day for male rats and 1.7, 4.3, 8.3, 16.9, 39.4, and 70.0 mg/kg bw/day acrylamide for female rats. Furthermore, male rats fed with 7.4, 18.5, 37, 74, 185, and 370 mg acrylamide per kg diet. These experiments showed that liver weights, liver to brain weight ratios were decreased in all rats administered 7.03 mM acrylamide in the drinking water for 14 days. Neoplastic findings were not detected in these animals. Dilation of the urinary bladder was determined in several male and female rats, and the lesion was determined microscopically. Furthermore, hind-leg paralysis was observed in the rats which had grossly dilation. Germinal epithelium degeneration in the seminiferous tubules of testes was observed microscopically in male rats administered 7.03 mM acrylamide in the drinking water. There were some proof that the lesions occurred in the reproductive tissues. The numbers of germinal cells showed a decline, and multinucleated spermatids existed in the lumens of seminiferous tubules [30].

Acrylamide has neurotoxic effects in mice, rats, guinea pigs, rabbits, cats, dogs, and monkeys [1,31–34]. The loss of motor function such as hind-limb splay and withered rotarod performance is the main neurotoxic response, whereas the ultrastructural and biochemical changes are the other overt responses. Axonal degeneration was increased in a dose-dependent manner in the sciatic nerve of both males and females following the administration of acrylamide to the rats. The dilation was microscopically observed in axons besides this foamy macrophages, and myelin debris were also observed [30].

Small-diameter nerves may be damaged by acrylamide intoxication. It has been shown that the response of sweat glands to pilocarpine which is a cholinergic agonist is dropped in acrylamide intoxicated mice [35]. Acrylamide can damage unmyelinated axons in the vagus nerve and the enteric nervous system [36,37]. Interference with axonal transport and the enhanced calcium input [38] are some mechanisms of acrylamide neurotoxicity [39,40]. Acrylamide intoxication leads to degeneration of epidermal nerves in the skin [41]. Acrylamide intoxication causes some degeneration in epidermal and dermal nerves. Typical epidermal nerves with a varicose appearance and some branching in the suprabasal layers are shown in Figure 1(a). Early stage of the acrylamide intoxication (in 5 days) epidermal nerves and dermal nerves are still readily demonstrated with the protein gene product (PGP) 9.5. Epidermal and dermal nerves have the same morphology with the control mice but the branching patterns of the epidermal nerves had become more elaborate at this stage (Figure 1(b)). In 21 days of the acrylamide intoxication (in the late stage), the number of epidermal nerves was reduced (Figure 1(c)) [41].

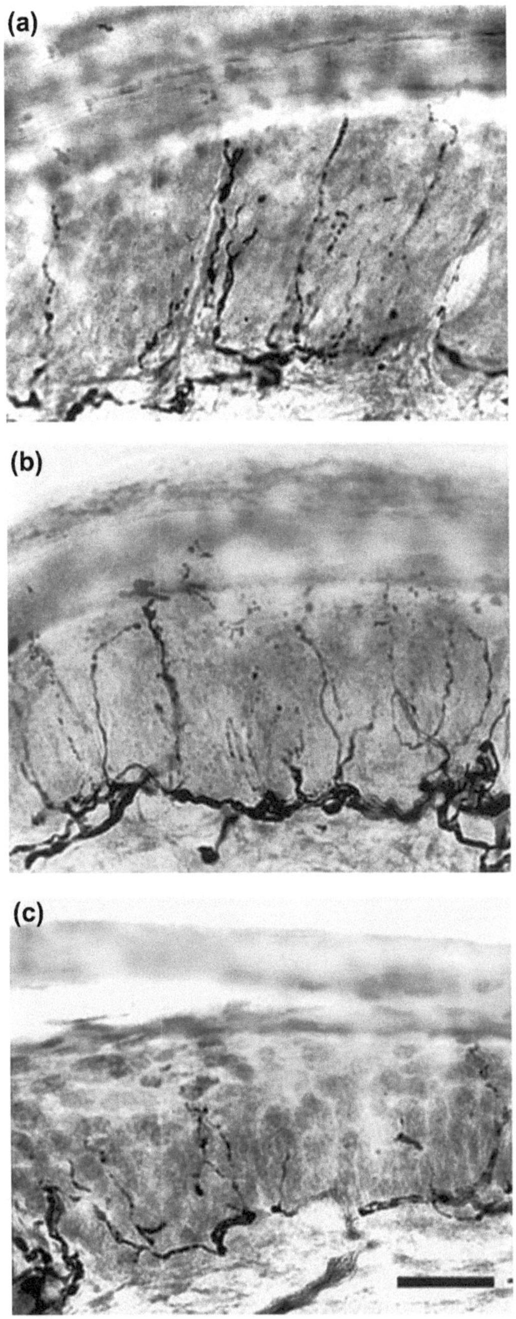

Figure 1 Skin innervation seen in acrylamide-intoxicated mice. This micrograph shows the innervation of the epidermis in control mice (a), in the initial stage (b), and in the late stage (c) of acrylamide intoxication. The skin of the hindfoot was immunocytochemically stained with the protein gene product 9.5 (PGP 9.5). (a) In the subepidermal nerve regions, PGP 9.5 (+) nerves are found and typical varicose appearance and some branching points are found in the epidermis. (b) The initial stage seems after the acrylamide intoxication, the branching patterns of PGP 9.5 (+) epidermal nerves have become prominent. (c) In the late stage after the acrylamide intoxication, the abundance of epidermal nerves significantly decreased. Bar = 20 μm [41].

After the acrylamide intoxication, the degeneration pattern of dermal nerves resembles to epidermal nerves. At the light microscopy, individual PGP 9.5 (+) dermal nerves have been shown with a linear pattern of similar axonal caliber (Figure 2(a)). After the acrylamide intoxication (in the late stage), the PGP 9.5 immunoreactivity of dermal nerves appeared as fragmented and globular in mice (Figure 2(b)). A single Schwann cell

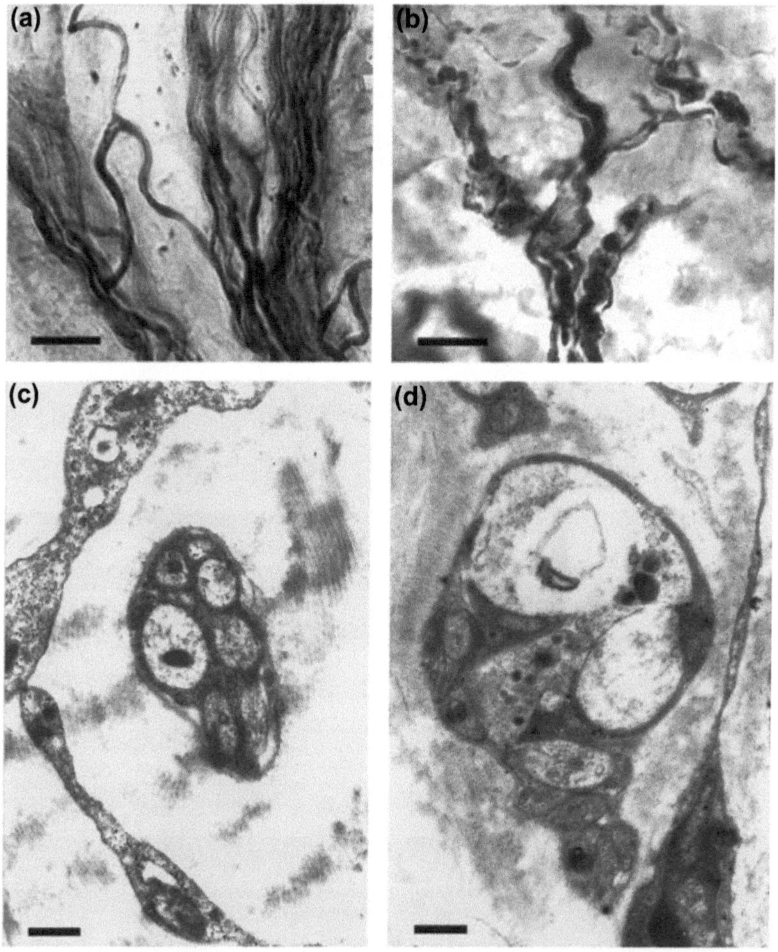

Figure 2 Dermal nerves seen in acrylamide-intoxicated mice. Skin sections of the control group (a and c) and 21 days in acrylamide intoxicated mice (the late stage, b and d) seem by immunostaining with protein gene product 9.5 (PGP 9.5) (a, b) and by regular electron microscopy (c, d). (a) PGP 9.5 (+) dermal nerves showing a discrete linear shape in control group. (b) Dermal nerves showing fragmented shape with a globular appearance in acrylamide-intoxicated mice. (c) Unmyelinated nerves showing as enclosed by a single Schwann cell in the dermis of control group. The axoplasm and organelles show a regular structure. (d) The proportion of intact axons is variable in acrylamide-intoxicated mice. Some axons seem like swollen shape, the corruption has become in organelles and the appearance of vacuoles. Bars = 10 μm (a, b) = 1 μm (c, d) [41].

enveloped the unmyelinated nerves at the ultrastructural level in the control mice (Figure 2(c)). Conversely, dermal axons became swollen with dissolution of organelles and the appearance of vacuoles in acrylamide-intoxicated mice (Figure 2(d)) [41].

Acrylamide administration to F334/N rats (for male and female) caused retinal degeneration in the eyes. The incidence of degeneration raised after the administration of 0.70 mM to the male rats, whereas it raised after the administration of 0.35 and 0.70 mM to the females. Some degeneration was recorded in the retina, including the loss of photoreceptors related with hypocellularity and thinning in other retinal layers [30].

In preputial glands, the prevalence of duct ectasia had been increased by administration of acrylamide (0.175, 0.35, and 0.70 mM) to the male rats. The dilation of the keratin-filled main ducts is a microscopical sign of this lesion. Furthermore, inflammation also usually exists in diverse intensity. Two non-neoplastic lesions, which were focal hypertrophy and diffuse cytoplasmic vacuolation, including adrenal cortex appeared with the incidence being significantly raised in the 0.70 mM of acrylamide-treated female rats. Focal hypertrophy referred to as the varied focal expansion of cortical cells in the zona glomerulosa or zona fasiculata of adrenal cortex had occurred by administration of acrlamide. The compression of adjacent tissue was not recorded. The cytoplasm was mostly eosinophilic and granular in nature, and some of them had several lipid vacuoles, although lipid accumulation was not the first reason of cellular enlargement. The cytoplasmic lipid vacuoles of the zona fasiculata or zona reticularis increased in a focal or diffuse type. In the spleen, the proliferation of hematopoietic cells elevated by 0.70 mM of acrylamide administration to the female rats. A slight enlargement of the spleen was usually recorded. The splenic enlargement derived from the increased hematopoietic activity had been microscopically shown. Bone marrow hyperplasia and ovarian atrophy, which were known as nonneoplastic lesions, significantly increased in the 0.35 and/or 0.70 mM groups of female rats [30].

Acrylamide rapidly distributed to tissues [17,42–44]. Acrylamide distribution appeared to be the highest in the muscle tissue ranging between 30% and 50% of the administered dose in dogs and pigs, after acrylamide administration of 1 mg/kg bw. Liver was the next with about 14% in dogs, whereas the gastrointestinal tract was the second with about 20% in pigs. In pigs, the absorption of acrylamide would be slower than dogs [45].

After administration of 120 mg/kg (2,3-14C) acrylamide to the male and 13.5–17.5 days pregnant female Swiss–Webster mice, radioactive-labeled acrylamide was detected in the gastrointestinal tract, liver, pancreas, testis, brain, and gallbladder, as well as epithelia of oral cavity, esophagus, and bronchi. The 13.5 days fetuses also had uniformly labeled acrylamide, whereas fetal brain had slightly increased level. In the tissues of 17.5 days fetuses, the distribution of labeled acrylamide resembled that in the maternal tissues but the increased level was reported in fetal skin [43].

The blood tissue had the majority of the absorbed dose at 24 h following a 6-h inhalation exposure to 3 ppm acrylamide in rats, followed by the skin, spleen, and lung. However, the distribution pattern was different although the same protocol was carried out in mice. The skin had the highest level of the absorbed dose followed by the subcutaneous fat, testes, and blood [44].

Acrylamide doses less than 15 mg/kg/day did not cause serious problems in clinical signs of toxicity or neurotoxicity, indices of reproductive performance (mating, pregnancy, and fertility) or the number of corpora lutea and the signs of histological or neuropathological changes. However, doses higher than 15 mg/kg/day resulted in the signs of neurotoxicity and changes in copulatory behavior. These higher doses also caused a decrease in fertility affecting sperm motility and morphology in both rats and mice [46,47].

CARCINOGENICITY IN ANIMAL STUDIES

FAO/WHO reported that acrylamide was carcinogenic in rats, producing increased incidences of benign and malignant tumors in some organs such as thyroid, adrenals, and tunica vaginalis. Hence, acrylamide induced lung and skin tumors, in a series of non-standard carcinogenicity bioassays in mice. Animal experiments have shown that acrylamide is a multiorgan carcinogen in mice and rats. However, none exhibited excess cancer risk at lower doses of acrylamide [48].

Two-year studies were done in F-344 rats receiving doses of about 0, 0.01, 0.1, 0.5, or 2 mg/kg/day acrylamide in drinking water. The incidence of tumors in mammary and clitoral glands, uterus, testes, brain and spinal cord, adrenal, pituitary, thyroid, and oral cavity increased according to histopathological results at the termination stage [3,6]. Another two-year study in Fischer F344/N male and female rats that had received drinking water at concentrations of 0.0875, 0.175, 0.35, and 0.70 mM (equivalent to 0.33, 0.66, 1.32, and 2.71 mg/kg bw/day for males and 0.44, 0.88, 1.84, and 4.02 mg/kg bw/day for females) was performed in NTP laboratories. At the end of the study, the incidence of cancer in the pancreatic islets and of malignant mesotheliomas increased in male rats, whereas the incidence of cancers in the clitoral gland, mammary gland, skin, liver, and tongue increased in female rats. The incidence of thyroid gland and heart tumors increased in both male and female rats receiving acrylamide [30].

A previous study demonstrated that acrylamide had carcinogenic properties. A/J mice (male and female) were administered with acrylamide by gastric intubation, intraperitoneal injection, and topically to the shaved back over 8 weeks starting from an age of 8 weeks. The dose levels were 12.5 and 25 mg/kg bw applied by gavage three times a week, or were 0, 1, 3, 10, 30, and 60 mg/kg bw applied three times a week by intraperitoneal injection. After the administration, lung adenomas appeared in both sexes in a dose-related manner. Female SENCAR mice administered with intraperitoneal injection of acrylamide and dermal application of TPA (i.e., the skin tumor promoter,

12-*O*-tetradecanoylphorbol-13-acetate) following an initiation promotion protocol. Combined acrylamide and TPA administration exhibited increased skin tumors in SENCAR mice [2]. In the same laboratory, female Swiss-ICR mice were administered with acrylamide orally six times over 2 weeks with total doses of 75, 150, and 300 mg/kg. At the end of one year, skin and lung tumors were observed in mice [49].

In a two-year study, B6C3F1 mice (96, males and females) received acrylamide, concentrations of 0.0875, 0.175, 0.35, and 0.70 mM (6.25, 12.5, 25, and 50 ppm acrylamide) in the drinking water ad libitum. Afterward, the incidence of the Harderian gland adenoma and combined Harderian gland adenoma or adenocarcinoma was raised significantly in all acrylamide dose groups of male B6C3F1 mice. Furthermore, the incidence of lung alveolar/bronchiolar adenoma and combined lung alveolar/bronchiolar adenoma or carcinoma was raised significantly at 0.175 and 0.70 mM acrylamide doses. A significant increase was observed in the incidence of stomach (forestomach) squamous cell papilloma and combined stomach (forestomach) squamous cell papilloma or carcinoma at 0.35 and 0.70 mM acrylamide doses. The incidence of Harderian gland adenoma was raised significantly for all doses in females. The combined incidence of mammary gland adenoacanthoma or adenocarcinoma was raised significantly at 0.175, 0.35, and 0.70 mM acrylamide, and the incidence of mammary gland adenocarcinoma was raised significantly at 0.175 and 0.70 mM acrylamide. It was observed that the incidences of lung alveolar/bronchiolar adenoma, combined lung alveolar/bronchiolar adenoma or carcinoma, and malignant mesenchymal skin tumors (fibrosarcoma, hemangiosarcoma, liposarcoma, myxosarcoma, neurofibrosarcoma, or sarcoma) were raised significantly at 0.35 and 0.70 mM acrylamide doses. The incidences of ovary granulosa cell tumor (benign) and mammary gland adenocarcinoma at 0.70 mM were also significantly increased in female B6C3F1 mice [30]. The incidence of the tunica vaginalis mesotheliomas in Fischer 344 rats can be increased by long-term acrylamide exposure. The mesotheliomas of tunica vaginalis testis of Fischer 344 rats are benign and equivalent to human benign mesothelioma of tunica vaginalis testis or adenomatoid tumors by reason of cellular uniformity [50,51]. In Beland's study, F344/N rats received 0, 0.0875, 0.175, 0.35, and 0.70 mM doses of acrylamide in the drinking water for 2 years, and it has been shown a dose-related increasing trend in follicular cell adenoma and carcinoma, and combined follicular cell adenoma or carcinoma in thyroid gland. Malignant mesothelioma was commonly appeared in the epididymis, and all rats having mesothelioma on the testicular tunics also had this neoplasm on the epididymis in males. In females, the prevalence of fibroadenomas in the mammary gland was significantly increased than the control group [51]. In the past 2-year carcinogenicity study, Wistar Han rats received 0.5–3.0 mg/kg/day doses of acrylamide in drinking water. There are significantly increased incidences of thyroid follicular neoplasms in males and females. Increased incidence of mammary fibroadenomas in females is observed but not statistically significant [52]. Two-year carcinogenicity study results in animals that received different doses of acrylamide are shown in Table 2.

Table 2 Summary of some 2-year carcinogenicity study results in animals receiving different doses of acrylamide

Test animal	Gender	Min and max dosage of acrylamide	Tumor type	References
F344 rats	Female	0.01–2.0 mg/kg bw/day	Mammary gland adenoma, fibroadenoma, or fibroma, thyroid gland follicular cell adenoma or carcinoma, oral cavity papilloma, uterine adenocarcinoma, clitoral gland adenoma, pituitary adenoma	Johnson et al. (1986) [3]
F344 rats	Male	0.01–2.0 mg/kg bw/day	Thyroid gland follicular cell adenoma, mesothelioma of the testes tunica albuginea, adrenal pheochromocytoma, benign	Johnson et al. (1986) [3]
F344 rats	Female	1.0–3.0 mg/kg bw/day	Mammary gland fibroadenoma, thyroid gland follicular cell adenoma or carcinoma	Friedman et al. (1995) [6]
F344 rats	Male	0.1–2 mg/kg bw/day	Thyroid gland follicular cell adenoma, mesothelioma of the testes tunica	Friedman et al. (1995) [6]
F344/N (Nctr) rat	Female	0, 0.0875–0.70 mM	Mammary gland fibroadenoma, clitoral gland carcinoma, oral cavity papilloma or carcinoma, skin fibroma or sarcoma	Beland et al. (2013) [51]
F344/N (Nctr) rat	Male	0, 0.0875–0.70 mM	Thyroid follicular cell carcinoma, cardiac schwannomas, testicular mesothelioma, pancreatic islet cell adenoma	Beland et al. (2013) [51]
B6C3F1/ Nctr mice	Female	0, 0.0875–0.70 mM	Harderian gland adenoma, lung alveolar/bronchiolar adenoma, mammary gland adenoacanthoma, mammary gland adenocarcinoma, ovarian granulosa cell tumor, skin sarcoma, hemangiosarcoma, liposarcoma, myxosarcoma, neurofibrosarcoma, or sarcoma	Beland et al. (2013) [51]
B6C3F1/ Nctr mice	Male	0, 0.0875–0.70 mM	Harderian gland adenoma, lung alveolar/bronchiolar adenoma, forestomach squamous cell papilloma	Beland et al. (2013) [51]
F344 rats	Female	0.44–4.02 mg/kg bw/day	Mammary gland fibroadenoma, thyroid gland follicular cell adenoma or carcinoma	NTP(2012) [30]
F344 rats	Male	0.33–2.71 mg/kg bw/day	Thyroid gland follicular cell adenoma or carcinoma, mesothelioma of the epididymis or testes tunica vaginalis	NTP(2012) [30]
B6C3F1 mice	Female	1.10–9.96 mg/kg bw/day	Harderian gland adenoma, mammary gland adenocanthoma and adenocarcinoma, lung alveolar, bronchiolar adenoma, ovary granulosa cell tumors (benign), skin, various types of sarcoma, stomach, foretomach squamous cell papilloma	NTP(2012) [30]
B6C3F1 mice	Male	1.04–8.93 mg/kg bw/day	Lung alveolar, bronchiolar combined adenoma and carcinoma, stomach squamous combined papilloma or carcinoma	NTP(2012) [30]
Wistar Han rats	Female	0.5–3.0 mg/kg bw/day	Mammary gland fibroadenoma, adenoma, adenocarcinoma, adenoma and/or adenocarcinoma, thyroid follicular cell adenoma, follicular cell adenocarcinoma, follicular cell adenoma and/or adenocarcinoma	Maronpot et al. (2015) [52]
Wistar Han rats	Male	0.5–3.0 mg/kg bw/day	Thyroid follicular cell adenoma, follicular cell adenocarcinoma, follicular cell adenoma and/or adenocarcinoma	Maronpot et al. (2015) [52]

In a study, F344 rats and B6C3F1 mice were exposed to acrylamide with drinking water containing acrylamide by repeat dosing [53]. The daily doses of acrylamide were about 2.59 ± 0.69 mg/kg bw/day for mice, 1.07 ± 0.28 mg/kg bw/day for female rats, and 0.96 ± 0.28 mg/kg bw/day for male rats. The analysis results of Hb adducts of acrylamide and glycidamide formation showed significant linear relationships between the acrylamide and glycidamide Hb adduct levels and the corresponding AUCs (area under the curves) [53]. In preliminary studies, it has been demonstrated that acrylamide caused fibroadenomas of the mammary gland and thyroid gland follicular tumors in rats [3,6].

IN VITRO AND IN VIVO GENOTOXICITY

In addition, acrylamide and glycidamide are potential germ cell mutagens [54,55]. Previous studies have shown that acrylamide and its metabolite glycidamide induce micronuclei in rodents [56,57]. Furthermore, it has been shown that they are mutagenic and induce micronuclei in big blue mice. Both acrylamide and glycidamide are genotoxic in mice and both of them induce the mutant frequencies and types of mutations in the liver of mice [58]. Several studies were conducted to explore DNA damage induced by acrylamide using the Comet assay. It was observed that there was genotoxic effects in human HepG2 cells treated with 2.5, 5, 10, and 20 mM of acrylamide [59].

Acrylamide has genotoxic potential in vivo in somatic cells and germ cells. Thus, acrylamide can induce heritable damage at gene and at chromosome level. It has strong affinity to proteins of the sperm [54,55,58]. However, a number of nongenotoxic mechanisms such as endocrine disruption and oxidative stress have been proposed to cause at least some of the tumorigenicity of acrylamide in rats [60].

Glycidamide which is a product of metabolized acrylamide is a reactive epoxide and forms DNA adducts. This epoxide occurred in the CYP2E1 detoxification system and is a primary pathway which is responsible for the genotoxity of acrylamide in germ cells [61,62]. The heritable translocations and specific-locus mutations recovered in the offspring of male mice exposed to acrylamide or glycidamide have been evaluated in several studies. Exposure of spermatozoa or spermatids to acrylamide or glycidamide increases the frequency of mutational events. Dominant lethal mutations, heritable translocations, and unscheduled DNA synthesis in germ cells were assayed in male mice treated with glycidamide. The metabolic conversion of acrylamide to glycidamide may be responsible for the mutagenicity of acrylamide [55,63]. Furthermore, a dominant lethal assay study using 1-aminobenzotriazole to inhibit the metabolic conversion of acrylamide to glycidamide by cytochrome P450 was performed in mice, and treatment with 1-aminobenzotriazole decreased the dominant lethal response which occurred by acrylamide exposure. The lethal effects of acrylamide were inhibited in CYP2E1 knockout mice [64,65]. The levels of male germ cell mutagenicity [64], micronuclei [62] and glycidamide DNA adducts decreased in CYP2E1 knockout mice after the administration of acrylamide [62].

The strains of *Salmonella typhimurium* (Ames test), *Escherichia coli*, and *Klebsiella pneumonia* are used to assay bacterial gene mutation induced by acrylamide [66]. Acrylamide generally did not induce reverse mutations in bacteria. Because of the absence of CYP2E1, capable of metabolizing acrylamide to glycidamide, mutagenic effects of acrylamide do not occur in some bacterial strains. However, acrylamide induced chromosomal aberrations, micronuclei (containing whole chromosomes or acentric fragments), sister chromatid exchanges, polyploidy, and aneuploidy with lack of metabolic activation in mammalian cells [59,67–72].

HUMAN CANCER COHORT STUDIES

A series of epidemiological studies have been carried out to evaluate the potential association between acrylamide intake and cancer. Differences in the acrylamide content in foodstuffs have caused the controversial results to estimate the actual intake by commonly using an FFQ. These differences led to various debates of accurately classifying persons as low or high acrylamide consumers.

The CTS began in 1995–1996, and 133,479 active and retired female teachers and administrators were recruited as a large prospective cohort. In this large cohort study, the association between the incidence of breast cancer and some air pollutants such as acrylamide, carbon tetrachloride, and chloroprene was examined. The baseline dietary intake was measured using California State Teachers Retirement System returned a completed baseline questionnaire. Some women not available for the study conditions such as unknown history of prior cancer were excluded from the study population. Thus, the resulting population was comprised of 112,378 women. During the follow-up time of 15 years, there has been 5676 cases of invasive breast cancer, and many data from this study present evidence for increased risk of several compounds such as acrylamide, carbon tetrachloride, chloroprene, 4,4'-methylene bis(2-chloroaniline), propylene oxide, and vinyl chloride. Estrogen receptor positive or progesterone receptor positive (ER+/PR+) tumors have been related to higher ambient levels of acrylamide benzidine, carbon tetrachloride, ethylidene dichloride, and vinyl chloride [73].

A serial of cohort studies began in Sweden to lighten the association between acrylamide intake and cancer [74,75]. A prospective cohort study in Swedish men started in 1997 and lasted for 9.1 year; 45,306 men who were cancer-free and completed an FFQ were included for the study. Dietary intake was assessed using the FFQ which had self-administered 96 food items. The mean of daily intake of acrylamide was $36.1 \pm 9.6\,\mu g$. Through a follow-up study, it was understood that coffee (23%), whole-grain soft bread (17%), whole-grain crisp bread (8%), white bread (7%), cookies/buns (7%), breakfast cereals (6%), wafers/crackers (6%), fried potato (6%), and potato crisps (4%) were the main contributors to acrylamide intake. During the study, 2696 prostate cancer cases were diagnosed and 1088 cases had localized prostate cancer, whereas 951 cases had advanced prostate cancer. Relative risk (RR): 0.88; 95% confidence interval (CI): 0.70–1.09. No significant association was found between acrylamide intake and prostate cancer [74]. Afterward, to analyze dietary acrylamide intake and colorectal cancer risk in Swedish men (number of populations: 45,306), the same cohort population data were used. RR: 0.95; 95% CI: 0.74–1.20. Statistically significant association was not found between the acrylamide intake and colorectal cancer [75].

To evaluate the association between dietary acrylamide intake and epithelial ovarian cancer, a population-based prospective study namely the Swedish Mammography Cohort was conducted in Swedish women; 61,057 women were enrolled for the study and dietary acrylamide intake was calculated by FFQ at baseline 1987–1990, and at follow-up in 1997. During the follow-up of 17.5 years, 368 cases of ovarian cancer were recognized. There was no relationship found between acrylamide intake and ovarian cancer: RR: 0.84; 95% CI: 0.62–1.14. The mean dietary acrylamide intake for Swedish women was $24.6 \pm 7.6\,\mu g/day$ [76]. To evaluate the relationship between acrylamide intake and cancer, the same Swedish cohort was used. Then, a total of 687 incident cases of endometrial carcinoma were recognized from 36,369 women. The values of hazard ratio (HR) and 95% CI, respectively, were 0.96 and 0.76–1.21. There was no relationship found between endometrial carcinoma and acrylamide intake [76].

Breast cancer incidence was obtained from the Swedish Mammography cohort. During the follow-up, 2952 invasive breast cancer cases were recognized and 2062 cases were available for information status of ER and PR from a total of 61,433 women. Acrylamide intake was reversibly correlated with ER+PR+ and total breast cancer risk, but not significant after modification of dietary and other risk factors of breast cancer. The RR of breast cancer for the highest versus the lowest quartile and 95% CI were, respectively: 0.91 and 0.80–1.03 [77].

A multicenter prospective cohort study "European Prospective Investigation into Cancer and Nutrition (EPIC)" including 10 European countries started to assess the relationship between the dietary intake of acrylamide and ductal adenocarcinoma of the exocrine pancreatic cancer risk in 1992. European men and women more than 500,000 were enrolled in the EPIC cohort study. After the exclusions from the cohort, Cox regression modeling was used to evaluate the data including 477,308 participants. At the end of the study, 856 pancreatic adenocarcinomas were diagnosed and included for analyses. During a mean follow-up time of 11 years, Denmark had the highest mean and median dietary acrylamide intake in men and women. The UK and the Netherlands had, respectively, the second and the third order, whereas Italy had the last for acrylamide intake. EPIC cohort study showed that at baseline, the mean dietary acrylamide intake was $26.22 \pm 14.79\,\mu g/day$, it was higher in men ($31.90\,\mu g/day$ and $0.40\,\mu g/kg$ bw/day) than in women ($23.81\,\mu g/day$–$0.37\,\mu g/kg$ bw/day). Furthermore, there was no association between acrylamide intake and pancreas cancer risk in EPIC (HR: 0.95% and 95%, CI: 0.89–1.01). The association between acrylamide intake and pancreas cancer risk did not have statistically significant heterogeneity by geographical location or by mean acrylamide intake (*p* values were found to be >0.23) [78]. The same EPIC cohort data were used to evaluate the association between esophageal carcinomas and acrylamide intake. After the follow-up, the number of identified esophageal carcinomas was 341, most of these carcinomas was esophageal adenocarcinoma (EAC). An increased risk of esophageal carcinomas was observed at the second and third quartiles. HR for Q2 vs Q1 were 1.75%, and 95% CI was 1.12–2.74; HR for Q3 vs Q1 was 1.66, 95% CI, 1.05–2.61. No risk was observed at the fourth quartile which is the highest [79].

The Netherlands Cohort Study appraised the association between several cancer types and dietary acrylamide intake [80–86]. Of 62,573 women and 58 279 men aged 55–69 years enrolled to the Netherlands Cohort Study at baseline in 1986. Breast (2225 cases), prostate (2246 cases), colorectum (2190 cases), lung (1895 cases), lymphatic (1233 cases), stomach (563 cases), pancreas (349 cases), renal cell (339 cases), endometrium (221 cases), brain (216 cases), esophagus (216 cases), ovary (195 cases), larynx (180 cases), oral cavity (101 cases), oropharynx (83 cases), and thyroid (66 cases) cancers were detected in this prospective study. Dutch Food and Consumer Product Safety Authority was data supplier to estimate acrylamide exposure. In women, the mean acrylamide intake was $21.1 \pm 11.9\,\mu g$/day, whereas $22.5 \pm 12.2\,\mu g$/day in men. The mean acrylamide intake of total study population was $21.8 \pm 12.1\,\mu g$/day (i.e., $0.30 \pm 0.18\,\mu g$/kg bw/day). The sources of acrylamide were chiefly coffee (47%), Dutch spiced cake (15%), cookies (13%), French fries (8%), and potato crisps (2%) in the subcohort. After the follow-up study of 16.3 years, HRs for lymphatic malignancies were evaluated. HRs of multiple myeloma and follicular lymphoma were 1.14 (95% CI: 1.01–1.27) and 1.28 (95% CI: 1.03–1.61) per 10 mg acrylamide/day increment in men. This cohort study indicated that acrylamide could increase the multiple myeloma and follicular lymphoma in men but not in women [85].

To determine the association between endometrial, ovarian, and breast cancer risk and dietary acrylamide intake in the Netherlands Cohort Study, 1835, 327, and 300 of cases were diagnosed with cancers of breast, endometrium, and ovary, respectively, at the end of the follow-up of 11.3 years. Significantly, an increased relation was found between acrylamide intake and ovarian cancer in all groups. HR value and 95% CI were 1.11 and 0.99–1.25, respectively, with an increment of acrylamide daily intake of 10 μg. Furthermore, statistically significant association was found in the never smoker group. HR, 95% CI: 1.17 (1.01–1.36). No relation was found between breast cancer and dietary acrylamide intake. An increased risk of postmenopausal endometrial cancer was observed related with increased dietary acrylamide intake among nonsmoker participants. In this subcohort, the daily acrylamide intake was found to be $21.0 \pm 11.9\,\mu g$ ($0.32 \pm 0.19\,\mu g$/kg bw/day). While Dutch spiced cake was the main source of diversity in acrylamide intake for this population, coffee, French fries, potato crisps, and cookies were the next stuffs [80].

A case–cohort study from the Netherlands Cohort Study was performed to clarify the association with key gen mutations in Kirsten-ras (*KRAS*) and adenomatous polyposis coli (*APC*) in the risk of colorectal carcinoma and acrylamide intake. The tumor tissues of 733 cases were examined to perform mutation analysis in codons 12–13 of exon 1 of KRAS gene and 1286–1520 of exon 15 of APC gene. It was observed that significantly positive correlation in men between the colorectal cancer risk with an active mutation in KRAS gene and acrylamide intake, whereas significantly reverse correlation were observed between tumor risk with truncating mutation in the APC gene in women. The HR value of fourth vs. first quartile and 95% CI were, respectively, 2.12 and 1.16–3.87; p 0.01 for KRAS mutation. HR value of fourth vs. first quartile and 95% CI were, respectively, 0.47 and 0.23–0.94; p 0.02 for APC mutation [86]. In this large prospective cohort study, 1986–2002, a positive correlation was not determined between the head–neck and thyroid cancer risk and dietary acrylamide intake in men. A decreased risk was observed for oral cavity and oro- and hypopharynx cancers and intake of dietary acrylamide in men. An increased risk was found for oral cavity cancer and dietary acrylamide intake in nonsmoker women only (case number were not enough). The study group combined never smokers and former smokers, quit smoking before 10 years; HRs of former smokers were higher than the never smokers but not statistically significant, and lower than the current smokers and former smokers, quit smoking just before [83].

A meta-analysis of 25 studies on dietary acrylamide intake and colorectal, breast, kidney, prostate, ovarian, bladder, and esophageal cancer risk has shown no association but kidney has increased risk [87].

There are some case–control studies in relationship with EAC and esophageal squamous cell carcinoma and acrylamide intake [87,88].

A preliminary cohort study was conducted among workers who were occupationally exposed to acrylamide in the USA and the Netherlands from 1925 to 1983. The cohort consisted of 2293 men and 6561 controls. Data of exposure to acrylamide were gathered by interviews, after that the estimated exposure level was found to be 0.3 mg/m^3 for total work day. The respiratory cancer mortality was recorded at one plant (SMR = 1.31). However, the authors did not comment any association with acrylamide exposure and cancer [89].

The Nurses' Health Study I (NHS I) was a prospective cohort that included 121,700 female nurses started in 1976, and then one more cohort was found as NHS II in 1989 and consisted of 116,671 female nurses. Usual dietary acrylamide intake was assessed by a semi-quantitative FFQ in 1991 with a follow-up every 4 year in this prospective cohort study. Wilson and coworkers recorded 1179 of invasive breast cancer cases among 90,628 of women in the premenopause phase aged 26–56 years. The average acrylamide intake of the lowest and the highest quintile were 10.8 μg/day and 37.8 μg/day, respectively. French fries (23%) were the first contributor of acrylamide, coffee (15%) was the second, and afterward cold breakfast cereal (12%), potato chips (9%), and other potatoes (baked, roasted, and mashed; 5%). The authors neither did nor found any association between dietary acrylamide intake and breast cancer (RR: 0.92; 95% CI: 0.76, 1.11) [90]. In NHS, the association between the dietary acrylamide intake and breast, endometrial, and ovarian cancers was evaluated among premenopausal–postmenopausal US women; 6301 cases of invasive breast cancer, 484 cases of invasive endometrial adenocarcinoma, and 416 cases of epithelial ovarian cancer were recorded from 1980 to 2006. Acrylamide intake of the lowest quintile and the highest quintile ranged 9–26 μg/day, respectively. Coffee (20%) was the major source of acrylamide whereas breakfast cereal (15%) and French fries (12%) were the second and the third order. An association was not found between acrylamide intake and breast cancer (RR: 0.95; 95% CI: 0.87–1.03). An elevated risk of endometrial cancer was found among highest acrylamide consumers (RR: 1.41; 95% CI, 1.01–1.97; $P=0.03$). There was no detected risk of ovarian cancer with acrylamide intake (RR: 1.25; 95% CI: 0.88–1.77) [91].

The UK Women's Cohort Study for breast cancer occurrence including 33,731 women was started in 1995. It was adopted as UK arm of EPIC study to assess dietary information [92]. The results of some cohort studies on acrylamide intake and cancer are summarized in Table 3.

KEY FACTS

- Acrylamide was firstly classified as a probable carcinogen by IARC in 1994 [1]. It is determined as a "reasonably anticipated to be a human carcinogen" by NTP in 2011 [30]. FAO/WHO (Food and Agriculture Organization of the United Nations/World Health Organization) estimated the acrylamide levels in food categories and the contribution to total exposure of the general public. The estimated average intake for the general population was 0.3–0.8 μg/kg body weight (bw)/day [7].
- Increased tumor risk at different sites was found in 2-year drinking water studies on experimental animals administered with acrylamide [3,6]. Two different $BMDL_{10}$ for acrylamide are estimated to be 0.31 mg/kg bw/day for the induction of mammary tumors in female rats and 0.18 mg/kg bw/day for Harderian gland tumors in male mice by the Joint FAO/WHO Expert Committee on Food Additives [27].
- Various approaches have been suggested to assess the quantitative risk of carcinogens, and MoE (margin of exposure) approach is the most favorite among them. According to the Joint FAO/WHO Expert Committee on Food Additives, MoE is estimated to be 200 for an average acrylamide consumer, whereas MoE is estimated to be 50 for a high acrylamide consumer [26].
- Acrylamide has neurotoxic effects on experimental animals such as rats, mice, and guinea pigs [31]. Axonal degenerations were observed in a dose-dependent manner in acrylamide-administered rats [30]. Reproductive tissues are particularly vulnerable to the administration of different doses of acrylamide to the rats [30].

Table 3 Results of cohort studies on dietary acrylamide intake and some cancer types

Cancer side	Country	Number of cases	Dietary acrylamide intake	Acrylamide sources	References
Colorectal, colon, proximal colon cancer, distal colon cancer	Swedish men	11,326 participants 676 cases	36.1 ± 9.6 μg	Coffee 23%, grain bread 17%, crisp bread 8%, white bread 7%, cookies/buns 7%, wafers/cracker/rusks 6%, fried potato 6%	Larsson et al. (2009) [75]
Breast cancer	Swedish women	61,433 participants 2952 cases	24.6 ± 7.6 μg	Coffee 29%, grain bread 13%, crisp bread 8%, white bread 7%, cookies/buns 6%, breakfast cereals/muesli 7%, fried potato 5%	Larsson et al. (2009) [77]
Endometrial carcinoma	Swedish women	61,226 participants 687 cases	24.6 ± 7.6 μg	–	Larsson et al. (2009) [76]
Breast cancer	California women	112,378 participants 5676 cases	–	BMI <25 alcohol <20 nonsmoker	Garcia et al. (2015) [73]
Breast cancer	UK womens	33,731 participants 1084 cases	Mean = 15 ± 0.23 μg/day Range = 10–21/day	Patoto chips 23%, bakery goods 17%, potato crisps 14%, bread 10%, biscuits 9%, coffee 8%	Burley et al. (2010) [92]
Head-neck and thyroid cancer, oral cavity, larenx, ora and hypopharyngeal cancer	Netherlends	120,852 participants	Mean = 21.8 ± 12.1 μg 22.5 ± 12.1 μg (male) 21.1 ± 11.9 μg (female)	Coffee 15%, Dutch spiced 57%, French fries 14%. cookies 3%, potato crisps 5%	Schouten et al. (2009) [83]

Continued

Table 3 Results of cohort studies on dietary acrylamide intake and some cancer types—cont'd

Cancer side	Country	Number of cases	Dietary acrylamide intake	Acrylamide sources	References
Pancreatic cancer	10 European countries (Denmark, France, Germany, Greece, Italy, Norway, Spain, Sweden, the Netherlands, UK (EPIC)	521,330 participants 865 cases	26.22 ± 14.79 μg/day 10–90th percentile range was 10.25–45.89 μg/day	-	Obon-Santacana et al. (2013) [78]
Esophageal cancer	European	-	Range 15.7–23.3 μg/day	Alcohol 3(0.3–12.3)	Barrosa (2014) [79]
Colorectal cancer	The Netherlands	120,852 participants 733 cases	–	Coffee, Dutch spice cake, cookies, potato chips, French fries	Hogervorst et al. (2014) [86]
Epithelial ovarian cancer	Swedish women	61057 participants 368 cases	24.6 ± 7.6 μg	Coffee 29%, grain bread 13%, crip bread 8%, breakfast cereals 7.5%, wafers 7%, cookies/buns 6%, fried potato 5%, boiled potato 4.5%, white bread 4%, bread crumbs 3.5%, potato crisps 2.5%, French fries 2%, oatmeal porridge 2.5%	Larsson et al. (2009) [76]

Lymphatic malignancies, multiple myeloma, diffuse large-cell lymphoma, chronic lymphoctic leukemia, follicular lymphoma, waldenström macrlobu-linemia-immunocytoma, mante cell lymphoma, T-cell lympho-mas, hemato-logical malignancies	The Netherlands	120,852 participants 170 cases (male), 153 cases (female) for multiple myeloma 159 cases (male), 100 cases (female) for diffuse large cell lymphoma 134 cases (male), 66 cases for chronic lymphocytic leukemia 42 cases (male), 47 cases (female) for follicular lymphoma 54 cases (male), 35 cases (female) for Waldenström–macroglobulinemia and immunocytoma	–	Coffee, Dutch spice cake, cookies, potato chips, French fries	Bongers et al. (2012) [85]
Endometrial ovarian, breast cancers	The Netherlands	62,573 participants 1835 cases of breast cancer; 211 cases of endometrial cancer. 195 cases of ovarian cancer	Daily: 21.0 ± 11.9 μg	Dutch spices cake	Hogervorst et al. (2007) [80]
Prostate cancer	Swedish men	45,306 participants 2696 cases	Mean: 23.7 = lowest μg/day 49.8 μg/day = highest	–	Larsson et al. (2009) [74]

- Acrylamide occurrence in foods and intake in different population groups have been studied extensively worldwide over the past decade. A few comprehensive case–cohort studies such as "European Prospective Investigation into Cancer and Nutrition," "Swedish Mammography Cohort," and the "Netherlands Cohort Study" were performed to define the relationship between acrylamide intake and cancer. Although the overall evidence is limited, epidemiological studies found no direct association between acrylamide intake and risk of cancer except for the renal cell, endometrial, and ovarian cancers.

MINI DICTIONARY

MoE MoE is margin of exposure and used by risk assessors to evaluate probable safety concerns that stem from the food or feed of the substances which are genotoxic and carcinogenic. It is a ratio of two factors that appraise a population the dose at which a small but measurable adverse effect is first observed and the range of exposure to the matter considered.

Cohort study Cohort is a large group of people who are linked in some way and share common features (e.g., exposed to a drug, chemical, or vaccine) were born in a particular period. They are followed over time for a research. Scientists observe these groups and uses correlations to define the risk of subject relation.

BMDL BMDL is a statistical lower confidence limit on the dose at the BMD (benchmark dose) which is described a dose that produces a predetermined change in the response rate of an adverse effect compared to the background. Risk assessors calculate the lower bound on the BMD (typically the 95% lower confidence limit). This dose is called the Benchmark Dose Lower bound Confidence Limit. This BMDL extended with reducing the sample size, resulting in a lower, more health protective risk value, and appropriately reflecting uncertainty in the study.

Epoxidation Epoxidation is a reaction that yields an epoxide. Epoxide is a cyclic ether with a three-atom ring which is equilateral triangle that makes it extremely strained. Epoxides are more reactive by the strained ring than the other ethers. Epoxidation occurs in the P450 cytochrome monooxygenase system. For example, acrylamide is metabolized to glycidamide, a highly reactive epoxide by the P450 cytochrome monooxigenase CYP2E.

F344 rat The Fischer F344 inbred rats originate from a nucleus colony obtained from the National Institutes of Health, Bethesda, Maryland. After an inbred strain has been divided into several branches for a number of years and if this separation happens at the early years of inbreeding, genetic variations result among the branches as a result of new mutations. It is called "substrains" and remarked by a slash and letters or some numbers. For example, F344/N is the substrain maintained by the National Institutes of Health, designated as "N."

SUMMARY POINTS

Acrylamide is a commodity chemical used in the industry and laboratory. It is also obtained from carbon hydrate and amino acid by heating the food, primarily coffee, the products of potato, and French fries. Acrylamide has been proved to have toxic effects in experimental animals. It has been shown as a toxic chemical to some tissues such as the reproductive tissues and retina besides neurotoxicity in rats and mice. It is converted to glycidamide which is a highly reactive epoxide by CYP2E in humans, then acquires genotoxic and carcinogenic properties to several organs. Epidemiological studies did not indicate any association between occupational and dietary exposure of acrylamide and incidence of cancer. A few studies have shown a relationship between acrylamide intake and the incidence of cancers in renal cells, endometrium, and ovary.

REFERENCES

[1] International Agency for Research on Cancer (IARC). Monographs on the evaluation of carcinogenic risks to humans. Some industrial chemicals. Lyon, France: IARC; 1994.
[2] Bull RJ, Robinson M, Laurie RD, Stoner GD, Greisiger E, Meier JR, et al. Carcinogenic effects of acrylamide in Sencar and A/J mice. Cancer Res 1984;44:107–11.
[3] Johnson KA, Gorzinski SJ, Bodner KM, Campbell RA, Wolf CH, Friedman MA, et al. Chronic toxicity and oncogenicity study on acrylamide incorporated in the drinking-water of Fischer 344 rats. Toxicol Appl Pharmacol 1986;85:154–68.
[4] Tareke E, Rydberg P, Karlsson P, Eriksson S, Tornqvist M. Analysis of acrylamide, a carcinogen formed in heated foodstuffs. J Agric Food Chem 2002;50:4998–5006. Technical training on risk analysis for Saarc countries FAO RAP, Bangkok, Thailand quality council of India Delhi, India, June 17–21, 2013.
[5] Fennell TR, Sumner SC, Snyder RW, Burgess J, Spicer R, Bridson WE, et al. Metabolism and hemoglobin adduct formation of acrylamide in humans. Toxicol Sci 2005;85:447–59.
[6] Friedman MA, Dulak LH, Stedham MA. A lifetime oncogenicity study in rats with acrylamide. Fundam Appl Toxicol 1995;27:95–105.
[7] European Union Risk Assessment Report. Luxembourg: Acrylamide; 2002. ISBN: 92-894-1250-X. 210 p. Available at: http://europa.eu.int.
[8] Sumner SC, Williams CC, Snyder RW, Krol WL, Asgharian B, Fennell TR. Acrylamide: a comparison of metabolism and hemoglobin adducts in rodents following dermal, intraperitoneal, oral, or inhalation exposure. Toxicol Sci 2003;75:260–70.
[9] Diembeck W, Dusing HJ, Akhiani M. Dermal absorption and penetration of acrylamide ([C14]-acrylamide as tracer) in different cosmetic formulations and polyacrylamide-solution after topical application to excised pig skin. Beiersdorf: Rep 7061/PEN.203 Acrylamide Monomer Prod Assoc 1998;9.
[10] Marty JP. In vitro percutaneous absorption of acrylamide across human skin. Research Unit in Dermopharmacology and Cosmetology. Paris, France: University of Paris; 1998. Report for SNF, SEPPIC, 75 Quai d'Orsay, 75321 Paris Cedex.
[11] Friedman M. Chemistry, biochemistry, and safety of acrylamide. A review. J Agric Food Chem 2003;51:4504–26.
[12] Dybing E, Farmer PB, Andersen M, Fennell TR, Lalljie SP, et al. Human exposure and internal dose assessments of acrylamide in food. Food Chem Toxicol 2005;43:365–410.
[13] Smith CJ, Perfetti TR, Rumple MA, Rodgman A, Doolittle DJ. "IARC Group 2A carcinogens" reported in cigarette mainstream smoke. Food Chem Toxicol 2000;38:371–83.
[14] EFSA (European Food Safety Authority). Update on acrylamide levels in food from monitoring years 2007 to 2010. EFSA J 2012;10(10):2938. 38 pp. http://dx.doi.org/10.2903/j.efsa.2012.2938.
[15] EFSA CONTAM Panel (EFSA Panel on Contaminants in the Food Chain), 20YY. Scientific opinion on acrylamide in food. EFSA J 2015;13(6):4104, 321pp. http://dx.doi.org/10.2903/j.efsa.2015.4104. Available at: www.efsa.europa.eu/efsajournal.
[16] Freisling H, Moskal A, Ferrari P, Nicolas G, Knaze V, Clavel-Chapelon F, et al. Dietary acrylamide intake of adults in the European prospective investigation into cancer and nutrition differs greatly according to geographical region. Eur J Nutr June 2013;52(4):1369–80.
[17] Miller MJ, Carter DE, Sipes IG. Pharmacokinetics of acrylamide in Fisher-344 rats. Toxicol Appl Pharmacol 1982;63:36–44.
[18] Hagmar L, Törnqvist M, Nordander C, Rosén I, Bruze M, Kautiainen A, et al. Health effects of occupational exposure to acrylamide using hemoglobin adducts as biomarkers of internal dose. Scand J Work Environ Health 2001;27:219–26.

[19] Fennell TR, Sumner SC, Snyder RW, Burgess J, Friedman MA. Kinetics of elimination of urinary metabolites of acrylamide in humans. Toxicol Sci 2006;93:256–67.
[20] Schettgen T, Rossbach B, Kütting B, Letzel S, Drexler H, Angerer J. Determination of haemoglobin adducts of acrylamide and glycidamide in smoking and non-smoking persons of the general population. Int J Hygene Environ Health 2004;207:531–9.
[21] Vesper HW, Caudill SP, Osterloh JD, Meyers T, Scott D, Myers GL. Exposure of the U.S. population to acrylamide in the National Health and Nutrition Examination Survey 2003-2004. Environ Health Perspect 2010;118:278–83.
[22] Claeys W, Baert K, Mestdagh F, Vercammen J, Daenens P, De Meulenaer B, et al. Assessment of the acrylamide intake of the Belgian population and the effect of mitigation strategies. Food Addit Contam Part a-Chemistry Analysis Control Expo Risk Assess 2010;27:1199–207.
[23] Vinci RM, Mestdagh F, De Meulenaer B. Acrylamide formation in fried potato products–present and future, a critical review on mitigation strategies. Food Chem 2012;133:1138–54.
[24] Pelucchi C, Bosetti C, Galeone C, La Vecchia C. Dietary acrylamide and cancer risk: an updated meta-analysis. Int J Cancer November 18, 2014. http://dx.doi.org/10.1002/ijc.29339. [Epub ahead of print].
[25] EFSA (European Food Safety Authority). Opinion of the scientific committee on a request from EFSA related to a harmonised approach for risk assessment of substances which are both genotoxic and carcinogenic. EFSA J 2005;282:1–31.
[26] JECFA. Joint FAO/WHO Expert Committee on food additives. Rome: WHO/FAO; 2005. p. 47.
[27] JECFA. Joint FAO/WHO Expert Committee on food additives. Rome: WHO/FAO; 2010. p. 2 http://www.who.int/foodsafety/chem/summary72_rev.pdf.
[28] Sakamoto J, Kurosaka Y, Hashimoto K. Histological changes of acrylamide-induced testicular legions in mice. Exp Mol Pathol 1988;48:324–34.
[29] Yang H, Lee S, Jin Y, Choi J, Han D, Chae C, et al. Toxicological effects of acrylamide on rat testicular gene expression profile. Reprod Toxicol 2005;19:527–34.
[30] NTP (National Toxicology Program). NTP technical report on the toxicology and carcinogenesis studies of acrylamide (CAS No. 79-06-1) in F344/N rats and B6C3F1 mice (feed and drinking water studies). NTP TR 575. NIH Publication No. 12-5917. National Institutes of Health. Public Health Service. U.S. Department of Health and Human Services; July 2012.
[31] International Agency for Research on Cancer. IARC Monographs on the evaluation of carcinogenic risks to humans. Some chemicals used in plastics and elastomers, vol. 39. Lyon: Acrylamide; 1986. pp. 41–66.
[32] LoPachin RM. Acrylamide neurotoxicity: neurological, morphological and molecular endpoints in animal models. Adv Exp Med Biol 2005;561:21–37.
[33] Exon JH. A review of the toxicology of acrylamide. J Toxicol Environ Health Part B 2006;9:397–412.
[34] Shipp A, Lawrence G, Gentry R, McDonald T, Bartow H, Bounds J, et al. Acrylamide: review of toxicity data and dose-response analyses for cancer and noncancer effects. Crit Rev Toxicol 2006;36: 481–608.
[35] Navarro X, Verdu E, Guerrero J, Buti M, Gonalons E. Abnormalities of sympathetic sudomotor function in experimental acrylamide neuropathy. J Neurol Sci 1993;114:56–61.
[36] Post EJ. Unmyelinated nerve fibers in feline acrylamide neuropathy. Acta Neuropathol (Berl.) 1978;42:19–24.
[37] Belai A, Burnstock G. Acrylamide-induced neuropathic changes in rat enteric nerves: similarities with effects of streptozotocin- diabetes. J Auton Nerv Syst 1996;58:56–62.
[38] LoPachin RM, Lehning EJ. Mechanism of calcium entry during axon injury and degeneration. Toxicol Appl Pharmacol 1997;143:233–44.
[39] Takahashi A, Mizutani M, Itakura C. Acrylamide-induced peripheral neuropathy in normal and neurofilament-deficient Japanese quails. Acta Neuropathol (Berl.) 1995;89:17–22.
[40] Sickles DW, Brady ST, Testino A, Friedman MA, Wrenn RW. Direct effect of the neurotoxicant acrylamide on kinesin-based microtubule motility. J Neurosci Res 1996;46:7–17.
[41] Ko MH, Chen WP, Hsieh ST. Neuropathology of skin denervation in acrylamide-induced neuropathy. Neurobiol Dis October 2002;11(1):155–65.

[42] Ramsey J, Young JD, Gorzinski SJ. Acrylamide: Toxicodynamics in the rat. Midland, MI: Dow Chemical Company; 1984.
[43] Marlowe C, Clark MJ, Mast RW, Friedman MA, Waddell WJ. The distribution of [14C] acrylamide in male and pregnant Swiss-Webster mice studied by whole-body autoradiography. Toxicol Appl Pharmacol December 1986;86(3):457–65.
[44] Sumner SCJ, Bahman A, Williams CC, Moore TA, Fennell TR. Acrylamide, metabolism, distribution, and hemoglobin adducts in male F344 rats and B6C3F1 mice following inhalation exposure and distribution and hemoglobin adducts following dermal application to F344 rats. Research Triangle Park, NC: CIIT; 2001.
[45] Ikeda G, Miller E, Sapienza P, et al. Comparative tissue distribution and excretion of fl-"C]acrylamide in beagle dogs and miniature pigs. Food Chem Toxic 1987;25(11):871–5.
[46] Zenick H, Hope E, Smith MK. Reproductive toxicity associated with acrylamide treatment in male and female rats. J Toxicol Environ Health 1986;17:457–72.
[47] Tyl RW, Marr MC, Myers CB, Ross WP, Friedman MA. Relationship between acrylamide reproductive and neurotoxicity in male rats. Reprod Toxicol 2000;14(2):147–57.
[48] Rice JM. The carcinogenicity of acrylamide. Mutat Res 2005;580:3–20.
[49] Bull RJ, Robinson M, Stober JA. Carcinogenic activity of acrylamide in the skin and lung of Swiss-ICR mice. Cancer Lett 1984;24:209–12.
[50] Damjanov I, Friedman MA. Mesotheliomas of tunica vaginalis testis of Fischer 344(F344) rats treated with acrylamide: a light and electron microscopy study. Invivo 1998;12:495–502.
[51] Beland FA, Mellick PW, Olson GR, Mendoza MC, Marques MM, Doerge DR. Carcino-genicity of acrylamide in B6C3F (1) mice and F344/N rats from a 2-year drinking water exposure. Food Chem Toxicol 2013;51:149–59.
[52] Maronpot RR, Thoolen RJ, Hansen B. Two-year carcinogenicity study of acrylamide in Wistar Han rats with in utero exposure. Exp Toxicol Pathol February 2015;67(2):189–95.
[53] Tareke E, Twaddle NC, McDaniel LP, Churchwell MI, Young JF, Doerge DR. Relationships between biomarkers of exposure and toxicokinetics in Fischer 344 rats and B6C3F1 mice administered single doses of acrylamide and glycidamide and multiple doses of acrylamide. Toxicol Appl Pharmacol 2006;217:63–75.
[54] Dearfield KL, Douglas GR, Ehling UH, Moore MM, Sega GA, Brusick DJ. Acrylamide: a review of its genotoxicity and an assessment of heritable genetic risk. Mutat Res 1995;330:71–99.
[55] Favor J, Shelby MD. Transmitted mutational events induced in mouse germ cells following acrylamide or glycidamide exposure. Mutat Res 2005;580:21–30.
[56] Paulsson B, Grawe J, Tornqvist M. Hemoglobin adducts and micronucleus frequencies in mouse and rat after acrylamide or N-methylolacrylamide treatment. Mutat Res Genetic Toxicol Environ Mutagen 2002;516:101–11.
[57] Paulsson B, Kotova N, Grawe J, Henderson A, Granath F, Golding B, et al. Induction of micronuclei in mouse and rat by glycidamide, genotoxic metabolite of acrylamide. Mutat Res Genetic Toxicol Environ Mutagen 2003;535:15–24.
[58] Manjanatha MG, Aidoo A, Shelton SD, Bishop ME, McDaniel LP, Lyn-Cook LE, et al. Genotoxicity of acrylamide and its metabolite glycidamide administeredin drinking water to male and female Big Blue mice. Environ Mol Mutagen 2006;47:6–17.
[59] Jiang L, Cao J, An Y, Geng C, Qu S, Jiang L, et al. Genotoxicity of acrylamide in human hepatoma G2 (HepG2) cells. Toxicol In Vitro December 2007;21(8):1486–92.
[60] Park J, Kamendulis LM, Friedman MA, et al. Acrylamide-induced cellur transformation. Toxicol Sci 2002;65:177–83.
[61] Sumner SC, Fennell TR, Moore TA, Chanas B, Gonzalez F, et al. Role of cytochrome P450 2E1 in the metabolism of acrylamide and acrylonitrile in mice. Chem Res Toxicol 1999;12:1110–6.
[62] Ghanayem BI, McDaniel LP, Churchwell MI, Twaddle NC, Snyder R, Fennell TR, et al. Role of CYP2E1 in the epoxidation of acrylamide to glycidamide and formation of DNA and hemoglobin adducts. Toxicol Sci 2005;88:311–8.
[63] Generoso WM, Sega GA, Lockhart AM, Hughes LA, Cain KT, Cacheiro NLA, et al. Dominant lethal mutations, heritable translocations, and unscheduled DNA synthesis induced in male mouse germ cells by glycidamide, a metabolite of acrylamide. Mutat Res 1996;371:175–83.

[64] Ghanayem BI, Witt KL, El-Hadri L, Hoffler U, Kissling GE, Shelby MD, et al. Comparison of germ cell mutagenicity in male CYP2E1-null and wild-type mice treated with acrylamide: evidence supporting a glycidamidemediated effect. Biol Reprod 2005;72:157–63.
[65] Ghanayem BI, Witt KL, Kissling GE, Tice RR, Recio L. Absence of acrylamide-induced genotoxicity in CYP2E1-null mice. evidence consistent with a glycidamide-mediated effect. Mutat Res 2005;578:284–97.
[66] Zeiger E, Anderson B, Haworth S, Lawlor T, Mortelmans K, Speck W. Salmonella mutagenicity tests: III. Results from the testing of 255 chemicals. Environ Mutagen 1987;9:1–109.
[67] Baum M, Fauth E, Fritzen S, Herrmann A, Mertes P, Merz K, et al. Acrylamide and glycidamide: genotoxic effects in V79-cells and human blood. Mutat Res 2005;580:61–9.
[68] Koyama N, Sakamoto H, Sakuraba M, Koizumi T, Takashima Y, Hayashi M, et al. Genotoxicity of acrylamide and glycidamide in human lymphoblastoid TK6 cells. Mutat Res Genetic Toxicol Environ Mutagen 2006;603:151–8.
[69] Martins C, Oliveira NG, Pingarilho M, da Costa GG, Martins V, Marques MM, et al. Cytogenetic damage induced by crylamide and glycidamide in mammalian cells: correlation with 8423 specific glycidamide-DNA adducts. Toxicol Sci 2007;95:383–90.
[70] Katic J, Cemeli E, Baumgartner A, Laubenthal J, Bassano I, Stolevik SB, et al. Evaluation of the genotoxicity of 10 selected dietary/environmental compounds with the in vitro micronucleus cytokinesis-block assay in an interlaboratory comparison. Food Chem Toxicol 2010;48:2612–23.
[71] ATSDR (Agency for Toxic Substances and Disease Registry). Toxicological profile of acrylamide. U.S. Department of Health and Human Services. Public Health Service; December 2012. Available at: http://www.atsdr.cdc.gov/toxprofiles/tp203.pdf.
[72] Pingarilho M, Oliveira NG, Martins C, Gomes BC, Fernandes AS, Martins V, et al. Induction of sister chromatid exchange by acrylamide and glycidamide in human lymphocytes: role of polymorphisms in detoxification and DNA-repair genes in the genotoxicity of glycidamide. Mutat Res 2013;752:1–7.
[73] Garcia E, Hurley S, Nelson DO, Hertz A, Reynolds P. Hazardous air pollutants and breast cancer risk in California teachers: a cohort study. Environ Health January 30, 2015;14(1):14.
[74] Larsson SC, Akesson A, Wolk A. Dietary acrylamide intake and prostate cancer risk in a prospective cohort of Swedish men. Cancer Epidemiol Biomarkers Prev 2009;18:1939–41.
[75] Larsson SC, Akesson A, Bergkvist L, Wolk A. Dietary acrylamide intake and risk of colorectal cancer in a prospective cohort of men. Eur J Cancer 2009;45:513–6.
[76] Larsson SC, Akesson A, Wolk A. Long-term dietary acrylamide intake and risk of epithelial ovarian cancer in a prospective cohort of Swedish women. Cancer Epidemiol Biomarkers Prev 2009;18:994–7.
[77] Larsson SC, Akesson A, Wolk A. Long-term dietary acrylamide intake and breast cancer risk in a prospective cohort of Swedish women. Am J Epidemiol 2009;169:376–81.
[78] Obón-Santacana M, Slimani N, Lujan-Barroso L, Travier N, Hallmans G, Freisling H, et al. Dietary intake of acrylamide and pancreatic cancer risk in the European Prospective Investigation into Cancer and Nutrition (EPIC) cohort. Ann Oncol 2013;24:2645–51.
[79] Lujan-Barroso L, González CA, Slimani N, Obón-Santacana M, Ferrari P, Freisling H, et al. Dietary intake of acrylamide and esophageal cancer risk in the European Prospective Investigation into Cancer and Nutrition cohort. Cancer Causes Control May 2014;25(5):639–46.
[80] Hogervorst JG, Schouten LJ, Konings EJ, Goldbohm RA, van den Brandt PA. A prospective study of dietary acrylamide intake and the risk of endometrial, ovarian, and breast cancer. Cancer Epidemiology. Biomarkers Prev 2007;16:2304–13.
[81] Hogervorst JG, Schouten LJ, Konings EJ, Goldbohm RA, van den Brandt PA. Dietary acrylamide intake is not associated with gastrointestinal cancer risk. J Nutr 2008;138(11):2229–36.
[82] Hogervorst JG, Schouten LJ, Konings EJ, Goldbohm RA, van den Brandt PA. Lung cancer risk in relation to dietary acrylamide intake. J Natl Cancer Inst 2009;101:651–62.
[83] Schouten LJ, Hogervorst JG, Konings EJ, Goldbohm RA, van den Brandt PA. Dietary acrylamide intake and the risk of head-neck and thyroid cancers: results from the Netherlands Cohort Study. Am J Epidemiol 2009;170:873–84.
[84] Pedersen GS, Hogervorst JG, Schouten LJ, Konings EJ, Goldbohm RA, van den Brandt PA. Dietary acrylamide intake and estrogen and progesterone receptor-defined ostmenopausal breast cancer risk. Breast Cancer Res Treat 2010;122:199–210.

[85] Bongers ML, Hogervorst JG, Schouten LJ, Goldbohm RA, Schouten HC, van den Brandt PA. Dietary acrylamide intake and the risk of lymphatic malignancies: the Netherlands Cohort Study on diet and cancer. PLoS One 2012;7:e38016.
[86] Hogervorst JGF, de Bruijn-Geraets D, Schouten LJ, van Engeland M, de Kok TMCM, Goldbohm A, et al. Dietary acrylamide intake and the risk of colorectal cancer with specific mutations in KRAS and APC. Carcinogenesis 2014;35:1032–8.
[87] Pelucchi C, Galeone C, Levi F, Negri E, Franceschi S, Talamini R, et al. Dietary acrylamide and human cancer. Int J Cancer 2006;118(2):467–71.
[88] Lin Y, Lagergren J, Lu Y. Dietary acrylamide intake and risk of esophageal cancer in a population-based case-control study in Sweden. Int J Cancer 2011;128(3):676–81.
[89] Collins J, Swaen G, Marsh G, Utidjian M, Carporossi J, Lucas L. Mortality patterns among workers exposed to acrylamide. J Occup Med 1989;31:614–7.
[90] Wilson KM, Mucci LA, Cho E, Hunter DJ, Chen WY, Willett WC. Dietary acrylamide intake and risk of premenopausal breast cancer. Am J Epidemiol 2009;169:954–61.
[91] Wilson KM, Mucci LA, Rosner BA, Willett WC. A prospective study of dietary acrylamide intake and the risk of breast, endometrial, and ovarian cancers. Cancer Epidemiol Biomarkers Prev 2010;10: 2503–15.
[92] Burley VJ, Greenwood DC, Hepworth SJ, Fraser LK, de Kok TM, van Breda SG, et al. Dietary acrylamide intake and risk of breast cancer in the UK women's cohort. Br J Cancer 2010;103:1749–54.

CHAPTER 5

Maternal Acrylamide and Effects on Offspring

Monika Hułas-Stasiak[1], Piotr Dobrowolski[1], Ewa Tomaszewska[2]
[1]Department of Comparative Anatomy and Anthropology, Maria Curie-Sklodowska University, Lublin, Poland; [2]Department of Animal Biochemistry and Physiology, Faculty of Veterinary Medicine, University of Life Sciences, Lublin, Poland

INTRODUCTION

Prenatal development takes place in a sequence of carefully timed events that progress from the cellular level to the formation of tissues, the organ system, and morphological structures. Perturbations caused by developmental toxicants may completely alter the normal developmental processes, leading to disruption of fetal organogenesis.

Developmental toxicology is defined as an ability of environmental toxicants to exert adverse effects on the developing organism. The major manifestations of developmental toxicity include death of the developing organism, malformation, growth retardation, and functional deficiency. These adverse effects can result from exposure prior to conception (of either parent), during the prenatal period, or postnatally to the time of sexual maturation [1].

It is well known that the first period of embryonic development from fertilization to the early postimplantation stage is the most sensitive for embryo lethal effects induced by environmental agents. The period of organogenesis includes intensive cell proliferation and tissue differentiation; therefore, the embryo is then most susceptible to dysmorphogenesis from developmental toxicants, and a majority of congenital malformations can be induced. During the fetal period, other more complex structural abnormalities can develop. Alterations induced at this stage may involve growth retardation, developmental delay, or functional disturbances, particularly neurobehavioral problems as the brain matures relatively late. The dose or magnitude of the exposure to a developmental toxicant may greatly affect the ensuing pathogenesis and generate a variety of fetal responses manifested by a broad spectrum of developmental disorders [2].

Agents that produce developmental toxicity at a dose that is not toxic to the maternal animal are of special concern because the developing organism is affected, but the toxicity is not apparent in the mother. However, a situation when adverse developmental effects are produced only at doses that cause minimal maternal toxicity is more common; in these cases, the developmental effects are still considered to represent developmental toxicity and should not be discounted as being secondary to maternal toxicity [3].

Acrylamide in Food
http://dx.doi.org/10.1016/B978-0-12-802832-2.00005-X

Maternal toxicity is defined as an adverse, indirect effect on the developing organism that occurs secondarily because of some effect on the pregnant mothers. At doses causing excessive maternal toxicity (which is significantly greater than the minimal toxic dose), information on developmental effects may be difficult to interpret and of limited value. Exposure of the developing individuals is often secondary via placental transfer or through breast milk. Thus, exposure of the fetus or child may not be the same as for a pregnant or lactating mother, and measurements of the toxicant in maternal or cord blood and in breast milk may provide a better estimate of the developmental exposure [3].

It is likely that fetal malformations, functional alterations, and even death can result from direct effects on the fetus, indirect (maternally mediated) effects, or a combination of the two. Moreover, whether developmental effects are secondary to maternal toxicity or not, maternal effects may be reversible while effects on the offspring may be permanent.

Preconceptional and utero toxicity may be enhanced in postnatal life. Exposure from breast milk, food, and the general environment may be associated with further postnatal developmental effects (e.g., changes in behavioral and cognitive function, or growth).

In recent years, acrylamide has aroused great interest in mass media and scientific publications because of its formation in heat-treated foods and its high levels detected in commonly consumed food [4–7]. Furthermore, the prenatal exposure to acrylamide is of particular concern since reproductive and developmental toxicity of acrylamide has been reported in humans and rodents [8–12].

Although the literature on the neurotoxic, mutagenic, and carcinogenic effects of acrylamide is huge, the data on its biochemical and morphological effects during embryological and early postnatal development are scarce. Sorgel et al. [13] found that 10–50% of dietary acrylamide in pregnant women was transferred via blood through the placenta to the fetus and up to 18.8 μg/L is contained in breast milk. Because water-soluble acrylamide and its metabolite glycidamide readily pass through the placenta, as well as mother's milk there is an increasing number of evidence that the maternal acrylamide exposure during the gestation and lactation period impairs organogenesis and further postnatal development in offspring (Figure 1).

This chapter attempts to collect and summarize the available literature on the adverse effects of prenatally and postnatally administered acrylamide on offspring development. A better understanding of the risks arising from the diet rich in acrylamide prompts, especially pregnant women and children to change their eating habits.

CENTRAL AND PERIPHERAL NERVOUS SYSTEM

The development of the central nervous system (CNS) is incredibly complex. It is clear that the developing CNS has a protracted period of vulnerability, which extends beyond organogenesis. This includes developmental processes that occur throughout the fetal, neonatal, and infancy period.

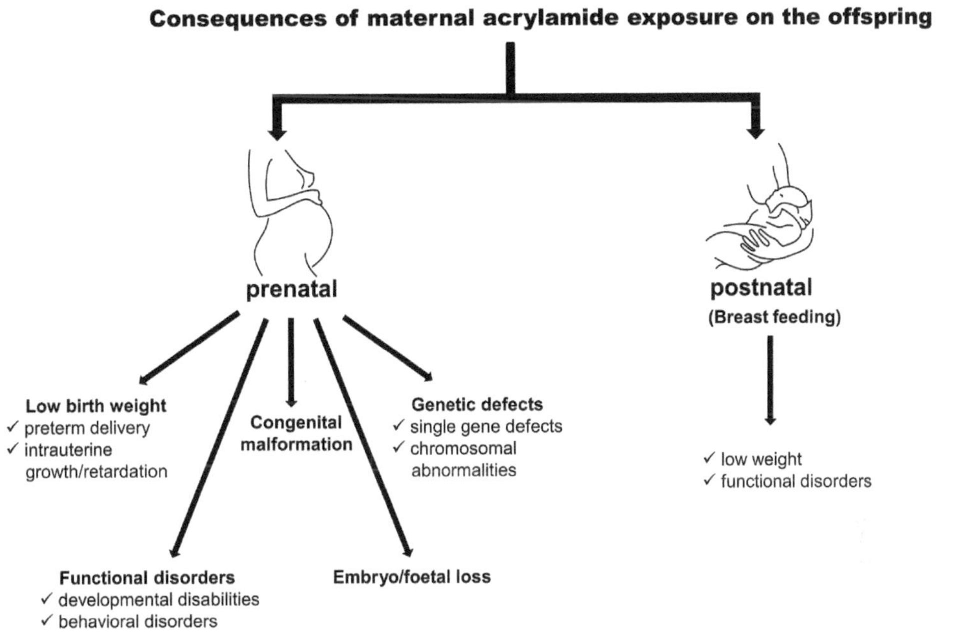

Figure 1 The figure highlights the adverse effects of maternal acrylamide exposure on the offspring. Prenatal exposure can induce genetic defects, embryo/fetal loss, congenital malformation, functional disorders, and low birth weight. These effects can be enhanced during postnatal life leading to further weight loss and functional disorders.

During brain maturation, there is an extensive reorganization of its structure including neurogenesis, neuronal differentiation, neuronal migration, arborization and synaptogenesis, synaptic organization, myelination, neuronal apoptosis, and gliogenesis, as well as glial migration and differentiation. The prenatal as well as postnatal exposure to a neurotoxicant such as acrylamide will likely affect the structure and function of the developing brain [14].

Allam and coworkers [15,16] examined acrylamide effects on the development of the cerebellum and medulla oblongata and on oxidative stress in newborn albino rats. Acrylamide at a dose of 10 mg/kg/day was administrated by gastric intubation to females from day 7 of gestation until birth (group B) as well as from day 7 of gestation until day 28 after birth (group C). The control pregnant female rats were given saline (group A). Signs of acrylamide toxicity were observed on the treated mothers postnatally and involved ataxia, splayed hind limb, weakness of hind-limb muscles, and finally paralysis, which caused alterations in maternal behavior. Newborns suffered from poor lactation and consequently malnutrition. At birth, the pups of all groups were hairless. There was a delay in fur appearing and ear and eye opening in newborns from the acrylamide-treated dams.

The results indicated that the acrylamide treatment during the prenatal and lactation periods induced a significant increase in oxidative stress and a marked suppression of the antioxidant defense system in the cerebellum and medulla oblongata of the rat offspring. Acrylamide induced a pronounced increase in lipid peroxidation while glutathione, total thiols, superoxide dismutase, and peroxidase activities were enormously depleted.

Allam et al. [17] also noted that the prenatal and further postnatal administration of acrylamide delayed proliferation, cellular migration, and differentiation in the external granular layer of cerebellum. According to Marcelo and Fahad [18] the external granular layer is the germinal layer that generates the granule cells, which then begin to migrate through the Purkinje cell layer to form the internal granular layer. Therefore, disappearance of the external granular layer is a normal part of the ontogeny of cerebellum. In acrylamide-treated animals this layer was thick in comparison with the control. These authors suggest that acrylamide impaired the migration of granule cells from the external to internal layer. The internal granular layer was affected by acrylamide-induced aberrations in the external granular layer. Ultrastructural studies of Purkinje cells in acrylamide-treated animals revealed microvacuolation and cell loss. The most striking features of acrylamide toxicity in the medulla oblongata and cerebral cortex were pyknosis and neurons chromatolysis which were observed at all investigated stages [16,17]. These observations clearly indicate that acrylamide or its metabolites disrupt the biochemical machinery, cause oxidative stress, and induce structural changes in the developing rat cerebellum, cerebrum, and medulla oblongata.

In another study, El-Bakry et al. [19] indicated histopathological and histochemical alterations in brachial and lumber regions of the spinal cord in newborn rats whose mothers were exposed to acrylamide during the gestation and the lactation periods. In the treated newborns, retardation in the development of sensorimotor reflexes was a manifestation of acrylamide neurotoxicity. Acrylamide induced neuropathy and neuronal loss, which lead to behavioral abnormalities. Behavioral abnormalities in the developing newborn rats depend on CNS regions where neuronal loss occurs. Acrylamide causes weakness, ataxia, and malformation in offspring activity. Late appearance and maturation of the hind-limb grasp reflex result from the dysfunction of spinal cord neurons. The neuronal loss in the spinal cord affects the spinal reflexes.

Moreover, El-Bakry et al. [19] showed a positive correlation between the intensity of Nissl granules and offspring activity, where granule intensity in the neurons refers to a high metabolic rate of these neurons. Therefore, the low activation of the newborns from the acrylamide-treated group was reflected by the low intensity of Nissl granules. The biochemical basis for neuropathy may involve modification of amino acids and proteins present in neurons and suppression of amino acid incorporation into proteins of the nervous system [20,21]. Acrylamide directly impairs protein synthesis in fetal tissues through its effect on RNA and DNA as well as total and subcellular protein contents [12,22]. The study presented by El-Bakry and coworkers confirmed chromatolysis in a majority of neurons after acrylamide treatment, which was caused by a high affinity of acrylamide and its metabolite glycidamide to DNA and formation of DNA adducts.

Studies by Ho et al. [23] reported that acrylamide was able to down-regulate the microtubular system and neurofilaments, thus blocking the intracellular transport in the receptor of the major neurotransmitter γ-aminobutyric acid in the CNS of chicken embryos.

El-Sayyad et al. [9] investigated the neurotoxic effects of acrylamide on postnatal development. In their study, female rats were treated with a dose of 30 mg/kg/day acrylamide during pregnancy or fed a standard diet (control group), and their offspring were examined. Pups were born with delayed growth and decreased body and brain weight. Light microscopic studies of the cerebellar cortex of the treated animals revealed a decrease in Purkinje cells and the internal granular layer. These neurons showed a different pattern of cell death from the control group. Ultrastructural analysis of Purkinje cells revealed changes in the endoplasmic reticulum, loss of the normal arrangement of polyribosomes, swollen mitochondria with abnormally differentiated cristae, and an abnormal Golgi apparatus.

The gastrocnemius muscle in the acrylamide-treated animals showed extensive degeneration of myofibrils, as evidenced by the poorly differentiated A, H, and Z bands. The study revealed that prenatally acrylamide exposure caused cerebellar cortical defects and myodegeneration of the gastrocnemius muscle during postnatal development.

Delayed pinnae detachment (a developmental landmark) and deficient negative geotaxis and rotarod performance results were reported in F344 rat pups that had been exposed prenatally to acrylamide at a dose of 10 mg/kg/day [24]. No such effects were seen at a dose of ≤5 mg/kg/day. A decrease in body weight was observed in pups exposed at doses of 1, 2.5, 5, or 10 mg/kg/day.

There is also evidence that acrylamide (ACR) (at a dose of 15 mg/kg/day) may affect the visual system [25]. These authors showed that the eye was sensitive to the cytotoxicity of acrylamide treatment during fetal and breast-feeding periods. The retina of ACR-treated offspring showed many histopathological changes such as loss of the building neuronal cells, massive alterations in retinal cell layers, and degeneration, vacuolization, and cell loss in the ganglion cell layer. Degenerated retina cell layers including pyknosis, convolution, and karyolysis of the nuclei were also observed. During the lactation period, acrylamide caused further drastic alteration in the retina of 7- and 14-day-old offspring.

The presented data undoubtedly prove that acrylamide ingestion by pregnant dams has a teratogenic effect on the offspring nervous system. Acrylamide consumption impairs motor coordination and behavioral activities that are regulated by functional integration of neuronal activity in various regions of the brain and spinal cord.

However, there are a few studies that do not confirm these observations. The study presented by Takahashi et al. [26] suggested that acrylamide exposure through maternal drinking water (at doses of 0, 50, 100, or 200 mg/L) did not cause irreversible damage to the nervous system of offspring, at least under the experimental conditions applied.

A decrease in food and water consumption as well as suppression of body weight were observed in the dams at ≥100 mg/L. Maternal neurotoxicity was evident at 100 mg/L. Acrylamide did not affect the gestation period, the number of implantations, or the live birth ratio of male pups. Body weights of pups in both sexes were significantly decreased from 100 mg/L. Maternal malnutrition was apparent at ≥100 mg/L during the lactation period. Therefore, maternal toxicity (food restriction during the gestation and lactation period) might suppress pups' body weight. Friedman et al. [27] reported increased mortality and reduced body weights in pups of Wistar rat dams at a dose of 25 mg/kg per day during lactation, but this effect was mediated by maternal toxicity.

REPRODUCTIVE SYSTEM

Reproductive toxicity involves the toxic effects of chemical and physical agents on the reproductive ability of an organism and development of its offspring. Reproductive toxicity is defined by the Globally Harmonized System [28] as adverse effects of chemicals on sexual function and fertility in adult males and females, as well as on developmental toxicity in the offspring [29].

Reproductive toxicity has many unique and challenging differences from toxicity to other systems. Whereas other forms of environmental toxicity typically involve development of disease in an exposed individual, reproductive toxicity will be expressed within a reproductive unit, or couple, since reproduction requires interaction between two individuals [30]. This unique, couple-dependent aspect, although obvious, makes reproductive toxicology distinct. For example, it is possible that exposure to a toxicant by one member of a reproductive couple (e.g., the male) will be manifested by an adverse reproductive outcome in its offspring [31,32].

There are other unique aspects that reflect the challenges of reproductive toxicology. Unlike renal, cardiac, or pulmonary function, reproductive function occurs intermittently. This means that occupational exposures can interfere with reproduction but go unnoticed during periods when fertility is not desired. This intermittent characteristic can make the identification of a reproductive toxicant in humans more difficult. Another unique characteristic of reproduction is that the complete assessment of the functional integrity of the reproductive system requires that the couple attempt pregnancy [30].

Reproductive health is a priority global health area. The latest data on human reproduction are disturbing. There are many infertile couples, fetal losses, malformed embryos, genetically defective children, and other kinds of abnormal reproductive outcomes [1]. According to Mascarenhes et al. [33], the number of couples suffering from infertility has increased from 42.0 million couples in 1990 to 48.5 million couples in 2010. Multiple factors such as infections, environment, genetic reasons, and even dietary can contribute to infertility. These factors may affect the female, the male, or both parents in a union, resulting in an inability to become pregnant or carry a child to term.

In considering a developing child's health and subsequent reproductive fitness as an adult, it is very important to take into account the health, genetic susceptibility, and exposure history of the maternal and parental germ cells and reproductive tracts.

Postconception toxicity to the pregnant mother may be a determinant of the health of her developing offspring in utero, neonatally, and after birth. Lesions arising in the germ cells of both parents before conception may also result in abnormal reproductive outcomes.

There is strong evidence suggesting that acrylamide has a negative impact on the reproductive ability. The compound has been evaluated for reproductive toxicity in multigenerational studies in rodents [22,34].

Acrylamide administered in drinking water or by gavage to rats or mice at doses ≥5 mg/kg/day resulted in reduced fertility, embryo implantation, and significant reduction in the number of live pups per litter [22,34]. These effects were seen in both the F0 and F1 generations. Crossover matings of treated males with untreated females indicated that observed fertility disorders could be attributed to effects on males [22,35].

Indeed, the affinity of acrylamide for germinal tissue was illustrated by studies on the male reproductive system. The data showed intense accumulation of acrylamide or its metabolite in testis, epididymis, and the glans penis. Acrylamide caused testicular atrophy, decreased sperm count, and increased abnormal sperm morphology [35–38]. Exposure of the male to acrylamide can diminish the quality and quantity of sperm produced.

Acrylamide is a germ cell mutagen inducing dominant lethal mutations and heritable chromosomal translocations in postmeiotic sperm of treated mice [12,22,39–41]. Paternally transmitted genetic defects or mutations can result in early or late fetal loss.

Animal data revealed that long-term, low-level acrylamide exposure (at doses of 0, 1 μg–2 mg/kg/day, equivalent to human dietary intake) generated DNA damage in germ cells of mice [42]. Long-term exposure to acrylamide (12 months) may have cumulative effects although no morphological alterations were observed in the testis.

Acrylamide and its metabolite glycidamide may cause chromosomal damage through alkylation of protamines, which are necessary proteins in DNA condensation in spermatids and spermatozoa [43]. Alternatively, they may deplete cellular glutathione levels essential for the deactivation of reactive oxygen species that can damage DNA [42].

The authors suggested that genetic damage induced by acrylamide exposure might accumulate in replicating spermatogonial stem cells, which are responsible for the regeneration of all subsequent germ cell types, and might be transmitted to future offspring. Therefore, acrylamide can impair offspring ability to produce fertile gametes and participate in normal reproduction when they are adults. Related studies by Holland et al. [44] report that acrylamide induces chromatin adducts in male germ cells of mice. Such adducts and the dose-dependent morphological abnormalities observed in preimplantation embryos indicate that acrylamide may elicit mutagenic effects in germ cells.

Thus, the genotoxicity of acrylamide should be taken into further consideration regarding male reproductive health in humans.

The effects observed in acrylamide studies have shown that the reproductive response is limited to the male and the female does not exhibit reproductive toxicity [12]. However, some data suggest that the exposure of female germ cells to acrylamide may continue its reproductive toxicity.

Gall et al. [45] described a drastic disorganization of cytokeratins and vimentin filaments presented in the foot processes crossing the zona pellucida, as well as isolation of the oocyte from its cumulus cells after acrylamide treatment. Electron microscopy examination revealed relocalization and aggregation of these filaments around the nucleus of granulosa cells. Gall et al. [45] suggest direct or indirect participation of intermediate filaments in the maintenance of cell contact integrity between the oocyte and granulosa cells. The isolation of immature oocytes may lead to their death.

Acrylamide is known as a cytoskeletal disrupting agent that selectively causes alterations in the distribution of intermediate filaments (IF), especially vimentin filaments, and does not affect other cytoskeletal elements [46,47]. Recent studies have indicated that vimentin, indeed, plays a functionally important role in adhesion and cell–cell interactions mainly by regulating integrin functions [48,49]. There is compelling evidence that integrins associated with vimentin form structures with unique mechanical and biological functions in different cell types [49]. It is well known that cell–cell interactions also play a key role during follicular development [50].

There are insufficient data describing the acrylamide effect on female offspring after prenatal exposure. Evaluation of the ovary is an important endpoint in toxicological assessment because toxicants that cause loss of oogonia, oocytes, or supportive somatic cells may have adverse effects on reproduction. None of the studies that have been conducted so far included a detailed morphological analysis of the ovary of newborn pups whose mothers were treated with acrylamide. Therefore, its toxic effect on ovarian function cannot be ruled out.

Indeed, Hulas-Stasiak et al. [51] have shown for the first time that prenatal exposure to acrylamide leads to a marked reduction of primordial and primary follicles in guinea pig offspring ovary. It is worth noting that the authors observed significant changes in the offspring ovary using a low dose of a toxicant (3 mg/kg/day), which did not cause toxicity to the mother. These authors observed a significant increase in the number of caspase 3 and TUNEL-positive oocytes in newborns. It is suggested that prenatally administered acrylamide enhanced follicular atresia mediated by oocyte apoptosis. Moreover, oocytes from all follicles from the treated group were negative for vimentin staining. Thus, the authors have speculated that acrylamide may disturb vimentin filaments causing defects in granulosa cells—oocyte interactions and finally leads to oocyte death. This may have a negative effect on the reproduction function, because depletion of the pool of primordial follicles can be manifested by female infertility.

DIGESTIVE SYSTEM

Recent data have indicated that maternal acrylamide treatment has a negative influence on small intestine histomorphometry of guinea pig offspring. Acrylamide increased the number of total, divided, and inactive crypts and the number of damaged villi in the duodenum and jejunum [52]. Furthermore, a decrease in villar epithelium thickness and the number of active crypts were observed. Acrylamide reduced the number of goblet cells and intact villi in the duodenum and mucosa thickness as well as the width of crypts in the jejunum. The intestine absorptive surface was also affected by acrylamide in the jejunum. Moreover, prenatal acrylamide exposure resulted in the reduction of the size of nerve plexuses and the jejunum wall thickness. These observations suggest that acrylamide disrupts the prenatal development of small intestine leading to its malfunction in offspring. Furthermore, strikingly a similar influence of acrylamide on the morphology of small intestine was observed in adolescent mice treated with acrylamide [53]. This might suggest a more general mechanism of action, independent of the route of exposure, since placental and oral exposure leads to similar effects in the small intestine. Additionally, taking into account breast-feeding and related exposure to acrylamide, the combined effect of prenatal and postnatal action of acrylamide cannot be ruled out.

Another study reported by Mohamed and Selim [54] showed the effect of prenatal acrylamide treatment (at a dose of 10 mg/kg/day) on the morphological changes in the liver of rat offspring. These authors described the loss of hepatocyte architecture, areas of hemorrhage and degeneration, and cellular infiltration. The most damaging effect of ACR was observed in 21-day-old pups (during the lactation period). The hepatocytes had shrunken dark stained nuclei and a vacuolated cytoplasm. According to these authors, the cytoplasmic vacuolization of the hepatic cells may be caused by progressive ischemia, hypoxia, and lipid accumulation in the cells. Moreover, bile ductular proliferation was evident in acrylamide-treated pups. The ductular reaction is a proliferative response to many types of liver injuries (caused by drugs or toxins) characterized by an increase in bile duct-like structures. Further, a significant increase in the number of apoptotic cells was observed. Acrylamide might alter the redox status of the cell and promote cell death. Moreover, Ito cells in the acrylamide-treated group exhibited irregular nuclei and areas of condensed nuclear chromatin. They were surrounded by collagen fibers and their cytoplasm was devoid of lipid droplets. It was reported that the hepatotoxic action of acrylamide was manifested by cell respiration disorders that interfere with oxidation and reduction mechanisms.

Similar effects were described by Allam and co-workers [55]. Acrylamide-induced biochemical and morphological changes were found in the liver of developing pups.

Prenatally and perinatally administered acrylamide (at a dose of 10 mg/kg/day) significantly increased lipid peroxidation and reduced glutathione and total thiol levels in liver. Significant inhibition of peroxidase and superoxide dismutase activities was observed in the liver tissue. Total lipids including cholesterol and triglycerides were significantly increased in the serum. The acrylamide treatment also increased serum alanine aminotransferase and aspartate aminotransferase activities and inhibited alkaline phosphatase activity. Microscopic studies showed acrylamide-induced fatty deposits, congested central vein, vacuolization, and chromatolysis in hepatocytes. The authors concluded that acrylamide induced structural changes in the liver, which may be caused by oxidative stress and perturbation of lipid and protein metabolism.

Recently, oxidative stress has been demonstrated to be one of the key mechanisms in many chemical-induced cell injuries. Oxidative stress in the cells or tissues refers to enhanced generation of ROS and/or depletion of the antioxidant defense system, leading to an imbalance between prooxidants and antioxidants. The delicate balance between the production and catabolism of oxidants is critical for maintenance of the biological function [56]. Developmental delay in this case is believed to be determined by inactivation of some respiratory enzymes, inhibition of some cofactors of these enzymes, and suppression of oxidative phosphorylation.

BODY WEIGHT

Diet during pregnancy is a well-recognized determinant of fetal growth [57]. A change in offspring body weight is a sensitive indicator of developmental toxicity. Reduced birth weight is associated with increased morbidity in infancy and is a risk factor for numerous adverse health effects in postnatal life (reduced stature, increased incidence of cardiovascular disease, diabetes, and osteoporosis) [58].

Decreased offspring body weight following maternal acrylamide exposure during gestation has been consistently observed in mice and rats [9,12]. Maternal consumption of foods rich in acrylamide, such as fried potatoes, in humans was associated with cord blood acrylamide adduct levels and with reduced birth weight [8,11]. Prenatal exposure to acrylamide was evaluated using hemoglobin adducts levels, which are well-established biomarkers for acrylamide exposure [59]. Birth weight decreased monotonically with increasing acrylamide and glycidamide adduct cord blood levels. These findings were consistent between countries and were shown in a full study population as well as in nonsmokers. However, the mechanisms underlying the negative effects of acrylamide on birth weight have not been fully elucidated. The ability of acrylamide to react readily with sulfhydryl and amino residues in proteins, including enzymes, receptors, and cytoskeletal proteins can affect a multitude of cellular processes. It has been suggested that a database of some of acrylamide toxic effects should be developed, which may contribute to the associations with birth outcomes observed in a human study population. The increase in head circumference is an important indicator of brain growth, and reduced birth head circumference has been associated with delayed neurodevelopment [60]. Besides, acrylamide readily passes through the placenta and can have negative effects on its function and, in particular, on the transport of nutrients to the fetus, which causes its growth retardation. Moreover, acrylamide passes through mother's milk to newborns during lactation [13,26]. Acrylamide affects the function of mammary glands because it leads to prolactin reduction in animals, thus, it impairs lactation [27,61]. Therefore, it causes malnutrition and consequently body weight loss.

CONCLUSIONS

According to Petersen [62] up to 40% of all foods consumed by humans contain acrylamide; therefore, there is a need to determine its role in human health.

The literature cited above clearly indicates that prenatally and postnatally administered acrylamide has harmful effects on offspring development. These toxic effects can result in long-term medical, cognitive, and behavioral problems. Thus, maternally ingested food and maternal as well as paternal health conditions have an important influence on fetus organogenesis and offspring postnatal life. To protect fetuses and newborns, pregnant and breast-feeding women should not consume products containing high concentrations of acrylamide.

This review also provides evidence supporting the need for changes in food production. Food manufacturers should start exploring methods to reduce acrylamide in their products.

KEY FACTS

In 1986, California voters approved an initiative to address their growing concerns about exposure to toxic chemicals. The initiative became the Safe Drinking Water and Toxic Enforcement Act of 1986, better known by its original name of Proposition 65. Proposition 65 requires the State to publish a list of chemicals known to cause cancer or birth defects or other reproductive harm. This list, which must be updated at least once a year, has grown to include approximately 800 chemicals since it was first published in 1987 [28].

The Office of Environmental Health Hazard Assessment (OEHHA) administers the Proposition 65 program. In February 2011, OEHHA added acrylamide to the Proposition 65 list as that causing reproductive and developmental effects because, in studies of laboratory animals, acrylamide affected the growth of offspring in the womb exposed to it and caused genetic damage that resulted in the death of mouse and rat embryos [28].

MINI DICTIONARY

Apoptosis It is the process of programed cell death (PCD) that may occur in multicellular organisms. Biochemical events lead to characteristic cell changes (morphology) and death. These changes include blebbing, cell shrinkage, nuclear fragmentation, chromatin condensation, and chromosomal DNA fragmentation.

Arborization It is terminal branching of nerve fibers in a treelike pattern.

Ataxia It is a neurological sign consisting of lack of voluntary coordination of muscle movements. Ataxia is a nonspecific clinical manifestation implying dysfunction of the parts of the nervous system that coordinate movement, such as the cerebellum. Several possible causes exist for these patterns of neurological dysfunction.

Chromatolysis It is the disintegration of Nissl bodies in the cell body of a neuron; dissolution and disintegration of the chromatin of cell nuclei.

Gliogenesis It is the generation of nonneuronal glial populations derived from multipotent neural stem cells.

Glutathione (GSH) It is an important antioxidant, preventing damage to important cellular components caused by reactive oxygen species such as free radicals and peroxides.

Hypoxia It is a condition in which the body or a region of the body is deprived of adequate oxygen supply.

Ischemia It is a restriction in blood supply to tissues, causing a shortage of oxygen and glucose needed for cellular metabolism (to keep tissues alive).

Ito cells The perisinusoidal fat-storing cells, stellate cells, or lipocytes of the liver are mesenchymal cells located in the perisinusoidal space (a small area between the sinusoids and hepatocytes, space of Disse). They are the main site of vitamin A storage in characteristic lipid droplets. Moreover, they demonstrate synthetic activity of collagens and other extracellular matrix proteins involved in hepatic fibrosis.

Karyolysis It is the complete dissolution of the chromatin of a dying cell due to the enzymatic degradation by endonucleases.

Neurogenesis It is the process by which neurons are generated from neural stem cells and progenitor cells. Most active during prenatal development, neurogenesis is responsible for populating the growing brain with neurons.

Oocyte It is a female gametocyte or germ cell involved in reproduction. An oocyte is produced in the ovary during female gametogenesis.

Oxidative stress It is defined as a disturbance in the balance between the production of reactive oxygen species (free radicals) and antioxidant defense.

Pyknosis It is the irreversible condensation of chromatin in the nucleus of a cell undergoing death.

Synaptogenesis It is the formation of synapses between neurons in the nervous system. An explosion of synapse formation occurs during early brain development.

Zona pellucida It is found in mammalian eggs. It is a glycoprotein membrane surrounding the plasma membrane of an oocyte.

SUMMARY POINTS

- Maternal acrylamide treatment causes developmental and reproductive toxicity in the offspring.
- Prenatal and postnatal exposure to acrylamide impairs offspring organogenesis.
- Acrylamide exposure leads to disturbances in the functioning of brain, ovary, testis, and digestive tract and causes depression in body weight.

- Genetic damage induced by maternal acrylamide treatment might accumulate in the offspring germ cells. It can reduce the quality and quantity of germ cells and lead to infertility in the offspring.
- The diet of pregnant women and maternal and paternal health conditions are important factors that determine the prenatal and postnatal development of the newborns.

REFERENCES

[1] Guidelines for developmental toxicity risk assessment. Published on December 5, 1991, Federal Register 56(234):63798–63826.
[2] Fawcett LB, Brent RL. Pathogenesis of abnormal development. Developmental and Reproductive toxicology: a practical approach. FL, USA: CRC Press, Taylor and Francis group; 2006. p. 61–81 [chapter 3].
[3] Hood RD, Miller DB. Maternally mediated effects on development. Developmental and Reproductive toxicology: a practical approach. FL, USA: CRC Press, Taylor and Francis group; 2006. p. 94–115 [chapter 4].
[4] Mottram DS, Wedzicha BL, Dodson AT. Acrylamide is formed in the Maillard reaction. Nature 2002;419:448–9.
[5] Tareke E, Rydberg P, Karlsson P, Eriksson S, Törnqvist M. Analysis of acrylamide, a carcinogen formed in heated foodstuffs. J Agric Food Chem 2002;50:4998–5006.
[6] Tareke E, Rydberg P, Karlsson P, Eriksson S, Törnqvist M. Acrylamide: a cooking carcinogen? Chem Res Toxicol 2000;13:517–22.
[7] JECFA (Joint FAO/WHO Expert Committee on Food Additives). Seventy-second meeting, Summary and Conclusions. Rome. 2010. p. 1–16 http://www.fao.org/ag/agn/agns/jecfa/JECFA72%20Summary%20Version%2016%20%20Mar%20FINAL%20rev%20%282%29.pdf.
[8] Duarte-Salles T, von Stedingk H, Granum B, Gutzkow K, Rydberg P, Tornqvist M, et al. Dietary acrylamide intake during pregnancy and fetal growth-results from the Norwegian mother and child cohort study (MoBa). Environ Health Perspect 2013;121:374–9.
[9] El-Sayyad HI, El-Gammal HL, Habak LA, Abdel-Galil HM, Fernando A, Gaur RL, et al. Structural and ultrastructural evidence of neurotoxic effects of fried potato chips on rat postnatal development. Nutrition 2011;27:1066–75.
[10] Manson J, Brabac MJ, Buelke-Sam J, Carlson GP, Chapin RE, Favor JB, et al. NTP-CERHR expert panel report on the reproductive and developmental toxicity of acrylamide. Birth Defects Res B Dev Reprod Toxicol 2005;74:17–13.
[11] Pedersen M, von Stedingk H, Botsivali M, Agramunt S, Alexander J, Brunborg G, et al. Birth weight, head circumference, and prenatal exposure to acrylamide from maternal diet: the European prospective mother–child study (NewGeneris). Environ Health Perspect 2012;120:1739–45.
[12] Tyl RW, Friedman MA. Effects of acrylamide on rodent reproductive performance. Reprod Toxicol 2003;17:1–13.
[13] Sorgel F, Weissenbacher R, Kinzig-Schippers M, Hofmann A, Illauer M, Skott A, et al. Acrylamide: increased concentrations in homemade food and first evidence of its variable absorption from food, variable metabolism and placental and breast milk transfer in humans. Chemotherapy 2002;48:267–74.
[14] Martin RP, Dombrowski SC. Prenatal central nervous system development. Prenatal exposures: Psychological and educational consequences for children. USA: Springer Science + Business Media, LLC; 2008. p. 15–25, [chapter 2].
[15] Allam A, El-Ghareeb A, Abdul-Hamid M, Bakery A, Sabri MI. Prenatal and perinatal acrylamide disrupts the development of cerebellum in rat: biochemical and morphological studies. Toxicol Ind Health 2011;27:291–306.
[16] Allam A, El-Ghareeb A, Ajarem J, Abdul-Hamid M, El-Bakry A. Effect of acrylamide on the development of medulla oblongata in albino rat: biochemical and morphological studies. Afr J Pharm Pharmacol 2013;7:1320–31.

[17] Allam A, Abdul-Hamid M, Bakery A, El-Ghareeb A, Ajarem J, Sabri M. Acrylamide disrupts the ontogeny of neurobehaviour in albino rats. Life Sci J 2013;10:1760–71.
[18] Marcelo RS, Fahad S. The cerebellum at birth in therian mammals, with special reference to rodents. Brain Behav 2002;59:101–13.
[19] El-Bakry AM, Abdul-Hamid M, Allam A. Prenatal and perinatal exposure of acrylamide disrupts the development of spinal cord in rats. World J Neurosci 2013;3:17–31.
[20] Hashimoto K, Ando K. Alteration of amino acid incorporation into proteins of the nervous system in vitro after administration of acrylamide to rats. Biochem Pharmacol 1973;22:1057–66.
[21] Hashimoto K, Ando K. Alteration of amino acid incorporation into nervous protein as an indicator of chemically induced neuropathy. Adverse effect of environmental chemicals and psychotropic drugs. Elsevier, Scientific Pub. Co; 1975. p. 185–188.
[22] Tyl RW, Friedman MA, Losco PE, Fisher LC, Johnson KA, Strother DE, et al. Rat two-generation reproduction and dominant lethal study of acrylamide in drinking water. Reprod Toxicol 2000;14: 385–401.
[23] Ho WH, Wang SM, Yin HS. Acrylamide disturbs the subcellular distribution of GABA receptor in brain neurons. J Cell Biochem 2002;85:561–71.
[24] Garey J, Ferguson SA, Paule MG. Developmental and behavioral effects of acrylamide in Fischer 344 rats. Neurotoxicol Teratol 2005;27:553–63.
[25] Sakr SA, Badawy GM, El-Sayyad HJ, Afify HS. Adverse effects of acrylamide on the developing retina of albino rats. J Basic Appl Sci Res 2011;1:706–12.
[26] Takahashi M, Shibutani M, Inoue K, Fujimoto H, Hirose M, Nishikawa A. Pathological assessment of the nervous and male reproductive system of rat offspring exposed maternally to acrylamide during the gestation and lactation periods–a preliminary study. J Toxicol Sci 2008;33:11–24.
[27] Friedman MA, Tyl RW, Marr MC, Myers CB, Gerling FS, Ross WP. Effects of lactational administration of acrylamide on rat dams and offspring. Reprod Toxicol 1999;13:511–20.
[28] A Guide to the globally harmonized system of classification and labelling of chemicals, United States Department of Labor, https://www.osha.gov/dsg/hazcom/ghs.html.
[29] UNECEF. part 3, http://www.unece.org/fileadmin/DAM/trans/danger/publi/ghs/ghs_rev05/English/03e_part3.pdf; 2013.
[30] Mattison DR. Introduction to male and female reproductive function, Encyclopaedia of occupational health and safety, http://www.ilocis.org/documents/chpt9e.htm.
[31] Hales BF, Robaire B. Paternally mediated effects on development. Developmental and Reproductive toxicology: a practical approach. FL, USA: CRC Press, Taylor and Francis group; 2006. p. 125–47, [chapter 5].
[32] Olshan AF, Faustman EM. Male-mediated developmental toxicity. Reprod Toxicol 1993;7:191–202.
[33] Mascarenhas MN, Flaxman SR, Boerma T, Vanderpoel S, Stevens GA. National, regional, and global trends in infertility prevalence since 1990: a systematic analysis of 277 health surveys. PLoS Med 2012:9–12. http://dx.doi.org/10.1371/journal.pmed.1001356.
[34] Chapin RE, Fail PA, George JD, Grizzle TB, Heindel JJ, Harry GJ, et al. The reproductive and neuronal toxicities of acrylamide and three analogues in Swiss mice, evaluated using the continuous breeding protocol. Fundam Appl Toxicol 1995;27:9–24.
[35] Sakamoto J, Hashimoto K. Reproductive toxicity of acrylamide and related compounds in mice–effects on fertility and sperm morphology. Arch Toxicol 1986;59:201–5.
[36] Wang H, Huang P, Lie T, Li J, Hutz RJ, Li K, et al. Reproductive toxicity of acrylamide-treated male rats. Reprod Toxicol 2010;29:225–30.
[37] Wang RS, McDaniel LP, Manjanatha MG, Shelton SD, Doerge DR, Mei N. Mutagenicity of acrylamide and glycidamide in the testes of big blue mice. Toxicol Sci 2010;117:72–80.
[38] Zhang X, Chen F, Huang Z. Apoptosis induced by acrylamide is suppressed in a 21.5% fat diet through caspase 3 independent pathway in mice testis. Toxicol Mech Methods 2009;19:219–24.
[39] Adler ID, Anderson D, Benigni R, Ehling UH, Laehdetie J, Pacchierotti F, et al. Synthesis report of the step project detection of germ cell mutagens. Mutat Res 1996;353:65–84.
[40] Dearfield KL, Abernathy CO, Ottley MS, Brantner JH, Hayes PF. Acrylamide: its metabolism, developmental and reproductive effects, genotoxicity and carcinogenicity. Mutat Res 1988;195:45–77.

[41] Ghanayem BI, Witt KL, El-Hadri L, Hoffler U, Kissling GE, Shelby MD, et al. Comparison of germ cell mutagenicity in male CYP2E1-null and wild-type mice treated with acrylamide: evidence supporting a glycidamide-mediated effect. Biol Reprod 2005;72:157–63.
[42] Nixon BJ, Stanger SJ, Nixon B, Roman SD. Chronic exposure to acrylamide induces DNA damage in male germ cells of mice. Toxicol Sci 2012;129:135–45.
[43] Sega GA, Generoso E, Brimer P. Acrylamide exposure induces a delayed unscheduled DNA synthesis in germ cells of male mice that is correlated with the temporal pattern of adduct formation in testis DNA. Environ Mol Mutagen 1990;16:137–42.
[44] Holland N, Ahlborn T, Turteltaub K, Markee C, Moore D, Wyrobek A, et al. Acrylamide causes preimplantation abnormalities in embryos and induces chromatin-adducts in male germ cells of mice. Reprod Toxicol 1999;13:167–78.
[45] Gall L, De Smedt V, Ruffini S. Co-expression of cytokeratins and vimentin in sheep cumulus-oocyte complexes. Alteration of intermediate filament distribution by acrylamide. Dev Growth Differ 1992;34:579–87.
[46] Arocena M. Effects of acrylamide on the cytoskeleton and apoptosis of bovine lens epithelial cells. Cell Biol Int 2006;30:1007–12.
[47] Sager PR. Cytoskeletal effects of acrylamide and 2,5-hexanedione: selective aggregation of vimentin filaments. Toxicol Appl Pharmacol 1989;97:141–55.
[48] Helfand BT, Chang L, Goldman RD. Intermediate filaments are dynamic and motile elements of cellular architecture. J Cell Sci 2004;117:133–41.
[49] Ivaska J, Pallari HM, Nevo J, Erikssn JE. Novel functions of vimentin in cell adhesin, migration and signaling. Exp Cell Res 2007;313:2050–60.
[50] Kidder GM, Mhawi AA. Gap junctions and ovarian folliculogenesis. Reproduction 2002;123:613–20.
[51] Hulas-Stasiak M, Dobrowlski P, Tomaszewska E, Kostro K. Maternal acrylamide treatment reduces ovarian follicle numer in newborn guinea pig offspring. Reprod Toxicol 2013;42:125–31.
[52] Tomaszewska E, Dobrowolski P, Puzio I, Prost L, Kurlak P, Sawczuk P, et al. Acrylamide-induced prenatal programming of intestine structure in guinea pig. J Physiol Pharmacol 2014;65:107–15.
[53] Dobrowolski P, Huet P, Karlsson P, Eriksson S, Tomaszewska E, Gawron A, et al. Potato fiber protects the small intestine wall against the toxic influence of acrylamide. Nutrition 2012;28:428–35.
[54] Mohamed DA, Selim SA. Effect of perinatal acrylamide exposure on the liver of albino rat offspring: light and electron microscopic study. Egypt J Hist 2012;35:371–82.
[55] Allam A, El-Ghareeb AW, Abdul-Hamid M, Bakery AE, Gad M, Sabri M. Effect of prenatal and perinatal acrylamide on the biochemical and morphological changes in liver of developing albino rat. Arch Toxicol 2010;84:129–41.
[56] Sridevi B, Reddy KV, Reddy SN. Effect of trivalent and hexavalent chromium on antioxidant enzyme activities and lipid peroxidation in a freshwater field crab, *Barytelphusa guerini*. Bull Environ Contam Toxicol 1998;61:384–90.
[57] Godfrey K, Barker D. Fetal nutrition and adult disease. Am J Clin Nutr 2000;71:1344–52.
[58] Gluckman PD, Hanson MA, Cooper C, Thornburg KL. Effect of in utero and early-life conditions on adult health and disease. N Engl J Med 2008;359:61–73.
[59] Vikström AC, Abramsson-Zetterberg L, Naruszewicz M, Athanassiadis I, Granath FN, Törnqvist MA. In vivo doses of acrylamide and glycidamide in humans after intake of acrylamide-rich food. Toxicol Sci 2011;119:41–9.
[60] Gale CR, O'Callaghan FJ, Bredow M, Martyn CN. The influence of head growth in fetal life, infancy, and childhood on intelligence at the ages of 4 and 8 years. Pediatrics 2006;118:1486–92.
[61] Uphouse L, Nemeroff CB, Mason G, Prange AJ, Bondy SC. Interactions between "handling" and acrylamide on endocrine responses in rats. Neurotoxicology 1982;3:121–5.
[62] Petersen B. Exposure and biomarkers. JIFSAN/NCFST acrylamide in food workshop. 2002. http://www.jifsan.umd.edu/acrylamide workshop.html.

CHAPTER 6

Metabolism of Acrylamide in Humans and Biomarkers of Exposure to Acrylamide

Tolgahan Kocadağlı, Vural Gökmen
Department of Food Engineering, Hacettepe University, Beytepe, Ankara, Turkey

INTRODUCTION

Acrylamide is a conjugated vinyl compound, and thus it is reactive to biological materials. The double bond in the acrylamide has a strong interest toward –SH groups in cysteine and interest toward $-NH_2$ groups in proteins/amino acids. Glycidamide, an oxidative metabolite of acrylamide, also reacts in the same manner via its epoxide ring, although to a higher extent with amino groups and bases in DNA. The in vivo reactions of both involve biological alkylation reactions of macromolecules, which were reviewed by Friedman [1]. A general knowledge on chemistry and biochemistry of the reactions is essential to better understand the metabolism and the undesirable consequences of dietary acrylamide. Many aspects of the reactions of proteins with electrophilic compounds were well described by Törnqvist et al. [2].

Bioavailability, metabolism, and usage of the metabolites as biomarkers of exposure are critically important parts of research on revealing the potential health risks of acrylamide found in foods. This chapter explains absorption, bioavailability, metabolism of acrylamide, and biomarkers of exposure and reviews human studies.

ABSORPTION AND BIOAVAILABILITY

Oral bioavailability can be simply defined as the fraction of the oral administered dose that reaches the systemic circulation. Since acrylamide is highly hydrophilic, it is expected to be absorbed completely. A number of studies have examined acrylamide bioavailability in humans and considerably in rodents. Orally administered acrylamide has revealed that it is absorbed rapidly in both rodents and humans.

Miller et al. demonstrated that about 90% of acrylamide was recovered from tissues, body fluids, urine, and feces after administration of 10 mg/kg body weight (bw) 2,3-^{14}C-labeled acrylamide by gavage to Fisher-344 rats [3]. Kadry et al. investigated the pharmacokinetics of acrylamide in rats (Sprague–Dawley) after oral administration of 50 mg/kg bw 1-^{14}C acrylamide in water [4]. Radioactivity was detected in blood after 5 min and the peak plasma

Acrylamide in Food
http://dx.doi.org/10.1016/B978-0-12-802832-2.00006-1

concentration occurred 38 min after administration and was equivalent to 47 μg/mL. Later on, Doerge et al. calculated the bioavailability in rats and mice by comparing oral administration with intravascular injection (both 0.1 mg/kg bw acrylamide) [5,6]. In Fisher-344 rats, the bioavailability of acrylamide in food and aqueous gavage was 32%–44% and 60%–98%, respectively [5]. In B6C3F1 mice, the acrylamide bioavailability was 23% from the diet and was 32%–52% from aqueous gavage [6]. The uptake of acrylamide from food was expectedly lower because of the interaction with the nucleophilic groups found in food components. Additionally, it was indicated that bioavailability is considerably different between species. On the contrary, Bjellaas et al. reported that acrylamide bioavailability was approximately complete in mice (C57BL/6JBomTac) fed with crisp bread [7]. The difference was attributed to the fact that Doerge et al. (2005) only measured the parent compound in plasma, and that they did not take into the account the first-pass metabolism of acrylamide to glycidamide [7].

Aureli et al. evaluated the difference in the absorption of acrylamide found in potato chips according to acrylamide dissolved in water in a feeding study of pigs [8]. The relative absorption of acrylamide from potato chips and water was the same in both low and high administration. However, it should be mentioned that the larger part of the diet of the pigs was composed of standard animal feed which is the main matrix in the gastrointestinal (GI) tract effecting bioaccessibility in both groups.

Sorgel et al. showed for the first time that acrylamide was absorbed from the GI tract in humans who ate crisp bread and potato products [9]. They indicated that the variability in urinary excretion of acrylamide between subjects was very high, which might be caused by high variability in oral bioavailability and/or hepatic metabolism. Boettcher et al. administered 1 mg d_3-acrylamide in water (13 μg/kg bw) to a male volunteer and on the first day 51% of the dose was recovered in urine as metabolites and on the second day the total recovered dose was 57% [10]. They concluded that the metabolic fate of acrylamide in humans was more similar to that in rats than in mice. Fennel et al. administered 1,2,3–$^{13}C_3$-acrylamide in water at doses of 0.5, 1, and 3 mg/kg to male volunteers [11]. The recovered urinary metabolites accounted for about 46%, 50%, and 40% of the doses 0.5, 1, and 3 mg/kg in 24 h, respectively. They have indicated that acrylamide was rapidly absorbed after administration. Kopp et al. administered 0.5 and 20 μg/kg bw $^{13}C_3$-acrylamide in water to three male and three female subjects [12]. No significant differences were observed for gender, and urinary metabolites were recovered close to 70% at both doses in 96 h. A rapid increase in acrylamide concentration of the serum was observed for both subjects.

Fuhr et al. evaluated the toxicokinetics of acrylamide in six young healthy volunteers after the consumption of potato chips containing 0.94 mg of acrylamide (12.4 Metin μg/kg bw) [13]. The mean cumulative recovery of acrylamide in urine accounted for 60 ± 11% of the dose in 72 h. They have concluded that most of the acrylamide ingested with food was absorbed in humans.

BIOTRANSFORMATION

A schematic diagram of absorption, distribution, metabolism, and excretion of acrylamide is given in Figure 1. After consumption, acrylamide is rapidly absorbed and distributed to all tissues in humans by systemic circulation. It can be also found in human breast milk and penetrates the human placenta [9]. Acrylamide is extensively converted to a range of metabolites that are excreted in urine. It can also interact with proteins and DNA resulting in covalent adducts.

The main pathways of acrylamide biotransformation are given in Figure 2. Acrylamide has an α,β-unsaturated structure with an amide group making it an electrophilic compound. Cellular nucleophiles react with acrylamide's double bond by the Michael addition. These nucleophiles include the –SH of cysteine, homocysteine, and glutathione; α-NH_2 groups of free amino acids and N-terminal amino acid residues of proteins, the ε-NH_2 of lysine, and the ring NH group of histidine [1]. Sulfhydryl groups in glutathione and proteins are the main targets for acrylamide, and amino groups found in proteins are also targets although to a lesser extent [14].

Acrylamide is partly converted to an epoxide glycidamide, which is also an electrophile, in vivo through oxidation by cytochrome P450 2E1 in phase I metabolism [15–17]. Therefore, like acrylamide, glycidamide covalently bonds to cellular components containing sulfhydryl and amino groups. Attack by sulfhydryl and amino groups or water results in the ring opening of the epoxide. Glycidamide can be further hydrolyzed by an epoxide hydrolase to glyceramide (2,3-dihydroxypropionamide) [15,18]. Nucleophilic nitrogens in DNA are also susceptible to form adducts especially with glycidamide, which is thought mainly to be responsible for the genotoxicity [19–23].

Orally ingested acrylamide in rodents and humans is extensively conjugated with glutathione in phase II metabolism. After degradation and acetylation of the conjugate, the resulting mercapturic acid *N*-acetyl-S-(2-carbamoylethyl)-L-cysteine (AAMA) is excreted with urine. Glutathione conjugation is of importance to reduce acrylamide reactivity against cellular components. Another important aspect of the conjugation of

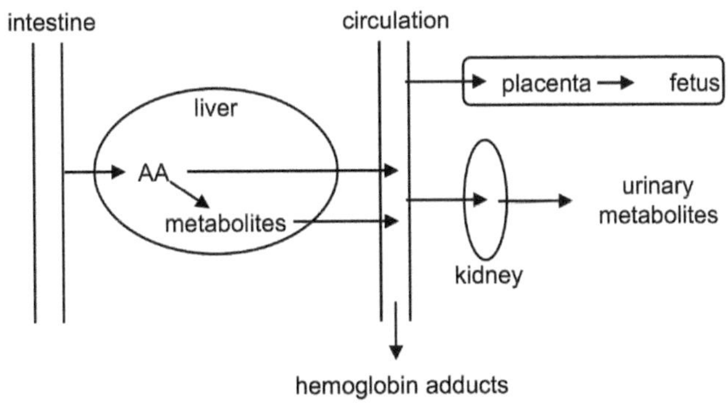

Figure 1 A schematic diagram of absorption, distribution, biotransformation, and excretion of acrylamide.

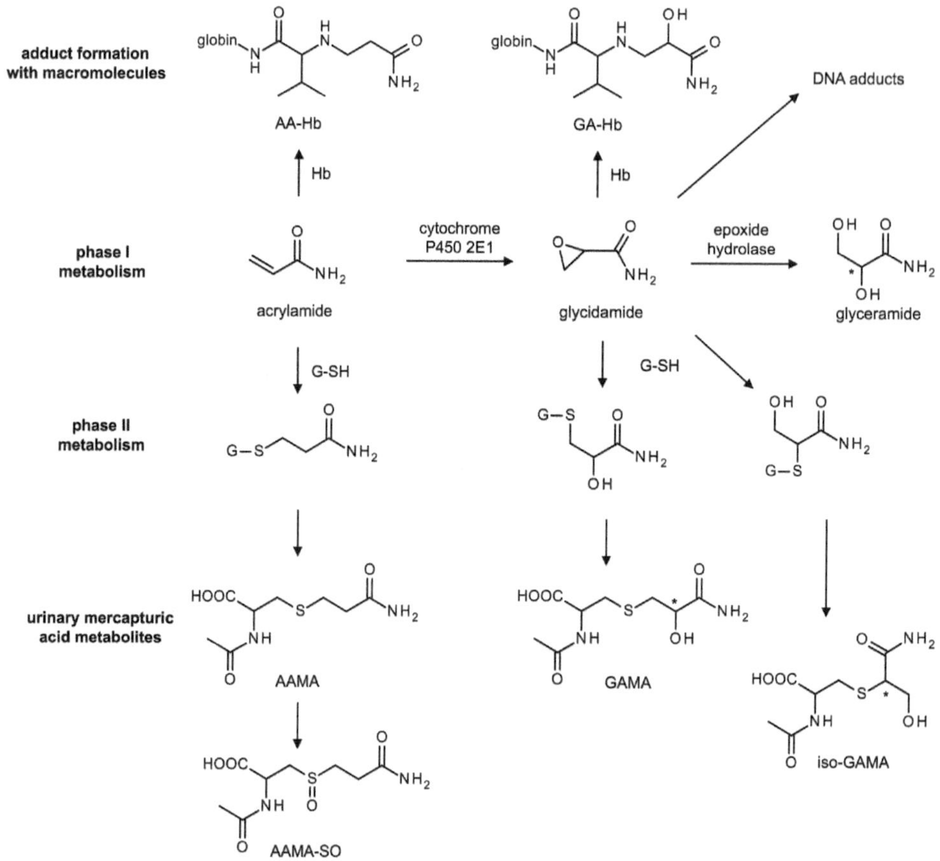

Figure 2 Biotransformation of acrylamide in human. * indicates chiral carbons.

acrylamide with glutathione is that it is less converted to glycidamide. Depletion of intracellular glutathione in rat hepatocytes and lung fibroblasts (V79) cells has shown to result in a significant induction of DNA strand breaks after incubation with 1 mM AA [24]. AAMA can be oxidized to its corresponding sulfoxide in humans but not in mice and rats [11,12]. It has been speculated that AAMA-sulfoxide may induce nephrotoxicity or bladder toxicity as suggested for acrolein-derived mercapturic acids [12,25].

Glycidamide can also be conjugated with glutathione and is excreted with urine after metabolic conversion to the mercapturic acids *N*-acetyl-*S*-(2-carbamoyl-2-hydroxyethyl)-L-cysteine (GAMA) and *N*-acetyl-*S*-(1-carbamoyl-2-hydroxyethyl)-L-cysteine (iso-GAMA) [18]. Glycidamide is more reactive than acrylamide toward DNA and proteins. However, conversion of acrylamide to glycidamide has been shown to be less important in humans than in rodents [13,26,27]. The amount of GAMA excreted in urine was about 9%–29% of the oral dose in rodents, and it was between 0.7% and 6% in human (Table 1). Hydrolysis of glycidamide is a less important pathway of detoxification alongside of glutathione conjugation.

Table 1 Comparison of urinary metabolites excreted within 24 h in humans

Study		Dose, mg/kg bw	% of dose excreted in urine						GAMA/AAMA	Total % of dose
			AA	AAMA	AAMA-SO	GA	GAMA	Glyceramide		
Fennel [26,27]	$^{13}C_3$-AA	3	nq	22.0 ± 5.30	4.2 ± 1.10	0.79 ± 0.24	nd	3.3 ± 1.10	–	34.0 ± 5.70
Boettcher [10,34]	d_3-AA	0.013	nd	45.1	nd	nd	2.8	nd	0.06	47.7
Fennel [11,29]	$^{13}C_3$-AA	0.5	4.7 ± 1.34	31.2 ± 6.5	8.3 ± 2.39	0.4 ± 0.20	0.8 ± 0.16	nd	0.03	45.6 ± 8.50
	$^{13}C_3$-AA	1	5.0 ± 1.65	34.4 ± 5.21	8.7 ± 1.21	0.6 ± 0.33	0.8 ± 0.11	nd	0.03	49.9 ± 6.30
	$^{13}C_3$-AA	3	3.2 ± 0.49	27.8 ± 7.99	7.3 ± 2.4	0.7 ± 0.33	0.7 ± 0.22	nd	0.03	39.9 ± 9.90
Fuhr [13]	Potato chips	0.0124	4.4 ± 1.5	50.0 ± 9.4	nd	nd	5.9 ± 1.2	nd	0.12	60.3 ± 11.2
Kopp [12,36]	$^{13}C_3$-AA	0.0005	nd	41.4 ± 3.47	7.2 ± 1.40	nd	3.8 ± 0.78	nd	0.09	52.4 ± 3.59
	$^{13}C_3$-AA	0.02	nd	37.4 ± 2.92	6.3 ± 1.77	nd	3.2 ± 0.69	nd	0.09	46.9 ± 3.70
Doroshyenko [37]	Potato chips	0.014	2.9	58	nd	nd	1.4	nd	0.024	63[a]

nd, not detected; nq, not quantified.
[a]Excretion over 72 h.

Several nucleophilic sites in proteins form adducts with acrylamide and glycidamide as mentioned above. Since albumin and hemoglobin are the most predominant proteins in blood, acrylamide and glycidamide readily form covalent adducts with the reactive amino acids of these proteins. Valine is the N-terminal amino acid residue of hemoglobin in all four globin subunits. The N-terminal in hemoglobin has been proposed as a major reactive nucleophilic site for many electrophiles, particularly in humans. In contrast to rat hemoglobin, there is no cysteine with high reactivity in human hemoglobin [14]. *N*-(2-carbamoylethyl)-valine (AA-Hb) and *N*-(2-carbamoyl-hydroxyethyl)-valine (GA-Hb) are formed by the addition of acrylamide and glycidamide, respectively, to the N-terminal valine residue in hemoglobin [15]. Hemoglobin adducts of acrylamide and glycidamide stay longer in the body rather than those of their free forms and the mercapturic acid derivatives. The mean lifetime of erythrocytes is 4 months in human, whereas it is only 2 months in rats. Measuring levels of long-lived hemoglobin adducts are indicative of a time-weighted average exposure.

BIOMARKERS OF EXPOSURE TO ACRYLAMIDE

Biomarkers, in general terms, are the indicators in biological materials that can be measured to monitor the presence of a chemical in the body or to assess biological responses, disease processes or adverse health effects. The measured quantities may indicate the exposure, dose, susceptibility, early biological and adverse health effects of a xenobiotic. This allows us to make decisions, which are more protective to health. Biomarkers are classified as biomarkers of exposure, effect, and susceptibility. Biomarkers of exposure can be defined as "an exogenous substance or its metabolite or the product of an interaction between a xenobiotic agent and some target molecule or cell that is measured in a compartment within an organism" [28]. Biomarkers of exposure involve measurements of parent compound, metabolites, and protein or DNA adducts. Biomarkers of exposure provide information on:

- exposure in individuals;
- bioavailability and metabolism;
- changes in levels over time;
- internal dose assessment for epidemiological studies;
- assessment of health risks;
- variability among species in risk assessment.

Free Acrylamide and Glycidamide in Biological Fluids

Acrylamide and glycidamide can be found in biological fluids of humans in their free form, although to a lesser extent than in the conjugated forms. Both substances have short half-life in plasma and serum, generally less than a few hours. They reflect very recent exposure and plasma levels are often below limits of detection. Therefore, plasma and serum levels of acrylamide and glycidamide are not a good choice as exposure biomarkers in humans. A number of studies were carried out to investigate the urinary levels of those.

Sorgel et al. investigated the excretion of acrylamide in urine of nine subjects who ate 0.5 kg of potato chips or crisp bread [9]. Total acrylamide excreted was found to be between 0.5 and 6 μg (corresponding to about 2–6 ng/mL) within 8 h. The variability in urinary excretion of acrylamide between subjects was indicated to be high, which might be caused by high variability in oral bioavailability and/or hepatic metabolism. The amount of acrylamide found in chips and crisps was not reported. In two subjects, they determined a half-life of 2.2 h for acrylamide in one case and 7 h in the other. They have concluded that the half-life of acrylamide in humans was short in some subjects and intermediate in others. In the same study, acrylamide was also found in the breast milk of two healthy mothers who ate 100 g of different kinds of potato chips (estimated exposures were mentioned as 0.8 and 1 mg of acrylamide). The concentration of acrylamide was found to be about 11 and 19 ng/mL in the 3rd and 4th h milk sample of the first subject and it was 5 and 3 ng/mL in the 4th and 8th h milk sample of the second.

Fennell et al. administered 3 mg/kg bw 1,2,3-^{13}C-acrylamide to six male volunteers and quantified glycidamide and glyceramide in urine along with other metabolites by using NMR spectroscopy [27]. Acrylamide was not quantified. Glycidamide and glyceramide concentrations were reported as 10 and 50 μM, their total amounts were 25 and 102 μmol, and ratios according to the total metabolites were 2.6% and 11% within 24 h, respectively. The percentages of the administered dose recovered as glycidamide and glyceramide were 0.8% and 3.3%, respectively. Fennell et al., in the second part of the study, performed LC-MS/MS analysis to define the kinetics of acrylamide elimination and its metabolites following oral administration of 0.5, 1, and 3 mg/kg bw 1,2,3-^{13}C-acrylamide to male volunteers (6 in each group) [29]. Total free acrylamide in urine was reported to be about 10-, 49-, and 95 μmol within 24 h, respectively. Unchanged acrylamide and glycidamide accounted for the 3%–5% and 0.4%–0.6% of the administered doses, respectively. The elimination of acrylamide in urine occurred at early time points, with the majority being excreted by 4 h of time. The rate of elimination (k_e) and the elimination half-life ($t_{1/2}$) were estimated to be between 0.21 and 0.26 h^{-1} and 3.1 and 3.5 h, respectively, in the administered doses. Glyceramide was not quantified in the second part of the study because of analytical limitations.

Fuhr et al. evaluated the toxicokinetics of acrylamide in six young healthy volunteers after the consumption of potato chips containing 0.94 mg (13.3 μmol) of acrylamide (12.4 μg/kg bw) [13]. Total recovered free acrylamide in urine samples were found to be between 0.3 and 0.8 μmol within 24 h, accounting for 4.4 ± 1.5% of the dose. The maximal urinary excretion rate of acrylamide was observed during 2–4 h sampling intervals. No concentrations of free glycidamide exceeding the limit of quantification (2.5 ng/mL) were reported during the entire sampling period (72 h).

Kopp et al. administered 20 μg/kg bw $^{13}C_3$-acrylamide in water to three male and three female subjects to monitor serum toxicokinetics of free acrylamide [12]. They reported that acrylamide was eliminated rapidly from blood, and it could only be quantified in the samples collected up to 2 h following exposure. Concentrations at 3 h after administration were between LOD (0.5 μg/L) and LOQ (1 μg/L). The time of maximum concentration (t_{max}) was reported as 0.94 h at 29.2 nM (C_{max}) and elimination half-life ($t_{1/2}$) was at 0.79 h with the elimination coefficient (k_e) being 0.96 h^{-1}. They have indicated that concentrations of free acrylamide in serum were extremely low (<0.5% of the dose), probably due to the higher affinity to blood specimens. On the other hand, 70% of the dose was recovered as urinary metabolites within 96 h.

Lee et al. investigated urinary free acrylamide in the South Korean adult population aged 18 to 69 years ($n = 1873$) [30]. The population-weighted geometric mean of urinary acrylamide concentrations were reported as 6.8 ng/mL (6.2 μg/g-creatinine) and were significantly different depending on sex, education, body mass index (BMI), and smoking status. The 50th and 95th percentiles of acrylamide levels were 8.1 and 26.0 ng/mL, respectively. The urinary acrylamide level of men was found to be 8.2 ng/mL, which was significantly higher than the value of 5.7 ng/mL found in women ($p < 0.001$). The urinary acrylamide level of overweight participants (BMI ≥ 25.0) was 7.8 ng/mL, and it was about 6.3 ng/mL for participants with a BMI value of 18.5–22.9.

Mercapturic Acid Metabolites

Free acrylamide, glycidamide, and glyceramide account for only a minor portion of the excreted substances as mentioned in the previous section. The main urinary metabolites are the mercapturic acids AAMA and GAMA (and iso-GAMA) derived from acrylamide and glycidamide, respectively. AAMA is the major and iso-GAMA is the minor urinary metabolite. Studies using labeled acrylamide with stable isotopes (1,2,3–$^{13}C_3$-acrylamide or d_3-acrylamide) to investigate urinary excretion of the metabolites administered to humans are listed in Table 1. Mercapturic acid metabolites are short-term biomarkers that can be used to estimate very recent exposures.

Calleman et al. investigated the level of AAMA in urine of humans exposed to acrylamide at work [31]. In the control group ($n = 10$), not exposed occupationally, AAMA was 3 μmol within 24 h. Urinary excretion of acrylamide metabolites (AAMA and GAMA) in German adults, who were not exposed occupationally, was studied by Boettcher et al. [32,33]. The median (range) concentration values of AAMA and GAMA in nonsmokers were found to be 29 (3–83) and 5 (0–14) μg/L, respectively. These values were about four times higher in smokers. The ratio of GAMA/AAMA in median (range) concentration was 0.2 (0.1–0.5) in nonsmokers. The ratio in all population was 0.12, and they found a good correlation between both the mercapturic acids ($R^2 = 0.76$). Moreover, their results indorsed that there was a background exposure to acrylamide for the general population, which was probably due to the diet.

Fennel et al. investigated the urinary metabolites after administration of 0.5, 1, and 3 mg/kg bw $^{13}C_3$-acrylamide in water to male volunteers [27,29]. The percentage of the dose as urinary metabolites is given in Table 1 with comparison to other studies. The major metabolite was AAMA and accounted for 68%–69% of the excreted metabolites measured in all dose groups. AMAA-SO was accounted for 17%–19% and, GAMA and iso-GAMA were accounted for less then 2% of the excreted metabolites measured in all dose groups. Sum of the metabolites excreted within 24 h was 145, 38.2, 3.1, and 0.7 μmol for AAMA, AAMA-SO, GAMA, and iso-GAMA, respectively, after administration of 0.5 mg/kg bw $^{13}C_3$-acrylamide. They indicated that the extent of metabolism of acrylamide to glycidamide in humans was considerably lower than it was in rats.

Boettcher et al. administered 1 mg d_3-acrylamide in water to a male volunteer (13 μg/kg bw) to investigate elimination kinetics of urinary metabolites [10]. From 1.5- to 8 h, d_3-AAMA concentrations increased linearly, and then a broad plateau between 8- and 18 h was observed and then began to decrease. GAMA excretion increased almost linearly similar to AAMA between 4 and 8 h after dosing, and the increase continued up to 22 h (t_{max}) with a lower rate and then began to decrease. They proposed two phases for elimination of both mercapturic acids, a faster first phase and a much slower second phase of excretion with half-lives of 3.5 and more than 10 h, respectively. Total recovery in urine after 48 h was about 52% for AAMA and 5% for GAMA. They have indicated that acrylamide was immediately absorbed, metabolized, and eliminated as AAMA. In a further study Boettcher et al. investigated the influence of fasting (acrylamide free diet) on urinary mercapturic acid metabolite excretion values in humans ($n=3$) [34]. Compared to the maximum values measured at the beginning of the fasting the mercapturic acid concentration has decreased by about 95%. In the previous study AAMA and GAMA levels measured 48 h after administration of d_3-acrylamide were also decreased by about 90% [10]. This confirmed that the diet is the main source of environmental AA exposure in humans, apart from smoking. Bjellaas et al. conducted a fasting study and investigated urinary metabolites ($n=6$) [35]. Starting with 24 h fasting (only water) all urine samples were collected during 48 h. Fasting during 1 day caused about 50% decrease in the total level of the metabolites (AAMA and GAMA), but after 1 day of a normal diet, the metabolite levels rose back to prefasting levels. The total amount of acrylamide in the form of urinary metabolites excreted over the period was estimated to be about 40 μg per day for the average nonsmoker, which is in accordance with the estimated daily dietary intake.

Fuhr et al. evaluated the toxicokinetics of acrylamide in six young healthy volunteers after the consumption of potato chips containing 0.94 mg (13.3 μmol) of acrylamide (12.4 μg/kg bw) [13]. Maximal urinary excretion rates of AAMA and GAMA were observed during the 6–8 and 12–16 h sampling intervals, respectively. The apparent terminal elimination half-lives for AAMA and GAMA were 17.4 ± 3.9 and 25.1 ± 6.4 h, respectively. Total mean (range) excretion of AAMA and GAMA within 72 h was about

6.3 (4.2–7.3) μmol and 0.7 (0.4–0.8) μmol, respectively. AAMA and GAMA accounted for urinary excretion of 50% and 6% of the dose, respectively.

Kopp et al. administered 0.5 and 20 μg/kg bw $^{13}C_3$-acrylamide in water to three male- and three female subjects and monitored mercapturic acids in urine [36]. Results were consistent in the low dose range reported in other studies (Table 1). Toxicokinetic analysis for urinary elimination of metabolites was reported for 20 μg/kg bw $^{13}C_3$-acrylamide administration. For AAMA, the time of maximum concentration (t_{max}) was reported as 10.2 h at 2.7 μmol (C_{max}) and elimination half-life ($t_{1/2}$) was 14.2 h with the elimination coefficient (k_e) of 0.05 h^{-1}. For AAMA-SO and GAMA, t_{max} was observed as 23 h at 0.45 and 0.29 μmol, respectively. AAMA was rapidly excreted, whereas concentrations of AAMA-SO and GAMA increased comparatively slower due to the time required for biotransformation. No significant differences were observed between female and male human subjects concerning toxicokinetics and biotransformation of acrylamide. Similar toxicokinetics and interindividual variations of acrylamide metabolites were observed in other studies in the urine of human volunteers [37,38].

In a study conducted with 47 nonsmoking human volunteers, the median (range) total excretion of acrylamide as mercapturic acid metabolites in urine was 16 (7–47) μg within 24 h [39]. The median (range) dietary intake of acrylamide estimated from the 24 h dietary recall was reported as 21 (3–178) μg. The total urinary metabolites correlated with the intake of aspartic acid, protein, starch, and coffee.

Brantsæter et al. assessed dietary exposure to acrylamide in 119 pregnant Norwegian women [40]. The dietary median (95th percentile) intake of AA was 0.48 (0.92) μg/kg bw per day as estimated by food frequency questionnaire, 0.41 (0.82) μg/kg bw per day as estimated by food diary, and 0.42 (0.70) μg/kg bw per day as estimated by probabilistic approach. The median (95th percentile) amount of acrylamide excreted as urinary metabolites was 0.16 (0.50) μg/kg bw per day in nonsmokers, corresponding to a dietary exposure of about 0.30 (0.91) μg/kg bw per day. A significant correlation was observed between the biomarker and the estimated dietary intake. The total urinary metabolites correlated with the intake of crisp bread, potato crisps, cooking oil, and garlic.

Hartmann et al. investigated the urinary metabolites in a German population ($n=91$) [41]. Concentrations ranged from <1.5 to 229 μg/L (median, 29 μg/L) for AAMA and from <1.5 to 85 μg/L (median, 7 μg/L) for GAMA. The median value for the GAMA/AAMA ratio was 0.3 within the range of 0.004–1.4. The median daily intake was 0.51 (<LOD to 2.32) μg/kg bw per day using mercapturic acids for calculations. The study showed that children take up about 1.5 times more acrylamide per kilogram body weight than do adults (estimation based on the sum of AAMA and GAMA). The ratio GAMA/AAMA was found to be significantly higher in the group of young children (6–10 years old, $n=12$) with a median level of 0.5 (range 0.2–1.3). The higher ratio in children and the large range probably reflect the differences in the enzymatic status of the individuals. A gender-related difference in internal exposure and metabolism was not observed.

The authors concluded that the relationship between internal exposure and age was a consequence of the fact that children consume more food per kilogram of body weight, as a function of their surface-to-volume ratio and hence higher caloric requirements.

In the study by Heudorf et al., the internal exposure to acrylamide through diet was studied in 110 children (5–6 year olds) [42]. Median (95th percentile) urinary levels were 36.0 (152.7) μg AAMA/L and 13.4 (55.9) μg GAMA/L. Based on the metabolite levels, the median (95th percentile) uptake of acrylamide in children was 0.54 (1.91) μg/kgbw per day, exceeding exposure in adults by 50%. The study provided evidence for a background exposure by nutrition. A number of associations with the consumption of French fries, various potato products, as well as fried cereals were found. Moreover, the ratio GAMA/AAMA was 0.4 (0.06–1.2), which was higher than the ratio reported in the study of Boettcher et al. [33]. The authors indicated that there could be a more potent oxidative metabolism in children [42].

In another study, AAMA levels were measured in Korean children ($n=31$, 10- to 13-years old), and their diets were surveyed [43]. The concentrations of metabolite AAMA in urine ranged between 15.4 and 196.3 μg/L, with a median level of 68.1 μg/L. The median (95th percentile) value of daily acrylamide was estimated as 1.04 (2.47) μg/kgbw per day.

Lee et al. investigated AAMA in the South Korean adult population aged 18-69 years ($n=1873$) [30]. The population-weighted geometric mean of AAMA concentrations was reported to be 29.9 μg/L and was significantly different depending on sex, BMI, and smoking status. The 50th and 95th percentiles of AAMA levels were 29.4 μg/L and 157.9 ng/mL, respectively. The AAMA level of men was 41.7 μg/L, which was significantly higher than that of women (21.5 μg/L) ($p<0.001$). The AAMA level of overweight participants (BMI ≥25.0) was 36.2 μg/L, and it was about 27 μg/L for participants with BMI of 18.5–22.9.

In a study on 195 nonsmoking adolescents in Canada, Brisson et al. investigated the urinary metabolites AAMA, AAMA-SO, and GAMA [44]. The most prevalent urinary metabolites were AAMA and AAMA-SO, with respective geometric mean concentrations of 81.7 and 39.7 μg/L. The daily intake of acrylamide was significantly correlated with the sum of AAMA and AAMA-SO urinary concentrations ($p<0.0001$).

Hemoglobin Adducts

To monitor cumulative exposure to a substance over a longer period, a marker should not be eliminated from blood at that time. Hemoglobin adducts of acrylamide and glycidamide stay longer in the body rather than those of their free forms and the mercapturic acid derivatives. Levels of hemoglobin adducts reflect the blood dose over the past 4 months, the lifespan of erythrocytes in human. Hemoglobin adducts in blood samples are measured to monitor exposure to several reactive chemicals, from both occupational and environmental levels, and hemoglobin adduct studies are conducted for quantitative cancer risk assessments [45]. Measurement of hemoglobin adducts is well established for the estimation of internal exposure to acrylamide and glycidamide in humans [46,47]. Because of the accessibility in large amounts and well-determined life span together with the absence of repair, hemoglobin (Hb) is preferred to DNA for dose monitoring [48]. Acrylamide-hemoglobin adduct (AA-Hb) may be used as a biomarker for average acrylamide exposure, while the glycidamide–hemoglobin adduct (GA-Hb) may be utilized as a biomarker for the internal glycidamide dose and the genotoxic impact [49].

The median background level of AA-Hb had been found as 30 pmol/g globin [45] and in several studies it was then reported to vary in a large range. A dose-dependent increase in the level of $^{13}C_3$-AA-Hb and $^{13}C_3$-GA-Hb adducts was observed in humans upon oral administration of $^{13}C_3$-acrylamide [27].

Schettgen et al. reported the mean (range) concentration of AA-Hb and GA-Hb in nonsmokers ($n=13$) as 19 (7–31) and 17 (9–23) pmol/g globin, respectively [50]. The conversion to glycidamide was indicated to be similar to that in rats, which is in accordance with the calculated values of urinary metabolites in other studies. Hagmar et al. investigated AA-Hb adducts in samples from the blood bank of Malmö Diet and Cancer Cohort [51]. In nonsmokers ($n=70$) the median (range) level was 31 (20–100) pmol/g globin. Significant differences between groups of men in high and low dietary exposure to acrylamide were observed, but not in women. Chevolleau et al. reported mean (range) AA-Hb and GA-Hb concentrations in the French population as 27 (9–70) and 22 (12–47) pmol/g globin [52].

In the EPIC (European Prospective Investigation into Cancer and Nutrition) study comprising nine European countries ($n=510$), AA-Hb and GA-Hb levels were investigated [53]. AA-Hb and GA-Hb values were found to range between 15 and 623 and 8 and 377 pmol/g hemoglobin, respectively. Both adducts differed significantly by country, sex, and smoking status. GA-Hb values were significantly lower in high alcohol consumers than in moderate consumers. With increasing BMI, GA-Hb in nonsmokers and AA-Hb in smokers decreased significantly.

In the US population ($n=7166$), Vesper et al. reported that AA-Hb and GA-Hb levels range from 3 to 910 and from 4 to 756 pmol/g hemoglobin, respectively [54]. Adjusted geometric mean levels were the highest in children of age 3-11 years and the lowest in adults aged over 60 years in nonsmokers. The higher levels in children were attributed to the known fact of larger intake of food per body mass in children, particularly acrylamide-rich foods such as French fries and potato chips. In addition, children had the highest GA-Hb to AA-Hb ratios, whereas adults over 60 years of age have the lowest ratios. This suggested differences in the acrylamide metabolism or metabolic rate in the age groups. The authors concluded that sex, age, and race/ethnicity do not strongly predict adduct levels. The large variability in adduct levels among nonsmoking individuals was likely a result of individual intake of acrylamide present in food and probably, to a lesser extent, due to passive smoking.

In order to reveal intra-individual variations of adduct levels, 13 nonsmokers were monitored during 20 months in the study of Vikström et al. [55]. Results showed that variation in AA-Hb and GA-Hb levels were up to 2- and 4-fold, respectively.

Inter-individual variations ($n = 68$, nonsmokers) were large being up to 6- and 8-fold for AA-Hb and GA-Hb, respectively. The intra-individual and inter-individual variations of GA-Hb/AA-Hb ratio were 3- and 11-fold, respectively. Daily intake estimation was found to be 0.52 and 0.67 μg/kg bw, which were calculated according to adduct levels and dietary history methodology, respectively. The authors concluded that for large cohorts the acrylamide adduct levels measured in a single blood sample from each subject could be used to calculate the mean intake and the variation in a cohort.

Food frequency questionnaires and the data of acrylamide content in foods are often used in intake estimation. This estimate also could be done by using hemoglobin adducts which reflects exposure over the past 4 months. Hemoglobin adduct concentrations can be used to derive the "area under the concentration–time curve" (AUC) in blood. This alternatively includes the absorption and metabolic rates into the estimation [2]. In a study by Vikström et al. AUC of acrylamide and glycidamide were determined according to adduct levels [56]. Acrylamide rich foods (2.5 or 11 μg acrylamide/kg bw/day for 30 and 4 days respectively) were given to nonsmokers and adduct levels were measured before and after exposure. The AA-Hb and GA-Hb adducts both increased about two-fold after the enhanced intake of acrylamide. The in vivo doses (AUC) were calculated in the high and medium groups for acrylamide 212 and 120 nM h/μg AA/kg bw and for glycidamide 49 and 21 nM h/μg AA/kg bw, respectively. The authors indicated that AUC values for acrylamide and glycidamide are about 5- and 2-fold higher in humans compared to rats. This can be used as a basis for interspecies extrapolation in cancer risk estimation.

In the study by Bjellaas et al., the relationship between acrylamide intake from food using food frequency questionnaires and the concentration of AA-Hb and GA-Hb as a measure of the internal exposure was investigated [49]. A significant positive correlation was reported between the AA-Hb adduct concentration and the intake of chips/snacks and crisp bread. GA-Hb adduct did not correlate with consumption of any of the main food groups. Neither AA-Hb nor GA-Hb adduct concentrations correlated with the total dietary intake of acrylamide as calculated from the reported food intake. On the contrary, Wirfält et al. found a significant association between AA-Hb and estimated acrylamide exposure from foods [57]. Kütting et al. showed a weak but significant correlation with the estimated dietary exposure in nonsmokers ($n = 828$) [58]. Outzen et al. studied the dietary determinants of AA-Hb and GA-Hb adduct levels in 537 nonsmoking women aged 50-65 years [59]. Results proposed that the intake of coffee and chips was associated with higher AA-Hb adduct levels and the intake of coffee and biscuits/crackers was associated with higher GA-Hb adduct levels. The study implied that dietary intake measured by a food frequency questionnaire explains only to a limited extent the variation in AA-Hb and GA-Hb concentrations. In the EPIC study, Ferrari et al. compared dietary estimates of acrylamide from questionnaires and from 24-h recalls to AA-Hb levels ($n = 510$) [60]. A weak correlation was observed between the self-reported diet and the biomarker. More details on estimating dietary acrylamide intake are found in Chapter 2 of this book.

Since ethanol and acrylamide are both substrates for cytochrome P450 2E1, Vikström et al. investigated the influence of alcohol consumption on acrylamide metabolism to glycidamide in 161 nonsmoking men [61]. Individual differences in metabolism were assessed by the ratio of GA-Hb to AA-Hb, and it was compared to alcohol intake provided by the questionnaire data. A negative, linear trend of the GA-Hb to AA-Hb ratio with increasing alcohol intake was observed and the strongest association was obtained in the group of men with the lowest adduct levels. A similar observation was also reported by Vesper et al. [62]. This demonstrated that there is a competitive effect between ethanol and acrylamide to cytochrome P450 2E1 as substrates. The authors emphasized that the influence of ethanol on metabolism of acrylamide to glycidamide partly explained earlier observations of only low-to-moderate associations between questionnaire data on dietary acrylamide intake and hemoglobin–adduct levels [61].

In the study by Schettgen et al., AA-Hb adducts were observed in the blood samples of 10 nonsmoking mothers and in the umbilical cord blood samples of the neonates [63]. The median (range) of AA-Hb adduct concentration was found to be 21 (18–34) pmol/g globin in mothers. The AA-Hb value in the blood of neonates was about half of the adduct level found in the blood of the mother. AA-Hb adduct levels in nonsmoking mothers and neonates showed a good correlation. Similar results were also reported by von Stedingk et al. [64]. It was indicated that the N-terminal amino acid residues of the two of globin chains are valine (other two chains have glycine) in fetal erythrocytes whereas all four chains have valine in adults. The procedures specifically measure N-terminal valine adducts. Therefore, the authors concluded that the fetus receives approximately the same in vivo dose of both acrylamide and its metabolite glycidamide as does the mother [64]. The studies showed that transplacental exposure of fetus to acrylamide might raise concerns. Maternal acrylamide and effects on offspring are found in Chapter 5 of this book.

DNA Adducts

Acrylamide reacts very slowly with DNA in vitro. However, its metabolite glycidamide is very reactive, about 100- to 1000-fold than acrylamide, toward DNA due to the electrophilic epoxide group [22]. DNA adducts of glycidamide can be used as markers of the biologically active dose of glycidamide that reaches the DNA [14]. The main adducts formed from glycidamide are *N*-7-(2-carbamoyl-2-hydroxyethyl)-guanine (N7-GA-Gua) and *N*-3-(2-carbamoyl-2-hydroxyethyl)-adenine (N3-GA-Ade) (Figure 3). The DNA adducts have shown to present in many tissues/organs and blood of rodents. The levels of N7-GA-Gua in tissues of mice were found to be about 100-fold higher than N3-GA-Ade [65]. A dose-dependent increase in DNA adduct levels is observed in experimental animals administered acrylamide or glycidamide. The DNA is repaired and the levels of adducts as biomarkers may only indicate recent exposure to acrylamide, about within a week.

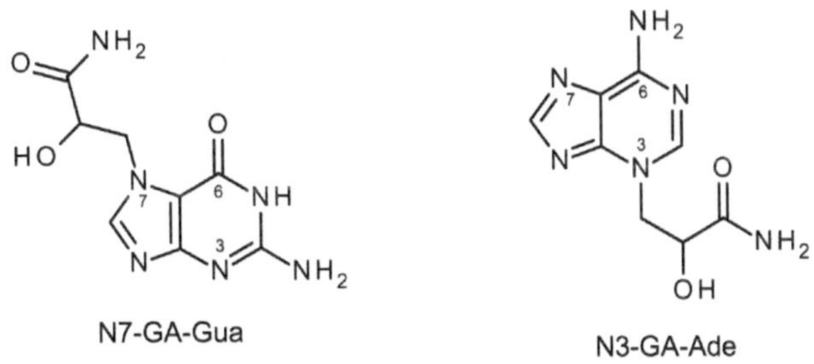

Figure 3 Guanine and adenine adducts of glycidamide.

Any of adducts have been identified in human tissues or blood. In a recent study urinary N7-GA-Gua was analyzed in humans [66]. The AAMA levels in urine were found to be significantly associated with N7-GA-Gua levels. The study indirectly demonstrated that DNA alkylation occured in nonsmokers, probably from acrylamide found in foods. It was concluded that the urinary N7-GA-Gua could be used as a biomarker for mutagenicity and in molecular epidemiology studies in human.

KEY FACTS

- The biological effects of acrylamide and glycidamide stem from their reactive nature. Acrylamide is a conjugated vinyl compound and thus it is reactive to biological materials. Amino ($-NH_2$) and sulfhydryl (–SH) groups react with the double bond of acrylamide via Michael-type nucleophilic addition reactions. The Michael addition is the reaction of a nucleophile (or a carbanion) with an α,β-unsaturated carbonyl compound. The nucleophile attacks the conjugated alkene at the β position, while electrons are withdrawn to the electronegative oxygen. This type of reaction of acrylamide results in the alkylation of biomacromolecules in vivo and in vitro.
- The structure of an epoxide contains oxygen atom attached to two adjacent carbon atoms (cyclic ether). Both the C–O bonds are polar due to the high electronegativity of the O atom and the ring is highly strained. The ring strain is relieved when it opens and thus the carbons in an epoxide group are very reactive electrophiles. Glycidamide, an oxidative metabolite of acrylamide, has an epoxide ring and thus reacts with nucleophiles in biological materials. Nucleophilic substitution of the glycidamide as an electrophile is faster than the double bond of the acrylamide. The alkylation of DNA occurs through the glycidamide, which is responsible for genotoxicity, not acrylamide itself. Therefore, it is important to assess internal exposure to glycidamide in cancer risk assessment by measuring its hemoglobin adducts and the mercapturic acid derivatives. It should be noted that conversion of acrylamide to glycidamide is higher in rodents, especially in mice, than in humans.
- Biotransformation is usually classified as phase I and II reactions. Phase I converts the molecule by adding a functional group that can be conjugated in phase II. Phase I

metabolism involves oxidation catalyzed by cytochrome P450 enzymes and phase II involves conjugation with glutathione. Therefore, acrylamide is converted to an epoxide glycidamide and both are conjugated with glutathione and excreted in urine as mercapturic acid derivatives. This is the predominant pathway of acrylamide detoxification.

- The measurement of AA-Hb and GA-Hb adducts was used to determine exposure to acrylamide in the occupational environment and from smoking. The study of Bergmark showed an unexpected high background of the acrylamide adduct found in the nonsmoking group, an indication of a regular exposure to acrylamide in the population [45]. It was stated in the study that the origin of this background was not known, and the estimated uptake from drinking water and environmental tobacco smoke was too low to account for these levels. Foods and beverages were suggested as possible origins had to be investigated. Later on, the preliminary estimation of the uptake, by researchers from Stockholm University, led to the conclusion that this exposure could lead to nonnegligible cancer risk [67]. In 2000, heated foods were proposed as a probable origin of the background exposure to acrylamide via measurement of hemoglobin adducts in rats fed fried feed [68]. However, the findings did not receive attention until Tareke et al. revealed that considerable amounts of acrylamide are formed during heating of a range of foods at high temperatures [69]. Then the origin/precursors, levels and the mitigation of acrylamide formation in foods were extensively studied worldwide.

MINI DICTIONARY

Glutathione An important tripeptide (γ-L-glutamyl-L-cysteinylglycine) preventing damage to important cellular components.

Toxicokinetics The quantitation of time course of a toxicant that will be absorbed, distributed, biotransformed, and excreted.

Hepatic metabolism The biochemical reactions that happen in the liver.

First-pass metabolism It is also known as first-pass effect. It is observed between the site of administration and the site of sampling for measurement of compound. The liver is the major site of first pass metabolism of orally-administered substances, which can be reduced before it reaches the systemic circulation.

Mercapturic acid The peptide bonds in glutathione conjugates are hydrolyzed enzymatically and subsequent acetylation of the amino group of cysteine yield mercapturic acid derivatives.

Cytochrome P450 A group enzyme that catalyzes oxidation of many chemicals.

SUMMARY POINTS

- This chapter describes the absorption, bioavailability, metabolism, and biomarkers of exposure to acrylamide.
- Most part of the acrylamide is absorbed and distributed to all tissues.

- Acrylamide is partly transformed to its genotoxic metabolite glycidamide.
- Acrylamide and glycidamide are conjugated with glutathione and thus detoxified.
- The conjugates are excreted in urine as mercapturic acid derivatives, which are biomarkers of short-term exposure to acrylamide.
- Acrylamide and glycidamide react with the N-terminal valine residue of hemoglobin forming covalent adducts.
- Hemoglobin adducts reflect the time-weighted exposure over 4 months.
- Hemoglobin adduct of acrylamide can be used in internal dose assessment and glycidamide adduct generally indicates the genotoxic effect.
- Nucleophilic nitrogens in DNA (bases) are also susceptible to form adducts especially with glycidamide.

REFERENCES

[1] Friedman M. Chemistry, biochemistry, and safety of acrylamide. A review. J Agric Food Chem 2003;51:4504–26.
[2] Tornqvist M, Fred C, Haglund J, et al. Protein adducts: quantitative and qualitative aspects of their formation, analysis and applications. J Chromatogr B: Anal Technol Biomed Life Sci 2002;778:279–308.
[3] Miller MJ, Carter DE, Sipes IG. Pharmacokinetics of acrylamide in Fisher-344 rats. Toxicol Appl Pharmacol 1982;63:36–44.
[4] Kadry AM, Friedman MA, Abdel-Rahman MS. Pharmacokinetics of acrylamide after oral administration in male rats. Environ Toxicol Pharmacol 1999;7:127–33.
[5] Doerge DR, Young JF, McDaniel LP, et al. Toxicokinetics of acrylamide and glycidamide in Fischer 344 rats. Toxicol Appl Pharm 2005;208:199–209.
[6] Doerge DR, Young JF, McDaniel LP, et al. Toxicokinetics of acrylamide and glycidamide in B6C3F(1) mice. Toxicol Appl Pharm 2005;202:258–67.
[7] Bjellaas T, Olstorn HBA, Becher G, et al. Urinary metabolites as biomarkers of acrylamide exposure in mice following dietary crisp bread administration or subcutaneous injection. Toxicol Sci 2007;100:374–80.
[8] Aureli F, Di Pasquale M, Lucchetti D, et al. An absorption study of dietary administered acrylamide in swine. Food Chem Toxicol 2007;45:1202–9.
[9] Sorgel F, Weissenbacher R, Kinzig-Schippers M, et al. Acrylamide: increased concentrations in homemade food and first evidence of its variable absorption from food, variable metabolism and placental and breast milk transfer in humans. Chemotherapy 2002;48:267–74.
[10] Boettcher MI, Bolt HM, Drexler H, et al. Excretion of mercapturic acids of acrylamide and glycidamide in human urine after single oral administration of deuterium-labelled acrylamide. Arch Toxicol 2006;80:55–61.
[11] Fennell TR, Sumner SCJ, Snyder RW, et al. Kinetics of elimination of urinary metabolites of acrylamide in humans. Toxicol Sci 2006;93:256–67.
[12] Kopp EK, Dekant W. Toxicoknetics of acrylamide in rats and humans following single oral administration of low doses. Toxicol Appl Pharm 2009;235:135–42.
[13] Fuhr U, Boettcher MI, Kinzig-Schippers M, et al. Toxicokinetics of acrylamide in humans after ingestion of a defined dose in a test meal to improve risk assessment for acrylamide carcinogenicity. Cancer Epidem Biomar 2006;15:266–71.
[14] Dybing E, Farmer PB, Andersen M, et al. Human exposure and internal dose assessments of acrylamide in food. Food Chem Toxicol 2005;43:365–410.
[15] Calleman CJ, Bergmark E, Costa LG. Acrylamide is metabolized to glycidamide in the rat: evidence from hemoglobin adduct formation. Chem Res Toxicol 1990;3:406–12.
[16] Settels E, Bernauer U, Palavinskas R, et al. Human CYP2E1 mediates the formation of glycidamide from acrylamide. Arch Toxicol 2008;82:717–27.

[17] Sumner SCJ, Fennell TR, Moore TA, et al. Role of cytochrome P450 2E1 in the metabolism of acrylamide and acrylonitrile in mice. Chem Res Toxicol 1999;12:1110–6.
[18] Sumner SC, MacNeela JP, Fennell TR. Characterization and quantitation of urinary metabolites of [1,2,3-13C]acrylamide in rats and mice using 13C nuclear magnetic resonance spectroscopy. Chem Res Toxicol 1992;5:81–9.
[19] Bandarra S, Fernandes AS, Magro I, et al. Mechanistic insights into the cytotoxicity and genotoxicity induced by glycidamide in human mammary cells. Mutagenesis 2013;28:721–9.
[20] Baum M, Fauth E, Fritzen S, et al. Acrylamide and glycidamide: genotoxic effects in V79-cells and human blood. Mutat Res-Gen Tox En 2005;580:61–9.
[21] Klaunig JE. Acrylamide carcinogenicity. J Agric Food Chem 2008;56:5984–8.
[22] Segerback D, Calleman CJ, Schroeder JL, et al. Formation of *N*-7-(2-carbamoyl-2-hydroxyethyl)guanine in DNA of the mouse and the rat following intraperitoneal administration of [14C] acrylamide. Carcinogenesis 1995;16:1161–5.
[23] Thielen S, Baum M, Hoffmann M, et al. Genotoxicity of glycidamide in comparison to (+/-)-anti-benzo[a]pyrene-7,8-dihydrodiol-9,10-epoxide and alpha-acetoxy-N-nitroso-diethanolamine in human blood and in mammalian V79-cells. Mol Nutr Food Res 2006;50:430–6.
[24] Puppel N, Tjaden Z, Fueller F, et al. DNA strand breaking capacity of acrylamide and glycidamide in mammalian cells. Mutat Res-Gen Tox En 2005;580:71–80.
[25] Ramu K, Fraiser LH, Mamiya B, et al. Acrolein mercapturates – synthesis, characterization, and assessment of their role in the bladder toxicity of cyclophosphamide. Chem Res Toxicol 1995;8:515–24.
[26] Fennell TR, Friedman MA. Comparison of acrylamide metabolism in humans and rodents. Adv Exp Med Biol 2005;561:109–16.
[27] Fennell TR, Sumner SC, Snyder RW, et al. Metabolism and hemoglobin adduct formation of acrylamide in humans. Toxicol Sci 2005;85:447–59.
[28] DeCaprio AP. Introduction to toxicologic biomarkers. In: Toxicologic biomarkers. New York: Taylor & Francis; 2006. p.6.
[29] Fennell TR, Sumner SC, Snyder RW, et al. Kinetics of elimination of urinary metabolites of acrylamide in humans. Toxicol Sci 2006;93:256–67.
[30] Lee JH, Lee KJ, Ahn R, et al. Urinary concentrations of acrylamide (AA) and *N*-acetyl-*S*-(2-carbamoylethyl)-cysteine (AAMA) and associations with demographic factors in the South Korean population. Int J Hyg Environ Health 2014;217:751–7.
[31] Calleman CJ, Wu Y, He F, et al. Relationships between biomarkers of exposure and neurological effects in a group of workers exposed to acrylamide. Toxicol Appl Pharmacol 1994;126:361–71.
[32] Boettcher MI, Angerer J. Determination of the major mercapturic acids of acrylamide and glycidamide in human urine by LC-ESI-MS/MS. J Chromatogr B: Anal Technol Biomed Life Sci 2005;824:283–94.
[33] Boettcher MI, Schettgen T, Kutting B, et al. Mercapturic acids of acrylamide and glycidamide as biomarkers of the internal exposure to acrylamide in the general population. Mutat Res 2005;580:167–76.
[34] Boettcher MI, Bolt HM, Angerer J. Acrylamide exposure via the diet: influence of fasting on urinary mercapturic acid metabolite excretion in humans. Arch Toxicol 2006;80:817–9.
[35] Bjellaas T, Janak K, Lundanes E, et al. Determination and quantification of urinary metabolites after dietary exposure to acrylamide. Xenobiotica 2005;35:1003–18.
[36] Kopp EK, Dekant W. Toxicokinetics of acrylamide in rats and humans following single oral administration of low doses. Toxicol Appl Pharmacol 2009;235:135–42.
[37] Doroshyenko O, Fuhr U, Kunz D, et al. In vivo role of cytochrome P450 2E1 and glutathione-S-transferase activity for acrylamide toxicokinetics in humans. Cancer Epidem Biomar 2009;18:433–43.
[38] Watzek N, Scherbl D, Feld J, et al. Profiling of mercapturic acids of acrolein and acrylamide in human urine after consumption of potato crisps. Mol Nutr Food Res 2012;56:1825–37.
[39] Bjellaas T, Stolen LH, Haugen M, et al. Urinary acrylamide metabolites as biomarkers for short-term dietary exposure to acrylamide. Food Chem Toxicol 2007;45:1020–6.
[40] Brantsaeter AL, Haugen M, de Mul A, et al. Exploration of different methods to assess dietary acrylamide exposure in pregnant women participating in the Norwegian Mother and Child Cohort Study (MoBa). Food Chem Toxicol 2008;46:2808–14.

[41] Hartmann EC, Boettcher MI, Schettgen T, et al. Hemoglobin adducts and mercapturic acid excretion of acrylamide and glycidamide in one study population. J Agric Food Chem 2008;56:6061–8.
[42] Heudorf U, Hartmann E, Angerer J. Acrylamide in children – exposure assessment via urinary acrylamide metabolites as biomarkers. Int J Hyg Environ Health 2009;212:135–41.
[43] Ji K, Kang S, Lee G, et al. Urinary levels of *N*-acetyl-*S*-(2-carbamoylethyl)-cysteine (AAMA), an acrylamide metabolite, in Korean children and their association with food consumption. Sci Total Environ 2013;456:17–23.
[44] Brisson B, Ayotte P, Normandin L, et al. Relation between dietary acrylamide exposure and biomarkers of internal dose in Canadian teenagers. J Expo Sci Environ Epidemiol 2014;24:215–21.
[45] Bergmark E. Hemoglobin adducts of acrylamide and acrylonitrile in laboratory workers, smokers and nonsmokers. Chem Res Toxicol 1997;10:78–84.
[46] Tornqvist M, Mowrer J, Jensen S, et al. Monitoring of environmental cancer initiators through hemoglobin adducts by a modified Edman degradation method. Anal Biochem 1986;154:255–66.
[47] von Stedingk H, Rydberg P, Tornqvist M. A new modified Edman procedure for analysis of N-terminal valine adducts in hemoglobin by LC-MS/MS. J Chromatogr B: Anal Technol Biomed Life Sci 2010;878:2483–90.
[48] Tornqvist M, Landin HH. Hemoglobin adducts for in vivo dose monitoring and cancer risk estimation. J Occup Environ Med 1995;37:1077–85.
[49] Bjellaas T, Olesen PT, Frandsen H, et al. Comparison of estimated dietary intake of acrylamide with hemoglobin adducts of acrylamide and glycidamide. Toxicol Sci 2007;98:110–7.
[50] Schettgen T, Rossbach B, Kutting B, et al. Determination of haemoglobin adducts of acrylamide and glycidamide in smoking and non-smoking persons of the general population. Int J Hyg Environ Health 2004;207:531–9.
[51] Hagmar L, Wirfalt E, Paulsson B, et al. Differences in hemoglobin adduct levels of acrylamide in the general population with respect to dietary intake, smoking habits and gender. Mutat Res-Gen Tox En 2005;580:157–65.
[52] Chevolleau S, Jacques C, Canlet C, et al. Analysis of hemoglobin adducts of acrylamide and glycidamide by liquid chromatography-electrospray ionization tandem mass spectrometry, as exposure biomarkers in French population. J Chromatogr A 2007;1167:125–34.
[53] Vesper HW, Slimani N, Hallmans G, et al. Cross-sectional study on acrylamide hemoglobin adducts in subpopulations from the European Prospective Investigation into Cancer and Nutrition (EPIC) Study. J Agric Food Chem 2008;56:6046–53.
[54] Vesper HW, Caudill SP, Osterloh JD, et al. Exposure of the US population to acrylamide in the National Health and Nutrition Examination Survey 2003–2004. Environ Health Perspect 2010;118:278–83.
[55] Vikstrom AC, Warholm M, Paulsson B, et al. Hemoglobin adducts as a measure of variations in exposure to acrylamide in food and comparison to questionnaire data. Food Chem Toxicol 2012;50:2531–9.
[56] Vikstrom AC, Abramsson-Zetterberg L, Naruszewicz M, et al. In vivo doses of acrylamide and glycidamide in humans after intake of acrylamide-rich food. Toxicol Sci 2011;119:41–9.
[57] Wirfalt E, Paulsson B, Tornqvist M, et al. Associations between estimated acrylamide intakes, and hemoglobin AA adducts in a sample from the Malmo Diet and Cancer cohort. Eur J Clin Nutr 2008;62:314–23.
[58] Kutting B, Uter W, Drexler H. The association between self-reported acrylamide intake and hemoglobin adducts as biomarkers of exposure. Cancer Causes Control 2008;19:273–81.
[59] Outzen M, Egeberg R, Dragsted L, et al. Dietary determinants for Hb-acrylamide and Hb-glycidamide adducts in Danish non-smoking women. Brit J Nutr 2011;105:1381–7.
[60] Ferrari P, Freisling H, Duell EJ, et al. Challenges in estimating the validity of dietary acrylamide measurements. Eur J Nutr 2013;52:1503–12.
[61] Vikstrom AC, Wilson KM, Paulsson B, et al. Alcohol influence on acrylamide to glycidamide metabolism assessed with hemoglobin-adducts and questionnaire data. Food Chem Toxicol 2010;48:820–4.
[62] Vesper HW, Sternberg MR, Frame T, et al. Among 10 sociodemographic and lifestyle variables, smoking is strongly associated with biomarkers of acrylamide exposure in a representative sample of the US Population. J Nutr 2013;143:995S–1000S.
[63] Schettgen T, Kutting B, Hornig M, et al. Trans-placental exposure of neonates to acrylamide – a pilot study. Int Arch Occup Environ Health 2004;77:213–6.

[64] von Stedingk H, Vikstrom AC, Rydberg P, et al. Analysis of hemoglobin adducts from acrylamide, glycidamide, and ethylene oxide in paired mother/cord blood samples from Denmark. Chem Res Toxicol 2011;24:1957–65.

[65] Gamboa da Costa G, Churchwell MI, Hamilton LP, et al. DNA adduct formation from acrylamide via conversion to glycidamide in adult and neonatal mice. Chem Res Toxicol 2003;16:1328–37.

[66] Huang CC, Wu CF, Shih WC, et al. Potential association of urinary N7-(2-carbamoyl-2-hydroxyethyl) guanine with dietary acrylamide intake of smokers and nonsmokers. Chem Res Toxicol 2015;28:43–50.

[67] Tornqvist M, Bergmark E, Ehrenberg L, et al. Risk estimation of acrylamide. Technical Report 7/98, pp. 28. Solna, Sweden: Swedish Chemicals Inspectorate; 1998.

[68] Tareke E, Rydberg P, Karlsson P, et al. Acrylamide: a cooking carcinogen? Chem Res Toxicol 2000;13:517–22.

[69] Tareke E, Rydberg P, Karlsson P, et al. Analysis of acrylamide, a carcinogen formed in heated foodstuffs. J Agric Food Chem 2002;50:4998–5006.

SECTION B

Acrylamide in Foods

CHAPTER 7

Acrylamide in Bakery Products

Marta Mesias, Francisco J. Morales
Institute of Food Science, Technology and Nutrition (ICTAN-CSIC), Madrid, Spain

INTRODUCTION

Bakery products comprise all food products in which cereal flour is the basic ingredient and which have been subjected to baking process. Baking is a process of cooking where heat is applied directly on the food product generally in an oven or heating appliance, reaching temperatures as high as 260 °C [1]. The term "bakery products" comprises bread and bread products, biscuits and cookies, crackers, cakes, pies and pastries, and other products such as croissants, muffins, pretzels, wafers, or waffles, all of them varying in the composition and in their manufacturing procedures.

Due to the high temperatures applied during baking and the low moisture levels reached, thermal processing of baked products triggers a series of chemical reactions between food components, with various consequences in the finished product. Some of these consequences involve an improvement of the textural and organoleptic characteristics of the food, affecting the flavor, color, and aroma, which are well appreciated by consumers. However, other negative consequences can occur, such as the natural formation of potentially toxic products such as acrylamide. Acrylamide is formed in bakery products during the thermal process as a consequence of the Maillard reaction, where reducing sugars react with asparagine amino acid at temperatures primarily above 120 °C [2], although formation of this compound can also occur at lower temperatures in some products [3]. Several factors will affect the acrylamide content in each product such as food composition, pH, water activity, and technological conditions during baking process [4].

Bakery products are widely consumed by the population, especially by children and adolescents. Therefore, acrylamide from these products significantly contribute to the total daily acrylamide intake and thus, it is a public health concern. This chapter is focused on the occurrence levels of acrylamide in bakery products and the main factors involved in the acrylamide formation in these foodstuffs. Compositional and technological processing factors as well as possible mitigation strategies to reduce acrylamide content in this kind of products are described.

Acrylamide in Food
http://dx.doi.org/10.1016/B978-0-12-802832-2.00007-3

OCCURRENCE LEVELS OF ACRYLAMIDE IN BAKERY PRODUCTS

Due to the consideration of acrylamide as a possible carcinogen and the association of its intake with a human health concern, the determination of acrylamide content in foods and the contribution of different food sources to the global acrylamide exposure has become an intensive focus of research. In order to create a database regarding acrylamide levels in relevant foodstuffs and to evaluate the consumption of these foods in certain population groups, research data are required to assess whether this toxic compound poses a risk for the population. In this regard, calculated exposure levels may be compared with a relevant toxicological reference value with the aim of estimating the potential risk from acrylamide exposure and monitor the effectiveness of any reduction strategies which may be implemented not only by industry but also at a domestic cooking scale.

Monitoring of the occurrence of acrylamide in foods and identifying the factors involved in the acrylamide formation, especially in certain food products, has been the objective of many research studies [3,5–7] and extensively revised [8]. Among the different food sources, potato products, bakery products, and coffee have been considered the most important contributors to acrylamide intake [9–11]. Therefore, it is necessary to evaluate the acrylamide levels and establish strategies to mitigate acrylamide formation in these kinds of products and, consequently, to reduce acrylamide exposure, especially in vulnerable sectors of the population.

Table 1 summarizes the acrylamide levels in bakery products according to data reported in the literature. Acrylamide content in bakery products will depend on the composition of the different products as well as on the thermal treatment conditions; it thus can vary from no detected values, as it is observed in bread crumbs [25] to more than 2000 μg/kg in products such as crackers [14]. In spite of the acrylamide levels in some bakery products not being too high, as in the case of bread, their daily consumption in large quantities explains their importance as a source of acrylamide exposure in the diet, especially among children and adolescents. According to data reported by European Food Safety Authority (EFSA) [26], bakery products may contribute to 20%–60% of the total mean acrylamide intakes, which involves exposures to acrylamide ranging from 0.13 to 0.31 μg/kg body weight/day in adults and reaching up to 0.55–0.75 μg/kg body weight/day in children and adolescents.

Recently the European Commission (EC) has adopted new indicative acrylamide values based on the EFSA monitoring data from 2007 to 2012 for the main sources of acrylamide exposure [27]. These recommendations involve reference values for acrylamide content in the major contributors to the intake of this compound. In cases where the level of acrylamide in a foodstuff exceeds the acrylamide indicative value set for the respective food category according to the recommendation, investigations into the production and processing methods used by food producers must be carried out. Thereby, risk associated with the acrylamide intake can be regulated and, consequently, controlled.

Table 1 Acrylamide levels in bakery products according to data reported in the literature

	Acrylamide (μg/kg)	References
Bread and bread products		
Bread	5–116	[10]
	40–90	[12]
	<10–152	[13]
	27–36	[9]
	35–110	[14]
	<30–160	[15]
Crisp bread	<30–430	[16]
	<68–1480	[17]
	65–1271	[14]
	152–1546	[18]
	422–908	[19]
	<30–151	[20]
Gingerbread	260–1410	[16]
	108–1697	[9]
	196–280	[13]
Toasted bread	<30–1430	[16]
	<10–107	[21]
	16–154	[19]
Biscuits and cookies		
Biscuits	<30–700	[16]
	20–1514	[9]
	<10–1060	[13]
	<68–1150	[17]
	47–370	[22]
	104	[23]
	<30–2085	[20]
	697	[24]
Cookies	400	[10]
	40–350	[12]
	42–250	[22]
Crackers	80–420	[16]
	566–2017	[14]
	197–328	[19]
	15–1040	[21]
	<30–296	[20]
	39–510	[22]
Cakes and pastries	12	[23]
	26	[24]
	20–200	[13]
	48–672	[14]

Continued

Table 1 Acrylamide levels in bakery products according to data reported in the literature—cont'd

	Acrylamide (μg/kg)	References
Other bakery products		
Croissants	<30–40	[16]
	22	[24]
Muffins	12–55	[19]
	<100	[15]
Pretzels	62–86	[21]
	53–83	[18]
Wafers and waffles	53–280	[22]
	42	[15]

Table 2 Indicative acrylamide values for bakery products based on the European Food Safety Authority (EFSA) monitoring data from 2007 to 2012 according to the European Commission of November 8, 2013. 2013/647/EU [27]

Foodstuff	Indicative value (μg/kg)
Soft bread	
1. Wheat-based bread	80
2. Soft bread other than wheat-based bread	150
Biscuits and wafers	500
Crackers with the exception of potato-based crackers	500
Crisp bread	450
Gingerbread	1000
Products similar to the other products in this category	500
Biscuits and rusks for infants and young children	200

Indicative acrylamide values for bakery products according to the latest EC Commission Recommendation of November 8, 2013 (2013/647/EU) are given in Table 2. Figure 1 summarizes the range of acrylamide content found in the different bakery products according to data reported by the literature (Table 1). Indicative acrylamide values for bakery products according to the EC Recommendation (Table 2) have been represented with a white square. Products that do not have assigned a specific indicative value have been included inside the category of "products similar to the other products in this category" and have been represented with an asterisk. As it can be observed, some of the baked products exhibit acrylamide levels above recommendations, which mean that strategies of mitigation should be performed to reduce the higher levels.

Strategies for acrylamide mitigation and monitoring programs on acrylamide levels in foods have been carried out in many countries worldwide by official organizations and companies producing acrylamide-containing foods [28,29]. In this sense, since 2003 the Food Drink Europe, formerly the confederation of food and drink industries in the UE, published the Acrylamide Toolbox in 2013 with the aim of providing descriptions of intervention steps which may prevent and reduce

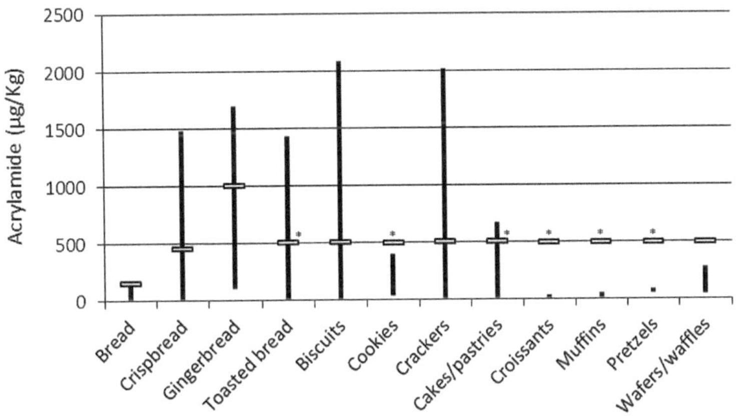

Figure 1 Range of acrylamide levels in bakery products according to data reported in the literature (see Table 1) and representation of the indicative acrylamide values based on the EFSA monitoring data from 2007 to 2012 (see Table 2). *For these products the indicative value for "products similar to the other products in this category" has been taken as reference.

formation of acrylamide in specific manufacturing processes and products. The FoodDrinkEurope Acrylamide Toolbox has for several years now been an invaluable resource for food and drink manufacturers who want to identify the best ways to reduce acrylamide in their products, and it has been largely implemented in the European food industry. The tools have been focused on the three main product categories with higher risk of acrylamide formation, such as potatoes, cereals, and coffee. Besides the application at industrial level, the strategies described may be useful for the domestic food preparation and in food service establishments [30]. The EFSA and the EC have also been actively involved in these efforts in recent years. The EFSA has monitored the levels of acrylamide in different food categories and several reports have been published. This information can be used for assessing both time trends in the acrylamide levels for the different categories and effectiveness of the mitigation strategies to reduce acrylamide content in the different foodstuffs. Moreover, groups of foods exceeding the indicative levels according to the recommendations can be identified and mitigation actions can be focused on them. Time trends of the mean of acrylamide levels in bakery products sampled from 2007 to 2010 based on data reported by the EFSA in 2012 [11] is represented in Figure 2. Acrylamide content in most of the products tended to decrease over the years although an increase has seen observed in 2010. According to EFSA, in 2010 alone 3%–7% of the cereal products exceeded the indicative limit, except for the food category "biscuits, crackers, crisp bread, and similar," where over 12% of the samples exceeded it. Therefore, special attention should be focused on mitigation actions to reduce acrylamide levels in these kinds of products.

As mentioned above, acrylamide formation in the food product will depend on the recipe and composition of the foodstuff and the technological process at which the

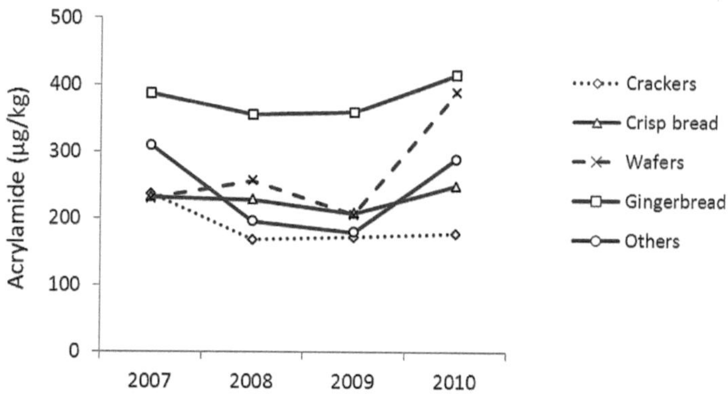

Figure 2 Time trends of acrylamide levels (mean) in bakery products sampled from 2007 to 2010 based on data reported by EFSA [11].

product will be subjected. All these factors have to be taken into account to understand the formation of this compound and to develop strategies for its mitigation. Factors affecting acrylamide formation in bakery products and strategies to mitigate acrylamide development in this foodstuff will be discussed in the following section.

FACTORS AFFECTING ACRYLAMIDE FORMATION IN BAKERY PRODUCTS

Basic ingredients of bakery products are flour, water, salt, and yeast or chemical leavening agents. Other ingredients such as fat and oils, sweeteners, eggs, dairy products, honey, fruits, nuts, spices, or chocolate will be either present or not depending on the recipe and the type of product. Initially a viscoelastic dough is prepared by mixing up of the different ingredients and after various steps of fermentation and baking, the bakery products are obtained. The differences in products will be the result of the different proportions of ingredients used in the formulation and the different treatments applied. It is well established that the amino acid asparagine and reducing sugars are the main precursors to the formation of acrylamide in foods. The ingredients will provide in a greater or lesser extent these precursors whereas the processing conditions will determine the degree of development of the Maillard reaction. Accordingly, all these factors will substantially affect the acrylamide levels in the baked product and the mitigation strategies will be focused on them. A scheme of the factors affecting acrylamide formation in bakery products is represented in Figure 3.

COMPOSITIONAL FACTORS

Asparagine

Asparagine is considered to be the main precursor for acrylamide in foods and its content has been confirmed to be the limiting factor for acrylamide formation [3].

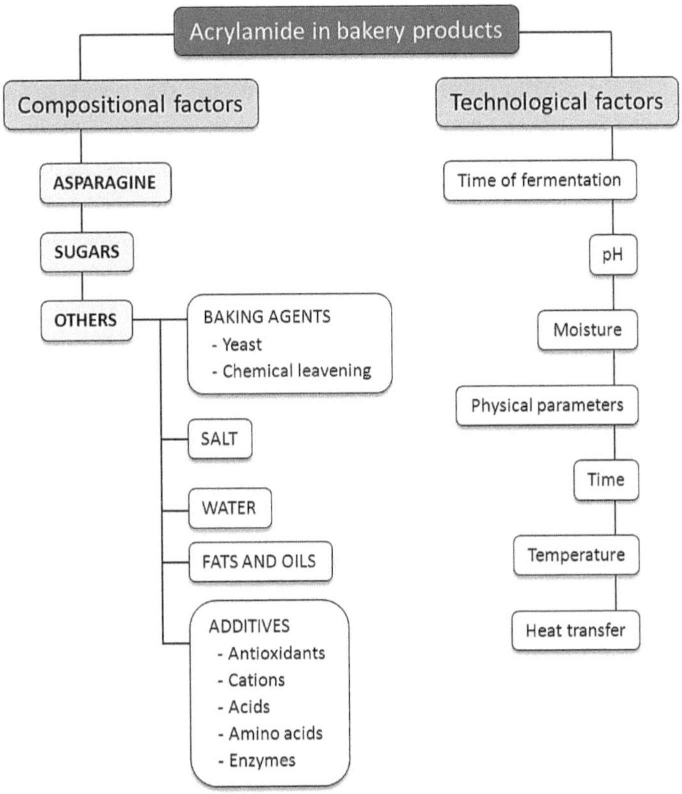

Figure 3 Factors affecting acrylamide formation in bakery products.

Several studies have demonstrated that addition of asparagine to the dough of different bakery products increased acrylamide formation, while added sugars did not significantly influence the acrylamide content in the final product. In this respect, Surdyk et al. [31] reported that asparagine added to the dough of yeast-leavened wheat bread in a range of 0.1–0.7 g/100 g of flour dramatically increased the content of acrylamide in crusts, reaching values up to 6000 μg/kg. The level of acrylamide in the isolated crumb was also increased, although to a lesser extent. In a similar way, Levine and Smith [5] and Amrein et al. [3] observed that addition of asparagine to the dough of crackers and gingerbread caused an increase in acrylamide levels. In all these studies addition of fructose or sucrose had no significant effect on either acrylamide formation or elimination.

In bakery products, the major source of free asparagine is mainly flour, although other ingredients such as honey and nuts could contribute to the total asparagine content in the dough. Several factors may condition the concentration of amino acids in the cereal flour [32], being responsible for the free asparagine content and, consequently, affecting the acrylamide levels in the baked food.

Cereal Varieties

Acrylamide formation will depend on the type of flour employed in the recipe of the bakery product, since asparagine content is different for the different cereal varieties. Flours mainly used for bakery products are from wheat but many types of flour from other grains such as rye, corn, rice, barley, oat or soy can also be used, contributing in different proportion to the asparagine content in the product. For instance, concentrations of free asparagine in the range of 1.74–19.05 mg/100 g have been described for wheat flour; 6.46–12.17 mg/100 g for spelt flour and 41.37–44.10 mg/100 g for rye flour [32].

Growth of Cereal Cultivars

According to Springer et al. [33], the cultivation of cereal through organic farming or conventional agriculture may affect the asparagine content flour, although controversial information has been reported about this concern.

Harvest Year

Harvest year has a crucial impact on asparagine content in cereals, probably due to the different weather conditions, since temperature and humidity can influence protein and amino acid composition [32].

Sprouting

It has been observed that asparagine levels increase in flour from sprouted wheat grains. This fact can be explained because sprouting of the grains increases enzyme activities, promoting the degradation of both starch and proteins and releasing precursors of acrylamide [32].

Fertilization

Claus et al. [32] reported that nitrogen fertilizers significantly increased amounts of asparagine in wheat grains and positively influenced acrylamide formation in bakery products, despite that reducing sugars in wheat were not affected by fertilization. Results for sulfur fertilizers, however, are controversial, since no effect [32] or decreased levels of asparagine [34] have been described in sulfur-fertilized cereals.

Milling

It is known that the concentration of free amino acids and sugars in cereal flours increases with a higher degree of extraction during milling. Springer et al. [33] suggested that the highest asparagine content is in the outer layers of the grain. Thereby, sifted flour has lower content of free asparagine (0.2 in wheat flour and 0.6 in rye flour) than the bran fraction (2 g/kg in wheat flour and 3 g/kg in rye flour). The highest content is present in germ wheat (5 g/kg in wheat) [35]. The fractionation efficiency of the milling process, therefore, will condition the asparagine content in the flour and, in consequence, it will affect the acrylamide levels in bread.

All these factors mean that a significant reduction of the acrylamide content in the bakery product could be achieved by choosing ingredients with a lower content of free asparagine [3]. Moreover, the decomposition of free asparagine prior to baking is suggested to be an additional procedure to mitigate acrylamide formation. For this purpose, other actions can be performed on the dough previous to the baking process, such as promoting fermentation or adding other ingredients as asparaginase or amino acids different from asparagine, which will be described below.

Sugars

As mentioned above, reducing sugars together with asparagine are the main precursors to the acrylamide formation in foods. Although asparagine is the limiting factor in bakery products, the type and concentration of sugars also play a crucial role in the development of acrylamide.

The relative reactivity of glucose, fructose, and sucrose with regard to acrylamide generation was evaluated by Hamlet et al. [18] in soft dough biscuits. Results revealed that addition of 1% fructose to the flour increased acrylamide formation by about 20%, showing much more reactivity than glucose and sucrose. Similarly, Graf et al. [36] evaluated the influence of the type of sugars on acrylamide formation in biscuits using a sucrose solution (nonreducing sugars) instead of inverted sugar syrup in the formulation, which resulted in an acrylamide content over 70% lower than the content in the standard product (46 mg/kg vs 170 mg/kg). These results suggest that replacing reducing sugars by sucrose is an effective way to significantly reduce the acrylamide content of sweet bakery. These findings are in agreement with those observed by Vass et al. [37] and Gökmen et al. [6], who concluded that the replacement of reducing sugars by sucrose reduced acrylamide content by 60% and 50% in crackers and cookies, respectively. However, contrary results have been found by Taeymans et al. [38] in experiments analyzing the acrylamide formation from reactions between asparagine and various sugars in a washed whole wheat flour matrix. In this case, fructose tended to be more effective than glucose but the reaction with sucrose resulted in amounts of acrylamide close to those recorded for fructose and glucose, probably because sucrose was hydrolyzed to form reducing sugars during thermal treatment.

Regarding the amount of sugars added into the formulation, different results have been reported. On the one hand, Surdyk et al. [31] indicated that addition of different amounts of fructose to yeast-leavened wheat bread did not influence the content of acrylamide and, on the other hand, Gökmen et al. [6] observed that increasing the amount of sucrose from 10 to 35 g in the recipe of cookies almost doubled the amount of acrylamide formed during baking.

Other Factors

Baking Agents

Baking agents or leavening agents are substances incorporated into the dough that cause or incorporate gases in the baked products, which must be retained in the product to promote an increase in volume and produce shape and crumb texture [1]. Different types of baking agents can be used in bakery products and they are known to strongly influence the acrylamide formation [3].

Yeast

Yeast (*Saccharomyces cerevisae*) is responsible for the fermentation process into the dough, by which sugars are transformed into carbon dioxide gas and alcohol. Gas will produce the leavening action in the yeast product while alcohol is evaporated during baking process. Asparagine is also utilized during the process of fermentation with yeast; therefore, extensive fermentation time may be one possible way to reduce acrylamide content in baked foods, and this is explained under the section Technological Factors.

Chemical Leavening

The use of chemical leavening is considered to be faster, more convenient and often a more consistent method than the use of yeast. The two major chemical leavening agents used in the manufacture of bakery products are sodium bicarbonate (called baking soda) and ammonium bicarbonate, both systems being decomposed into carbon dioxide gas when exposed to heat [1].

It has been shown that ammonium bicarbonate enhances acrylamide formation in bakery products. Amrein et al. [3] investigated the influence of this baking agent on acrylamide formation in gingerbread. In its absence almost no acrylamide was formed, whereas the addition of 0.4, 0.8, and 1.6 g/100 g dough led to acrylamide levels of 170, 501, and 880 μg/kg, respectively. Authors concluded that the effect of ammonium bicarbonate on the formation of acrylamide might be promoted indirectly by providing more reactive carbonyl compounds such as glyoxal and methylglyoxal originated from the reaction of ammonia with reducing sugars. Glyoxal and methylglyoxal are formed from glucose and fructose in Maillard reaction models and may react more quickly with amino acids than with reducing sugars [39]. In agreement with these results, Marconi et al. [40] found that acrylamide content increased more than twice (from 76 to 162 μg/kg) when 11.9 g instead of 3.0 g of ammonium bicarbonate were added to the dough of shortbreads.

An alternative to ammonium bicarbonate is the use of sodium bicarbonate, whose application seems to reduce acrylamide content by raising pH in the system. The partial or total replacement of ammonium bicarbonate by sodium bicarbonate has been studied by several authors. Amrein et al. [3] observed a decrease in acrylamide levels by one-third when sodium bicarbonate was used in gingerbread instead of ammonium bicarbonate. Similarly, the complete replacement of NH_4HCO_3 by $NaHCO_3$ reduced the acrylamide content by over 70% in biscuits. This effect was not observed if ammonium bicarbonate was not fully replaced by sodium bicarbonate [36]. A decrease in acrylamide levels in crackers was also observed by Levine and Smith [5] when 5 g of $NaHCO_3$

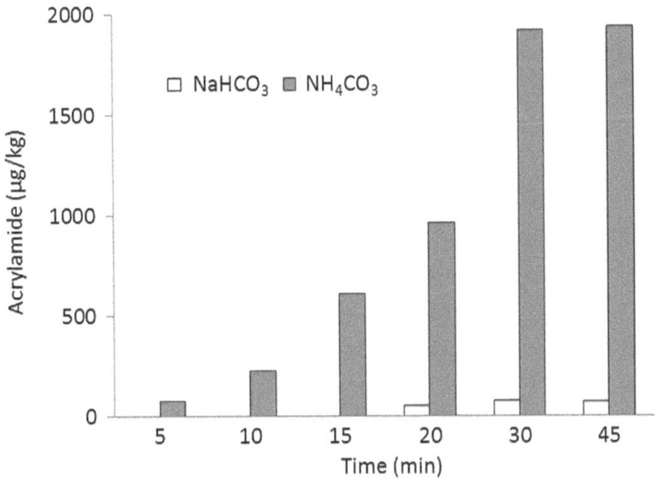

Figure 4 Acrylamide formation in crackers with added $NaHCO_3$ or NH_4CO_3 during baking at 180 °C. *Adapted from Levine and Smith [5].*

instead of 5 g NH_4CO_3 was added to the formulation (Figure 4). According to these authors, besides decreasing the acrylamide formation, $NaHCO_3$ has been demonstrated to promote acrylamide elimination, which is another positive aspect of using sodium bicarbonate instead of ammonium bicarbonate.

Salt

Salt has been traditionally used during the manufacture of baked products as it causes several important changes in rheological, technological, and sensory parameters. Its role in baking is not only seasoning or as a flavor enhancer but salt is also involved in the structure and formation of the gluten matrix, increasing stability, flexibility, and resistance of the dough and decreasing stickiness and water absorption [41]. In addition, sodium chloride inhibits yeast growth in the dough controlling then the fermentation process [42].

Salt has a considerable inhibitory effect on acrylamide formation. However, it is important to control the amount of NaCl added to the dough, since elevated levels may override this inhibitory effect. In this respect, addition of 3% NaCl instead of 1% to the bread rolls dough has been demonstrated to decrease the acrylamide content due to lower amylase activity and, therefore, lower reducing sugars' content [4]. In this experiment asparagine also tended to decrease when 3% salt was added to the dough. Conversely, higher concentrations of salts significantly increased the amount of acrylamide in the product, probably due to an inhibition of the yeast growth. Van Der Fels-Klerx et al. [43] also observed that the presence of NaCl in the biscuit formulation lowered acrylamide concentrations when baked at 180 °C and 190 °C but not at 200 °C.

Water

Water plays an important role in the baking process since it may control factors such as quality, texture, taste, smell, volume, flavor, and mouthfeel of bakery products [44]. The involvement of water in relation to acrylamide formation will be discussed in the technological section.

Fats and Oils

Fats and lipids are added to many bakery products to enhance sensorial and rheological characteristics, favor leavening conditions, provide structure, and improve technological parameters [1]. Type of vegetable oil or fat used in the formulation and their composition may affect the amount of acrylamide produced. In this regard, Capuano et al. [45] observed higher acrylamide levels in fat-rich model systems containing sunflower oil than in systems made with palm oil. An important aspect in the composition of different oils is the content in antioxidant compounds, which may affect the higher or lower acrylamide formation in the final baked product. Arribas-Lorenzo et al. [46] evaluated this fact selecting three types of virgin olive oils according to their content in phenolic compounds (low, intermediate, and high) and used them in the formulation of three cookie doughs to investigate the effect of the olive oil phenolic profile on the formation of acrylamide during baking. No significant differences among recipes were found for short baking times. However, formulations with high phenolic compounds content presented acrylamide levels 20% lower than those with low phenolic compounds content when prolonging the baking time up to 16 min.

Another point to consider is the relationship between lipid oxidation and acrylamide formation. According to Zamora et al. [47], carbonyl compounds deriving from lipid oxidation during food processing can degrade asparagine to acrylamide, suggesting that lipids can contribute to acrylamide formation during heat treatment of fat-rich foods or during frying processes. This statement was supported by Capuano et al. [45], who showed a positive influence of lipid oxidation on acrylamide formation in fat-rich model systems. The effect of oil oxidation level on acrylamide formation was more evident when sugar was removed from the formulation, since lipids become the main source of carbonyl groups. Investigations from Arribas-Lorenzo et al. [46] also supported the hypothesis that secondary lipid oxidation products may actively participate in the formation of acrylamide in fat-rich foods. These authors evaluated acrylamide levels in cookies formulated with heated and unheated sunflower oil and concluded that oxidized oil in the formulation led to a huge increase in the formation of acrylamide during baking. On the contrary, the contribution of lipid oxidation on acrylamide formation when bakery products are poor in fat, as in the case of bread or bread crisp, is practically negligible [48].

Additives

Antioxidants

Several research studies suggest that addition of antioxidant compounds to the dough prior to the baking process may be successfully used to reduce acrylamide formation not only by the inhibition of the formation of acrylamide, but also by the inhibition of the overall rate of the Maillard reaction. Zhang and Zhang [49] described significant acrylamide reductions in fried bread sticks upon an addition of antioxidants from bamboo leaves and green tea. In a similar way, Hedegaard et al. [50] reported that the addition of antioxidants from aqueous rosemary extract, rosemary oil, and dried rosemary leaves to wheat dough decreased the acrylamide levels in bread by 62%, 67%, and 57%, respectively. These effects could be explained due to the carbonyl compound trapping capacity of antioxidant compounds, which prevent the Maillard reaction development and, consequently, acrylamide generation. In line with this affirmation, Levine and Smith [5] reported that addition of ferulic acid, an antioxidant found in the seeds and leaves of many plants, reduced the net acrylamide formed in a cracker model system. Authors concluded that ferulic acid could react with acrylamide precursors or intermediates avoiding acrylamide formation.

On the contrary, it has also been declared that the effectiveness of antioxidant compounds to reduce acrylamide formation may be limited. Likewise, Açar and Gökmen [51] found that grape seed extract added to wheat flour had no effect on acrylamide formation in bread. Capuano et al. [48] also discussed a limited and uncertain effect of antioxidant compounds to decrease acrylamide content in bread. These authors added an antioxidant extract from green tea at different concentrations to a wheat formulation and tested the effect in reducing acrylamide formation. When the extract was added at a concentration of 0.1% and 0.02% acrylamide formation was not inhibited, but at a concentration of 0.5%, a reduction was observed. However the possible relation of acrylamide formation and residual moisture content in bread was discussed, since moisture was much higher in the presence of extracts than in those without the antioxidant. This could explain the lower acrylamide content detected since acrylamide formation is reported to be strictly dependent on final moisture content [52].

The role of antioxidants on acrylamide inhibition can also be disclosed in fat systems. In this respect, Capuano et al. [45] evaluated the impact of the presence of catechin in fat-rich model systems subjected to different oxidation levels. Catechin reduced acrylamide formation in systems containing unheated oil and partially oxidized oil whereas no significant lowering effect was observed when highly oxidized oil was used. It could be deduced that catechins reduced acrylamide formation presumably by trapping carbohydrates and/or preventing lipid oxidation. An inverse relationship between the total phenolic compounds content added to the formulation of cookies baked for 16 min and acrylamide formation has also been found by Arribas-Lorenzo et al. [46].

Cations

Cations have long been known to cause reduction in acrylamide formation. The higher effectiveness of divalent cations (Ca^{2+} and Mg^{2+}) as inhibitors of acrylamide formation in comparison with monovalent ones (Na^{+} and K^{+}) has been previously stated in model systems and food models [53,54]. Thereby, decreases of around 30% in acrylamide content have been described in bread [55] and crackers [56] when Ca^{2+} was added to the formulation. In a similar way, reductions in acrylamide formation have been reported when formulations of bakery products were prepared with calcium chloride [51,57]. Among inorganic salts, $CaCl_2$ has been suggested to be the most effective acrylamide formation inhibitor in cereal matrices [58]. In accordance with that, results from Hamlet [18] showed a higher acrylamide inhibition when calcium chloride instead of calcium carbonate was added to soft dough biscuit.

According to Mestdagh et al. [59], the addition of metal cations to certain foods may cause a decrease in the pH which, in agreement with Sadd et al. [57], may induce reduction of the acrylamide content. In this respect, Quarta and Anese [54] found that Ca^{2+} was capable of inducing significant pH reduction in model systems. For this reason, it has been probed whether the inhibition of the formation of acrylamide when Ca^{2+} is added to the system is due to the ions or to the drop in the pH value [59].

Acids

The addition of acids, by means of lowering pH, results in moderate Maillard reaction and, in consequence, decreases the amount of acrylamide formed in foods during heating [3,31,59]. In bakery products, acidity can minimize acrylamide content by both the reduction of acrylamide formation and increase of the degradation of acrylamide formed. The protonation of the α-amino group of asparagine hinders the formation of the N-substituted glycosylamine, which may explain the reduced acrylamide content.

Amrein et al. [3] carried out various experiments checking the ability of citric acid to reduce the acrylamide content in gingerbread. Addition of 0.5 and 1.0 g of citric acid/100 g of dough resulted in drops of pH to 5.6 and 5.0, respectively, and in a reduction of the acrylamide concentration by factors of 4 and 40, respectively (Figure 5). These results are

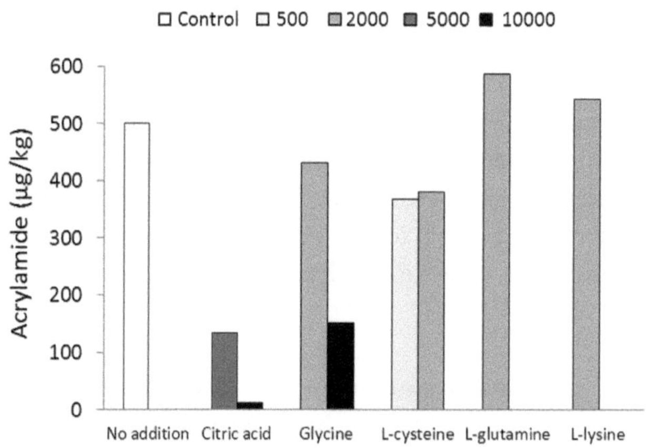

Figure 5 Effect of the addition of organic acids and amino acids (g/kg dough) on the acrylamide content in gingerbread. *Adapted from Amrein et al. [3].*

well in line with those reported by Graf et al. [36], who added tartaric acid in different proportions to the dough of bread leading to a decrease of acrylamide formation, although these results were also affected by a partial replacement of ammonium bicarbonate by sodium bicarbonate. Levine and Smith [5] added ascorbic acid to cracker models and found effects on both acrylamide formation and elimination, thus significantly decreasing acrylamide contents. In summary, the addition of some organic acids is another way to significantly reduce the acrylamide content of bakery products. However, the amount of acid to be added may be limited, mainly for sensory reasons [3]. Decreases in the pH value can also be achieved by the addition of metal cations, as mentioned above.

Amino Acids

The addition of certain amino acids to the formulation of bakery products has been demonstrated to reduce acrylamide content. Amino acids may compete with asparagine in the Maillard reaction avoiding the acrylamide formation but, moreover, may react with acrylamide through Michael addition after its formation, promoting its elimination [52].

Amrein et al. [3] checked the ability of different amino acids to reduce the acrylamide content in gingerbread. Results of this experiment revealed that a moderate addition of L-glutamine, L-lysine, or glycine (2 g/kg of dough) did not lower the acrylamide content, whereas larger addition of glycine (10 g/kg of dough) reduced the acrylamide content to one-third (Figure 5). These results are in good agreement with those from Capuano et al. [7], who reported that the addition of glycine (5 g/kg of flour) to the basic formulation of bread crisp significantly reduced acrylamide formation, observing declines up to 60% at different toasting times. In a similar way, Brathen et al. [60] observed that the application of glycine to the dough prior to fermentation and production of breads and flat breads reduced acrylamide by up to 80% and 90% for flat breads and breads, respectively. Acrylamide levels were also reduced by 16% when the dough surface of yeast-leavened bread was sprayed with a 10% glycine solution prior to baking [61]. Glycine is known to strongly enhance browning and to react with α-dicarbonyls. Therefore, the effect could be due to the competition of the amino acids for the reactive carbonyls and/or the elimination of formed acrylamide by a reaction with glycine.

Regarding L-cysteine, opposite data have been reported. Some experiments showed a trend or even a decrease of acrylamide formation when this amino acid was added to the formulation. Likewise, application of cysteine tended to reduce the acrylamide content in gingerbread [3] (Figure 5) and decreased acrylamide formation in crackers by almost 50% [5]. Claus et al. [4] also reported lower acrylamide levels in bread whose dough was treated with cysteine but, on the contrary, its application as a solution to the bread surface showed no effect, probably due to the limited mobility of cysteine in the dry crust. Flückiger and Salih [62] neither observed any effect when cysteine was added directly to the dough of crisp bread.

Enzymes

One of the main strategies to decrease acrylamide formation is reducing precursors in the food product. Therefore, the decomposition of free asparagine prior to baking is supposed to result in a decrease in acrylamide content. For this purpose, the application of asparaginase during dough preparation has been tested [3]. This enzyme catalyzes the hydrolysis of asparagine in aspartic acid and ammonia.

The effectiveness of asparaginase in the reduction of acrylamide formation has been demonstrated in several studies. Capuano et al. [7] assessed the acrylamide content in bread crisp toasted at different times and concluded that asparaginase added to the dough of bread led to a reduction of up to 80% in the samples toasted for 25 min and to a reduction of about 67% in the sample toasted for 22 min. Moreover, acrylamide started to form later, not being detected after 18 min of toasting. This result was even more effective than this observed with the addition of glycine to the formulation. Vass et al. [37] also described a decrease of acrylamide levels by at least 70% when asparaginase was added to wheat cracker production and Kumar et al. [63] recently reported that acrylamide content was reduced when bread was prepared with asparaginase treated dough; higher the concentration of the enzyme, higher the reduction of acrylamide. Thereby, acrylamide levels decreased from 729 to 24 μg/kg in bread crust and from 143 to 41 μg/kg when 300 U of enzyme was added to the dough (Figure 6). Anese et al. [64], however, reported that the effectiveness of asparaginase in the reduction of acrylamide formation is highly influenced by the formulation of bakery products, both by the matrix composition and structure. These authors observed that short dough biscuits with high water content promoted acrylamide development by favoring precursor mobility and enzyme capability to reducing acrylamide formation. On the contrary, the presence of fat significantly

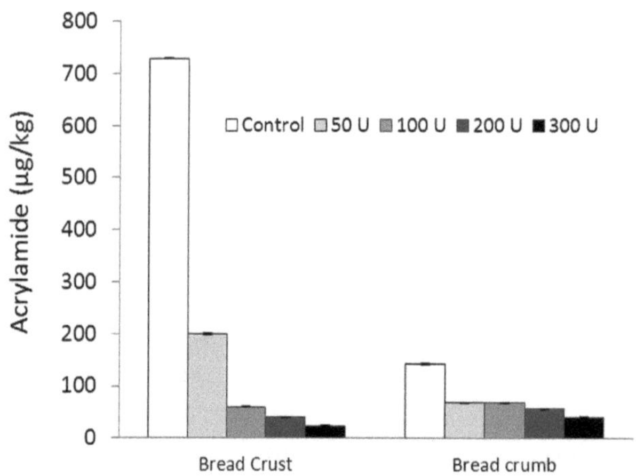

Figure 6 Acrylamide concentration in bread crust and crumb in L-asparaginase-treated breads. *Adapted from Kumar et al. [63].*

reduced acrylamide development and enzyme activity as compared with the fat-free formulation.

Although asparaginase seems to be very efficient for acrylamide mitigation, it is rather expensive compared with other strategies, which limits its use in the production of low-price foodstuffs, such as some bakery products.

TECHNOLOGICAL FACTORS

Time of Fermentation

As previously mentioned, acrylamide content of bakery products can be reduced by prolonged fermentation of the dough. Fredriksson et al. [65] demonstrated that long fermentation time reduced acrylamide content in bread made with whole grain wheat (87%) and rye bran (77%) when compared with short fermentation time. In this experiment, most of the asparagine was utilized after 2 h of fermentation with yeast. Similar results have been described by Claus et al. [4], showing a decrease of 60% in the acrylamide content in wheat bread when fermentation time was prolonged. However, excessive time is not suitable for this purpose, since gluten could be degraded and the capacity of yeast for asparaginase metabolization limited [61,65]. In this regard, Fredriksson et al. [65] considered that prolongation of fermentation time to at least 1 h was sufficient for acrylamide reduction in industrial bread production.

pH

The effects of pH on acrylamide formation in bakery products have also been investigated. As mentioned before, addition of organic acids or metal cations lead to decreased pH value, which reduce Maillard reaction development and, consequently, minimize acrylamide generation. However, a more alkaline pH does not necessarily imply higher acrylamide content in the baked product [3].

Moisture and Water Activity

Acrylamide in bakery products is specially affected by moisture content of the product and by its water activity. Depending on the water content in the foodstuff, water may have an inhibitory or an accelerating impact on the Maillard reaction and, therefore, on acrylamide formation. Acrylamide is formed when water activity is reduced to below 0.8, and in low water activity foods, maximum acrylamide formation takes place when water activity is about 0.4. Additional reduction of water activity tends to decrease the amount of acrylamide. However, it is well known that moisture content is related to thermal processing conditions and it is difficult to separate the effect of water content from that of temperature or time of heating on acrylamide formation [25]. In this regard, Taeymans et al. [38] conducted a trial on a model biscuit system to evaluate acrylamide formation in relation to the temperature and moisture content. At 10% moisture, no acrylamide was

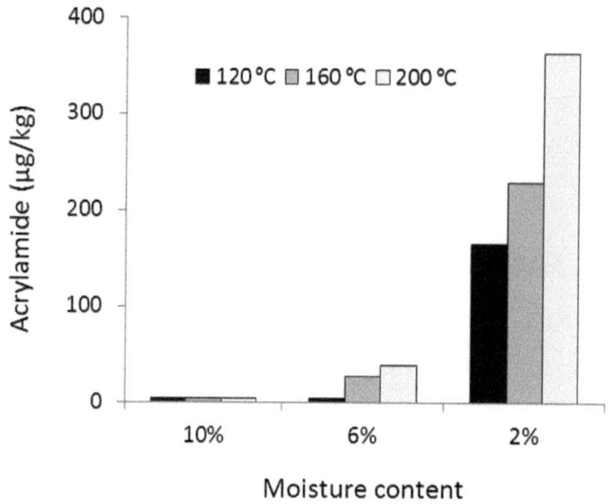

Figure 7 Acrylamide concentrations in model biscuits in relation to baking temperature and final moisture content. *Adapted from Taeymans et al. [38].*

observed under the conditions employed (120, 160, and 200 °C of baking temperatures). At 6% moisture, acrylamide was detected at 160 and 200 °C, but at moderate amounts. At a moisture content of 2%, acrylamide ranged from 165 to 363 µg/kg, depending on the baking conditions (Figure 7). In a similar way, Ahrné et al. [25] studied the effect of crust temperature and water content on acrylamide formation during the baking of white bread. Acrylamide concentrations increased when water content decreased down to 4%, which was related to the increase in the temperature. Thereafter, a trend toward decreased acrylamide concentration was observed, due to higher temperatures in the crust and lower water content (from 4% to 2%), reaching the lower acrylamide concentration in the crust having a 2.7% water content. These findings were in agreement with those reported by Elmore et al. [66], who cooked wheat cakes in an oven at 180 °C for 5–60 min. Acrylamide formation in cakes was relatively low when the moisture levels were above 5%. A decrease of the moisture was observed along the baking time and acrylamide was formed to a large degree since the moisture contents fell below 5%. Acrylamide reached the maximum level at 30 min of heating and then started to decline (Figure 8), possibly due to secondary reactions between acrylamide and other food components, or evaporation of acrylamide from the surface of the cakes.

Physical Parameters

Physical parameters such as dimensions of the bakery product are another factor to affect acrylamide formation. According to Açar and Gökmen [51] the product thickness is an important parameter that determines the rate of drying and, consequently, the rate of acrylamide formation during baking. These authors reported that in the dough having thickness of 1 mm, acrylamide concentration increased rapidly at 180 °C, reaching an

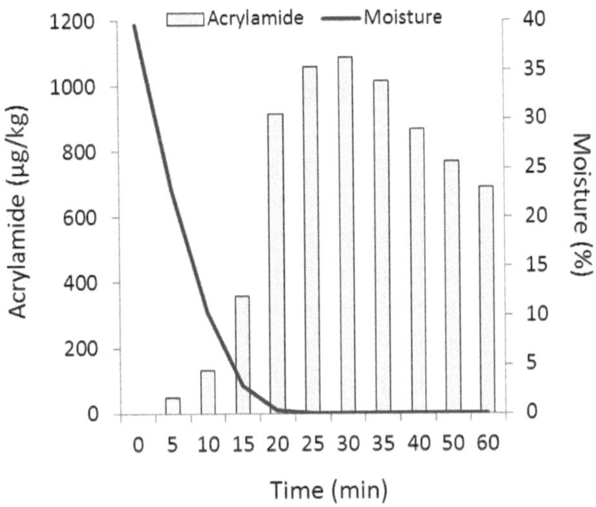

Figure 8 Acrylamide concentrations and moisture content of wheat cakes as influenced by the time of baking at 180 °C. *Adapted from Elmore et al. [66].*

apparent maximum of 411 µg/kg within baking time of 8 min. Then acrylamide levels decreased exponentially. Conversely, no acrylamide was detected in the dough having thickness of 10 mm within 15 min of heating under the same baking conditions, because the moisture content was above 10%. After this time, acrylamide concentration tended to increase linearly, reaching a value of 217 µg/kg after 30 min of heating.

Mustafa et al. [35] evaluated the effect of the height of rye crisp bread on acrylamide formation. Breads with height ranging from 4.5 to 7.2 mm were baked for 3–15 min at temperatures of 207°–282 °C. Bread heights, together with the time of heating and its interaction with the temperature, had significant effect on the acrylamide levels detected in the final baked product.

Time and Temperature of Baking

Many researches have indicated that acrylamide content is affected by the time and temperature of thermal treatment during baking [31,53,60,66,67]. According to Açar et al. [68], formation and degradation of acrylamide occur simultaneously during heating at elevated temperatures. At lower temperatures acrylamide is formed along the time until reaching an upper level. However, at higher temperatures, maximum acrylamide formation is reached in a short period of time and then acrylamide levels decrease due to a high rate of degradation. This affirmation was in agreement with the information reported by Friedman [2], who pointed out that temperature of about 120 °C are required for the formation of acrylamide; increasing to about 200 °C and decreasing above that temperature because elimination mechanisms begin to predominate.

All these hypotheses have been demonstrated in several studies. For instance, Mogol and Gökmen [69] reported that increasing baking temperature or time significantly increased the amounts of acrylamide formed in biscuits during conventional baking. As expected, at a constant temperature (200 °C), acrylamide levels increased from 39 to 211 µg/kg when duration of treatment changed between 8 and 15 min, respectively. In parallel, at a constant time (13 min), acrylamide levels increased from 21 to 191 µg/kg when temperature of baking changed from 180 to 200 °C. Amrein et al. [3] investigated the influence of temperature and time on acrylamide formation in gingerbread baked under different conditions. These authors observed an increase of acrylamide levels in the products baked at both 180 and 200 °C for 20 min and a decrease when time of heating was prolonged. Ahrne et al. [25] reported that the highest concentration of acrylamide was observed in bread baked at 260 °C for 15 min, whereas levels were lower prolonging the time of baking upto 20 min. Similarly, Marconi et al. [40] observed lower acrylamide levels in shortbread baked at 300 °C for 3 min when compared with those baked at 240 °C for 7 min.

As mentioned before, temperature and time of heating may be inversely related with moisture content in the foodstuff and, in turn, the interaction among these three factors will affect the final concentration of the acrylamide in the baked product. In accordance with Taeymans et al. [38], at a certain temperature, acrylamide levels in biscuits increased when moisture conditions were lower whereas at certain moisture conditions, increase of acrylamide levels occurred when temperatures were raised. Differences were not observed when 10% moisture was evaluated because in these conditions acrylamide formation is not promoted (Figure 7). A relationship between moisture and baking conditions were also observed by Elmore et al. [66] (Figure 8). In a similar way, higher temperatures and time of baking have been demonstrated to decrease moisture content in the baked bread, promoting the formation of crust at the surface in a higher proportion, from 30% when baked at 170 °C for 17 min to 45% when baked at 270 °C for 32 min. Both baking temperature and baking time increased the content of acrylamide in the crust [31].

Heat Transfer

During baking, heat and mass transfers occur simultaneously, causing the formation of a temperature gradient within the product, which will affect the rates of the chemical reactions occurring in the baked product [68]. Both baking temperature and heat transfer will determine heat and water content inside the product and, consequently, will condition acrylamide formation.

Oven type and different baking processes have been tested in order to find a relationship with the acrylamide formation in the baked product. Claus et al. [4] studied the differences between deck and convection ovens using wheat bread baked at 220 and 260 °C for 60 min. Deck ovens are characterized by heat transfer via conduction and radiation, whereas convection ovens use forced air circulation. The results revealed that at 200 °C, acrylamide levels were 60% higher in breads baked in a multi-deck oven (109.6 μg/kg) than in a convection oven (173.9 μg/kg); at 260 °C these differences were only 35%, probably due to a higher acrylamide formation rate at this temperature. Authors suggested that the higher levels in breads baked in a convection oven may be because forced air circulation leads to faster and more intense drying of the bread crust, enhancing acrylamide formation. Therefore, deck ovens seem useful to mitigate acrylamide levels in bakery products.

Mogol and Gökmen [69] compared conventional baking and vacuum baking and evaluated their effects on acrylamide content of biscuits. The results indicated that the levels of acrylamide attained during vacuum baking (11–51 μg/kg) were significantly lower than those attained during conventional baking (39–211) at 200 °C for 8–15 min. Compared with conventional baking, heating under lower pressure provided lower time—temperature profile with slightly accelerated drying rate of biscuits due to evaporation of water in dough. In addition, since oven air was partially removed in vacuum baking, convective heating was limited, and acrylamide formation decreased.

Ahrne et al. [25] compared the traditional baking with steam and falling temperature baking. The use of steam during baking of white bread reduced the concentration of acrylamide compared with levels found in baking without steam, probably because steam both increased the water content and decreased temperature in the product, avoiding acrylamide formation. In the falling temperature baking, oven temperature was lowered after 5, 10, and 15 min of baking, which reduced the acrylamide concentration by 67%, 36%, and 35%, respectively, compared with that produced by constant temperature in the traditional baking. These results are in agreement with those reported by Isleroglu et al. [70] and Sakin-Yilmazer et al. [71], who observed lower acrylamide levels in cookies and muffins baked in a steam-assisted oven when compared with those baked in natural and forced convection ovens.

Alternatively, Anese et al. [72] proposed the application of radiofrequency heating in the last stages of the baking process, which resulted to be a promising strategy to keep low the acrylamide levels in the bakery products. Radiofrequency heating allows to rapidly and uniformly heat foods and it has been used as a post-baking process to reduce the residual moisture of bakery products and the baking times. The application of this treatment in combination at the final step of conventional heating was demonstrated to reduce acrylamide levels in biscuits. In this case, biscuits subjected to this alternative baking process had about 35% acrylamide content lesser than those obtained by using only the conventional heating.

SYNOPSIS OF ANALYTICAL TECHNIQUES

Sample Extraction

Initially, the sample should be finely ground in order to improve the extraction yield. Water at room temperature is used to extract acrylamide in bakery products in most published analytical methods. Other sample preparations also include the addition of heated water (60°–70 °C) or a mix of water–methanol or water–acetonitrile. Some procedures include precipitation of co-extractives with Carrez reagent (potassium ferricyanide (II) and zinc (II) sulfate (VI)). Extraction is carried out by vigorous shaking and it can be intensified by ultrasound. In order to control the recoveries and to take into account possible losses occurring during the whole sample pre-treatment, an internal standard is added to the sample matrix after homogenization and before extraction. Most published studies use carbon-labeled (^{13}C) or deuterium-labeled (d_3) acrylamide as internal standard. Finally centrifugation of the samples is performed to obtain the supernatant. Samples can be protected from the light during the extraction.

A QuEChERS (Quick, Easy, Cheap, Effective, Rugged, and Safe) method based on water extraction and partition with acetonitrile by salting out with $MgSO_4$ and NaCl is also reported. In this case homogenized sample and spiked d_3-acrylamide is extracted with a mixture of hexane–water–acetonitrile containing $MgSO_4$ and NaCl. The salt combination enhances separation of the water and acetonitrile layers. Acrylamide is mainly concentrated into the acetonitrile layer. In addition, the use of acetonitrile serves for deproteinization, and a simple centrifugation step is sufficient to clarify the extract. In a second step, acetonitrile extract is transferred to a tube containing the sorbent for dispersive solid-phase extraction and anhydrous $MgSO_4$. The final extract is analyzed either by liquid chromatography–tandem mass spectrometry or by gas chromatography–mass spectrometry.

Purification and Cleanup Step

Different solid-phase extraction (SPE) cartridges are applied as an effective cleanup step to eliminate interfering compounds. The cleanup step can be achieved by using single or multiple SPE with different sorbents (silica, polymer, or carbon-based), or combinations of these (mixed mode). Different sorbent and commercially available cartridges are used, such as Oasis HLB (poly(divinyl-benzene-co-*N*-vinylpyrrolidone)), Isolute Multimode (mixed mode containing nonpolar (C18)), and Bond Elut Accucat cartridges (C18 hydrophobic silica-based encapped sorbent).

Chromatographic Separation and Detection

The literature describes two different methods for the analysis of the extracts: liquid chromatography coupled to tandem mass spectrometry (LC-MS/MS) and gas chromatography coupled to tandem mass spectrometry (GC–MS/MS). Determination of the acrylamide concentration is most often assayed by LC–MS/MS, which enables omitting the time-consuming step of derivatization that is necessary before GC–MS separation.

For the GC–MS/MS method a derivatization of acrylamide such as bromination or the formation of a stable thioether by reaction with 2-mercaptobenzoic acid is included.

The product of bromination is 2,3-dibromopropionamide, which can be back-extracted from the aqueous bromination solution into a solvent adequate for gas chromatographic analysis, or treated with triethylamide to form the more stable 2-bromopropionamide. Following removal of the ethyl acetate by evaporation, triethylamine is added to partially debrominate, after which the sample extracts are injected onto a GC–MS system for quantification.

According to the literature, several LC columns can be employed for the chromatographic separation such as Hypercarb (graphite carbon column), Inertsil ODS, Alltima C18, Kinetex C18 or Luna ODS2. The GC columns most commonly employed for the analysis of the bromo-derivative are midpolar to polar columns. Several GC capillary columns are proposed, such as DB17 (50%-phenyl-dimethylpolysiloxane), ZB50 (50%-phenyl-dimethylpolysiloxane), Optima WAX (olyethyleneglycol 20 kDa), or RTX-200 Crossbond (trifluoropropylmethylpolysiloxane). Dimensions should be: 30 m of length and 0.25 mm of internal diameter with 0.25 μm of thickness.

The application of tandem mass spectrometry working in multiple reactions monitoring mode (MRM) is recommended for acrylamide analysis. MRM allows monitoring the transition from a precursor ion to a product ion. The main ions observed for acrylamide are m/z 72, corresponding to the protonated molecular ion; m/z 55, corresponding to a loss of NH_2; and m/z 27, corresponding to a subsequent loss of CO. For acrylamide quantitation, mass transition 72 > 55 is selected due to its high intensity. Respectively, transition of 75 > 58 is described for isotopic-labeled acrylamide, d_3-acrylamide, and $^{13}C_3$-acrylamide. For the brominated extracts acrylamide is quantified using the product ion m/z 135 (precursor ion m/z 152) for the analyte, 2,3-dibromopropionamide and m/z 137 (precursor ion m/z 155) for internal standard (d_3-acrylamide) brominated to 2,3-dibromo(d_3)propionamide.

Quantification is performed with standard solutions spiked with the internal standard. The peak identity is confirmed by comparison of the peak area ratios for the transitions of the ions from sample extracts and standard solutions, respectively.

KEY FACTS

- **Baking fermentation.** Fermentation in food processing is a metabolic process that converts carbohydrates to organic acids, carbon dioxide and/or alcohol. For the production of bakery products, fermentation can be performed by using yeast. Gas will produce the leavening action in the product whereas alcohol is evaporated during baking process.
- **Cereal milling.** Milling is the process of grinding the cereal grain with the objective of separating the floury edible endosperm from the other kernel portions.

- **Convection oven.** A convection oven is an oven that has fans to circulate air around food. Whereas conduction is a static process, convection is a more efficient method of heat transfer, since it heats food faster resulting in more even cooking and shorter cooking times. Convection oven also tends to increase the browning of food by concentrating more heat on the food's outer surface.
- **Deck oven.** Deck ovens or traditional oven use conduction heat to bake products. Conduction heating is a static process in which heat travels directly from a hot stone or deck, to the product being baked.
- **Steam oven.** Steam ovens boil water from an inbuilt reservoir to create a cloud of steam within the oven chamber. Significant benefits are associated with this procedure such as higher retention of flavor, texture, and color, less losses of vitamins and minerals and less drying. However, it also presents certain disadvantages, since steam oven is unable to brown the outer skin of the baked product.

MINI DICTIONARY

Bakery products Bakery products comprise all food products in which cereal flour is the basic ingredient and which have been subjected to the process of baking.

Baking agents Baking agents or leavening agents are substances incorporated into the dough that cause or incorporate gases in the baked products, which must be retained in the product to promote an increase in volume and produce shape and crumb texture.

Baking Baking is a process of cooking food where dry heat is applied directly on the food product generally in an oven or heating appliance, reaching temperatures as high as 260 °C.

EFSA The European Food Safety Authority (EFSA) is the keystone of European Union (EU) risk assessment regarding food and feed safety.

Yeast Yeast (*Saccharomyces cerevisae*) is responsible for the fermentation of the dough, by which sugars are transformed into carbon dioxide gas and alcohol.

SUMMARY POINTS

- Acrylamide is formed in bakery products during the thermal process as a consequence of the Maillard reaction, where reducing sugars react with asparagine amino acid at temperatures primarily above 120 °C.
- Bakery products together with potato products and coffee have been considered the most important contributors to acrylamide intake.
- Acrylamide formation in the bakery product depends on the recipe and composition of the foodstuff and the technological process to which the product is subjected.
- The content of asparagine, sugars, and other ingredients such as baking agents, salt, water, fats and oils, and other additives, will condition the acrylamide formation during the baking process.
- Technological factors such as time of fermentation, pH, moisture, physical parameters, time/temperature, and heat transfer will also affect the final levels of acrylamide in the bakery products.

REFERENCES

[1] Lai HM, Lin TC. Bakery products: science and technology. In: Hui HY, editor. Bakery products. Science and technology. Iowa (USA): Blackwell Publishing Professional; 2006. p. 3–65.

[2] Friedman M. Chemistry, biochemistry, and safety of acrylamide. A review. J Agric Food Chem 2003;51(16):4504–26.

[3] Amrein TM, Schönbächler B, Escher F, et al. Acrylamide in gingerbread: Critical factors for formation and possible ways for reduction. J Agric Food Chem 2004;52(13):4282–8.

[4] Claus A, Mongili M, Weisz G, et al. Impact of formulation and technological factors on the acrylamide content of wheat bread and bread rolls. J Cereal Sci 2008;47(3):546–54.

[5] Levine RA, Smith RE. Sources of variability of acrylamide levels in a cracker model. J Agric Food Chem 2005;53(11):4410–6.

[6] Gökmen V, Açar ÖÇ, Köksel H, et al. Effects of dough formula and baking conditions on acrylamide and hydroxymethylfurfural formation in cookies. Food Chem 2007;104(3):1136–42.

[7] Capuano E, Ferrigno A, Acampa I, et al. Characterization of the Maillard reaction in bread crisps. Eur Food Res Technol 2008;228(2):311–9.

[8] Friedman M, Levin CE. Review of methods for the reduction of dietary content and toxicity of acrylamide. J Agric Food Chem 2008;56(15):6113–40.

[9] Matthys CM, Bilau Y, Govaert Y, et al. Risk assessment of dietary acrylamide intake in Flemish adolescents. Food Chem Toxicol 2005;43(2):271–8.

[10] Biedermann M, Grundböck F, Fiselier K, et al. Acrylamide monitoring in Switzerland, 2007–2009: results and conclusions. Food Addit Contam 2010;27(10):1352–62.

[11] EFSA (European Food Safety Authority). Update on acrylamide levels in food from monitoring years 2007 to 2010. EFSA J 2012;10(10):2938. http://www.efsa.europa.eu/efsajournal. [accessed 31.10.14].

[12] El-Ziney MG, Al-Turki AA, Tawfik MS. Acrylamide status in selected traditional Saudi foods and infant milk and foods with estimation of daily exposure. Am J Food Technol 2009;4(5):177–91.

[13] Pugajeva I, Zumbure L, Melngaile A, et al. Determination of acrylamide levels in selected foods in Latvia and assessment of the population intake. In: Foodbalt conference proceedings. 2014. p. 111–6.

[14] Mojska H, Gielecinska I, Szponar L, et al. Estimation of the dietary acrylamide exposure of the Polish population. Food Chem Toxicol 2010;48(8–9):2090–6.

[15] Svensson K, Abramsson L, Becker W, et al. Dietary intake of acrylamide in Sweden. Food Chem Toxicol 2003;41(11):1581–6.

[16] Konings EJM, Baarsb AJ, van Klaverenc JD, et al. Acrylamide exposure from foods of the Dutch population and an assessment of the consequent risks. Food Chem Toxicol 2003;41(11):1569–79.

[17] Eerola S, Hollebekkers K, Hallikainen A, et al. Acrylamide levels in Finnish foodstuffs analysed with liquid chromatography tandem mass spectrometry. Mol Nutr Food Res 2007;51(2):239–47.

[18] Hamlet CG, Sadd PA, Liang L, et al. Exploiting processing conditions to reduce acrylamide in cereal-based foods. Report No. C021. http://www.foodbase.org.uk/results.php?f_report_id=145; 2007. [accessed 30.10.14].

[19] Cressey P, Thomson B, Ashworth M, et al. Acrylamide in New Zealand food and updated exposure assessment. MAF Technical Paper No: 2011/19. Risk Assessment Studies. Report No 43. http://www.foodsafety.govt.nz/elibrary/industry/acrylamide-in-nz-food-updated-exposure-assessment.pdf; 2012 [accessed 3.11.14].

[20] Rufián-Henares JA, Arribas-Lorenzo G, Morales FJ. Acrylamide content of selected Spanish foods: survey of biscuits and bread derivatives. Food Addit Contam 2007;24(4):343–50.

[21] Normandin L, Bouchard M, Ayotte P, et al. Dietary exposure to acrylamide in adolescents from a Canadian urban center. Food Chem Toxicol 2013;57:75–83.

[22] Centre for Food Safety of the Food and Environmental Hygiene Department. Risk assessment studies Report No. 43 Dietary exposure to acrylamide of Hong Kong adult population, http://www.cfs.gov.hk/english/programme/programme_rafs/files/programme_rafs_fc_01_25.pdf; 2012. [accessed 11.11.14].

[23] Saleh SI, El-Okazy AM. Assessment of the mean daily dietary intake of acrylamide in Alexandria. J Egypt Public Health Assoc 2007;82(3–4):331–45.

[24] Sirot V, Hommet F, Tard A, et al. Dietary acrylamide exposure of the French population: results of the second French Total Diet Study. Food Chem Toxicol 2012;50(3–4):889–94.

[25] Ahrné L, Claes-Göran A, Floberg P, et al. Effect of crust temperature and water content on acrylamide formation during baking of white bread: steam and falling temperature baking. LWT Food Sci Technol 2007;40(10):1708–15.
[26] EFSA (European Food Safety Authority). Results on acrylamide levels in food from monitoring years 2007–2009. EFSA J 2011;9(4):2133. http://www.efsa.europa.eu. [accessed 31.10.14].
[27] European Commission. Commission Recommendation of 8 November 2013 on investigations into the levels of acrylamide in food. 12-11-13 Off J Eur Union 2013;L301:315–7, http://www.boe.es/doue/2013/301/L00015-00017.pdf. [accessed 29.10.14].
[28] Codex Alimentarius. Code of Practice for the Reduction of acrylamide in foods. Rome (Italy), http://www.codexalimentarius.net/download/standards/11258/CXP_067e.pdf; 2009. [accessed 17.11.14].
[29] Lineback DR, Goughlin JR, Stadler RH. Acrylamide in foods: a review of the science and future considerations. Annu Rev Food Sci Technol 2012;3:15–35.
[30] FDE (Food and Drink Europe). Food drink Europe acrylamide toolbox 2013. http://ec.europa.eu/food/food/chemicalsafety/contaminants/toolbox_acrylamide_201401_en.pdf; 2013. [accessed 29.10.14].
[31] Surdyk N, Rosén J, Andersson R, et al. Effects of asparagine, fructose and baking conditions on acrylamide content in yeast-leavened wheat bread. J Agric Food Chem 2004;52(7):2047–51.
[32] Claus A, Schreiter P, Weber A, et al. Influence of agronomic factors and extraction rate on the acrylamide contents in yeast-leavened breads. J Agric Food Chem 2006;54(23):8968–76.
[33] Springer M, Fischer T, Lehrack A, et al. Acrylamidbildung in Backwaren. Getreide Mehl Brot 2003;57:274–8.
[34] Muttucumaru N, Halford NG, Elmore JS, et al. The formation of high levels of acrylamide during the processing of flour derived from sulfatedeprived wheat. J Agric Food Chem 2006;54(23):8951–5.
[35] Mustafa A, Anderson R, Kamal-Eldin A, et al. Fortification with free amino acids affects acrylamide content in yeast leavened bread. In: Preedy, Watson R, Patel VB, editors. Flour and breads and their Fortification in health and disease prevention. London (UK): Academic Press; 2011. p. 325–36.
[36] Graf M, Amrein TM, Graf S, et al. Reducing the acrylamide content of a semi-finished biscuit on industrial scale. LWT Food Sci Technol 2006;39(7):724–8.
[37] Vass M, Amrein TM, Schönbächler B, et al. Ways to reduce the acrylamide formation in cracker products. Czech J Food Sci 2004;22(special issue):19–21.
[38] Taeymans D, Wood J, Ashby P, et al. A review of acrylamide: an industry perspective on research, analysis, formation, and control. Crit Rev Food Sci Nutr 2004;44(5):323–47.
[39] Piloty M, Baltes W. Investigations on the reaction of amino acids with R-dicarbonyl compounds. I. Reactivity of amino acids in the reaction with R-dicarbonyl compounds. Z Lebensm Unters Forsch 1979;168:368–73.
[40] Marconi O, Bravi E, Perretti G, et al. Acrylamide risk in food products: the shortbread case study. Anal Methods 2010;2(11):1686–91.
[41] Beck M, Jekle M, Becker T. Impact of sodium chloride on wheat flour dough for yeast-leavened products. I. Rheological attributes. J Sci Food Agric 2012;92(3):585–92.
[42] Hutton T. Sodium: technological functions of salt in the manufacturing of food and drink products. Br Food J 2002;104(2):126–52.
[43] Van Der Fels-Klerx HJ, Capuano E, Nguyen HT, et al. Acrylamide and 5-hydroxymethylfurfural formation during baking of biscuits: NaCl and temperature–time profile effects and kinetics. Food Res Int 2014;57:210–7.
[44] Zhou W, Therdthai N, Hui YH. Introduction to baking and bakery products. In: Zhou W, editor. Bakery products science and technology. Iowa (USA): Wiley-Blackwell; 2014. p. 3–65.
[45] Capuano E, Oliviero T, Açar Ö, et al. Lipid oxidation promotes acrylamide formation in fat-rich model systems. Food Res Int 2010;43:1021–6.
[46] Arribas-Lorenzo G, Fogliano V, Morales FJ. Acrylamide formation in a cookie system as influenced by the oil phenol profile and degree of oxidation. Eur Food Res Technol 2009;229(1):63–72.
[47] Zamora R, Gallardo E, Hidalgo FJ. Strecker degradation initiated by 2,4-decadienal or methyl 13- oxooctadeca-9,11-dienoate in model systems. J Agric Food Chem 2007;55(4):1308–14.
[48] Capuano E, Ferrigno A, Acampa I, et al. Effect of flour type on Maillard reaction and acrylamide formation during toasting of bread crisp model systems and mitigation strategies. Food Res Int 2009;42(9):1295–302.

[49] Zhang Y. Study on reduction of acrylamide in fried bread sticks by addition of antioxidant of bamboo leaves and extract of green tea. Asia Pac J Clin Nutr 2007;16(Suppl. 1):131–6.
[50] Hedegaard RV, Granby K, Frandsen H, et al. Acrylamide in bread: effect of prooxidants and antioxidants. Eur Food Res Technol 2008;227(2):519–25.
[51] Açar OC, Gökmen V. Investigation of acrylamide formation on bakery products using a crust-like model. Mol Nutr Food Res 2009;53(12):1521–5.
[52] Stadler RH. The formation of acrylamide in cereal products and coffee. In: Skog K, Alexander J, editors. Acrylamide and other hazardous compounds in heat-treated foods. Cambridge (UK): Woodhead Publishing; 2006. p. 23–40.
[53] Gökmen V, Senyuva HZ. Effects of some cations on the formation of acrylamide and furfurals in glucose–asparagine model system. Eur Food Res Technol 2007;225(5):815–20.
[54] Quarta B, Anese M. The effect of salts on acrylamide and 5-hydroxymethylfurfural formation in glucose-asparagine model solutions and biscuits. J Food Nutr Res 2010;49(2):69–77.
[55] CIAA. The CIAA acrylamide "toolbox," rev 9. http://www.ciaa.be; 2006. [accessed 14.11.14].
[56] Elder VA, Fulcher JG, Leung H, et al. Method for reducing acrylamide in thermally processed foods. Patent US20040058045; 2004.
[57] Sadd PA, Hamlet CG, Liang L. Effectiveness of methods for reducing acrylamide in bakery products. J Agric Food Chem 2008;56(15):6154–61.
[58] Kukurová K, Ciesarová Z, Bednáriková A, et al. Effect of inorganic salts on acrylamide formation in cereal matrices. Czech J Food Sci 2009;27:S425–8.
[59] Mestdagh F, Maertens J, Cucu T, et al. Impact of additives to lower the formation of acrylamide in a potato model system through pH reduction and other mechanisms. Food Chem 2008;107(1):26–31.
[60] Brathen E, Kita A, Knutsen SH, et al. Addition of glycine reduces the content of acrylamide in cereal and potato products. J Agric Food Chem 2005;53(8):3259–64.
[61] Fink M, Andersson R, Rosén J, et al. Effect of added asparagine and glycine on acrylamide content in yeast-leavened bread. Cereal Chem 2006;83(2):218–22.
[62] Flückiger R, Salih E. Method to limit acrylamide in heated foods. Patent WO2006017526; 2006.
[63] Kumar NSM, Shimray CA, Indrani D, et al. Reduction of acrylamide formation in sweet bread with L-asparaginase treatment. Food Bioprocess Tech 2014;7(3):741–8.
[64] Anese M, Quarta B, Peloux L, et al. Effect of formulation on the capacity of L-asparaginase to minimize acrylamide formation in short dough biscuits. Food Res Int 2011;44(9):2837–42.
[65] Fredriksson H, Tallving J, Rosen J, et al. Fermentation reduces free asparagine in dough and acrylamide content in bread. Cereal Chem 2004;81(5):650–3.
[66] Elmore S, Koutsidis G, Dodson AT, et al. Measurement of acrylamide and its precursors in potato, wheat, and Rye model systems. J Agric Food Chem 2005;53(4):1286–93.
[67] Mustafa A, Andersson R, Roseän J, et al. Factors influencing acrylamide content and color in rye crisp bread. J Agric Food Chem 2005;53(15):5985–9.
[68] Açar OC, Gökmen V. A new approach to evaluate the risk arising from acrylamide formation in cookies during baking: total risk calculation. J Food Eng 2010;100(4):642–8.
[69] Mogol BA, Gökmen V. Mitigation of acrylamide and hydroxymethylfurfural in biscuits using a combined partial conventional baking and vacuum post-baking process: preliminary study at the lab scale. Innov Food Sci Emerg Tech 2014;26:265–70.
[70] Isleroglu H, Kemerli T, Sakin-Yilmazer M, et al. Effect of steam baking on acrylamide formation and browning kinetics of cookies. J Food Sci 2012;77(10):257–63.
[71] Sakin-Yilmazer M, Kemerli T, Isleroglu H, et al. Baking kinetics of muffins in convection and steam assisted hybrid ovens (baking kinetics of muffin). J Food Eng 2013;119(3):483–9.
[72] Anese M, Sovrano S, Bortolomeazzi R. Effect of radiofrequency heating on acrylamide formation in bakery products. Eur Food Res Technol 2008;226(5):1197–203.

CHAPTER 8

Acrylamide in Fried Potato Products

Pieternel Luning[1], Maimunah Sanny[2]
[1]Department of Agrotechnology and Food Sciences, Wageningen University, Wageningen, The Netherlands; [2]Faculty of Food Science and Technology, Putra University, Serdang, Malaysia

ACRYLAMIDE IN FOOD AND ITS TOXICITY

Since ancient times, foods have been heat treated to modify and to preserve its organoleptic and nutritional properties for consumption. However, heat treatment can also generate undesired effects that may exhibit mutagenic and carcinogenic properties, such as heterocyclic amines, 5-hydroxymethylfurfural, and polycyclic aromatic hydrocarbons [1–4]. Acrylamide has been added to the list of food-borne toxicants since 2002 [5,6]. Acrylamide is used worldwide to synthesize polyacrylamide and applied as a soil conditioner, in wastewater treatment, in cosmetics, paper, and textile industries [7]. Acrylamide is neurotoxic in animals and humans, and it has shown to be a reproductive toxicant, genotoxic, and carcinogen in rodents [4,7,8]. Acrylamide is classified by the International Agency for Research on Cancer [9] as a probable human carcinogen on the basis of positive results with bioassays in rodents, supported by evidence that acrylamide is metabolized in mammalian tissues into glycidamide [10], which is reactive toward DNA (genotoxic), and thus can form several DNA adducts, which is believed to be the key process in the carcinogenicity of acrylamide [4,7,11].

Extensive epidemiological studies have been conducted to assess the cancer risk because of dietary acrylamide intake in human subjects. Most of these studies found no association between estimated dietary acrylamide intake and increased cancer risk [12–15]. However, more recent studies reported statistically significant relations between estimated dietary intake of acrylamide and risk on some cancer types for some populations [16–18].

ACRYLAMIDE FORMATION

The major mechanistic pathway for acrylamide formation in foods is the Maillard reaction with free asparagine as main precursor [19–22]. Asparagine can be thermally decomposed by deamination and decarboxylation, but the yield of acrylamide from asparagine is much higher when a carbonyl source is present [19,23,24], which explains the high level of acrylamide detected in foods rich in reducing sugars and free asparagine such as fried potatoes [19,23,24]. Moreover, acrylamide can be generated by deamination of

Acrylamide in Food
http://dx.doi.org/10.1016/B978-0-12-802832-2.00008-5

3-aminopropionamide (3-APA) [25], which can be formed by enzymatic decarboxylation of free asparagine and can yield acrylamide upon heating even in absence of a carbonyl source [26,27]. The sugar–asparagine adduct *N*-glycosylasparagine is a direct precursor of acrylamide [20]. High temperature and reduction in water content in food-processing systems enhance the formation of the *N*-glycoside [20], and the ability to form an open-chain of *N*-glycosylasparagine favors the formation of the decarboxylated Amadori product of asparagine. This compound can be formed under mild conditions but requires elevated temperatures to cleave the carbon–nitrogen covalent bond and to produce acrylamide [23]. Different authors showed that acrylamide formation is more sensitive to temperature changes than browning at low-moisture contents [28,29]. Therefore, lowering of the frying temperature toward the end of the frying process decreases acrylamide formation more strongly as compared to browning [29].

Furthermore, acrylamide is an intermediate of the Maillard reaction rather than an end product because it is subjected to a degradation reaction [30]. The level of acrylamide in heated foodstuffs, therefore, is the net result of complex reactions leading to the formation and degradation of this compound [31,32].

ACRYLAMIDE IN FRIED POTATO PRODUCTS

Acrylamide is found in a wide range of heat-treated foods prepared commercially or cooked in the home including bread, crisp bread, bakery wares, breakfast cereals, potato products, chocolate, and coffee [33]. High levels of acrylamide have been detected in fried potato products, such as French fries and potato crisps [34–36]. The Joint FAO/WHO Expert Committee of Food Additives (JECFA) and other risk assessment bodies have recommended that appropriate efforts to reduce acrylamide concentrations in food should continue [10]. The European Commission issued "indicative" levels for acrylamide in food in early 2011 [37], for French fries and potato crisps these levels have been set at 600 and 1000 μg/kg, respectively [37]. Investigations are recommended if acrylamide levels exceed prescribed indicative values [35].

The EFSA [38] released a report on the update on acrylamide levels in a variety of foods from the monitoring years 2007–2010 (Table 1). The study reported the analysis of 13,162 samples in 25 countries over the 4-year period, covering 10 different food categories. The trend analysis indicated only few changes in acrylamide levels from 2007 to 2010. In the different categories of food, 6–17 % of the samples exceeded the indicative values for acrylamide, the highest proportion of samples exceeding the level was in the subcategory "potato crisps," followed by "French fries, sold as ready-to-eat" [38]. However, Powers and coauthors [39] reported a significant downward trend for mean levels of acrylamide in potato crisps, ranging from 763 ± 91.1 μg/kg in 2002 to 358 ± 2.5 μg/kg in 2011, which is a substantial decrease of $53\% \pm 13.5\%$ (Table 2). They used a larger dataset from 20 European countries over the years 2002–2011.

Table 1 Number of observations (*n*), mean, and 95% quantile (Q95) for acrylamide in samples of fried potato products from 2007 to 2010 reported by EFSA

	2007			2008			2009			2010		
Food category	*n*	Mean (μg/kg)	Q95 (μg/kg)	*n*	Mean (μg/kg)	Q95 (μg/kg)	*N*	Mean (μg/kg)	Q95 (μg/kg)	*n*	Mean (μg/kg)	Q95 (μg/kg)
French fries, sold as ready-to-eat	648	356	1114	563	277	767	501	342	900	256	338	1024
French fries from fresh potatoes	204	237	606	301	251	679	158	278	810	196	325	921
French fries from potato dough	–	–	–	1	406	406	0	–	–	1	150	150
Unspecified French fries	444	410	1378	261	306	830	343	371	1226	59	382	1377
Potato crisps	293	551	1589	532	580	1853	414	639	2125	242	675	2080
Potato crisps from fresh potatoes	100	570	1395	104	541	1460	132	619	1833	173	758	2193
Potato crisps from potato dough	26	402	928	39	361	788	23	409	1350	19	435	1000
Unspecified potato crisps	167	564	1673	389	612	1900	259	670	2336	50	481	1389
Precooked French fries/potato products for home cooking	137	306	1144	254	223	777	249	270	1000	117	331	1159
Deep-fried fries	32	395	1450	34	229	710	44	220	627	64	198	681
Unspecified potato products for home cooking	97	272	796	99	213	763	134	253	914	25	270	928

Adapted from EFSA [38].

Table 2 Number of observations (*n*), mean, standard error of mean (SE), variance, minimum, median (Q50), 95% quantile (Q95), and maximum levels for acrylamide in samples of potato crisps from 2002 to 2011 reported by Powers and coauthors [39]

Year	*n*	Mean (µg/kg)	SE	Variance	Minimum (µg/kg)	Median (µg/kg)	Q95 (µg/kg)	Maximum (µg/kg)
2002	42	763	91.1	348,149	70	493	2080	2500
2003	136	573	27.3	101,440	160	475	1118	2080
2004	321	624	26.6	227,186	130	490	1580	4450
2005	123	621	21.3	103,875	100	565	1210	1780
2006	1151	577	11.9	163,776	22	460	1350	2830
2007	3206	570	7.0	157,220	30	476	1270	5900
2008	5692	472	5.1	145,005	30	367	1170	4300
2009	6493	500	5.8	215,229	40	350	1400	6000
2010	10,971	435	4.1	182,087	22	330	1020	12,000
2011	12,213	258	2.5	74,727	17	280	870	3090

They claimed that their dataset more likely represents the actual situation compared to the EFSA dataset. Nevertheless, even in 2011, a small proportion of samples still contained too high levels of acrylamide, with 0.2% exceeding 2000 μg/kg.

VARIABILITY IN ACRYLAMIDE LEVELS

Dietary intake studies observed large differences in acrylamide levels both between single foodstuffs (different brands), within particular food categories, and within batches of products processed under the same conditions [40–43]. Similarly, considerable differences have been observed even on repetition of a specific mitigation measure under controlled conditions [44,45]. Different authors stated that the large variation in acrylamide levels complicates the estimation of actual dietary exposure to acrylamide [4,46]. Furthermore, Grob [47] showed that high acrylamide levels combined with high degree of consumption could increase the dietary exposure for an individual by a factor of 5, from an average exposure of 2.8 μg/day. The FAO/WHO [45] identified the observed significant variation and emphasized that the causes of variation need to be understood to develop effective measures.

FACTORS CONTRIBUTING TO VARIATION IN ACRYLAMIDE LEVEL

Due to the high risk on acrylamide formation and wide consumption of French fries, its preparation in Food Service Establishments (FSE) should be well controlled to reduce dietary exposure (Sanny et al. [83]). Major factors influencing acrylamide formation in French fries preparation are glucose and fructose, and actual frying conditions (time and temperature). Table 3 shows a compilation of literature data indicating the possible contribution of variable glucose and fructose and process conditions to variation in acrylamide levels, at the different steps in French fries production. A typical flow chart of French fries preparation in FSE is shown in Figure 1.

Receiving Supplied Materials

The first step of French fries production is the receipt of usually frozen par-fried potato strips [48]. For potatoes, a strong correlation between acrylamide formation and available reducing sugars was found [49–52], and they were indicated as the limiting factor for acrylamide formation in French fries [53–55]. Since the levels of reducing sugars vary widely among potato varieties, it affects the variation in the actual formation of acrylamide. To illustrate, Amrein and coauthors [49] found that the level of glucose ranged from 0.04 to 2.7 g/kg, with the lowest values found in samples of variety Lady Claire and Marlene, and the highest in Naturella and Nicola. Levels of fructose varied similarly but were generally lower than of glucose, again with the highest values in Naturella and Nicola. Some varieties are inherently more prone to higher levels of

Table 3 Technological factors and conditions contributing to variation in acrylamide level

Steps in French fries production	Factors and conditions affecting variation in acrylamide level	Actual product properties and process conditions	Observed acrylamide levels	Reference(s)
Receiving supplied materials	Initial level of glucose and fructose in par-fried potato strips	• Glucose level ranged from 0.04 to 2.7 g/kg, a 68-fold difference between 17 potato cultivars (the lowest value in Lady Claire and Marlene, and the highest in Naturella and Nicola) • Fructose level ranged from 0.033 to 1.537 g/kg, a 47-fold difference between 17 potato cultivars (the lowest value in Markies and Marlene, and the highest values in Naturella and Nicola)	Grated potato (placed in a preheated oven at 120 °C for 40 min) of cultivar Nicola had the highest potential of acrylamide formation (maximum, 2020 μg/kg), followed by Charlotte (maximum, 1700 μg/kg), while the cultivar Panda exhibited the lowest mean potential (80 μg/kg). The difference between the extremes corresponds to a factor of 25	Amrein et al. [49]
		• Glucose level ranged from 1.1 to 34.7 mmol/kg fresh weight, a 31.3-fold difference between 31 potato cultivars (the lowest value in Jelli and the highest in Yukon Gold B) • Fructose level ranged from 1.73 to 33.6 mmol/kg fresh weight, a 19.5-fold difference between 31 cultivars (the lowest value in Fingerling Ozette and the highest in red)	No acrylamide formation was analyzed	Vivanti et al. [60]
		• Glucose level ranged from 1.8 to 9.8 g/kg, a 5.4-fold difference between five potato cultivars (the lowest value in Saturna in year 2005 and the highest in Bintje in year 2004) • Fructose level ranged from 1.1 to 6.2 g/kg, a 5.6-fold difference between five potato cultivars (the lowest value in Saturna in year 2005 and the highest in Saturna in year 2006)	Potato chips of cultivar Bintje had the highest potential of acrylamide formation (maximum, 7600 μg/kg), while the cultivar SW 91 102 exhibited the lowest mean potential (1800 μg/kg). The difference between the extremes corresponds to a factor of 4.2	Viklund et al. [52]

		• Glucose level ranged from 2.22 to 48.33 mol/kg, a 21.8-fold difference between five potato cultivars (the lowest value in Saturna and the highest in Estima) • Fructose level ranged from 2.22 to 37.78 mol/kg, a 17-fold difference between five potato cultivars (similar to glucose, the lowest value in Saturna, and the highest in Estima)	Potato chips of cultivar Estima had the highest potential of acrylamide formation (4.12 log[acrylamide]), while the cultivar Saturna exhibited the lowest mean potential (3.3 log[acrylamide]). The difference between the extremes corresponds to a factor of 6.6	Williams [51],
		• The mean glucose concentration in Lady Claire in November 2009, for example, was 1.17 mmol/kg, whereas in King Edward it was 17.38 mmol/kg	Acrylamide formed in heated flour from Lady Claire tubers was 1660 µg/kg, whereas in flour from Pentland Dell tubers it reached 7110 µg/kg. The difference between the extremes corresponds to a factor of 4.2.	Halford et al. [109],
Storage and thawing	Not expected	No studies found	No studies found	
Frying	Frying time–temperature condition	Frying time–temperature conditions range: 7 min at 150 °C and 3.5 min at 190 °C	Acrylamide level of potato slices increased from 500 to 4500 µg/kg, a 9-fold difference	Pedreschi et al. [65]
		Frying time–temperature conditions range: 10 min at respectively 170 °C and 190 °C	Acrylamide level in French fries increased from 800 to 3700 µg/kg, a 5-fold difference	Matthäus et al. [61]
		Time–temperature conditions range: 22 min at 140 °C and 17 min at 180 °C	Acrylamide level in a glucose–asparagine model system increased 450–2900 µg/kg a 6.4-fold difference	Claeys et al. [32]
		• Time–temperature conditions range: 15 and 45 min at 140 °C • Time–temperature conditions range: respectively 2, 4, 9, and 45 min at 180 °C	• Acrylamide level in a glucose–asparagine ranged from 0.8 to 1.4 mmol/l • Acrylamide level in a glucose–asparagine model system increased from 0.85, 2.5, to 3, and decreased to 1.4 mmol/l (after 45 min)	Knol et al. [30]
		Time–temperature conditions range: respectively 1.5, 4, and 12 min at 160 °C (temperature constant after about 1.5–2.0 min frying at initial oil temperature of 180 °C)	Acrylamide level in the Lady Rosetta potato crisps increased from 0.2 to 12 and decreased to 10 mg/kg (after 12 min)	Knol et al. [71]

Continued

Table 3 Technological factors and conditions contributing to variation in acrylamide level—cont'd

Steps in French fries production	Factors and conditions affecting variation in acrylamide level	Actual product properties and process conditions	Observed acrylamide levels	Reference(s)
		Frying time–temperature conditions range: 3 min at 150 °C and 5 min at 190 °C	Acrylamide level in French fries increased from 1995 to 7079 μg/kg (3.3 − 3.85 log([acrylamide])), a 3.5-fold difference	Williams [51]
		Frying time–temperature conditions range: respectively 4, 5, 7 and 9 min at 120 °C	Acrylamide level in French fries increased from 50, 68, 208 and 830 μg/kg	Romani et al. [72]
		Frying time–temperature conditions range: from 5 to 60 min at 180 °C, a rapid increase between 10 and 20 min, reaching a maximum value between 20 and 35 min, followed by a slow linear decrease	Maximum levels for acrylamide in potato, rye, and whole wheat cakes were 6800, 3200, and 1100 μg/kg, respectively	Elmore et al. [73]
	Portion size	Portion-frying time–temperature drop conditions range: 30 g potato strips to 2 l of oil and a temperature drop to 168 °C (from an initial temperature of 170 °C) required frying time of 3:30 min, 100 g potato strips and a temperature drop to 158 °C required frying time of 5:15 min, 150 g potato strips and a temperature drop to 144 °C required frying time of 6:30 min, and 200 g potato strips and a temperature drop to 136 °C required frying time of 8 min.	Acrylamide level in French fries decreased from 75, 45, 35, and 25 μg/kg	Fiselier et al. [74]
		Portion size–temperature drop conditions range (at a constant frying time and an initial temperature of 180 °C): 50 g potato strips to 1 l of oil (5%) and 100 g potato strips (10%) and a temperature drops by 20–35 °C	Acrylamide level in French fries decreased from 710 to 530 μg/kg, a 1.3-fold difference	Grob et al. [70]
Holding	Not expected	No studies found	No studies found	

Updated from Sanny et al. [83]

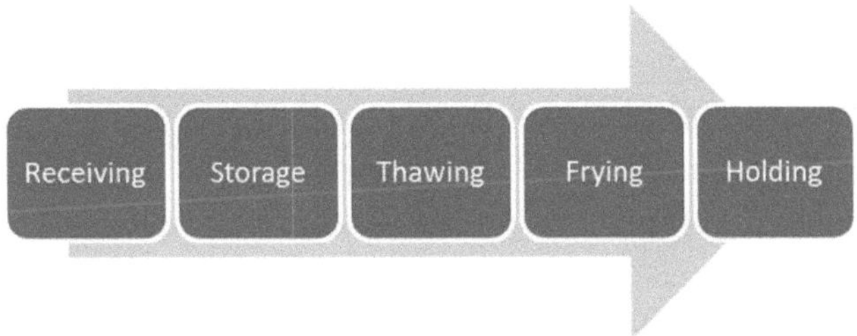

Figure 1 Steps in the preparation of French fries in FSE.

reducing sugars than others, and should therefore be avoided if at all possible for high-temperature cooking processes because they are a potential source of high and variable acrylamide levels [56]. However, considerable differences in reducing sugars among potatoes of the same cultivar suggested that other factors, such as storage, may have an even stronger influence on variation [57]. Cooling of potato tubers below 8–10 °C causes a large increase in reducing sugars that are released from starch, the well-known low-temperature sweetening [58,59]. This effect can be minimized by reconditioning after storage [60]. The levels of reducing sugars may vary over storage time as well, and its level at a specific time may not always reflect levels present at earlier or later time periods [60]. For example, a mean value of 0.82 g/kg fresh weight (FW) was found after harvesting and may increase to 1.26 g/kg FW after storage at 8 °C [61]. Frozen par-fried potato strips can thus largely vary in reducing sugars with a considerable variation between potatoes, potentially contributing to high and variable acrylamide levels if no specific measures are taken earlier in the chain.

Storage of Frozen Par-Fried Potato Strips

The frozen par-fried potato strips are subsequently stored at −18 °C until use. At frozen state, the molecular mobility strongly reduces and rates of all chemical and biochemical reactions slow down. In fact, almost no chemical activity was observed [62]. Therefore, this step does not seem to contribute to variation.

Thawing

The sacks with frozen French fries are removed from the freezer storage, prior to the deep-frying process. In practice, the sacks of French fries usually stand for 2 or more hours at room temperature, until thawed to a point where they can be deep fried [63]. To our knowledge, no studies have been performed on the effect of thawing on changes in glucose and fructose, and or acrylamide formation.

Frying

Deep frying of par-fried potato strips is the most important processing step. The formation of acrylamide starts at temperatures slightly above 120 °C, and reaches a maximum around 170–180 °C, depending on the model system studied and the duration of heating [6,19,64]. Pedreschi and coauthors [65] reported that the acrylamide level of potato slices was about 500 µg/kg after frying for 7 min at 150 °C as opposed to about 4500 µg/kg after frying for 3.5 min at 190 °C. Matthaus and coauthors [66] found that the level of acrylamide in French fries was about 800 µg/kg after frying for 10 min at 170 °C, and it increased to about 3700 µg/kg after 10 min at 190 °C. However, at higher temperatures (180 and 200 °C), the drastic increase was followed by a fast decrease due to degradation of acrylamide [30,67,68]. The increase in acrylamide level with increasing frying temperature followed an exponential function [29,61,69,70], which implies that small deviations in frying temperature can have large consequences for acrylamide formation. Therefore, frying temperature is expected to be an important factor that determines the extent of variation in acrylamide levels.

However, not only temperature is crucial, but also the time of frying is important, because both determine the kinetics of acrylamide formation [71]. Williams [51] found that most acrylamide formation occurs during the early stages of frying at the higher temperatures (175 °C), but sufficient quantities of precursors still remained for acrylamide formation at the early stage when frying at lower temperature (150 °C for 3 min). Romani and coauthors [72] observed that the increase of time became a key factor in acrylamide formation in French fries after around 4 min of frying when the temperature of potato surface reached 120 °C and the temperature of oil bath reached 140 °C. Elmore and coauthors [73] cooked potato flakes, whole meal wheat, and whole meal rye at 180 °C, from 5 to 60 min, and observed an exponential increase in acrylamide formation between 10 and 20 min, reaching a maximum value between 20 and 35 min, followed by a slow linear decrease. Above studies clearly demonstrated the profound effect of time as an important causal factor of variation, because minor deviations in frying time can result in large variation (and high values) of acrylamide levels in French fries.

Portion size is another factor that may contribute to variation, because it affects the actual temperature and time that is necessary to make French fries with optimal quality attributes (like crispness and flavor). According to Fiselier and coauthors [74], an isothermal frying at 160 °C is required, which can be achieved after a temperature drop from an initial frying temperature of 170–175 °C. However, bigger portion sizes will result in larger temperature drops and require longer frying times to achieve an optimum frying quality [70,74]. For example, Fiselier and coauthors [74] demonstrated that adding 30 g potato strips to 2 L of oil, dropped the temperature to 168 °C from an initial temperature of 170 °C. It required 3:30 min (210 s) before optimal quality of French fries was achieved and the fries contained 75 µg/kg acrylamide. When 150 g potato strips were added, temperature dropped to 144 °C, and it took 6:30 min (390 s) to obtain French fries with optimal quality, but acrylamide level was 35 µg/kg. Acrylamide level was reduced more than half, whereas the culinary quality was still good. The studies indicated that small deviations in portion size can result in variable acrylamide levels, and portion size seems therefore the factor contributing to variation.

Different authors have observed an increased acrylamide formation rate at low water content, whereas with increasing water content the reaction rate decreased. The increased acrylamide formation at lower water content, was explained by the temperature: moisture gradient within the food [29,75], which is dependent on reaction time and on distance from the center of the French fries [67,75]. As a result, the formation of acrylamide is concentrated at the crust region rather than core region [69,76].

Other studies demonstrated that the type of oil used in the frying process has no influence on acrylamide formation [51,61,77], and it is not affected by oil oxidation nor oil hydrolysis [78].

Holding

French fries are consumed hot and once the French fries are in the holding bin, they are distributed to the consumers as per their orders [63]. No systematic study on the effect of holding step on acrylamide formation has been reported [79], but it is unlikely that the holding step has an effect because the temperature is much below the temperature at which formation and degradation takes place.

Therefore, if no specific control measures are taken at FSE, it is expected that variable levels of reducing sugars in the par-fried potato strips at receipt and variable frying regimes (due to set temperature, time, and portion size) are important causes of variable and high acrylamide levels (Table 3).

ROLE OF FOOD HANDLERS IN ACRYLAMIDE FORMATION IN FRIED POTATOES PRODUCTS

Most studies on acrylamide formation focus on investigating the influence of technological factors on the formation of acrylamide [80,81]. However, the preparation of French fries in FSE is highly relying on food handlers and their impact on acrylamide should not be ignored [82,83]. Decisions of managers and food handlers (i.e., the managerial factors) at critical points of control in the preparation of French fries were systematically analyzed based on a "food quality decision model" [84]. The conceptual model in Figure 2 shows the possible technological and managerial causes of variation, how they can interact and contribute to variation in acrylamide level in French fries prepared in FSE [85]. Technological causes of variation include those that relate to the variation in product properties (product factor) and the variation in technological conditions, such as equipment and process conditions (process factor). Managerial causes of variation refer to variation in (daily) control decisions that aim at keeping the product and the process status within acceptable tolerances by measuring, comparing, and taking corrective actions if necessary. Similarly, it aims at keeping peoples' actual

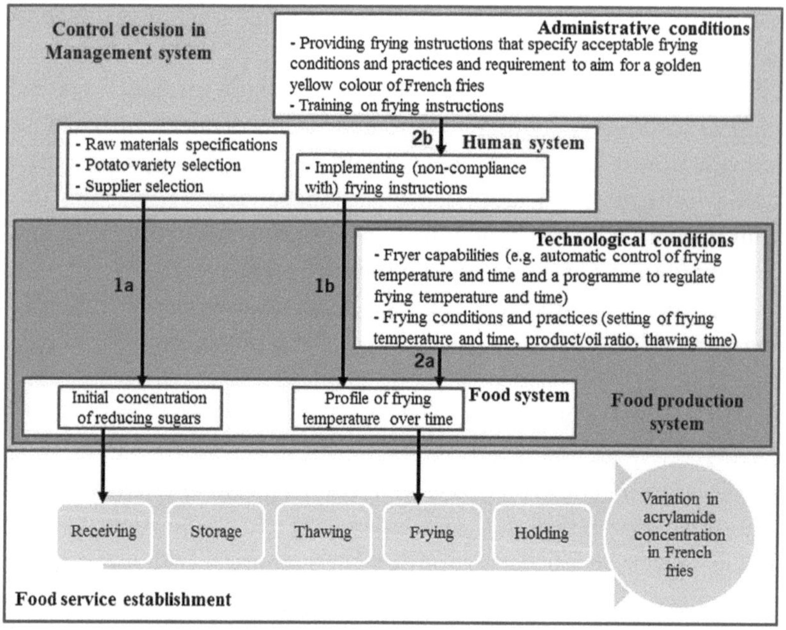

Figure 2 A conceptual model showing possible technological and managerial causes of variation in acrylamide level in French fries prepared in FSE: (1a) control decisions on food dynamics, (1b) control decisions on human dynamics, (2a) control decisions on technological conditions affecting food dynamics, and (2b) control decisions on administrative conditions affecting human dynamics [85].

behavior (e.g., executing certain tasks) within certain desired tolerances [84,86]. Managers are usually the owner of an FSE, and food handlers are the ones who do the actual frying in an FSE, they daily take control decisions. Managers typically take decisions on resources (also called resource control), such as recruiting people with adequate competences, purchasing appropriate equipment, selecting specific suppliers, providing procedures and instructions, and so on. Food handlers typically take decisions on out-of-tolerance situations of product properties and process conditions (also called product and process controls) [84]. Literature reviewing on FSE practices indicated that manager's decisions to control the appropriate supplies and suppliers are restricted because potatoes with low level of reducing sugars are not widely available [87]. Manager's decision to select appropriate fryers is also restraint because fryers normally used and available for FSE have no features for regulating a temperature profile that is needed to reduce acrylamide formation [74]. However, managers can decide to select fryers equipped with automatic control of frying temperature and time because this type of fryer is nowadays widely available and used in chain fast-food services [88,89]. Managers can also decide to provide procedures and instructions to support food handler's daily decisions. Procedures on standardized control of frying conditions, for example, require food handlers to avoid frying above 175 °C and to take corrective actions such as reducing frying time when frying smaller portion of frozen par-fried potato strips [44]. The actual compliance to food safety practices is, however, affected by various factors such as training, motivation, and adequate resources, and an appropriate management culture can limit the actual compliance to food safety practices [90,91]. Furthermore, daily decisions of food handlers in this type of (often) small and medium enterprises are expected to be less controlled because of time constraint as food handlers need to prepare and supply food to order rather than from stock [83,90,91]. These factors are expected to affect food handler's daily decisions on the setting of frying conditions and controlling them between acceptable tolerances. Food handler's variable and inadequate decisions in controlling the frying conditions could thus affect the variation in acrylamide levels. Nevertheless, researchers commonly dedicate causes of variation in acrylamide level to technological factors such as frying time–temperature regimes and levels of reducing sugars and amino acids [80,81].

An observational study in three types of FSE, namely chain fast-food services (CFS), institutional caterers (IC), and restaurants (R) found that the least variation in acrylamide level was found in French fries prepared in the CFS compared with the IC and the R, although the mean level of acrylamide among the three FSE types was not significantly different. The mean level of acrylamide was 231 μg/kg for the CFS, 254 μg/kg for the IC, and 354 μg/kg for the R. In addition, the highest level of acrylamide (1023 μg/kg) was found in the R type of FSE [89]. Comparing the frying practices among the three FSE types revealed that the setting of frying temperature of 177 °C for 2 min and 45 s of digitally controlled fryers in the CFS apparently resulted in a small range of actual frying temperature (170–177 °C) and time (150–165 s). Correspondingly, the lowest variation in acrylamide level (i.e., 150–392 μg/kg, a 2.6-fold difference) was found in the CFS type. In the R type, frying pans were used in which the temperature and time cannot be regulated. The food handlers visually inspected the oil to estimate its temperature to start frying and checked the color of French fries to end frying, which apparently resulted in a large range of actual frying temperature (148–215 °C) and time (199–694 s). Correspondingly, the largest variation in acrylamide levels (i.e., 152–1023 μg/kg, a 6.7-fold difference) was found in the R type. The study concluded that lack of standardized control of temperature and time (due to inadequate frying equipment) and variable frying practices of food handlers seem to contribute to large variation and high acrylamide level in French fries prepared in the R type compared with the other FSE types. Various authors discussed that acrylamide level increases exponentially with increasing frying temperature [29,61,69], which might explain why the large range of frying temperature and time as recorded in restaurants corresponded with the largest variation in acrylamide.

MEASURES TO MITIGATE ACRYLAMIDE FORMATION IN FRENCH FRIES PREPARED IN FSE

The final preparation of French fries in FSE is most critical to be controlled properly because acrylamide is formed toward the end of the frying process [74,92]. A number of control measures to mitigate acrylamide formation were proposed and tested in controlled laboratory conditions [87,93,94]. However, the results of these studies can differ from field studies [95,96], and they must be validated under actual food-processing conditions [44,97]. Although French fries are widely prepared in FSE, data on the effects of control measures in a real situation of French fries preparation are limited [97,98], compared to the data available for industrial production [99,100]. Sanny and coworkers [101] conducted a study to evaluate the effect of lowering reducing sugars level in par-fried potato strips (product factor) on the level and variation of acrylamide in French fries. French fries using par-fried potato strips with lower (LRS) and higher level of reducing sugars (HRS) were compared with the normal level (NRS). The food handlers in the CFS and the IC prepared French fries under standardized frying conditions, in which frying temperature and time of fryer were set at 177 °C and 2 min 45 s, respectively. The study found that the use of par-fried potato strips with lower levels of reducing sugars than the commonly used potato strips is an effective measure to reduce acrylamide levels in French fries. The lowest mean acrylamide level in French fries was found for the par-fried potato strips with LRS (322 μg/kg) and was *significantly* different ($p < 0.05$) from the strips with NRS (491 μg/kg) and HRS (928 μg/kg) [101]. A strong correlation between reducing sugars level and acrylamide level ($r = 0.758$; $n = 216$; $p < 0.05$) was found. However, the variation in reducing sugars level in the par-fried potato strips was not reflected in the variation in acrylamide level in French fries. The dominant effect of frying temperature–time above the effect of reducing sugars was attributed to the fact acrylamide is sensitive to temperature changes (at low moisture contents) and its level increases exponentially with increasing frying temperature [28,68,102].

Another intervention study was conducted to obtain insight into the effect of frying instructions on food handlers' control decisions (process factor) in the restaurant type of FSE, and the impact on the level and variation of acrylamide in French fries [103]. The frying instructions that specified the acceptable frying conditions and practices were provided and compliance with the frying instructions was observed. The study found that frying instructions supported food handlers' decisions to start frying when the oil temperature reached 175 °C, but an inconsistent effect of the instructions on the food handlers' decisions to end frying was observed. Although providing instructions did not result in a significant difference in the mean level of acrylamide in the restaurants as a group, data analysis for each restaurant showed that implementing frying instructions reduced the level of acrylamide in French fries. When instructions were strictly followed, the acrylamide level was significantly lowered. However, in case of noncompliance, even higher means of acrylamide levels were found compared with before instructions.

The study demonstrated the difficulty of managerial interventions as robust way to mitigate the risk of acrylamide.

CONCLUDING REMARKS

Quantifying the impact of food properties and technological conditions and the interference of food handlers in the preparation of French fries on the variation of acrylamide could lead to a broader insight into which factors contribute more to large variation and high level of acrylamide that are found in practice. The estimation of actual dietary exposure to acrylamide deals with ambiguity because of the variable level of acrylamide [104], but researchers often neglect its presence and they regularly consider variability as given fact [105,106]. The occurrence of high acrylamide levels that are reported in the literature [107,108] could be attributed to the lack of control over frying conditions in the food production chain, may be more than to the level of reducing sugars alone. In the assessment of acrylamide risk, the impact of people in preparing potato-based products should not be ignored as an important risk factor.

SYNOPSIS ANALYTICAL TECHNIQUE

The empirical studies described in this chapter used an acrylamide analysis method as described by Sanny and coauthors [89] and briefly summarized below.

For extraction, grounded samples of French fries were added to water containing 1000 ng 13C3-labeled acrylamide as the internal standard (final concentration = 100 ng/ml) were added. Standards contained acrylamide at concentrations of 10, 25, 50, 100, 250, 500, and 1000 ng/ml and isotopically labeled acrylamide at 100 ng/ml. The mixture was shaken and centrifuged, filtrated, and elute was collected. Elutes and standards were treated with bromination reagent overnight at 4 °C and excess bromide was decomposed by adding sodium thiosulfate solution. After phase separation, the lower aqueous layer was discarded and the organic phase was transferred into a centrifuge tube containing calcinated sodium sulfate anhydrous and centrifuged at 10,956 RCF (*g*) for 10 min. The liquid phase was decanted through glass wool. Pooled fraction was evaporated to dryness under vacuum at 40 °C. The residue was re-dissolved in ethyl acetate and triethylamine was added to convert 2,3-dibromopropionamide to 2-bromopropenamide. The resulting solution was transferred into an insert in an amber vial and stored in a freezer at −18 °C until analysis by gas chromatograph–time-of-flight-mass spectrometry (GC–TOF-MS).

The brominated sample extracts and calibration standards were analyzed on a GC–TOF-MS detector with positive electron impact ionization. The GC column was a capillary column (30 m, 0.25 mm i.d., 0.25 m film thickness), and the carrier gas was helium at 1.6 ml/min. Following injection, the column was held at 65 °C for 1 min, then programmed at 15 °C/min to 170 °C, 5 °C/min to 200 °C, followed by 40 °C/min to

250 °C, and held for 15 min at 250 °C. Injections were made in split-less mode with a purge activation time of 1.0 min and an injection temperature of 250 °C. The GC–TOF-MS interface transfer line was held at 260 °C. Ions monitored were m/z 70, 149, and 151 for 2-bromopropenamide, and m/z 110 and 154 for 2-bromo(13C3)propenamide.

Acrylamide in French fries samples was quantified using the ion at m/z 149 for 2-bromopropenamide, and the ion at m/z 154 for 2-bromo(13C3)propenamide. Other ions at m/z 70, 110, and 151 were considered only for confirmation purposes. A calibration graph was constructed by plotting peak area ratios (149/154) against the corresponding ratios of analyte amounts. Acrylamide concentrations in sample extracts were calculated from the calibration slope and intercept value. Calibration curves were linear ($r^2 > 0.999$). The limit of detection was 1 µg/kg and the recoveries were in the range between 102% and 110%. The intraday and interday precisions (expressed as the RSD) were 6.4% and 3.1%, respectively.

KEY FACTS

- The major mechanistic pathway for acrylamide formation in foods is the Maillard reaction with free asparagine as main precursor.
- Asparagine can thermally decompose by deamination and decarboxylation, but the yield of acrylamide from asparagine is much higher when a carbonyl source is present.
- Acrylamide is an intermediate of the Maillard reaction rather than an end product because it is subjected to a degradation reaction.
- Formation of acrylamide starts at temperatures slightly above 120 °C, and reaches its maximum around 170–180 °C, depending on model system studied and duration of heating.
- Increase in acrylamide with increasing frying temperature follows an exponential function.
- Minor deviations in frying time result in large variation and high values of acrylamide in French fries.
- Food handlers frying practices in FSE contribute to variability in acrylamide in French fries.
- Data on effects of control measures to mitigate risk of acrylamide formation in French fries production in real situations are still limited.

MINI DICTIONARY

Food Service Establishments (FSE) These are establishments that prepare and serve food for consumption on premises. FSE can be categorized into restaurants, home caterer (only using standard kitchen equipment), institutional caterer (food preparation and servicing in hospitals, schools, and so on), limited operation caterer (prepare and serve ready-to-eat food for off-premise consumption), commercial caterer (catering operation based in a store, food stand, mobile unit, pushcart, or other commercial establishment, or a home kitchen equipped with commercial equipment).

Food handler Any person employed in a food enterprise/establishment, who may be involved in the manufacturing, preparation, or packing of food for sale.

SUMMARY POINTS

- High acrylamide concentrations have been detected in foods rich in reducing sugars and free asparagine, such as fried potatoes.
- Many empirical studies found large variation in glucose and fructose concentrations in raw potatoes and in acrylamide levels in fried potato based products.
- Large variation in acrylamide complicates estimation of dietary exposure to acrylamide.
- Acrylamide studies mainly focused on technological factors under well-controlled laboratory conditions, but preparation of French fries in FSE is highly relying on food handlers.
- Major sources of variation in French fries preparation are variable levels of reducing sugars in par-fried potato strips and variable frying regimes.
- Lack of standardized control of temperature and time due to inadequate frying equipment and variable frying practices by food handlers contributed to large variation and high acrylamide level in French fries prepared; variation was higher in restaurants compared with chain fast-food services, and institutional caterers.
- Use of potatoes with low reducing sugar concentrations as mitigation measure lowered acrylamide concentration but variation remained high due to variable frying practice in FSE.
- Systematic introduction of frying protocols in FSE resulted in variable and high acrylamide concentrations due to variable compliance to producers, which is a common problem in food safety control.
- Acrylamide risk assessment should consider the impact of people in preparing potato-based products as an important risk factor.

REFERENCES

[1] Park JH, Penning TM. Polyaromatic hydrocarbons. In: Stadler RH, Lineback DR, editors. Process-induced food toxicants: occurence, formation, mitigation, and health risks. New Jersey: John Wiley & Sons, Inc; 2009.

[2] Turesky RJ. Heterocyclic aromatic amines. In: Stadler RH, Lineback DR, editors. Process-induced food toxicants: occurrence, formation, mitigation, and health risks. New Jersey: John Wiley & Sons, Inc.; 2009.

[3] van Boekel M, et al. A review on the beneficial aspects of food processing. Mol Nutr Food Res 2010;54(9):1215–47.

[4] Capuano E, Fogliano V. Acrylamide and 5-hydroxymethylfurfural (HMF): a review on metabolism, toxicity, occurrence in food and mitigation strategies. LWT – Food Sci Technol 2011;44(4):793–810.

[5] Tareke E, et al. Acrylamide: a cooking carcinogen? Chem Res Toxicol 2000;13(6):517–22.

[6] Tareke E, et al. Analysis of acrylamide, a carcinogen formed in heated foodstuffs. J Agric Food Chem 2002;50(17):4998–5006.
[7] Friedman M. Chemistry, biochemistry, and safety of acrylamide. A review. J Agric Food Chem 2003;51(16):4504–26.
[8] LoPachin RM. The changing view of acrylamide neurotoxicity. NeuroToxicol 2004;25(4 Spec. Iss.):617–30.
[9] IARC. Some industrial chemicals, Lyon: IARC. IARC Monogr Eval Carcinog Risk Chem Hum 1994;60:435–53.
[10] FAO/WHO. Joint FAO/WHO expert commitee on food additives, sixty-fourth meeting, summary and conclusions. 2005. Available from: http://www.who.int/ipcs/food/jecfa/summaries/summary_report_64_final.pdf. [13 May 2008].
[11] Rice JM. The carcinogenicity of acrylamide. Mutat Res – Genet Toxicol Environ Mutagen 2005;580(1–2):3–20.
[12] Mucci LA, et al. Dietary acrylamide and cancer of the large bowel, kidney, and bladder: absence of an association in a population-based study in Sweden. Br J Cancer 2003;88(1):84–9.
[13] Mucci LA, et al. Acrylamide intake and breast cancer risk in Swedish women. J Am Med Assoc 2005;293(11):1326–7.
[14] Pelucchi C, et al. Fried potatoes and human cancer. Int J Cancer 2003;105(4):558–60.
[15] Pelucchi C, et al. Dietary acrylamide and human cancer. Int J Cancer 2006;118(2):467–71.
[16] Hogervorst JG, et al. A prospective study of dietary acrylamide intake and the risk of endometrial, ovarian, and breast cancer. Cancer Epidemiol Biomarkers Prev 2008;16(11):2304–13.
[17] Wilson KM, et al. A prospective study on dietary acrylamide intake and the risk for breast, endometrial, and ovarian cancers. Cancer Epidemiol Biomarkers Prev 2010;19(10):2503–15.
[18] Olesen PT, et al. Acrylamide exposure and incidence of breast cancer among postmenopausal women in the Danish diet, cancer and health study. Int J Cancer 2008;122(9):2094–100.
[19] Mottram DS, Wedzicha BL, Dodson AT. Acrylamide is formed in the Maillard reaction. Nature 2002;419(6906):448–9.
[20] Stadler RH, et al. Acrylamide from Maillard reaction products. Nature 2002;419(6906):449–50.
[21] Becalski A, et al. Acrylamide in foods: occurrence, sources, and modeling. J Agric Food Chem 2003;51(3):802–8.
[22] Zyzak DV, et al. Acrylamide formation mechanism in heated foods. J Agric Food Chem 2003;51(16):4782–7.
[23] Yaylayan VA, Wnorowski A, Locas CP. Why asparagine needs carbohydrates to generate acrylamide. J Agric Food Chem 2003;51(6):1753–7.
[24] Weißhaar R, Gutsche B. Formation of acrylamide in heated potato products – model experiments pointing to asparagine as precursor. Dtsch Lebensm 2002;98(11):397–400.
[25] Granvogl M, Schieberle P. Thermally generated 3-aminopropionamide as a transient intermediate in the formation of acrylamide. J Agric Food Chem 2006;54(16):5933–8.
[26] Granvogl M, et al. Quantitation of 3-aminopropionamide in potatoes – a minor but potent precursor in acrylamide formation. J Agric Food Chem 2004;52(15):4751–7.
[27] Granvogl M, Schieberle P. Quantification of 3-aminopropionamide in cocoa, coffee and cereal products: correlation with acrylamide concentrations determined by an improved clean-up method for complex matrices. Eur Food Res Technol 2007;225(5–6):857–63.
[28] Eichner K, Laible R, Wolf W. The influence of water content and temperature on the formation of Maillard reaction intermediates during drying of plant products. In: Simatos D, Multon JL, editors. In properties of water in foods in relation to quality and stability. NATO Advanced Science Institute Serie Martin Nijhoff Publ; 1985. p. 191–210.
[29] Amrein TM, et al. Influence of thermal processing conditions on acrylamide generation and browning in a potato model system. J Agric Food Chem 2006;54(16):5910–6.
[30] Knol JJ, et al. Toward a kinetic model for acrylamide formation in a glucose-asparagine reaction system. J Agric Food Chem 2005;53(15):6133–9.
[31] De Vleeschouwer K, et al. Role of precursors on the kinetics of acrylamide formation and elimination under low moisture conditions using a multiresponse approach – Part I: effect of the type of sugar. Food Chem 2009;114(1):116–26.

[32] Claeys WL, De Vleeschouwer K, Hendrickx ME. Quantifying the formation of carcinogens during food processing: acrylamide. Trends Food Sci Technol 2005;16(5):181–93.
[33] Mills C, Mottaram DS, Wedzicha BL. Acrylamide. In: H. SR, Lineback DR, editors. Process-induced food toxicants: occurrence, formation and health risks. Hoboken, New Jersey: John Wiley & Sons, Inc.; 2009.
[34] Claeys WL, et al. Assessment of the acrylamide intake of the Belgian population and the effect of mitigation strategies. Food Addit Contam – Part A Chemistry, Analysis, Control, Expo Risk Assess 2010;27(9):1199–207.
[35] EFSA. Results on acrylamide levels in food from monitoring years 2007–2009 and exposure assessment. Eur Food Saf Auth J 2011;9(4):2133.
[36] Biedermann M, et al. Acrylamide monitoring in Switzerland, 2007–2009: results and conclusions. Food Addit Contam – Part A Chemistry, Analysis, Control, Expo Risk Assess 2010;27(10):1352–62.
[37] European-Commission. Commission recommendation of 10 January 2011 on investigations into the levels of acrylamide in food. 2011. Available from: http://ec.europa.eu/food/food/chemicalsafety/contaminants/recommendation_10012011_acrylamide_food_en.pdf. [December 8, 2014].
[38] EFSA. Update on acrylamide levels in food from monitoring years 2007 to 2010. Eur Food Saf Auth J 2012;10(10):2938.
[39] Powers SJ, et al. Acrylamide concentrations in potato crisps in Europe from 2002 to 2011. Food Addit Contam – Part A Chem Analysis, Control, Expo Risk Assess 2013;30(9):1493–500.
[40] Svensson K, et al. Dietary intake of acrylamide in Sweden. Food Chem Toxicol 2003;41(11):1581–6.
[41] Matthys C, et al. Risk assessment of dietary acrylamide intake in Flemish adolescents. Food Chem Toxicol 2005;43(2):271–8.
[42] Dybing E, Sanner T. Risk assessment of acrylamide in foods. Toxicol Sci 2003;75(1):7–15.
[43] Konings EJM, et al. Acrylamide exposure from foods of the Dutch population and an assessment of the consequent risks. Food Chem Toxicol 2003;41(11):1569–79.
[44] CIAA. The CIAA acrylamide toolbox. 2006. Available from: http://www.ciaa.be/documents/brochures/CIAA_Acrylamide_Toolbox_Oct2006.pdf. [13 May 2008].
[45] FAO/WHO. Joint FAO/WHO Food Standards Programme CODEX Committee on Contaminants in Foods. Proposed draft code of practice for the reduction of acrylamide in food. 2007. Available from: ftp://ftp.fao.org/codex/cccf1/cf01_15e.pdf. [13 May 2008].
[46] Lineback DR, Coughlin JR, Stadler RH. Acrylamide in foods: a review of the science and future considerations. Annu Rev Food Sci Technol 2012;3(1):15–35.
[47] Grob K. Options for legal measures to reduce acrylamide contents in the most relevant foods. Food Addit Contam 2007;24(supp 1):71–81.
[48] Zhang L, et al. U.S. opportunities in China's frozen french fry market. Am J Potato Res 1999;76(5):297–304.
[49] Amrein TM, et al. Potential of acrylamide formation, sugars, and free asparagine in potatoes: a comparison of cultivars and farming systems. J Agric Food Chem 2003;51(18):5556–60.
[50] Becalski A, et al. Acrylamide in french fries: influence of free amino acids and sugars. J Agric Food Chem 2004;52(12):3801–6.
[51] Williams JSE. Influence of variety and processing conditions on acrylamide levels in fried potato crisps. Food Chem 2005;90(4):875–81.
[52] Viklund GAI, et al. Impact of harvest year on amino acids and sugars in potatoes and effect on acrylamide formation during frying. J Agric Food Chem 2008;56(15):6180–4.
[53] Matthaus B, Haase NU. Acrylamide – still a matter of concern for fried potato food? Eur J Lipid Sci Technol 2014;116(6):675–87.
[54] Bethke PC, Bussan AJ. Acrylamide in processed potato products. Am J Potato Res 2013;90(5):403–24.
[55] Pedreschi F, Mariotti MS, Granby K. Current issues in dietary acrylamide: formation, mitigation and risk assessment. J Sci Food Agric 2014;94(1):9–20.
[56] De Wilde T, et al. Influence of storage practices on acrylamide formation during potato frying. J Agric Food Chem 2005;53(16):6550–7.
[57] Biedermann M, et al. Experiments on acrylamide formation and possibilities to decrease the potential of acrylamide formation in potatoes. Mittl Aus Leb Hyg 2002;93(6):668–87.

[58] Foot RJ, et al. Acrylamide in fried and roasted potato products: a review on progress in mitigation. Food Addit Contam 2007;24(Suppl. 1):37–46.
[59] Granda C, Moreira RG, Castell-Perez E. Effect of raw potato composition on acrylamide formation in potato chips. J Food Sci 2005;70(9):E519–25.
[60] Vivanti V, Finotti E, Friedman M. Level of acrylamide precursors asparagine, fructose, glucose, and sucrose in potatoes sold at retail in Italy and in the United States. J Food Sci 2006;71(2):C81–5.
[61] Matthäus B, Haase NU, Vosmann K. Factors affecting the concentration of acrylamide during deep-fat frying of potatoes. Eur J Lipid Sci Technol 2004;106(11):793–801.
[62] Archer DL. Freezing: an underutilized food safety technology? Int J Food Microbiol 2004;90(2):127–38.
[63] Glenn OR, et al. Automated French fry cooking apparatus. United States: Burger King Corporation; 1993.
[64] Claeys WL, De Vleeschouwer K, Hendrickx ME. Kinetics of acrylamide formation and elimination during heating of an asparagine-sugar model system. J Agric Food Chem 2005;53(26):9999–10005.
[65] Pedreschi F, Kaack K, Granby K. Reduction of acrylamide formation in potato slices during frying. Lebensmittel-Wissenschaft Und-Technologie – Food Sci Technol 2004;37(6):679–85.
[66] Avermaete T, et al. Determinants of product and process innovation in small food manufacturing firms. Trends Food Sci Technol 2004;15(10):474–83.
[67] De Vleeschouwer K, et al. Investigation of the influence of different moisture levels on acrylamide formation/elimination reactions using multiresponse analysis. J Agric Food Chem 2008;56(15): 6460–70.
[68] De Vleeschouwer K, et al. The kinetics of acrylamide formation/elimination in asparagine-glucose systems at different initial reactant concentrations and ratios. Food Chem 2008;111(3):719–29.
[69] Gökmen V, Palazoğlu TK, Şenyuva HZ. Relation between the acrylamide formation and time-temperature history of surface and core regions of French fries. J Food Eng 2006;77(4):972–6.
[70] Grob K, et al. French fries with less than 100 μg/kg acrylamide. A collaboration between cooks and analysts. Eur Food Res Technol 2003;217(3):185–94.
[71] Knol JJ, et al. Kinetic modelling: a tool to predict the formation of acrylamide in potato crisps. Food Chem 2009;113(1):103–9.
[72] Romani S, et al. Effect of frying time on acrylamide content and quality aspects of French fries. Eur Food Res Technol 2008;226(3):555–60.
[73] Elmore JS, et al. Measurement of acrylamide and its precursors in potato, wheat, and rye model systems. J Agric Food Chem 2005;53(4):1286–93.
[74] Fiselier K, et al. Influence of the frying temperature on acrylamide formation in French fries. Eur Food Res Technol 2006;222(3–4):414–9.
[75] De Vleeschouwer K, et al. Kinetics of acrylamide formation/elimination reactions as affected by water activity. Biotechnol Prog 2007;23(3):722–8.
[76] Franke K, Sell M, Reimerdes EH. Quality related minimization of acrylamide formation-an integrated approach. In: Friedman M, Mottram D, editors. In chemistry and safety of acrylamide in food. New York: Springer Science Business Media, Inc.; 2005.
[77] Kotsiou K, et al. Acrylamide formation and colour development in low-fat baked potato products as influenced by baking conditions and oil type. Eur Food Res Technol 2013;236(5):843–51.
[78] Mestdagh F, De Meulenaer B, Van Peteghem C. Influence of oil degradation on the amounts of acrylamide generated in a model system and in French fries. Food Chem 2007;100(3):1153–9.
[79] Castle L. Analysis of acrylamide in foods. In: Skog K, Alexander J, editors. Acrylamide and other hazardous compounds in heat-treated foods. Cambridge, England: Woodhead Publishing Limited; 2006.
[80] Boon PE, et al. Calculations of dietary exposure to acrylamide. Mutat Res – Gene Toxicol Environ Mutagen 2005;580(1–2):143–55.
[81] Birlouez-Aragon I, et al. The health and technological implications of a better control of neoformed contaminants by the food industry. Pathol Biol 2010;58(3):232–8.
[82] Luning PA, Marcelis WJ. A techno-managerial approach in food quality management research. Trends Food Sci Technol 2006;17(7):378–85.
[83] Sanny M, et al. Impact of control behaviour on unacceptable variation in acrylamide in French fries. Trends Food Sci Technol 2010;21(5):256–67.
[84] Luning PA, Marcelis WJ. A conceptual model of food quality management functions based on a techno-managerial approach. Trends Food Sci Technol 2007;18(3):159–66.

[85] Sanny M. Variation in acrylamide concentration in French fries: effects of control measures in food service establishments. Wageningen University; 2012.
[86] Luning PA, Marcelis WJ. A food quality management research methodology integrating technological and managerial theories. Trends Food Sci Technol 2009;20(1):35–44.
[87] Fiselier K, Grob K. Legal limit for reducing sugars in prefabricates targeting 50 μg/kg acrylamide in French fries. Eur Food Res Technol 2005;220(5–6):451–8.
[88] Al-Kahtani HA. Survey of quality of used frying oils from restaurants. J Am Oil Chem Soc 1991;68(11):857–62.
[89] Sanny M, et al. Possible causes of variation in acrylamide concentration in French fries prepared in food service establishments: an observational study. Food Chem 2012;132(1):134–43.
[90] Griffith CJ. Food safety: where from and where to? Br Food J 2006;108(1):6–15.
[91] Clayton DA, et al. Food handlers' beliefs and self-reported practices. Int J Environ Health Res 2002;12(1):25–39.
[92] Amrein TM, et al. Occurrence of acrylamide in selected foods and mitigation options. Food Addit Contam 2007;24:13–25.
[93] Palazoğlu TK, Gökmen V. Reduction of acrylamide level in French fries by employing a temperature program during frying. J Agric Food Chem 2008;56(15):6162–6.
[94] Ishihara K, et al. Examination of conditions inhibiting the formation of acrylamide in the model system of fried potato. Biosci Biotechnol Biochem 2006;70(7):1616–21.
[95] Snyder PO, Matthews ME. Percent retention of vitamin C in whipped potatoes after pre-service holding. J Am Dietetic Assoc 1983;83(4):454–8.
[96] De Graaf C, et al. A comparison between liking ratings obtained under laboratory and field conditions: the role of choice. Appetite 2005;44(1):15–22.
[97] Stadler RH. Acrylamide formation in different foods and potential strategies for reduction. In: Friedman M, Mottram DS, editors. In chemistry and safety of acrylamide in food. New York: Springer Science Business Media, Inc.; 2005. p. 157–69.
[98] Taeymans D, et al. A review of acrylamide: an industry perspective on research, analysis, formation and control. Crit Rev Food Sci Nutr 2004;44(5):323–47.
[99] Vinci RM, et al. Implementation of acrylamide mitigation strategies on industrial production of French fries: challenges and pitfalls. J Agric Food Chem 2011;59(3):898–906.
[100] Vinci RM, et al. Effective quality control of incoming potatoes as an acrylamide mitigation strategy for the French fries industry. Food Addit Contam – Part A Chem Analysis, Control, Expo Risk Assess 2010;27(4):417–25.
[101] Sanny M, et al. Is lowering reducing sugars concentration in French fries an effective measure to reduce acrylamide concentration in food service establishments? Food Chem 2012;135(3):2012–20.
[102] Gökmen V, Palazoğlu TK. Acrylamide formation in foods during thermal processing with a focus on frying. Food Bioprocess Technol 2008;1(1):35–42.
[103] Sanny M, et al. Effect of frying instructions for food handlers on acrylamide concentration in French fries: an explorative study. J Food Prot 2013;76(3):462–72.
[104] Besaratinia A, Pfeifer GP. A review of mechanisms of acrylamide carcinogenicity. Carcinogenesis 2007;28(3):519–28.
[105] Smith MK. Common mistakes in using statistics: spotting and avoiding them. 2011. Available from: http://www.ma.utexas.edu/users/mks/statmistakes/terminologyrevariability.html. [27 November 2011].
[106] Röös E, Sundberg C, Hansson PA. Uncertainties in the carbon footprint of food products: a case study on table potatoes. Int J Life Cycle Assess 2010;15(5):478–88.
[107] Mills C, Mottram DS, Wedzicha BL. Acrylamide. In: Stadler RH, Lineback DR, editors. Process-induced food toxicants: occurence, formation, mitigation, and health risks. New Jersey: John Wiley & Sons, Inc.; 2009.
[108] IRMM. Joint Research Centre: acrylamide monitor database. 2006. Available from: http://irmm.jrc.ec.europa.eu/activities/acrylamide/Pages/database.aspx. [11 July 2011].
[109] Halford NG, et al. Concentrations of free amino acids and sugars in nine potato varieties: effects of storage and relationship with acrylamide formation. J Agric Food Chem 2012;60(48):12044–55.

CHAPTER 9

Acrylamide in Coffee and Coffee Substitutes

Monica Anese
Department of Food Science, University of Udine, Udine, Italy

INTRODUCTION

Coffee together with tea is the most popular beverage in the world, because of its attractive aroma and taste, health promoting, and pharmaceutical potentials. The biological effects of coffee are attributed to actions of caffeine as well as hundreds of biologically active compounds, including polyphenols and Maillard reaction products. It has been recently reported that moderate doses of coffee (three to four cups per day) not only may have beneficial effects on alertness, vigilance, and cognitive abilities but also may reduce the risk of type 2 diabetes mellitus and hypertension, as well as other conditions associated with cardiovascular risk, such as obesity and depression [1]. Additionally, the potential risks of coffee consumption (mostly associated to its high caffeine content) include anxiety, insomnia, tremulousness, and palpitations, as well as bone loss and possibly increased risk of fractures [2]. For this reason, a variety of non-coffee products (i.e. without caffeine) have been developed as coffee substitutes. Currently, the most used ingredients are malted barley, roasted chicory root, and roasted rye, while roasted rice, sugar beet, peas, soybean, carob, almonds, and figs are used to a lesser extent.

Coffee and coffee substitutes are prepared by brewing roasted and ground coffee beans and non-coffee ingredients [3]. During roasting, the matrix is subjected to temperatures higher than 200 °C [4,5] which are responsible for inducing complex chemical reactions, that is, the Maillard reaction, caramelization, and pyrolysis. As a consequence, important changes of chemical, physical, and sensory properties of the coffee beans and non-coffee ingredients occur [4] (Figure 1).

Among these, color and flavor development are the ones mostly affecting beverage acceptability as they are perceived by senses. Besides, other important "invisible" modifications generally occur. These are relevant to the formation of desired and undesired heat-induced components. It has been extensively demonstrated that coffee has strong antioxidant activity mainly attributable to the Maillard reaction products that form during the roasting process [6–8]. Recently, it has been found that roasting may induce the formation of toxic molecules, that is, acrylamide as well as furan and 5-hydroxymethylfurfural [9,10]. According to the International Agency for Research

Acrylamide in Food
http://dx.doi.org/10.1016/B978-0-12-802832-2.00009-7

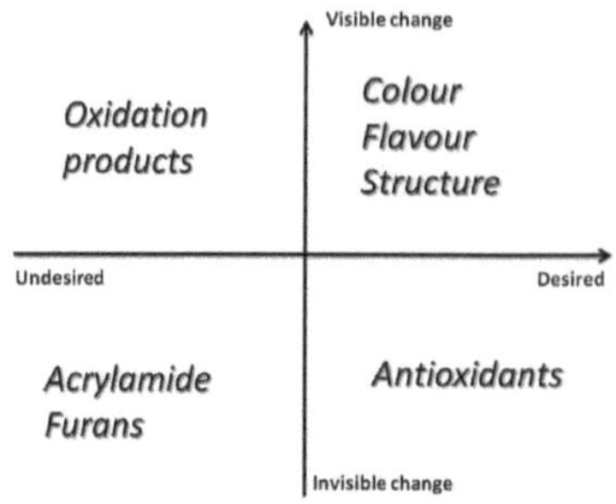

Figure 1 Consequences of roasting of coffee and coffee substitutes.

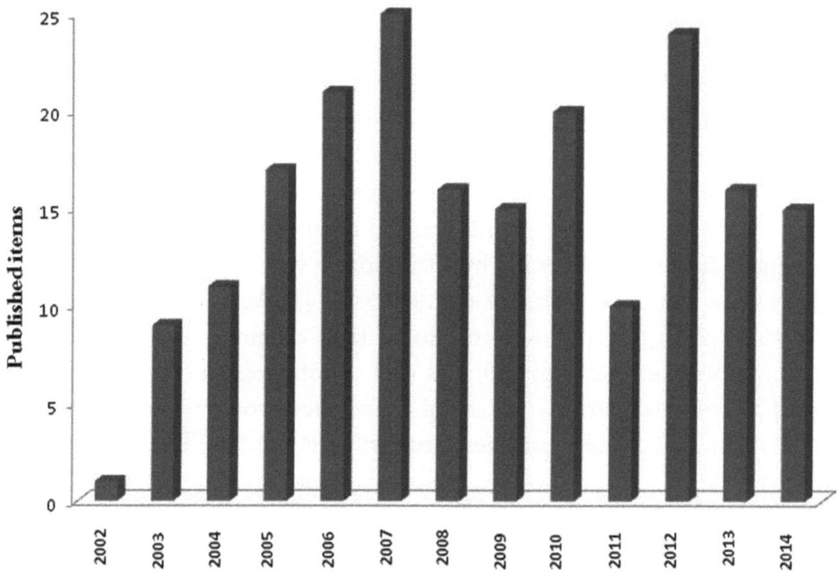

Figure 2 Published items on acrylamide in coffee and coffee substitutes per year. *Source: Web of Science.*

on Cancer, acrylamide is known to be "possibly carcinogenic to humans" [11]. Results of the survey carried out by the EFSA Panel on Contaminants in the Food Chain (CONTAM) on over 8000 food samples collected from 24 European countries since 2010 and analyzed for acrylamide content, evidenced that coffee and coffee substitutes contain highest levels of acrylamide [12].

It must be pointed out that, although more than 5000 articles have been published on acrylamide in all food categories since 2002, only about 200 are relevant to acrylamide in coffee and coffee substitutes (Figure 2). They include both scientific papers and patents on methods of analysis, mechanism of formation, quantification studies, dietary acrylamide intake, and mitigation interventions.

MECHANISMS OF ACRYLAMIDE FORMATION IN COFFEE AND COFFEE SUBSTITUTES, AND INFLUENCING FACTORS

To the best of our knowledge, a few studies specifically deal with the mechanisms of acrylamide formation in coffee and coffee substitutes [13–15]. Nevertheless, more than one route of generation of the toxic molecules have been suggested [16]. As for other foods subjected to heating, the reaction between asparagine and carbohydrates or carbonyl compounds represents the major pathway for the formation of acrylamide in coffee and coffee surrogates [17,18]. In particular, asparagine has been described as the limiting factor for acrylamide formation in coffee [19]. A positive correlation between the formation of acrylamide and asparagine concentration was actually found in coffee and chicory-based substitutes [20,21].

Possible additional or alternative mechanisms of acrylamide formation in coffee may involve intermediate products of thermally induced reactions, such as acrolein, acrylic acid, and 3-aminopropanamide (3-APA) [22–25]. Acrolein and acrylic acid may further generate acrylamide following reaction with ammonia. Although Granvogl et al. [25] demonstrated that the 3-APA is an efficient precursor of acrylamide in model systems and potatoes, where it can be generated also through the biochemical decarboxylation of asparagine, it was neither detected in raw nor in roasted coffee [13]. Due to the relatively high roasting temperatures, pyrolytic reactions can plausibly represent a further pathway for acrylamide formation in coffee, analogously to what has been already observed in bread [26]. It has been recently suggested that 5-hydroxymethylfurfural (HMF), which is also generated during roasting, could play a role in acrylamide formation [14,15]. HMF could act as a more efficient precursor than sugars to form acrylamide. Therefore, any ingredient/component able to promote HMF formation would increase acrylamide generation [15].

Different factors may affect acrylamide concentration in coffee and coffee substitutes. These are relevant to the raw material; that is, vegetable type (i.e., coffee or non-coffee ingredients), coffee species, and to process interventions (i.e., roasting, formulation, and storage).

According to the results of the evaluation carried out by the EFSA Panel on Contaminants in the Food Chain (CONTAM) on more than 1500 coffee products collected and analyzed from 2010 to 2013, an average acrylamide concentration of 578 ng/g was found in roasted coffee, whereas instant coffee and cereal-based coffee substitutes showed higher values [12]. These data are consistent with the results of recent studies showing that the acrylamide content in soluble coffee varied from 200 to 645 ng/g with a median concentration of 472 ng/g, and in roasted barley grains it ranged from 50 to 410 ng/g [27,28]. The chicory-based substitutes showed the highest levels, sometimes exceeding the indicative value of 4000 for this food category by the European Commission [29]. These high values for chicory derivatives have been attributed to high levels of fructose in chicory that result in higher rates of acrylamide formation [27]. Moreover, chicory cultivar as well as level of nitrogen applied during fertilization resulted to affect the amount of free asparagine and consequently acrylamide levels in roasted chicory [21].

Acrylamide formation is also affected by the degree of roast, which is the time of roasting. In fact, the amount of acrylamide during roasting time increases exponentially, reaches a maximum and then decreases rapidly [9,16,19,20,30]. This has been also confirmed by the results of the CONTAM survey [12] showing average concentrations of acrylamide of 374, 266, and 187 μg/kg in light, medium, and dark roasted coffee, respectively. The decrease in acrylamide at high degrees of roast occurs when the rate of degradation exceeds the rate of formation, probably due to the reaction of acrylamide with other reactive species present in the coffee [31]. Recently, Pastoriza et al. [32] stated that acrylamide may react with coffee melanoidins during roasting, leading to a net decrease in acrylamide content.

It has been demonstrated that also coffee species may affect the acrylamide level in roasted beans and coffee brews [19,20,33,34]. In fact, significantly higher amounts of acrylamide form in Robusta (*Coffea canephora robusta*) than in Arabica (*Coffea arabica*). Bagdonaite et al. [20] found that the mean acrylamide concentration in Robusta and Arabica coffees, that were roasted in a small-scale convection roaster at 240 °C for 7.5 min, was 708 ± 77 ng/g and 374 ± 86 ng/g, respectively. Similarly, Summa et al. [33] reported that acrylamide formation in Robusta coffee after 9 min of hot air roasting at 236 °C was about 250 ng/g, while in Arabica it was 150 ng/g. These differences in acrylamide formation are attributed to differences in species' chemical composition. In fact, the concentration of free asparagine in Robusta was found to be almost twice (797 ± 23 μg/g) than in Arabica (486 ± 97 μg/g) [20]. Alves et al. [34] observed that not only Robusta was richer in acrylamide than Arabica is but also the concentration of the undesired molecule decreased proportionally with increasing the degree of roast in both ground coffee and espresso brew. Provided the same roasting conditions are used, acrylamide formation in Robusta is 45%–55% higher than in Arabica, independently of the coffee's geographical origin [20] and roasting degree [34] (Figure 3). This proportion is maintained also in the cup [34]. In fact, as acrylamide is freely soluble in water, the transfer of acrylamide from roasted beans to the beverage is almost quantitative. Lower rates of transfer were observed for espresso coffee due to the lower extraction time as compared with other extraction methods [34].

Mitigation of Acrylamide Concentration

The exposure to acrylamide from coffee and coffee substitutes contributes substantially to the total dietary acrylamide intake [35,36]. According to EFSA [12], the consumers of coffee (both as beverage and as ingredient of a recipe) represent from 34% up to 96% of the population across the surveys in adults (18–64 years), elderly (65–74 years) and very elderly (≥75 years) age groups. The highest consumption levels were observed in the elderly age group, with a median average and 95th percentile consumption levels at,

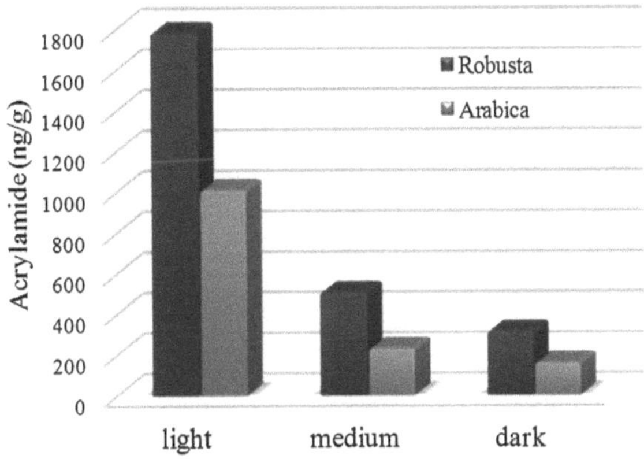

Figure 3 Acrylamide concentration in light, medium, and dark roasted ground coffee. *Adapted from Alves et al. [34].*

respectively, 14 and 32 g dry equivalent per day, followed by the very elderly and adults (median averages of 11 and 10 g dry equivalent per day, and 95th percentile consumption levels at 28 and 26 g dry equivalent per day, respectively). Acrylamide exposure from coffee largely varies from country to country depending on consumer age and gender, degree of roast, coffee volume, and so on. For instance, in Northern European countries (e.g., Norway and Sweden) and Switzerland coffee is the largest contributor to the total acrylamide intake, accounting for 29–40%, while in most countries located in the south and west of Europe it is the second largest (range 14–28%) [37–39]. Also in the United States coffee is among the foods mainly contributing to acrylamide intake [40]. Brewed coffee is ranked sixth in the top 20 foods by mean acrylamide intake (0.027 μg/kg bw day) and first for acrylamide per portion (1.9 μg for 240 g portion size) [41].

As acrylamide is supposed to have negative consequences to human health [11], its widespread presence in foods has generated great concern and many efforts have been carried out in recent years to find possible strategies able to mitigate its presence in food and keep its levels as low as reasonably achievable. Information on potential interventions to be acquired by industry operators to reduce acrylamide levels in foods, including that in coffee and coffee substitutes, is given in the so-called acrylamide Toolbox, developed by FoodDrinkEurope [42]. Reduction of acrylamide in coffee is actually a challenging issue. Indeed, a limiting factor for the exploitation of the mitigation strategies is the existence of a close relationship between the formation of the undesired molecule and the development of the desired sensory properties of coffee. As the formation of this heat-induced molecule is concomitant with that of color, flavor, and texture, it is very difficult to minimize its formation without affecting the coffee sensory acceptability. Roasters are still looking for strategies that can be effectively applied to reduce acrylamide in coffee. Moreover, these strategies should not only exert low impact on the coffee sensory properties but also meet regulatory compliance and be economically sustainable.

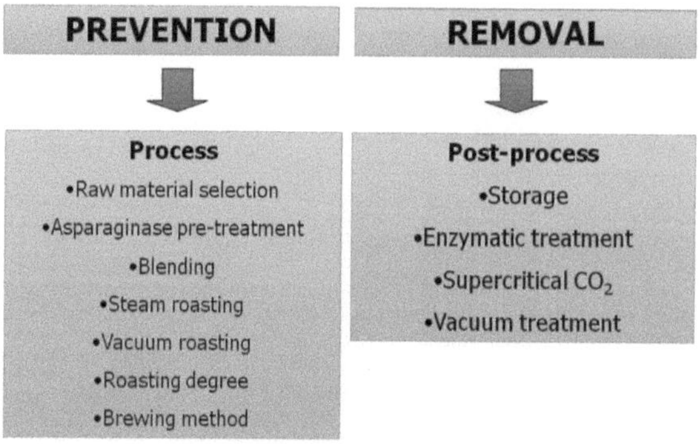

Figure 4 Technological strategies to reduce acrylamide levels in coffee and coffee substitutes. *Modified from Anese et al. [43].*

In broad terms, the ways of mitigating acrylamide levels in food can be regarded as preventive and removal strategies [43]. The preventive interventions are aimed to keep as low as possible the formation of acrylamide during roasting, by acting on precursors and mechanistic pathways, thus creating less favorable reaction conditions. The removal strategies are intended to remove or decompose the already formed acrylamide from the finished product. As the removal interventions are thought as post-process interventions, in principle, they are not affected by the previous process steps. Relevant technological strategies of acrylamide mitigation in coffee and coffee substitutes are summarized in Figure 4. They can be applied in one or more steps of coffee production, from raw material selection, through pretreatment, roasting, and postprocess interventions, to brew preparation.

Here acrylamide mitigation strategies are discussed, focusing on their pros and cons with respect to their potential applicability at the industrial level.

Preventive Strategies

Pretreatment

Raw coffee can be subjected to enzymatic pretreatment to reduce the levels of precursor asparagine. In particular, the use of enzymes that hydrolyze the amide group on the side chain of asparagine has been suggested, to form aspartic acid and ammonia [44]. The most studied enzyme capable of hydrolyzing the amide group of free asparagine to prevent acrylamide formation is asparaginase, which has been claimed to reduce acrylamide levels up to 70% without impairing the sensory properties of roasted coffee [45]. The main issues related to the use of these enzymes are relevant to their efficiency in reducing the undesired molecule when exploited at the industrial scale. Besides the ineluctable batch-to-batch variability due to a non-constancy of raw materials, acrylamide prevention in coffee by means of the use of asparaginase can be affected by an insufficient or inhomogeneous delivery of the enzyme to the substrate. In fact, factors limiting the contact of the enzyme with the substrate might be responsible for incomplete hydrolysis of asparaginase and partial reduction of acrylamide formation. Among these are coffee structure and composition.

In fact, the "tight" coffee structure might hamper enzyme diffusion through the coffee matrix and its contact with the substrate. Moreover, the presence of fat may create hydrophobic zones, where the activity of the aqueous enzymatic suspension is limited [46]. Asparaginase efficiency can be improved by subjecting coffee beans to any pretreatment that favors enzyme diffusion through the matrix and thus its contact with asparagine. These include drying, hydrating, steaming, and heating; eventually in combination with reduced or increased pressure [47]. Each pretreatment would create a driving force for the enzyme solution to migrate into the beans. Also, particle size reduction could greatly improve enzyme efficiency and thus acrylamide reduction [47]. However, this strategy cannot be realistically applied at the industrial level as coffee beans have to be roasted whole. Further issues affecting the use of asparaginase on the industrial scale are relevant to process conditions, that is, temperature and incubation time as well as enzyme concentration. A proper choice of these variables is essential to effectively reduce acrylamide levels. In particular, the use of optimal dose would limit costs and avoid flavor generation due to an excess of ammonia formed from the hydrolysis of asparagine. However, this may depend on precursor natural variability associated to coffee species and variety [16,46,47].

Roasting Step

As already pointed out, the final acrylamide concentration depends on the roasting degree. Acrylamide decreases in coffee beans upon darker roasting, reaching levels up to 20%–30% of the maximum one [9,19,30]. However, this may be not always a suitable strategy as the roasting degree must be compatible with the coffee flavor profile, which affect consumer acceptability. Medium roasted coffee with lower acrylamide concentration can be obtained by means of roasting under reduced pressure conditions [48]. The authors subjected coffee beans to roasting process at either atmospheric or low (1.1. Torr) pressure conditions and evaluated acrylamide formation and degradation during roasting. Different acrylamide levels were obtained, provided the same time–temperature combination was received by the beans at each roasting time (Figure 5).

At a medium degree of roast, accounting for 7% weight loss [4], the vacuum roasted coffee had about 50% less acrylamide than the conventionally roasted beans, while they did not differ in color. Moreover, the conventionally and vacuum medium roasted coffee beans were not perceived as different by a panel of trained assessors. It was suggested that the low pressure generated inside the roaster could prevent acrylamide from being accumulated. In particular, the reduced pressure would exert a stripping effect toward water as well as water soluble compounds such as acrylamide, and/or promote acrylamide sublimation. As long as acrylamide is generated during the roasting process, the low pressure inside the roaster would cause its transition from the solid phase to the vapor one. The latter would leave the coffee matrix and easily move out from the oven.

Steam roasting has been also studied as a potential alternative to conventional roasting [49]. Coffee beans were roasted at temperatures ranging from 200 to 240 °C in a roaster equipped with a steam injector. Results showed that steam roasting allowed 10%

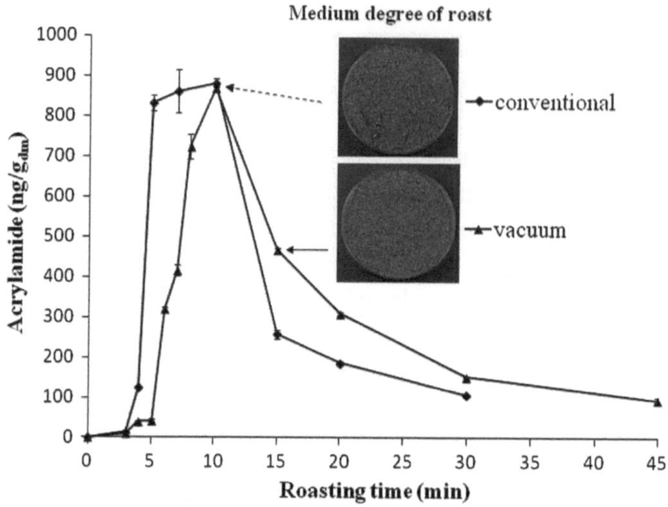

Figure 5 Acrylamide evolution during conventional and vacuum roasting of coffee. Acrylamide concentrations of medium roasted coffees obtained by means of the conventional and vacuum roasting are indicated by the continuous and dashed arrows, respectively. Images of the ground medium roasted coffees are also shown. *Adapted from Anese et al. [48].*

acrylamide reduction but to reach a comparable color end point longer process time was required with a negative impact on coffee taste and aroma.

Formulation Step

This can be regarded as coffee species selection and blending. It has already been mentioned that the coffee species affects the acrylamide content in roasted beans and coffee brews [20,33,34], acrylamide amounts being higher in Robusta than in Arabica. Provided the same roasting conditions are used, acrylamide formation in Robusta is 40%–50% higher than in Arabica [20,34] and this proportion is maintained also in the brewed cup [34]. Reduction of acrylamide levels in coffee can thus be achieved by selecting coffee blends in higher percentages of Arabica and darker in degree of roasting. Blending coffee with other vegetable matrices that are not a source of acrylamide could represent an alternative way to minimize the levels of the toxic molecule. As described in a recent patent, the addition of a herbal blend (e.g., *Sambucus* and *Hibiscus*) to roasted coffee before or after brewing could reduce the amount of acrylamide in the brewed coffee product by at least 10% [50].

Removal Strategies

Coffee Storage

It has been reported that acrylamide concentration decreases during storage of roasted coffee and barley-based coffee substitutes, possibly due to interactions with the matrix components [19,28,51,52]. Acrylamide losses were affected by the storage temperature.

In particular, a temperature dependence of acrylamide loss was observed by Lantz et al. [19] for vacuum packed roast and ground coffee stored at temperatures from −18 to 37 °C for 12 months. These authors reported that the reduction of acrylamide followed a second order reaction kinetics and had an activation energy level of 73 kJ/mol. This suggests that acrylamide concentration in the "market coffee" could vary considerably depending on its actual "age." It must be pointed out that during storage coffee also undergoes flavor decay. As flavor is one of the key parameters affecting coffee quality, storage does not represent a mitigation opportunity.

Biochemical Removal

Acrylamide can be removed from the roasted coffee by metabolic conversion operated by yeasts or enzymes [53,54]. By mixing brewed coffee added with sucrose with *Saccharomyces cerevisiae*, acrylamide concentration exponentially decreased during fermentation [54]. Acrylamide removal was affected by sucrose and yeast concentration. A maximum decrease of about 65% of the toxic molecule was achieved after 24 h by employing 2% w/v yeast and 10% sucrose. Cha [53] suggested the possibility of removing acrylamide from coffee brews by using the bacterial enzyme acrylamidase, which is able to promote acrylamide hydrolysis to acrylic acid and ammonia. This enzyme was able to degrade up to 100% acrylamide in coffee maintained at 70 °C for at least 30 min. As keeping coffee hot and stirring are conditions needed for the catalytic activity, the coffee filter machine was suggested as the only extraction technique suitable for this application. However, the main constraints to the application of this intervention are represented by the release of ammonia, that may affect coffee flavor, and acrylic acid, that is known to be toxic and corrosive for skin and mucosa. Finally, it is noteworthy that the use of enzyme in foods for human consumption must meet the current regulations.

Supercritical CO_2 Removal

Banchero et al. [55] suggested the use of supercritical CO_2 to remove acrylamide from roasted coffee. In the supercritical state ($T \geq 31.1$ °C; $P \geq 7.37$ MPa), CO_2 has liquid-like density and gas-like viscosity that confer high diffusivity and high extraction power. Moreover, as CO_2 is cheap, non-flammable and a non-toxic solvent, supercritical CO_2 extraction is an environmentally friendly technique. In addition, industries are already acquainted with this technology as it is widely used in the coffee decaffeination process. The authors investigated the effect of different process conditions (i.e., pressure, process time and temperature, co-olvent addition) on extraction efficiency of acrylamide. Temperature was the variable that mostly affected the extraction efficiency; 100 °C, 200 MPa and 9.5% ethanol as co-solvent being the best working conditions. Under these experimental conditions, the highest extent of acrylamide removal (70%) was obtained with process duration of about 22 h. It is noteworthy that high-temperature treatments for long time may cause further Maillard reaction development and consequently (undesired) color and flavor changes. For these reasons, currently supercritical CO_2 removal is likely to be hardly applicable at the industrial level.

Vacuum Treatment of Hydrated Roasted Beans

It could be argued that acrylamide can be removed from foods by exploiting its physical chemical properties (e.g., low boiling point) and that removal might be favored by reduced pressure conditions. However, it has been observed that acrylamide removal from roasted coffee is hindered, due to the glassy state of the coffee matrix, the glass transition temperature of roasted coffee being about 65 °C [56]. It has been demonstrated that acrylamide removal by means of the application of a vacuum treatment is possible only for foods with a high water activity value [43]. A hydration step of the roasted coffee to water activity higher than 0.8 prior the vacuum treatment could allow acrylamide to be removed. However, the vacuum treatment of hydrated coffee would promote not only acrylamide escape but also release of flavor compounds, and thus an overall decrease of the odor intensity of coffee. Further studies are therefore needed in order to elucidate optimal process conditions to obtain coffee with reduced acrylamide and good sensory properties. This could be done by properly modulating the process parameters (pressure, temperature, and time) to maximize acrylamide removal while minimizing volatile loss, or alternatively by flavor enrichment of low acrylamide-containing (vacuum-treated) coffee.

In conclusion, despite a number of preventive and removal strategies being reported in the literature, currently there are no industrially applicable interventions for mitigating acrylamide in coffee and coffee substitutes. A feasible opportunity is represented by the selection of commercial blends high in Arabica content, which should be roasted to a dark degree to prepare short beverages. It is noteworthy that coffee prepared as such should meet the consumers' preference. Although further research has to be conducted at pilot and industrial scales, roasting under reduced pressure could represent an interesting strategy to obtain medium-roasted coffee with low acrylamide concentration and minimal impact on the sensory properties. Currently, the most promising mitigation intervention is represented by the use of asparaginase. However, for this strategy to be reliable for exploitation, further investigation has to be performed taking into account the inevitable non-constancy of the raw material composition and quality as well as the process parameters.

SYNOPSIS OF ANALYTICAL TECHNIQUES

Different analytical techniques are currently used to determine acrylamide in coffee and coffee substitutes. Acrylamide concentration is generally quantified by means of mass spectrometer coupled with a gas (GC–MS) or liquid chromatograph (LC–MS), using water as extraction solvent. According to Soares et al. [57], 11 LC-MS and 2 GC–MS methods have been developed to determine acrylamide in coffee products. Even more are the clean up procedures. Coffee is actually is one of the most complex matrices, due to the presence of interfering compounds [51,58]. Purification steps include precipitation of co-extractives with organic solvents or Carrez reagent, defatting with *n*-hexane or dichloromethane, partition with ethyl acetate, single or multiple solid-phase extraction (SPE) with different sorbents, acrylamide derivatization (bromination), formation of a stable thioether by reaction with 2-mercaptobenzoic acid. Also, a QuEChERS (Quick, Easy, Cheap, Effective, Rugged and Safe) method based on water extraction and partition with acetonitrile by salting out with $MgSO_4$ and NaCl has been developed [59]. Soares et al. [57] proposed a purification method based on matrix solid-phase dispersion of the coffee sample with a conditioned C18 phase followed by extraction through a bilayered (C18/Isolute Multimode) SPE column. Wenzl et al. [60] developed a method based on water extraction and simultaneous defatting with *n*-hexane, and subsequent purification of the aqueous phase by passage through Isolute Multimode followed by Env + SPE columns. More recently, a rapid and reliable purification method has been developed [61], based on an in-house made single mixed SPE column. The SPE column, consisting of a mixture of C18, strong cation (SCX) and anion exchange (SAX) sorbents in the ratio 2/1.5/1.5 (w/w/w), allowed to retain the majority of the chemically different interfering coffee components such as melanoidins, trigonelline, chlorogenic acids, and caffeine in one step, thus obtaining a clean sample amenable to HPLC-MS/MS analysis.

KEY FACTS

- Discovery of acrylamide in foods dates back to April 2002, when the Swedish National Food Administration highlighted its presence in a wide variety of heated foods.
- At high concentrations, acrylamide is a neurotoxic and potentially carcinogenic molecule in humans. As acrylamide can accumulate in the organism, a low dietary acrylamide exposure might not be negligible because also a low daily exposition to acrylamide potentially can induce toxicological risk in a long-term period.
- Physical and chemical properties of acrylamide. Acrylamide is a reactive di-functional monomer containing a reactive electron deficient double bond and an amide group, which may undergo reactions typical of these functionalities. At ambient temperature and pressure acrylamide is a colorless and odorless crystalline solid with a melting point of 84.5 °C and a molecular mass of 71.08 g/mol. Acrylamide is soluble in water and low-molecular-weight alcohols and is insoluble in nonpolar organic solvents.
- Furans are a wide class of heterocyclic, low-molecular-weight molecules that are formed as products or intermediates in heat-induced reactions and can significantly contribute to the sensory properties of heated foods. The common feature of furans is the furan ring that can present one or more substitutes. Among furans, furan and HMF have been reported to be toxic and probably carcinogenic to humans. High levels of furan and

HMF have been detected in a large variety of food commodities for adults and infants; both industrially processed or prepared at a domestic level. These include canned or jarred foods, cereal derivatives, fruit juices, dried foods, honey, milk, and coffee.

MINI DICTIONARY

Roasting It is a unit operation that uses heated air to alter the sensory properties of foods (i.e., improve palatability, develop aromas and modify the texture) as well as to destroy enzymes and microorganisms.

Roasted coffee It is obtained by subjecting the green beans to a roasting process which is generally carried out at temperatures higher than 200 °C.

The degree of roast It is the way to describe the external color of roasted coffee. Roasted coffee can be characterized by light, to medium to dark degree of roast. Coffees with different degree of roast are consumed worldwide, the medium-roasted coffee being preferred in the American and North European markets, whereas the dark-roasted type mostly used in the Mediterranean countries.

Instant coffee It is known also as soluble coffee and coffee powder and is made from roasted and ground beans that are extracted with hot water. The coffee extract is subsequently dried by spray-drying or freezing-drying.

SUMMARY POINTS

- Acrylamide has been reported to be toxic and probably carcinogenic to humans and may form during heating of a variety of foods, including coffee and coffee substitutes.
- Coffee is the major source of acrylamide for regular coffee consumers in many countries.
- Acrylamide intake depends on many variables, such as the vegetable (i.e., coffee or non-coffee), matrix, coffee species, degree of roast, beverage preparation method, and storage.
- Reduction of acrylamide levels in coffee and coffee substitutes is a challenging issue, because it is very difficult to mitigate its levels without affecting coffee acceptability.
- Currently, there are no industrially applicable interventions for mitigating acrylamide in coffee and coffee substitutes.
- Feasible opportunities to reduce acrylamide levels in coffee and coffee substitutes are represented by the selection of commercial blends high in Arabica content, which should be roasted to a dark degree to prepare short beverages, and the use of asparaginase. Regarding the latter opportunity, further investigation is needed at both the pilot and industrial scales.

REFERENCES

[1] Nehlig A. Physiological effects of coffee and human health. A review. Cah Agric 2011;21:197–207.
[2] O'Keefe JH, Bhatti SK, Patil HR, et al. Effects of habitual coffee consumption on cardiometabolic disease, cardiovascular health, and all-cause mortality. J Am Coll Cardiol 2013;62:1043–51.
[3] Buffo RA, Cardelli-Freire C. Coffee flavor: an overview. Flavor Fragr J 2004;19:99–104.
[4] Clarke RJ. Roasting and grinding. In: Clarke RJ, Macrae R, editors. Coffee. Technology, vol. 2. London: Elsevier Applied Science Publ. Ltd; 1987. p. 73–107.

[5] Eggers R, Pietsch A. Technology I: roasting. In: Clarke RJ, Vizthum OG, editors. Coffee: recent developments (chapter 4). London: Blackwell Science Ltd; 2001. p. 90–107.
[6] Nicoli MC, Anese M, Manzocco L, et al. Antioxidant properties of coffee brews in relation to the roasting degree. Lebensm Wiss Technol 1997;30:292–7.
[7] Borrelli CR, Visconti A, Mennella C, et al. Chemical characterization and antioxidant properties of coffee melanoidins. J Agric Food Chem 2002;50:6527–33.
[8] Del Castillo MD, Gordon MD, Ames JM. Peroxyl radical-scavenging activity of coffee brews. Eur Food Res Technol 2005;221:471–7.
[9] Taeymans D, Wood J, Ashby P, et al. A review of acrylamide: an industry perspective on research, analysis, formation and control. Crit Rev Food Sci Nutr 2004;44:323–47.
[10] Claeys WL, de Vleeschouer K, Hendrickx ME. Quantifying the formation of carcinogens during food processing: acrylamide. Trends Food Sci Technol 2005;16:181–93.
[11] IARC. Monographs on the evaluation of carcinogenic risks to humans, vol. 60. Lyon, France: International Agency for Research on Cancer; 1994.
[12] EFSA CONTAM Panel. EFSA panel on contaminants in the food chain. Scientific Opinion on Acrylamide in Food. 2014. http://www.efsa.europa.eu/it/consultations/call/140701.pdf.
[13] Bagdonaite K, Viklund G, Skog K, et al. Analysis of 3-aminopropionamide: a potential precursor of acrylamide. J Biochem Biophys Methods 2006;69:215–21.
[14] Kocadagli T, Goncuoglu N, Hamzalioglu A, et al. In depth study of acrylamide formation in coffee during roasting: role of sucrose decomposition and lipid oxidation. Food Funct 2012;3:970–5.
[15] Cai Y, Zhang Z, Jiang S, et al. Chlorogenic acid increased acrylamide formation through promotion of HMF formation and 3-aminopropionamide deamination. J Hazar Mat 2014;268:1–5.
[16] Guenther H, Anklam E, Wenzl T, et al. Acrylamide in coffee: review of progress in analysis, formation and level reduction. Food Add Contam 2007;26:60–70.
[17] Mottram DS, Wedzicha BL, Dodson AT. Acrylamide is formed in the Maillard reaction. Nature 2002;419:448.
[18] Stadler RH, Blank I, Varga N, et al. Acrylamide from Maillard reaction products. Nature 2002;419:449–50.
[19] Lantz I, Ternité R, Wilkens J, et al. Studies on acrylamide levels in roasting, storage and brewing of coffee. Mol Nutr Food Res 2006;50:1039–46.
[20] Bagdonaite K, Derler K, Murkovic M. Determination of acrylamide during roasting of coffee. J Agric Food Chem 2008;56:6081–6.
[21] Loaëc G, Niquet-Leridon C, Henry N, et al. Effects of variety, agronomic factors, and drying on the amount of free asparagine and crude protein in chicory. Correlation with the acrylamide formation during roasting. Food Res Int 2014;63(part C):299–305.
[22] Yasuhara A, Tanaka Y, Hengel M, et al. Gas chromatographic investigation of acrylamide formation in browning model systems. J Agric Food Chem 2003;51:3999–4003.
[23] Becalski A, Lau BPY, Lewis D, et al. Acrylamide in foods: occurrence, sources, and modeling. J Agric Food Chem 2003;52:3801–6.
[24] Yaylayan VA. Stadler RH Acrylamide formation in food: a mechanistic perspective. J AOAC Int 2005;88:262–7.
[25] Granvogl M, Jezussek M, Koehler P, et al. Quantitation of 3-aminopropionamide in potatoes—a minor but potent precursor in acrylamide formation. J Agric Food Chem 2004;52:4751–7.
[26] Claus A, Weisz GM, Schieber A, et al. Pyrolytic acrylamide formation from purified wheat gluten and gluten-supplemented wheat bread rolls. Mol Nutr Food Res 2006;50:87–93.
[27] Loaëc G, Jacolot P, Helou C, et al. Acrylamide, 5-hydroxymethylfurfural and Nε-carboxymethyl-lysine in coffee substitutes and instant coffees. Food Add Contam Part A 2014;31:593–604.
[28] Mizukami Y, Yoshida M, Isagawa S, et al. Acrylamide in roasted barley grains: presence, correlation with colour and decrease during storage. Food Add Contam Part A 2014;31:995–1000.
[29] European Commission. SANCO/10680/2014-revision 5. Draft commission recommendation of on investigations into the levels of acrylamide in food. 2013.
[30] Şenyuva HZ, Gökmen V. Study of acrylamide in coffee using an improved liquid chromatography mass spectrometry method: investigation of colour changes and acrylamide formation in coffee during roasting. Food Add Contam 2005;22:214–20.

[31] Biedermann M, Biedermann-Brem S, Noti A, et al. Methods for determining the potential of acrylamide formation and its elimination in raw materials for food preparation, such as potatoes. Mitt Lebensm Hyg 2002;93:653–67.
[32] Pastoriza S, Rufián-Henares JÁ, Morales FJ. Reactivity of acrylamide with coffee melanoidins in model systems. LWT-Food Sci Technol 2012;45:198–203.
[33] Summa CA, de la Calle B, Brohee M, et al. Impact of the roasting degree of coffee on the in vitro radical scavenging capacity and content of acrylamide. Lebensm Wiess Technol 2007;40:1849–54.
[34] Alves RC, Soares C, Csal S, et al. Acrylamide in espresso coffee: influence of species, roast degree and brew length. Food Chem 2010;119:929–34.
[35] Mucci LA, Dickman PW, Steineck G, et al. Dietary acrylamide and cancer risk: additional data on coffee. Br J Canc 2003;89:774–6.
[36] Granby K, Fagt S. Analysis of acrylamide in coffee and dietary exposure to acrylamide from coffee. Anal Chim Acta 2004;520:177–82.
[37] Svensson K, Abramsson L, Becker W, et al. Dietary intake of acrylamide in Sweden. Food Chem Toxicol 2003;41:1581–6.
[38] Swiss Federal Office of Public Health. Assessment of acrylamide intake by duplicate diet study. 2002. http://www.bfr.bund.de/cm/343/assessment_of_acrylamide_intake_by_duplicate_diet_study.pdf.
[39] Freisling H, Moskal A, Ferrari P, et al. Dietary acrylamide intake of adults in the European Prospective Investigation into cancer and nutrition differs greatly according to geographical region. Eur J Nutr 2013;52:1369–80.
[40] FDA, Food and Drug Administration. Available from: http://www.fda.gov/Food/FoodborneIllness Contaminants/ChemicalContaminants/ucm053549.htm.
[41] Di Novi M. The 2006 Exposure assessment for acrylamide. 2006. Available from: http://www.fda.gov/downloads/Food/FoodborneIllnessContaminants/UCM197239.pdf.
[42] Food Drink Europe. Food drink Europe acrylamide toolbox. 2013. Available from: http://www.fooddrinkeurope.eu/uploads/publications_documents/AcrylamideToolbox_2013.pdf. 2006.
[43] Anese M, Suman M, Nicoli MC. Technological strategies to reduce acrylamide levels in heated foods. Food Eng Rev 2009;1:169–79.
[44] Zyzak DV, Sanders RA, Stojanovic M, et al. Acrylamide formation mechanism in heated foods. J Agric Food Chem 2003;51:4782–7.
[45] Anonymous. Enzyme can help reduce acrylamide levels in coffee. Coffee & Cocoa 2011;38(29):31.
[46] Anese M, Quarta B, Peloux L, et al. Effect of formulation on the capacity of L-asparaginase to minimize acrylamide formation in short dough biscuits. Food Res Int 2011;44:2837–42.
[47] Dria GJ, Zyzak DV, Gutwein RW, et al. Method for reduction of acrylamide in roasted coffee beans, roasted coffee beans having reduced levels of acrylamide, and article of commerce. US Patent 7220440 B2. 2007.
[48] Anese M, Nicoli MC, Verardo G, et al. Effect of vacuum roasting on acrylamide formation and reduction in coffee beans. Food Chem 2014;145:168–72.
[49] Theurillat V, Leloup V, Liardon R, et al. Impact of roasting conditions on acrylamide formation in coffee, Montpellier, France: ASIC 21st International Conference, September 11–15, 2006.
[50] Fasullo Nachtrieb C. Coffee composition for brewing roasted coffee, and for reducing the level of acrylamide in brewed roasted coffee product, comprises roasted unbrewed coffee, cacao, and second herbal component, e.g. sambucus, rosehips, and hibiscus. US Patent US2013189417–A1. 2013.
[51] Delatour T, Périsse A, Goldmann T, et al. Improved sample preparation to determine acrylamide in difficult matrixes such as chocolate powder, cocoa, and coffee by liquid chromatography tandem mass spectroscopy. J Agric Food Chem 2004;52:4625–31.
[52] Hoenicke K, Gaterman R. Studies on the stability of acrylamide in food during storage. AOAC Int 2005;88:268–73.
[53] Cha M. Enzymatic control of the acrylamide level in coffee. Eur Food Res Technol 2013;236:567–71.
[54] Akillioglu HG, Gökmen V. Mitigation of acrylamide and hydroxymethyl furfural in instant coffee by yeast fermentation. Food Res Int 2014;61:252–6.
[55] Banchero M, Pellegrino G, Manna L. Supercritical fluid extraction as a potential mitigation strategy for the reduction of acrylamide level in coffee. J Food Eng 2013;115:292–7.

[56] Anese M, Manzocco L, Maltini E. Effect of coffee physical structure on volatile release. Eur Food Res Technol 2005;221:434–8.
[57] Soares CMD, Alves RC, Casal S, et al. Development and validation of a matrix solid-phase dispersion method to determine acrylamide in coffee and coffee substitutes. J Food Sci 2010;75:T57–63.
[58] Riediker S, Stadler RH. Analysis of acrylamide in food by isotope-dilution liquid chromatography coupled with electrospray ionization tandem mass spectrometry. J Chrom A 2003;1020:121–30.
[59] Mastovska K, Lehotay SJ. Rapid sample preparation method for LC−MS/MS or GC−MS analysis of acrylamide in various food matrices. J Agric Food Chem 2006;54:7001–8.
[60] Wenzl T, Szilagyi S, Rosen J, et al. Validation by collaborative trial of an isotope dilution liquid chromatographic tandem mass spectrometric method to determine the content of acrylamide in roasted coffee. Food Add Contam 2009;26:1146–52.
[61] Bortolomeazzi R, Munari M, Anese M, et al. Rapid mixed mode solid phase extraction method for the determination of acrylamide in roasted coffee by HPLC-MS/MS. Food Chem 2012;35:2687–93.

CHAPTER 10

Acrylamide in Soybean Products, Roasted Nuts, and Dried Fruits

Slađana Žilić
Department of Food Technology and Biochemistry, Maize Research Institute, Belgrade, Serbia

INTRODUCTION

Thermal processes have a strong impact on the final food quality: its nutritional, sensory, and functional characteristics. The latest developments in food processing have been aimed at substantial reduction of damages to nutrients and sensory components by applying the optimal heating conditions. In spite of that, the formation of compounds that are not naturally present in foods and that reveal harmful effects such as cytotoxic, carcinogenic, and mutagenic effects (neo-formed contaminants, NFC) is the major concern arising from heating processes [1]. These compounds can develop during baking, frying, and roasting due to various chemical reactions such as the Maillard reaction (MR), caramelization, and lipid oxidation. Acrylamide is one of the neo-formed contaminants that has acquired much attention due to its high toxicological potential and wide occurrence in foods. Shortly after the discovery of acrylamide by Swedish researchers [2], it has been clearly established that the major pathway for its formation in foods is the MR with free asparagine and reducing sugars as main reactants [3,4]. However, a number of theoretical mechanisms have been proposed for the formation of acrylamide in heated foods. In addition to *N*-glycosylasparagine (Schiff base), that is a direct precursor of acrylamide [5], 3-aminopropionamide (3-APA) [4,6], hydroxycarbonyls, dicarbonyls [4,7], alkadienals from lipid oxidation [8], and 5-hydroxymethyl-2-furfural (HMF) [9] can be involved in the acrylamide formation as important transient intermediates. Acrylamide has been classified as a potential carcinogen to humans by the IARC [10]. Also, many studies indicate that acrylamide is neurotoxic in animals and humans and that it is a reproductive toxicant and germ-cell mutagen in animals [11–13].

ACRYLAMIDE IN SOYBEAN PRODUCTS

In many countries, soybeans are principally used in the forms of soybean protein products, such as flours, grits, flakes, isolates, concentrates, and textured soya proteins, and besides are employed as ingredients in foods. Many soybean-containing products have been designated as functional foods due to the several health benefits attributed to

Acrylamide in Food
http://dx.doi.org/10.1016/B978-0-12-802832-2.00010-3

soybean, such as lowering of plasma cholesterol, prevention of cancer, diabetes, and obesity [14]. However, given that soybean must be processed before consumption due to the complex structure of its proteins and the high content of antinutritional components [15,16], the MR is the main chemical event occurring in this food ingredient during thermal processing [17]. The principal markers of the MR development in soybean products are furosine, as well as acrylamide, HMF, and carboxymethyl-lysine [18]. Results of Žilić et al. [18] indicate that heat generation and its transfer through the raw material, which are specific to each of the thermal processes, caused significant differences in behavior and concentration of heat-induced contaminants in the soybean products. In spite of all, the database of the neo-formed contaminants has not taken into account the huge number of soybean-based products which are widely present in the diet. Therefore, the data presented in this chapter contribute to the improvement of the knowledge of potential risks of infrared and microwave radiation, extrusion, fermentation, and baking conditions in soybean flakes, roasted kernels, grits, fermented soybean products (e.g., miso and soya sauces), and soybean-containing products, respectively.

Soybeans and the Maillard Reaction Reactants

Many studies indicate that the MR occurs in raw ingredients of food [19,20]. Rufián-Henares et al. [20] explained that the amount of neo-formed contaminants in non-heated cereals and soybean samples indicates *initial thermal damage* that depends on the amount and type of proteins. In the case of soybean, the high protein content (about 40%) could explain the higher acrylamide generation coming from the MR. Like many globulins from other species, soybean proteins characteristically have a good balance of the essential amino acids [21]. Soybeans are deficient in sulfur-containing amino acids and rich in acidic amino acids, such as aspartic acid and asparagine that correlate with the acrylamide formation. Average asparagine values ranged from 9.57% to 11.27% of the total residues [22,23]. In addition, the sugar content of soybean (60 and 110 mg/100 g of glucose and fructose, respectively) [20] may increase the acrylamide formation. Acrylamide may be formed in raw soybeans during storage depending on the humidity and temperature conditions. Soybean dried at 38 °C for 48 h prior to thermal processing was found to contain 22.36 ng/g of acrylamide. Under the stated conditions of drying, the initial moisture content of beans was reduced from 15.0% to 7.3% inducing the MR to a certain extent [18].

Roasted Soybean Products

Microwave and infrared (micronization) radiation are widely used for heating soybean. Similar to conventional heating methods such as boiling or frying, microwave and micronization processing can in the same way facilitate MRs, resulting in food browning. There is little information about microwave and infrared heating effects on acrylamide formation in soybean so far. These processes differ from each other in heat generation and heat transfer through the soybean. Infrared waves are emitted by the heating element and absorbed by the surface of soybeans. Heat is transferred through the soybeans by conduction. In addition, heated soybeans are subjected to pressure force of a flake roller. Unlike the micronization method, after microwaves' penetration, heat is generated within the product. The ability to absorb microwave energy is governed by the dielectric properties of soybeans. Water in soybeans is often the primary component responsible for dielectric heating [18]. According to results of Žilić et al. 's study [18] during microwave heating the acrylamide concentration rapidly increased to the maximum (59.82 ng/g) at 90 °C for 2 min. Extensive heating (≥115 °C for ≥3 min) resulted not only in darker color in soybeans, but also in lower acrylamide contents (Figure 1). These results suggest that acrylamide is formed as an intermediate product during the MR and its concentration begins to decrease as the rate of degradation exceeds the rate of formation during heating, which is in accordance with the results obtained by Gökmen and Şenyuva [24]. By comparison, Becalski et al. [25] analyzed a few food samples that were subjected to roasting and found that roasted soybeans contained 25 ng/g of acrylamide. Furthermore, according to the data of US Food and Drug Administration, the concentration of acrylamide in roasted soya nuts was 91 ng/g [26]. The increasing trend of acrylamide formation during micronization indicates the intensity of the heat treatment. The applied conditions of infrared radiation significantly increased the concentration of acrylamide in soya flakes. The acrylamide concentration was found to be 69.49 ng/g in soybeans infrared heated at 140 °C for 100 s, which was about 1.6 times higher than that found in soybean flakes obtained at 100 °C heated for 40 s [18] (Figure 2). However, infrared radiation much more influenced the acrylamide formation in cereals. On comparison, acrylamide concentration was found to be

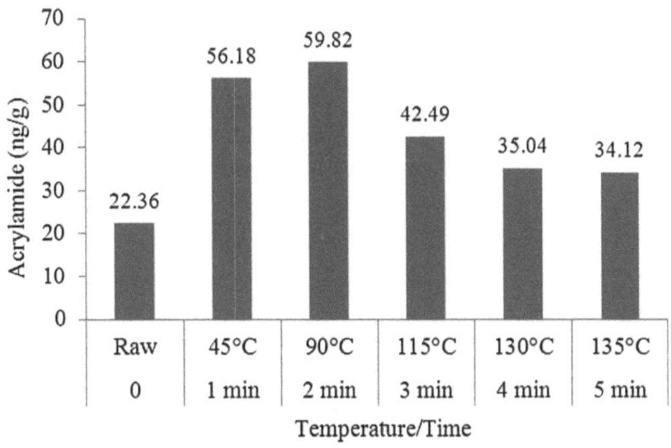

Figure 1 Acrylamide concentration in microwaved soybeans. *Source: Žilić et al. [18].*

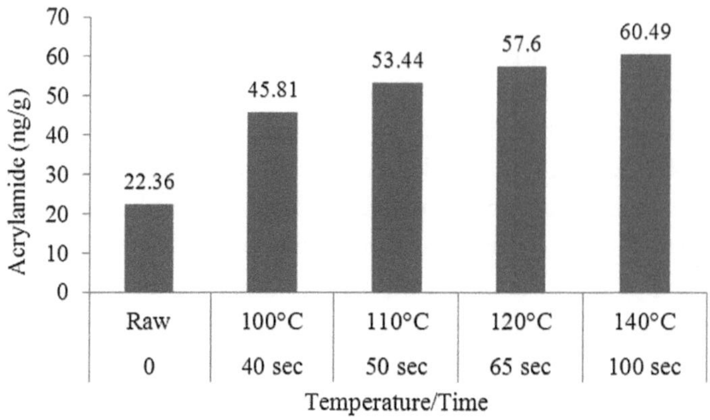

Figure 2 Acrylamide concentration in soybean flakes obtained by infrared heating (micronization). *Source: Žilić et al. [18].*

704.51 ng/g in maize kernels, which were infrared heated at 140 °C for 100 s. [27]. In addition, during infrared heating of soybean, the acrylamide formation greatly increases with the decreasing moisture content especially at the end of the heating process. An exponential increase in acrylamide concentration of fried potatoes due to reduced moisture conditions in the final stage of the frying process has been also reported in the study of Rydberg et al. [28].

Hydrothermal-Treated Soybean Products

Extrusion is a type of a hydrothermal treatment, which is often used for processing soybeans. The extruded soybean product usually changes texture because of the reduction of forces and release of moisture and heat. The cooking process takes place within the extruder where the material produces its own friction and heat due to the pressure generated. Results of Žilić et al. 's study [18] showed that the acrylamide content in different heated soybean products depends not only on the temperature, but also on the time of heating and the moisture content. However, the heating temperature appeared to have a greater effect on the acrylamide formation during extrusion. At a constant heating time and moisture content, increasing the temperature load results in the significantly increased amount of acrylamide formed in soybean reaching 62.62 ng/g during extrusion at 130 °C for 20–30 s [18]. However, acrylamide levels were similar in the soybean grits obtained by extrusion cooking at 130 °C and 140 °C for 20–30 s (Figure 3). Significantly higher acrylamide concentration (118.77 ng/g) was found in extruded soybean analyzed by Palermo et al. [29].

Fermented Soybean Products

The fermented soybean foods usually have high free amino acid contents. Regardless, fermented foods, including miso and soya sauce, rarely contained acrylamide at

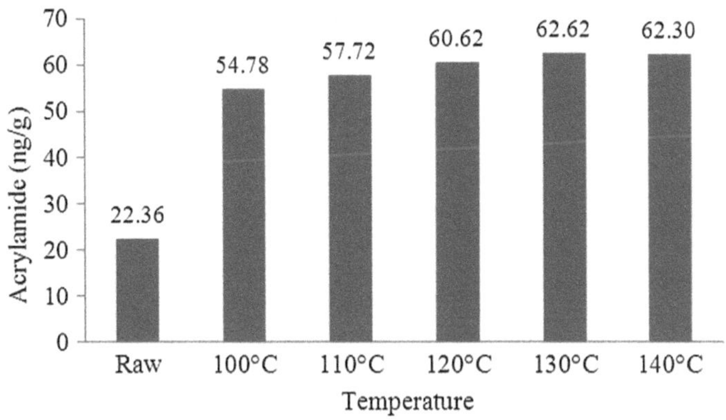

Figure 3 Acrylamide concentration in soybean grits obtained by extrusion. *Source: Žilić et al. [18].*

quantifiable levels. Though these foods are considered essential foods in the Asian diet, due to their use as seasonings in cooking, their intake is not as great as of cereal products. Therefore, from a health risk viewpoint, the acrylamide levels in those foods are of minimal concern. Four out of 50 analyzed samples of soya sauce were found to have low acrylamide concentrations (4.0–6.0 ng/g) [30]. At the same time, all 50 of the miso samples were found to contain no acrylamide. According to data available from the US Food and Drug Administration, acrylamide is not detected in soya sauce [31].

Soybean-Containing Products

Soybean-containing foods are widely consumed in our diet. Palermo et al. [29] have analyzed the concentration of acrylamide in nine products containing soya, such as crisp bread, shortbread, wholemeal cookies, toasted bread, breakfast cereals, crackers, dry cookies, butter cookies, and cereal bars, and these results are compared with the values in the conventional products (Table 1). The concentration of acrylamide in these products ranged from 48.45 ng/g in soya-containing toasted bread to 734.78 ng/g in soya-containing butter cookies. In most of the cases, the products containing soybean showed a higher acrylamide content compared to conventional products. It was interesting that all soya products except one had a higher water content. On the other hand, conventional product designated "dry cookies" behaved as expected, having a lower water content and also a higher acrylamide concentration than the corresponding soybean product (272.14 vs 63.16 ng/g) [29].

Okara is a by-product in the production of soybean cheese and can be used as a protein and as a dietary fiber supplement in various bakery products. However, according to Palermo et al. [29] okra cookies were found to have significantly higher acrylamide concentrations (588.84 ng/g) compared to that in wheat cookies (361.88 ng/g). Okara samples had an asparagine concentration in line with those usually present in wheat flour. On the other hand, in reducing sugar okara was richer than wheat flour (3830 vs 1700 mg/100 g) causing the formation of acrylamide to a large extent. In addition, soya is a common ingredient in infant

Table 1 Acrylamide concentrations determined in commercial soybean-containing samples and acrylamide concentration reported by the European Food Safety Agency (EFSA)

	Conc. (ng/g d.m.) in commercial samples		Conc. (ng/g d.m.) in EFSA ref.
	Soybean-containing	Conventional (control)	
Crisp bread	198.36	ND	5–2838
Shortbread	341.25	189.04	5–2949
Wholemeal cookies	496.33	169.58	15–499
Toasted bread	48.45	ND	10–1430
Breakfast cereals	64.42	90.04	10–440
Crackers	97.27	ND	5–830
Dry cookies	63.16	272.14	10–4256
Butter cookies	734.78	62.22	5–3324
Cereal bars	112.04	85.47	96

ND, not detected
Source: Palermo et al. [29].

formulas. Soya-based formulas are a nutritious alternative to cow's milk-based food for millions of infants worldwide. So far, according to data available from the US Food and Drug Administration, acrylamide has not yet been detected in these products [31]. Soya crisps can be similar in appearance and texture to potato chips and tortilla chips, but according to the data of the US Food and Drug Administration, it has a considerably lower content of acrylamide (17 ng/g). Potato and tortilla chips contained acrylamide in the range of 117 and 2762 ng/g and a value of 117 ng/g, respectively [31]. A total of 123 samples classi fied into 15 categories were analyzed for amounts of acrylamide by Chen et al. [32]. Among samples tested the acrylamide contents were the lowest (4.08–24.40 ng/g) in soybean paste.

ACRYLAMIDE IN ROASTED NUTS

Nuts are energy-dense foods, rich in oil and unsaturated fatty acids. The favorable fatty acid profile of consumed nuts probably contributes to their beneficial effects observed in epidemiologic studies (prevention of coronary heart disease and diabetes) and feeding trials (cholesterol lowering) [33]. Besides oil, nuts contain many bioactive compounds: vegetable protein, fiber, minerals, tocopherols, and phenolic compounds. Nuts also can contain antinutritional factors, such as tannins and phytates, which have to be inactivated by the heat treatment before human consumption [34]. Further, heat treatments inactivate enzymes and destroy undesirable microorganisms. Roasting is the main step of nut processing. Flavor, color, texture, and the appearance of nuts are significantly enhanced by the roasting process. These changes are mainly related to the MR and a large number of volatile components have been identified [35]. However, comparatively little information is available on acrylamide in roasted nuts, although the acrylamide content of roasted almonds can exceed 1000 ng/g [25]. This chapter includes the literature data about the acrylamide concentration in almond, tropical almond, hazelnut, chestnuts, pistachios, and peanuts.

Almonds

Compared to all other nuts, almonds have the highest content of nutrients and beneficial components. Sweet almonds are edible, while bitter almonds are used for oil production. Almonds contain acrylamide precursors, asparagine and reducing sugars, in high quantity. The content of free asparagine has been reported to be in the range of 98–641 mg/100 g [36], while sucrose was reported to be the main constituent of carbohydrates in almonds. The sucrose content has ranged from 3100 to 4680 mg/100 g, while the total content of glucose and fructose ranged from 70 to 134 mg/100 g [37]. The high content of precursors, as well as the roasting temperature and the physical form (whole, sliced, and cut) of the almond kernel strongly influences the acrylamide formation. According to results of Amrein et al. [38] almonds of the European origin contained significantly less free asparagine and formed significantly less acrylamide during roasting as compared to almonds from the United States. The roasted almonds contained a very high amount of acrylamide. Becalski et al. [25] detected acrylamide in roasted almonds in the concentration of 260 ng/g, while the highest concentration of 2147 ng/g was found in the dark roasted almonds used as an ingredient in biscuit production [39]. In marzipan, where cooked, blanched, and raw almonds are usually used, no acrylamide (limit of detection 20 ng/g) or very low amounts were found. In the first 11 min of roasting at 130 °C acrylamide was not detected. However, after heating for 15 min at 150 °C 35 times more acrylamide was found than after heating for 16.5 min at 130 °C (715 vs 44 ng/g) [39]. The same authors found that glucose and fructose react much faster than free asparagine upon heating of almond kernels. For example, after roasting at 180 °C for 7 min only 4%, that is, 7% of the initial content of glucose, that is, fructose, was found, respectively, whereas the free asparagine content had decreased only by 27%. According to Amrein et al. [40] a decrease in the acrylamide content was found after roasting at above 180 °C. Nevertheless, these samples were over-roasted. The measurement of the kernel temperature during roasting showed that the onset of the acrylamide formation in almonds corresponded to that moment when the kernel temperature exceeded ~140 °C [41]. Acrylamide is not stable in roasted almonds. The acrylamide contents decreased in samples stored at room temperature over a period of up to 300 days. These contents decreased significantly by 44–62% between the first and the final analyses [38]. Although the level of acrylamide in roasted almonds is high, the low rate of consumption and infrequent consumption would suggest that the average daily intake of acrylamide by consumers of roasted almonds would not exceed 1.0 μg/d [42]. The acrylamide content of various almond-containing products or prefabricates was determined by Amrein et al. [39]. The median acrylamide content of roasted almonds was 4-fold higher than that of bakery products with almonds. For the bakery products, only eight values exceeded 200 ng/g. Nonetheless, there was one biscuit product that contained an extraordinary high acrylamide content of 1307 and 1574 ng/g. The presence of ammonium hydrogen carbonate (E 503), and a large amount of reducing sugar could be the reason for this high value.

Tropical Almonds

Tropical almond nuts can be consumed roasted or incorporated into many recipes. However, little information is available on the acrylamide content and its reduction in the roasted tropical almond nuts. According to the available data, although tropical almond has a relatively high content of reducing sugars, sucrose (3030 mg/100 g), fructose, and glucose (300 mg/100 g) [43], low levels of acrylamide (8–86 ng/g) were obtained in the roasted nuts [44].

Hazelnuts

Hazelnuts are an important commercial crop in many countries. However, Turkey is the largest producer of hazelnuts in the world with about 73% of the worldwide production. Roasted hazelnuts are broadly used as ingredients in the preparation of chocolate and bakery products, but there is little information on the acrylamide content of roasted hazelnuts. According to results of Amrein et al. [38], roasted hazelnuts contained very little acrylamide (16–56 ng/g) because of the low content of free asparagine in the raw nut. Higher acrylamide content was detected in Chinese roasted hazelnuts at a concentration of about 150 ng/g [45]. Hazelnuts contained about 40 times less free asparagine as compared to that in almonds, while the sugar contents were alike. The free asparagine contributed only 2%–7.5% to the pool of free amino acids [38,46]. Under the same conditions (165 °C for 12.5 min), the content of acrylamide in hazelnuts was 15 times less compared to the content in roasted almonds [38]. According to the results of the same authors, during hazelnut roasting the sugars were consumed in a similar manner as compared to that in almonds. After 10 min at 150 °C, over 80% of the reducing sugars were degraded, while the free asparagine content had decreased only by 30%.

Chestnuts

Nowadays, food products obtained from chestnuts have become increasingly important due to their healthy and organoleptic properties. The carbohydrate content of chestnuts is comparable with that of wheat and rice. In addition, sucrose is the main sugar present in chestnuts. The mean concentration of sucrose was 7390 mg/100 g, which points out to the important variation between the varieties of chestnuts analyzed (3110–9940 mg/100 g) [47]. De la Montaña Míguelez et al. [48] determined that glucose and fructose concentrations varied between 0 and 310 mg/100 g for both monosaccharides, respectively, whereas, in comparison to these data, Künsch et al. [49] found higher concentrations of fructose (370–690 mg/100 g), and similarly of glucose (100–120 mg/100 g). As far as the amino acids, aspartic acid, asparagine, and glutamic acid are predominantly free amino acids in chestnuts. According to De Vasconcelos et al. [50] the asparagine content varied from 62.0 to 149.5 mg/100 g. The variations in the acrylamide precursor contents, as well as the conditions of the thermal treatment, result in high variability of the acrylamide content in the roasted chestnut samples. The highest content of commercial roasted chestnuts was determined in one sample originating from Spain (1278 ng/g) [51]. However, this seems to be an extreme value. The mean acrylamide content of the commercial roasted chestnut samples was determined to be about 90 ng/g. According to results of Karasek et al. [51], acrylamide concentrations in roasted chestnuts, excluding samples from Spain, were in the range of <9–828 ng/g, whereas only low amounts (<4–159 ng/g) were found in chestnut products (Tables 2 and 3). These foods (puree, flour, pasta, and glazed chestnuts) are of minor relevance with regard to the acrylamide intake by humans via food, due to their low content, and probable low quantities consumed.

The acrylamide content in roasted Chinese chestnut was detected at concentrations of about 140–160 ng/g [45]. After 35 min of roasting at 200 °C, when chestnut has become deeply dark, the acrylamide content was around 9000 ng/g. After 40 min the acrylamide content decreased [51].

Pistachios

A large percentage of pistachios are consumed as salted and as roasted snacks and confections. Pistachios are also used in food industries for the production of cakes, biscuits, bread, candies, chocolates, ice creams, fermented meat, sauces, pudding, and desserts.

Table 2 Acrylamide content in roasted chestnuts of different origin

Origin	Acrylamide (ng/g)	Number of samples
Austria-Gratz	46–203	9
Austria-Vienna	<LOQ–74	3
Czech Republic-Prague	88–315	2
France-Provence	828	1
Germany-Aachen	23–265	4
Germany-Cologne	32–100	2
Spain-Segovia	1278	1

LOQ–9 μg/kg.
Source: Karasek et al. [51].

Table 3 Acrylamide content in chestnut-based products of different origin

	Origin	Acrylamide (ng/g)
Cooked chestnut	France	36–38
Chestnut puree	France, Hungary, Italy	<LOQ–9
Chestnut flour	Italy	159
Chestnut pasta	Italy	4
Sweet chestnut	Italy	12

LOQ–4 μg/kg.
Source: Karasek et al. [51].

Roasting is the key unit operation in pistachio processing and improves the characteristic aroma, color, texture, and appearance of all nuts, including pistachio [52]. However, there is no detailed information about the effect of roasting on the acrylamide concentration in this nut. Özer et al. [53] detected acrylamide in pistachios in the concentration of 462 and 318 ng/g, while the lowest concentration of 300 and 83 ng/g was found in baklava-containing pistachios cooked at temperatures 240 °C and 220 °C, respectively. According to Ötles and Ötles [54] the acrylamide level in pistachio garland bun was <30 ng/g.

Peanuts

Peanuts are a good source of protein, lipid, and fatty acids for human nutrition. A larger proportion of peanut production in the world is destined to foods such as peanut butter and snack products. Peanuts are roasted to achieve the desired color and flavor characteristics in peanut products, especially in peanut butter. When it comes to precursors in the synthesis of acrylamide, the results of Venkatachalam and Sathe [55] showed that peanuts had a higher content of asparagine but a lower content of sugars in comparison with other nuts. For this reason, similar to hazelnuts, roasted peanuts have a relatively low content of acrylamide. According to the available data, the content of acrylamide in roasted peanuts does not exceed 55 ng/g [56,57]. The US Food and Drug Administration has tested 15 samples of roasted peanuts, and levels of acrylamide ranged from non-detectable to 28 ng/g [31]. Having this in mind, it is understandable that the contribution of roasted peanuts to the overall acrylamide intake is marginal. According to the data of the Office of Environmental Health Hazard Assessment, the lower (population-based) and upper (everyday eaters) bounds on acrylamide intake from roasted peanuts did not exceed 1.0 μg/d [42]. A somewhat higher content of acrylamide was detected in peanut products. In accordance with the data of Health Canada [58], the concentrations of acrylamide in peanut butter available in the Canadian market ranged from 60 to 133 ng/g. The data of the US Food and Drug Administration for the acrylamide content were similar, and it ranged from 76 to 125 ng/g in the creamy peanut butter samples [31]. On the other hand, the South Australia Department of Health undertook a limited survey to examine concentrations of acrylamide in non-carbohydrate-based foods and found that the concentrations of acrylamide in peanut butter were only in trace amounts [57]. The Office of Environmental Health Hazard Assessment established that the upper (everyday eaters) bound on acrylamide intake from peanut butter exceeds 1.0 μg/d by 3-fold, while the lower (population-based) bound does not exceed 1.0 μg/d. This means that persons who consume peanut butter would have to eat peanut butter once every third day to achieve an acrylamide intake of 1.0 μg/d [42]. In addition to peanut butter, acrylamide was detected in peanut paste. A survey of acrylamide levels in foods from the Turkish market showed the presence of 45–63 ng/g of acrylamide in the peanut paste samples [59].

ACRYLAMIDE IN DRIED FRUITS

The use of fresh fruits in the diet is a great way to get needed nutrients and vitamins. However, in recent years, special attention has been paid to the quality of dehydrated foods, such as dried fruits. Processing and drying of fruits significantly affect their color, texture, flavor, density, and sorption characteristics. As a result of drying, the fruit loses moisture thereby increasing the concentration of nutrients in the remaining mass. However, degradation of heat-labile bioactive compounds and vitamins can occur. Ascorbic acid, as well as thiamine is heat-sensitive, but other water-soluble vitamins are more stable to heat and oxidation, and according to Fellows [60] losses during drying rarely exceed 5–10%. A number of natural volatile flavor compounds of fruits are reduced by heat treatments. On the other hand, heat treatments induce neo-formed volatile flavor compounds through the autoxidation of unsaturated fatty acids, thermal decomposition, and initiation of the MR. However, if the fruits are sulfured, enzymatic browning can be inhibited, and the MR retarded. The dietary exposure to acrylamide from the dried fruit is often underestimated. However, for instance, the additional exposure from dried fruits is estimated to be less than 0.3% of total exposure [61] (mean level in dried fruits: 47 ng/g [62] and its average consumption amount: 0.62 g/person/day). Due to this reason, details on the content of acrylamide in dried fruit, such as dried plums, pears, apricots, banana, raisins, and figs, is covered in this chapter. Generally, data on acrylamide in dried fruit are scarce in the literature.

Dried Plums (Prunes)

After harvest, plums that will be used to make prunes are dehydrated in hot air at 85–90 °C for 18 h. Following dehydration, plums are further processed into prune juice, puree, or other prune products. Prunes are considered to be exemplar of functional foods because of their sweet flavor and established mild laxative effects. Dried plums contain about 6100 mg/100 g of dietary fiber, while their laxative action could be explained by high sorbitol content (14,700 mg/100 g). Prunes do not mediate a rapid rise in blood sugar concentration, even though they are a good source of energy in the form of simple sugars [63]. In addition to sugar, the data of Amrein et al. [40] demonstrate that asparagine is present in large amounts in plums, which partly explains the presence of acrylamide in the corresponding dried products. The analyses of free amino acids showed that fresh plums contained 179 mg/100 g of free asparagine. Furthermore, the data presented by Amrein et al. [40] reveal that dried plums may contain substantial amounts of acrylamide, although the temperatures applied during drying usually do not exceed 90 °C. A sample of dried plums contained from 19 to 146 ng/g of acrylamide. The levels of acrylamide in pitted prunes in the United States ranged from 31 to 87 ng/g [64]. In addition, the concentration of acrylamide in dried plums and prunes without pits, as published in the studies of Health Canada [58], amounted to <10 to 101 and 332 ng/g, respectively. According to the results of Becalski et al. [65], there were significant differences in acrylamide concentration in dried plums between different brands. For example, dried plums of brand B contain, on average, more than 5.7 times the amount of acrylamide found in dried plums of brand D. The concentration in these samples ranged from 58 to 332 ng/g. In order to examine the effect of additional heating, Amrein et al. [40] exposed dried plums to a moderate heat treatment (120 °C, 40 min), which corresponded to the potential of acrylamide formation as used for potatoes. The values of acrylamide in re-heated plums ranged from 1010 to 2240 ng/g, that is, on average, 19.7 times higher than in prunes. Becalski et al. [65] analyzed the acrylamide concentration in 24 samples of baby foods based on dried plums and 18 other prune products. The presence of high levels (about 900 ng/g) of acrylamide in one organic brand of prune nectar is rather surprising. The acrylamide contents of 75–261 ng/g and 33–61 ng/g were reported for baby foods and apple prune juice blend, respectively.

Dried Pears

Pears contain acrylamide precursors, asparagine and reducing sugars, in a relatively high concentration. According to Fourie et al. [66] fructose was the major sugar present in pears and comprised between 41.0% and 61.0% of the total sugars, or between 4990 and 5690 mg/100 g. The glucose content varied from 1000 to 2800 mg/100 g, while preliminary analyses of free amino acids in fresh fruit samples showed that fresh pears contained ~69 mg/110 g of free asparagine [40]. The same authors observed the contents of acrylamide in dried pears samples. The highest concentration was found in dried large, dark whole pears ("Weichspeckbirnen" in German). The average amount of acrylamide of three "Weichspeckbirnen" dried pears categories was 1457 ng/g. In addition, the values of single fruits varied greatly. For example, the single pears of whole, dark dried pears ("Weichspeckbirnen") from the same lot contained 500, 990, 1070, 1430, 2760, and 4710 ng/g acrylamide. However, the values of the other pear samples, not belonging to the "Weichspeckbirnen" category, amounted to less than 10 ng/g. The concentration of acrylamide in these pears coincides with the results of Health Canada [58].

Dried Apricots, Figs, Banana, and Raisins

In the literature, data on acrylamide in dried apricots, figs, banana, and raisins are very scarce. The only data are found in the study of Amrein et al. [40]. A sample of dried apricots and dried banana contained 180 ng/g and 50 ng/g of acrylamide, respectively. In dried figs and raisins, only traces of acrylamide were detected (LOQ 10 ng/g).

SYNOPSIS OF ANALYTICAL TECHNIQUES

The presence of acrylamide in soybean products, roasted nuts, and dried fruits is determined using gas chromatography–mass spectrometry (GC–MS) after derivatization with bromine, or by a liquid chromatography–mass spectrometry (LC–MS) directly. The bromination procedure, which is a derivatization step used in the GC–MS method, is complex and the derivatized products may be thermally unstable and adsorptive due to remaining amide groups. Yamazaki et al. [30] have improved GC–MS method using xanthydrol as the derivatization reagent that are classically used for the identification of urea. This reagent reacts with the amide group of acrylamide to form a stable xanthyl compound used in subsequent GC–MS. Quantification of acrylamide by LC–MS using water or 10 mM formic acid as extraction solvent is relatively simple. The extracted samples require purification. The coextracted colloids are precipitated by adding 0.68 M potassium hexacyanoferrate (II) trihydrate solution (Carrez I) and 2 M zinc sulfate heptahydrate solution (Carrez II). Purification steps also include defatting with n-hexane, partition with ethyl acetate, solid phase extraction (SPE) with different sorbents (elution through a preconditioned MCX or HLB Oasis cartridge). This procedure provides a clean sample amenable to LC-MS/MS analysis.

KEY FACTS

- Acrylamide (2-propenamide) is a colorless and odorless crystalline solid that is used as a cement binder and in the synthesis of polymers and gels.
- Since 2002 acrylamide has been added to the list of food-borne toxicants.
- The major pathway for acrylamide formation in foods is the MR with free asparagine as the main precursor. In addition to N-glycosylasparagine (Schiff base), 3-aminopropionamide (3-APA), hydroxycarbonyls, dicarbonyls, alkadienals from lipid oxidation and 5-hydroxymethyl-2-furfural (HMF) can be involved in the acrylamide formation as important transient intermediates.
- Acrylamide has two reactive sites—the electron-deficient vinylic double bond and the amide group. The double bond can participate in nucleophilic reactions with active functional groups (the SH group, the α-NH_2 groups, the ε-NH_2 the ring NH group), Dielse–Alder additions and radical reactions.
- The International Agency for Research on Cancer (IARC) has classified acrylamide as "probably carcinogenic to humans" (Group 2A). Many studies indicate that acrylamide is neurotoxic in humans and that it is a reproductive toxicant, a germ-cell mutagen and a carcinogen in animals.
- Avoidance or elimination of foods containing acrylamide is impossible. One the most promising tool to control acrylamide content in heat-treated foods is the addition of the enzyme asparaginase, as well as the control of thermal treatment conditions.

MINI DICTIONARY

Microwave radiation It is a unit operation based on the effects of microwaves that are placed in the near infrared spectrum with a wavelength ranging from one mm to several cm.

Micronization It is a unit operation based on the effects of near-infrared rays of a wavelength ranging from 1.8 to 3.4 μm. Kernels were fed onto the vibratory conveyor via a vibrating feeder providing a uniform exposure of kernels to infrared energy. The heated kernels were subjected to pressure force of a flake roller.

Extrusion It is a type of a hydrothermal treatment. This procedure is based on the temperature impacts appearing during friction of the material that is squeezed through a cylinder by a specially designed volute.

Fermented soybean These are foods produced by the action of microorganisms. The natural bacteria feed on sugar and starch from food creating lactic acid.

Roasting It is a cooking method that uses dry heat, whether by an oven, or through other heat sources.

Drying It is a mass transfer process consisting of the removal of water by evaporation from a solid food.

SUMMARY POINTS

- The major contributors to acrylamide intake are potato crisps and chips, bread, and coffee. Other foods, including inter alia soybean products, roasted nuts, and dried fruits, can account for up to 10% of the total intake.
- Despite the mostly marginal contribution of roasted nuts and dried fruits to the overall acrylamide intake, the concentration of this contaminant therein can be very high.
- In addition to moisture, heat generation, and its transfer through the raw material, which are specific to each of the thermal processes, caused significant differences in behavior and concentration of acrylamide in the soybean products.
- The fermented soybean foods rarely contained acrylamide at quantifiable levels.
- In most of the cases, the products containing soybean showed a higher acrylamide content compared to conventional products.
- The content of acrylamide precursors, asparagine and reducing sugars, in high quantity, strongly influence the acrylamide formation in almond kernel.
- Acrylamide is not stable in roasted almonds.
- Roasted hazelnuts contained very little acrylamide because of the low content of free asparagine in the raw nut.
- The variations in the acrylamide precursor contents result in high variability of the acrylamide content in the roasted chestnut samples.
- Due to a higher content of asparagine but a lower content of sugars in comparison with other nuts, roasted peanuts have a relatively low content of acrylamide.
- Dried plums may contain substantial amounts of acrylamide, although the temperatures applied during drying usually do not exceed 90 °C.
- The values of acrylamide in re-heated plums are, on an average, 19.7 times higher than in prunes.
- There were significant differences in acrylamide concentration in dried plums and pears between different brands.
- When the fruits are sulfured, enzymatic browning can be inhibited, and the MR retarded.

REFERENCES

[1] Capuano E, Fogliano V. Acrylamide and 5-hydroxymethylfurfural (HMF): a review on metabolism, toxicity, occurrence in food and mitigation strategies. LWT-Food Sci Technol 2011;44(4):793–810.

[2] Tareke E, Rydberg P, Karlsson P, Eriksson S, Törnqvist M. Analysis of acrylamide, a carcinogen formed in heated foodstuffs. J Agri Food Chem 2002;50(17):4998–5006.

[3] Mottram DS, Wedzicha BL, Dodson AT. Acrylamide is formed in the Maillard reaction. Nature 2002;419(6906):448–9.

[4] Zyzak DV, Sanders RA, Stojanovic M, Tallmadge DH, Eberhardt BL, Ewald DK, et al. Acrylamide formation mechanism in heated foods. J Agri Food Chem 2003;51(16):4782–7.

[5] Stadler RH, Robert F, Riediker S, Varga N, Davidek T, Devaud S, et al. In-depth mechanistic study on the formation of acrylamide and other vinylogous compounds by the Maillard reaction. J Agri Food Chem 2004;52(17):5550–8.

[6] Yaylayan VA, Stadler RH. Acrylamide formation in food: a mechanistic perspective. J AOAC Int 2005;88(1):262–7.

[7] Amrein TM, Schonbachler B, Escher F, Amadò R. Acrylamide in gingerbread: critical factors for formation and possible ways for reduction. J Agri Food Chem 2004;52(13):4282–8.

[8] Zamora R, Hidalgo FJ. Contribution of lipid oxidation products to acrylamide formation in model systems. J Agri Food Chem 2008;56(15):6075–80.

[9] Gökmen V, Kocadağlı T, Goncuoğlu N, Atac Mogol B. Model studies on the role of 5-hydroxymethyl-2-furfural in acrylamide formation from asparagine. Food Chem 2012;132(1):168–74.

[10] IARC. Monographs on the evaluation of the carcinogenic risks of chemicals to humans, vol. 60. Lyon: International Agency for Research on Cancer (IARC); 1994.

[11] Friedman M. Chemistry, biochemistry, and safety of acrylamide. A review. J Agri Food Chem 2003;51(16):4504–26.

[12] LoPachin RM. The changing view of acrylamide neurotoxicity. Neurotoxicology 2004;25(4):617–30.

[13] Exon JH. A review of the toxicology of acrylamide. J Toxicol Env Heal B 2006;9(5):397–412.

[14] Friedman M, Brandon DL. Nutritional and health benefits of soy proteins. J Agri Food Chem 2001;49(3):1069–86.

[15] Žilić S, Božović I, Savić S, Mladenović-Drinić S, Bekrić V. Effects of microwave radiation of content of soluble proteins, urease and trypsin inhibitor. Food Sci Biotechnol 2002;11(6):590–5.

[16] Žilić S, Božović I, Hadži-Tašković Šukalović V. Thermal inactivation of soybean bioactive proteins. Int J Food Eng 2012;8(4). http://www.degruyter.com/view/j/ijfe. [23.10.12].

[17] Žilić S, Božović I, Savić S, Šobajić S. Heat processing of soybean kernel and its effect on lysine availability and protein solubility. Cent Eur J Biol 2006;1(4):572–83.

[18] Žilić S, Ataç Mogol B, Akıllıoğlu G, Serpen A, Delić N, Gökmen V. Effects of extrusion, infrared and microwave processing on Maillard reaction products and phenolic compounds in soybean. J Sci Food Agri 2014;94(1):45–51.

[19] Guerra-Hernandez E, Corzo N, Garcia-Villanova B. Maillard reaction evaluation by furosine determination during infant cereal processing. J Cereal Sci 1999;29(2):171–6.

[20] Rufián-Henares AJ, Delgado-Andrade C, Morales JF. Assessing the Maillard reaction development during the toasting process of common flours employed by the cereal products industry. Food Chem 2009;114(1):93–9.

[21] Žilić S, Barać M, Pešić M, Ignjatović-Micić D, Srebrić M, Mladenović Drinić S. Characterization of proteins from kernel of different soybean varieties. J Sci Food Agri 2011;91(1):60–7.

[22] Zarkadas CG, Voldeng HD, Yu ZR, Choi VK. Determination of the protein quality of three new northern adapted cultivars of common and miso type soybeans by amino acid analysis. J Agri Food Chem 1997;45(4):1161–8.

[23] Zarkadas CG, Gagnon C, Gleddie S, Khanizadeh S, Cober ER, Guillemette RJD. Assessment of the protein quality of fourteen soybean [Glycine max (L.) Merr.] cultivars using amino acid analysis and two-dimensional electrophoresis. Food Res Int 2007;40(1):129–46.

[24] Gökmen V, Şenyuva HZ. Study of colour and acrylamide formation in coffee, wheat flour and potato chips during heating. Food Chem 2006;99(2):238–43.

[25] Becalski A, Lau PYB, Lewis D, Seaman WS. Acrylamide in foods: occurrence, sources, and modelling. J Agri Food Chem 2003;51(3):802–8.
[26] FDA. Survey data on acrylamide in food: Individual food products. Data through October 7, 2004. United States Food and Drug Administration; 2004. Available from: http://www.fda.gov/food/foodborneillnesscontaminants/chemicalcontaminants/ucm053549.htm#u1102.
[27] Žilić S, Ataç Mogol B, Akıllıoğlu G, Serpen A, Babić M, Gökmen V. Effects of infrared heating on phenolic compounds and Maillard reaction products in maize flour. J Cereal Sci 2013;58(1):1–7.
[28] Rydberg P, Eriksson S, Tareke E, Karlsson P, Ehrenberg L, Törnqvist M. Investigations of factors that influence the acrylamide content of heated foodstuffs. J Agri Food Chem 2003;51(24):7012–8.
[29] Palermo M, Fiore A, Fogliano V. Okara promoted acrylamide and carboxymethyl-lysine formation in bakery products. J Agri Food Chem 2012;60(40):10141–6.
[30] Yamazaki K, Isagawa S, Kibune N, Urushiyama T. A method for the determination of acrylamide in a broad variety of processed foods by GC–MS using xanthydrol derivatization. Food Addit Contam 2012;29(5):705–15.
[31] FDA. Survey data on acrylamide in food: Individual food products. Data through November 15, 2002. United States Food and Drug Administration; 2002. Available from: http://www.fda.gov/food/foodborneillnesscontaminants/chemicalcontaminants/ucm053549.htm#u1102.
[32] Chen YH, Xia EQ, Xu XR, Ling WH, Li S, Deng GF, et al. Evaluation of acrylamide in food from China by a LC/MS/MS method. Int J Environ Res Publ Health 2012;9(11):4150–8.
[33] Ros E. Nuts and novel biomarkers of cardiovascular disease. Am J Clin Nutr 2009;89(Suppl. 5): 1649S–56S.
[34] Xu Y, Hanna MA. Nutritional and anti-nutritional compositions of defatted Nebraska hybrid hazelnut meal. Int J Food Sci Technol 2011;46(10):2022–9.
[35] Nicolotti L, Cordero C, Bicchi C, Rubiolo P, Sgorbini B, Liberto E. Volatile profiling of high quality hazelnuts (Corylusavellana L.): chemical indices of roasting. Food Chem 2013;138(2–3):1723–33.
[36] Seron LH, Poveda EG, Moya MSP, Carratalá MLM, Berenguer-Navarro V, Grané-Teruel N. Characterisation of 19 almond cultivars on the basis of their free amino acids composition. Food Chem 1998;61(4):455–9.
[37] Fourie PC, Basson DS. Sugar content of almond, pecan and macadamia nuts. J Agri Food Chem 1990;38(1):101–4.
[38] Amrein TM, Lukac H, Andres L, Perren R, Escher F, Amadò R. Acrylamide in roasted almonds and hazelnuts. J Agri Food Chem 2005;53(20):7819–25.
[39] Amrein TM, Andres L, Schonbachler B, Conde-Petit B, Escher F, Amadò R. Acrylamide in almond products. Eur Food Res Technol 2005;221(1–2):14–8.
[40] Amrein TM, Andres L, Escher F, Amadò R. Occurrence of acrylamide in selected foods and mitigation options. Food Addit Contam 2007;24(Suppl. 1):13–25.
[41] Lukac H, Amrein TM, Perren R, Conde-Petit B, Amado R, Escher F. Influence of roasting conditions on the acrylamide content and the color of roasted almonds. J Food Sci 2007;72(1):C033–8.
[42] OEHHA. Characterization of acrylamide intake from certain foods. Reproductive and Cancer Hazard Assessment Section, Office of Environmental Health Hazard Assessment, California Environmental Protection Agency. 2005. Available from: http://oehha.ca.gov/prop65/law/pdf_zip/acrylamideintake Report.pdf.
[43] Ezeokonkwo CA. Comparative effects of dry- and moist-heating treatments on the biochemical characteristics of *Terminalia catappa* seed. Food Sci Technol Int 2007;13(2):165–71.
[44] Lasekan O, Abbas K. Analysis of volatile flavour compounds and acrylamide in roasted Malaysian tropical almond (*Terminaliacatappa*) nuts using supercritical fuid extraction. Food Chem Toxicol 2010;48(8–9):2212–6.
[45] Zhuang H, Zhang T, Liu J, Yuan Y. Detection of acrylamide content in traditional Chinese food by high-performance liquid chromatography tandem mass spectrometry method. CyTA–J Food 2012;10(1):36–41.
[46] Alasalvar C, Shahidi F, Liyanapathirana CM, Ohshima T. Turkish tombul hazelnut (*Corylus avellana* L.). Compositional characteristics. J Agri Food Chem 2003;51(13):3790–6.
[47] Suárez HM, Galdón RB, Mesa RD, Romero DC, Rodríguez RE. Sugars, organic acids and total phenols in varieties of chestnut fruits from Tenerife (Spain). Food Nutr Sci 2012;3(6):705–15.

[48] De La Montaña Míguelez J, Míguez Bernárdez M, Garcia Queijeiro JM. Composition of varieties of chestnuts from Galicia (Spain). Food Chem 2004;84(3):401–4.
[49] Künsch U, Schärer H, Patrian B, Hurter J, Conedera M, Sassella A, et al. Quality assessment of chestnut fruits. Acta Hort 1999;494:119–27.
[50] De Vasconcelos MCBM, Bennett RN, Rosa EAS Cardoso JVF. Primary and secondary metabolite composition of kernels from three cultivars of Portuguese chestnut (*Castanea sativa* Mill.) at different stages of industrial transformation. J Agri Food Chem 2007;55(9):3508–16.
[51] Karasek L, Wenzl T, Anklam E. Determination of acrylamide in roasted chestnut and chestnut-based foods by isotope dilution HPLC-MS/MS. Food Chem 2009;114(4):1555–8.
[52] Hojjati M, Calín-Sánchez Á, Hadi Razavi S, Carbonell-Barrachin ÁA. Effect of roasting on colour and volatile composition of pistachios (*Pistaciavera* L.). Int J Food Sci Technol 2013;48(3):437–43.
[53] Özer MS, Kola O, Altan A, Duran H, Zorlugenç B. Acrylamide content of some Turkish traditional desserts. J Food Agric Environ 2012;10(1):74–7.
[54] Ötles S, Ötles S. Acrylamide in food: chemical structure of acrylamide. Electron J Environ Agric Food Chem 2004;3(5):1–8.
[55] Venkatachalam M, Sathe SK. Chemical composition of selected edible nut seeds. J Agri Food Chem 2006;54(13):4705–14.
[56] Hoenicke K, Gatermann R. Studies on the stability of acrylamide in food during storage. J AOAC Int 2005;88(1):268–73.
[57] Eckert P. Acrylamide. A survey of acrylamide in non-carbohydrate based foods. Government of South Australia: Survey Report. 2006. p. 1–8.
[58] Health Canada. Health Canada's revised exposure assessment of acrylamide in food. Health Canada; 2012. Available from: www.hc-sc.gc.ca.
[59] Ölmez H, Tuncay F, Özcan N, Demirel S. A survey of acrylamide levels in foods from the Turkish market. J Food Compos Anal 2008;21(7):564–8.
[60] Fellows P. Processing by application of heat. In: Food processing technology; Principles and practice. New York:VC; 1988. p. 306–10.
[61] FEHD. The first Hong Kong total diet study: Acrylamide. Hong Kong: Centre for Food Safety of the Food and Environmental Hygiene Department; 2013.
[62] FAO/WHO. Safety evaluation of certain contaminants in food. In: WHO food additives series 63/FAO JECFA, monographs 8 of 72nd Meeting of the Joint FAO/WHO Expert Committee on Food Additives (JECFA). 2011. Available from: http://whqlibdoc.who.int/publications/2011/9789241660631_eng.pdf.
[63] Stacewicz-Sapuntzakis M, Bowen PE, Hussain EA, Damayanti-Wood BI, Farnsworth NR. Chemical composition and potential health effects of prunes: a functional food? Crit Rev Food Sci Nutr 2001;41(4):251–86.
[64] FDA. Survey data on acrylamide in food: Individual food products. United States Food and Drug Administration Electronic citation; 2009. Available from: http://www.fda.gov/Food/FoodSafety/FoodContaminantsAdulteration/ChemicalContaminants/Acrylamide/ucm053549.htm/.
[65] Becalski A, Brady B, Feng S, Gauthier BR, Zhao T. Formation of acrylamide at temperatures lower than 100 °C: the case of prunes and a model study. Food Addit Contam 2011;28(6):726–30.
[66] Fourie PC, Hansmann CF, Oberholzer HM. Sugar content of fresh apples and pears in South Africa. J Agri Food Chem 1991;39(11):1938–9.

CHAPTER 11

Acrylamide in Tea Products

Fang Chen[1], Jie Liu[2], Daotong Li[1], Pengpu Wang[1]
[1]College of Food Science and Nutritional Engineering, National Engineering Research Centre for Fruits and Vegetables Processing, Key Laboratory of Fruits and Vegetables Processing, Ministry of Agriculture, China Agricultural University, Beijing, China; [2]College of Grain, Oil and Food, Henan University of Technology, Zhengzhou, China

INTRODUCTION

Aside from water, tea may be the oldest drink of the world. The high demand for tea is not only due to its desirable flavor but also because it contains many bioactive substances beneficial to health, and is particularly rich in polyphenol compounds, such as catechins, which are very important factor for quality control in the tea industry [1]. The benefits of tea have been evidenced by in vitro and in vivo studies in the last few decades. Many articles have discussed its role against cancer, cardiovascular diseases, congenital impairments, neurodegenerative diseases, depression, and many other diseases [2,3]. The process of oxidation and fermentation in harvested tea leaves plays an important role in the formation of their attractive taste, color, aroma, and nutrition values.

Unfortunately, different amounts of acrylamide in tea and tea drinks were discovered by the survey of acrylamide levels in Asian foods [4–13]. The Maillard reaction has been shown to be the main pathway for acrylamide formation in a wide range of foods containing important precursors, mainly asparagine and reducing sugars, processed at high temperatures [14–16]. Thus, the level of asparagine in the tea material and the roasting conditions (roasting temperature and time) strongly influence the process of acrylamide formation [10]. This chapter addresses the acrylamide levels in tea, analytical techniques for its quantification, and the effect of roasting treatments on its formation. This chapter would contribute to the improvement of the process for the reduction of acrylamide in tea.

ACRYLAMIDE LEVELS IN TEA

Acrylamide levels in tea products and infusion discovered by researchers from China, Korea, and Japan were summarized in Table 1. The acrylamide content in various tea products was reported by the Ministry of Agriculture, Forestry and Fisheries of Japan [13], and systematically investigated by Mizukami et al. [10]. As shown in Table 1, they reported that the amounts of acrylamide in roasted green tea such as Houjicha were 190–570 μg/kg and 247–1880 μg/kg, respectively. The levels of acrylamide in green tea

Acrylamide in Food
http://dx.doi.org/10.1016/B978-0-12-802832-2.00011-5

Table 1 Acrylamide content in tea products and tea drinks

Type of tea	Average		Content		Number of samples	Reference	Area
	Product (μg/kg)	Infusion (μg/L)*	Product (μg/kg)	Infusion (μg/L)*			
Green tea	–	–	ND, Tr		2	9	Japan
	45	0.5	ND, 100	ND 3	6	13	Japan
	56	1.2	27–110	0.6–2.1	12	10	Japan
	48		ND, 94		10	17	China
	34		ND, 95		10	4	China
	30		ND, 94		10	5	China
Roasted green tea	541		519–567		3	9	Japan
	385	7	190–570	4–11	4	13	Japan
	681	14.9	247–1880	4.8–41.2	13	10	Japan
Oolong tea	60		ND, 142		4	9	Japan
	75	1	60–90	ND 2	2	13	Japan
	57	2.6	31–85	0.9–4.8	3	10	Japan
	6		ND, 18		3	17	China
	6		ND, 19		3	4	China
Black tea	13	ND	ND, 20	ND	3	13	Japan
	21	ND	18–25	ND	3	10	Japan
	5		ND, 9		2	17	China
	5		ND, 12		2	4	China
	17		14–20			11	China
Pu-erh	28		ND, 85		13	4	China
			ND		1	9	Japan
Raw Pu-erh	26		ND, 83		10	17	China
	18		ND, 92		17	18	China
Ripened Pu-erh	17		ND, 50		3	17	China
	18		ND, 87		33	18	China
White tea	ND		ND		1	17	China
Yellow tea	17		17		1	17	China
Tea drink	ND		ND		3	4	China
Tea products	245		ND, 889		25	7	Korea

*, The condition to prepare tea infusion was as follows: 10 g, 90 °C, 430 ml, 1 min for green tea; 15 g, 90 °C, 650 ml, 0.5 min for roasted green tea; 15 g, 90 °C, 650 ml, 0.5 min for Oolong tea; 5 g, boiling water, 360 ml, 1.5–4 min or 5 g, 90 °C, 360 ml, 4 min for black tea, respectively.

ND, Not detected (below the detection limit of 0.2 ng/ml in tea infusion by Mizukami et al. [10], 0.26 μg/kg by Qu et al. [11], 1 μg/kg by Liu et al. [17] and by Chen et al. [4], 2 μg/L in infusion and 20 μg/kg in product in MAFFJ [13], 3.2 μg/kg by Lee et al. [7], 9 μg/kg by Takatsuki et al. [9], and 10 μg/kg by Zhuang et al. [5], respectively.); Tr, Trace ($9 \leq tr < 30$ μg/kg).

were far less than roasted green tea (less than 100 μg/kg and 110 μg/kg, respectively). The acrylamide content in the analyzed black tea and oolong tea samples were all less than 100 μg/kg. Thirty tea samples including green tea, oolong tea, black tea, white tea, yellow tea, and Pu-erh tea in China were analyzed for the acrylamide contents by Liu et al. [17]. The acrylamide levels of all samples were less than 100 μg/kg. Samples obtained from four classes of tea products including black tea, oolong tea, white tea, and yellow tea had a quite low acrylamide concentration (<20 μg/kg). The acrylamide contents detected only in Pu-erh tea were performed by Li et al. [18]. The acrylamide content was not detected in seven samples, accounting for 14% of the total 50 samples and the acrylamide contents of 26 samples were less than 10 μg/kg. The acrylamide contents in all of tea samples ranged from 2 to 92 μg/kg. Taiwan researchers applied an improved liquid chromatography-tandem mass spectrometry (LC–MS/MS) method for acrylamide analysis in foods, and the acrylamide contents in green, oolong, and black tea drink were all less than 3 μg/kg [6]. There was a lower amount of acrylamide in unroasted or unfried tea samples (17 ± 3 μg/kg) by electropolymerization of single-walled carbon nanotubes composited with polypyrrole as a solid-phase microextraction fiber and GC with electron-capture detection [11].

Based on the available data, it could be identified that acrylamide levels in green, oolong, Pu-erh, white, and yellow tea were quite lower than that in roasted green tea. Except roasted green tea, acrylamide contents were almost lower than 100 μg/kg, and there were always the samples from every type of tea with no detection of acrylamide. Only a few samples of oolong tea had slightly higher acrylamide concentration up to 142 μg/kg. High acrylamide levels occurred in roasted green tea produced in Japan (190–1880 μg/kg). Relatively high levels of acrylamide have been found in each type of tea from Japan and the surveyed tea products from Korean market (Table 1). Fortunately, the amounts of acrylamide in tea infusion were far lower than that in tea products, although the former was dependent on the latter. The acrylamide levels in the Chinese tea products were in the following order: white tea < yellow tea < black tea < oolong tea < Pu-erh tea < green tea. The acrylamide levels in the Japanese tea products were in the order of black tea < Pu-erh tea < green tea < oolong tea < roasted green tea.

FACTORS AFFECTING ACRYLAMIDE LEVELS IN TEA

A great variability in acrylamide levels between different types of tea could be attributed to factors such as variability of acrylamide precursors in raw leaves, difference in process methods and conditions. Furthermore, concentrations of acrylamide present in tea are the result of simultaneously occurring formation and elimination mechanisms. Zhang et al. [19] have investigated the effect of the extract of green tea (EGT) on the kinetics of acrylamide formation in asparagine–glucose model system under microwave and low moisture heating conditions. The results indicated that addition of EGT could significantly reduce the formation rate constant (k_F) of acrylamide but could not significantly affect the elimination rate constant (k_E) of acrylamide. Hence, the antioxidants coexisting with acrylamide precursors would reduce the acrylamide levels more effectively in tea matrix. It might explain that except roasted green tea, there were always samples in every type of tea with no detection of acrylamide.

Processing Methods

The acrylamide levels in tested Chinese tea products [17] were much lower than that in Japanese tea, especially in roasted green tea such as Houjicha [10]. This might ascribe to the higher processed temperature (170–200 °C) used in Japanese roasted green tea than that (100–150 °C) in Chinese tea. For the most Japanese green tea, the fresh leaves are usually steamed or pan-firing to inactivate polyphenol oxidase prior to drying steps, which are performed at temperatures below 80 °C. Green tea is finally manufactured by roasting the leaves from 120 to 140 °C for about 20 min. Houjicha is distinguished from others because it is roasted in a porcelain pot over charcoal at a high temperature. The process was first performed in Kyoto, Japan in the 1920s and its popularity persists today. While black and oolong teas are manufactured by drying the leaves at temperatures below 100 °C to preserve their attractive flavors. Thus, the processing methods used by different regions dramatically affect the acrylamide contents in tea.

In addition, different processing technique is another important factor. For example, Pu-erh tea is a unique tea manufactured in Yunnan Province of China and is mainly consumed by people in the Southeast Asia. Pu-erh tea can be categorized, depending upon the way it is manufactured, as either raw or ripened. After picking appropriate tender leaves, the first step in making raw or ripened Pu-erh is converting the leaf to rough tea. Rough tea can undergo wet-piling for several months prior to being compressed to produce ripened Pu-erh, or be directly compressed to produce raw Pu-erh. The difference in acrylamide levels between raw and ripened Pu-erh tea might be related to the wet-piling used for the ripened Pu-erh tea. Based on the dependence analysis among amino acids, sugar content, and the amount of acrylamide, it is showed that acrylamide in Pu-erh tea is mainly generated through acrolein way during fermentation, and not through asparagine way [18]. Wet-piling might inhibit the formation of acrylamide as the sugar content declined during fermentation and the gallic acid and tea polyphenols also showed an inhibitory effect on the acrylamide formation.

When the filamentous fungi routinely used in the food and beverage industries was incubated in acrylamide-containing roasted green tea for 3 days at 30 °C, there were no glycidamide generated, although acrylic acid as an intermediate was formed during acrylamide degradation [20]. However, there was no significant impact on the flavor, color, and other sensory properties. *Aspergillus oryzae* could be successfully used for reducing the acrylamide concentration in roasted green tea due to induction of acrylamide-degrading amidase.

Processing Conditions

Similar patterns of time- and temperature-dependent acrylamide formation were observed in roasted tea leaves with model system and other food matrix. The acrylamide concentration in the product increased with an increase in the roasting time from 100 up to 160 °C. The 180 °C for the roasting time of 10 and 20 min was found to be the most appropriate temperature for the formation of acrylamide in roasted green tea; and the highest level was observed at 180 °C for 10 min (Figure 1(a)). In the tea industry, roasted green tea (Houjicha) is processed from 170 to 200 °C for about 10 min and contains a considerable amount of acrylamide (about 250–1880 ng/g) [10]. The decrease in acrylamide content at higher temperatures was assumed due to the physical properties of acrylamide, which decomposes and polymerizes on melting at temperatures of >175 °C [16].

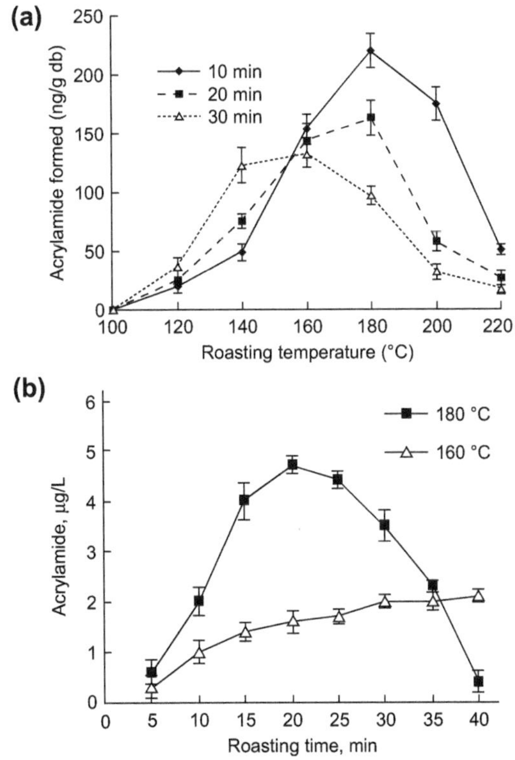

Figure 1 Effects of roasting temperature and time on acrylamide formation in roasted products (a) and tea infusion *(Mizukami et al., 2006 [10])* (b). Tea leaves weighing 100 g were placed on an aluminum plate and roasted in a laboratory oven. The roasting treatment, acrylamide extraction process, and preparation for tea infusion were repeated in triplicate. Each value is presented as mean ± SD ($n=9$) *(Mizukami et al., 2008 [21])*.

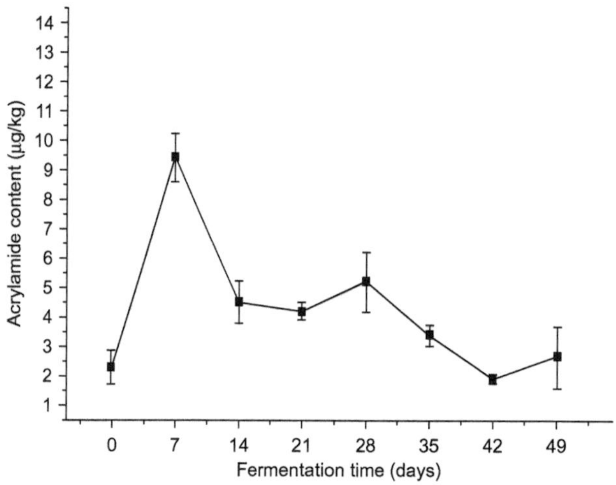

Figure 2 The trends of acrylamide contents during wet-piling of Pu-erh tea. *Source: Li et al. (2012) [18].*

In addition, for the roasting treatments of 160 °C for 30 min and 180 °C for 15 min, 2.0 and 4.0 μg/L of acrylamide were detected in tea infusion, yielding 1316 and 1057 mg/L of tea catechins, respectively (Figure 1(b)). On this condition, the desirable flavor evaluated by tea experts was obtained [21]. The deep roasting treatment induced a decrease in the levels of tea catechins by way of oxidation, decomposition, and polymerization [22]. The maximum amount of acrylamide in tea infusion was 4.7 μg/L when the tea leaves were roasted at 180 °C for 20 min. The acrylamide content in tea infusion slowly increased to 2.1 μg/L at the roasting temperature of 160 °C for 5–40 min. Thus, the roasting temperature of 160 °C is strongly recommended for the mitigation of acrylamide without compromising the sensory quality [21]. As shown in these experiments, the roasting time and temperature were covariant parameters for controlling the acrylamide levels in roasted green tea.

Changes of acrylamide levels during the wet-piling in Pu-erh tea process showed that unfermented tea leaves had little acrylamide content. However, the acrylamide content increased rapidly at the first stage of fermentation and reached its highest level on the 7th day of fermentation. Then, the acrylamide content decreased as the fermentation went on (Figure 2). It indicated that wet-piling could inhibit or decrease acrylamide formation in Pu-erh tea [18].

Soaking experiments of Pu-erh tea showed that no acrylamide was detected in tea infusion when the soaking time was from 1 to 9 min (Figure 3) [18]. Acrylamide dissolution increased gradually when the soaking time was from 15 to 60 min. When soaking at 60 min, the highest content of acrylamide (19.13 μg/L) had reached and the ratio of acrylamide dissolution is 20.80%. Tea soaking temperature, solid–liquid ratio, and soaking number had little effect on the dissolution of acrylamide in tea samples. Therefore, it is not recommended to soak tea for a long time in boiling water and the soaking time of <15 min is recommended for the tea soaking.

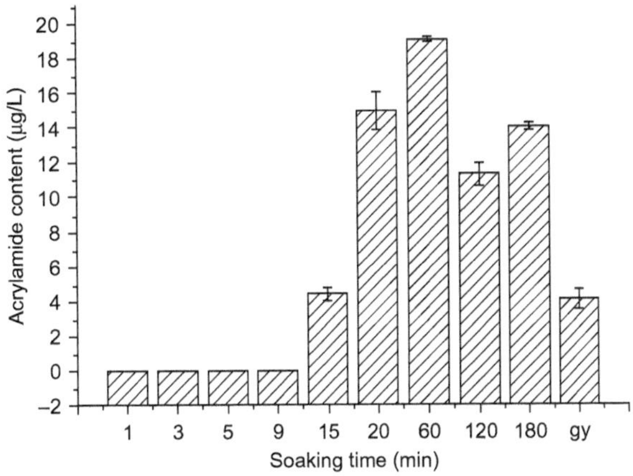

Figure 3 The influence of soaking time on acrylamide content in tea infusion. *Source: Li et al. (2012) [18].*

Precursors

Green tea samples prepared by the conventional method were roasted in a laboratory oven for 10, 20, or 30 min, and the acrylamide content measured by the GC–MS method was plotted against roasting temperature (Figure 4). The highest acrylamide concentration is about 180 μg/kg, and this seems to be the maximum potential of acrylamide accumulation in this tea sample [12]. However, some roasted green tea samples showed values above 500 μg/kg [8–10]. Therefore, not only roasting time and temperature but also components of tea leaves should affect the acrylamide level in tea products. Elucidating the relationship among these components and the formation of acrylamide in tea would contribute to the improvement of the process of reducing the acrylamide in tea.

Six cultivars of tea leaves were roasted at 180 °C for 10 min to explore the effects of the components of tea leaves on acrylamide formation. Their acrylamide concentration significantly varied from 49 to 584 μg/kg. On a molecular basis, the asparagine content was lower than that of the total reducing sugar, as the level of asparagine was <0.61 mg/g and the levels of fructose and glucose varied from 0.81 to 17.22 g/kg [10]. Therefore, the asparagine content in the leaves determined the acrylamide level in the tea products. A correlation was observed between the acrylamide level in the roasted product and the asparagine level in the tea leaves ($r = 0.806$) (Figure 5(a)). In contrast, no correlation was observed between acrylamide and glucose ($r = -0.196$) (Figure 5(b)) or between acrylamide and fructose ($r = -0.114$) (Figure 5(c)). Theanine, the most abundant amino acid in tea samples, was not a significant precursor in acrylamide formation.

The same result was reported for roasted almonds [23]. However, the asparagine content is generally more constant and substantially higher than the reducing sugar content in potato tubers [24–26]. Thus, the reducing sugar levels in the presence of asparagine determine acrylamide formation in heated potatoes, and a substantial amount of acrylamide in roasted tea products was mainly formed from asparagine in the presence of reducing sugars.

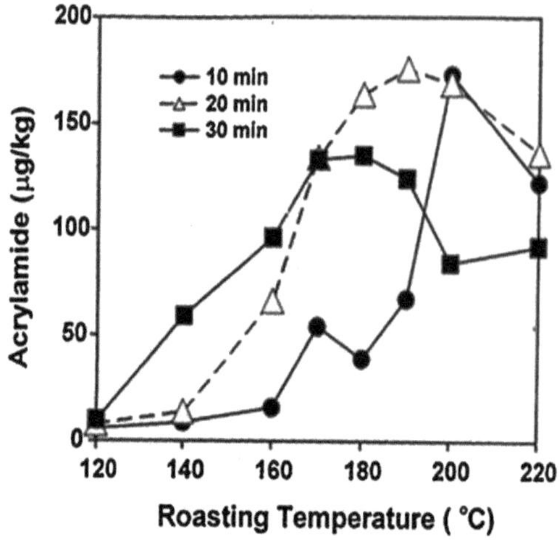

Figure 4 Effect of roasting temperature on acrylamide content in tea leaves. *Source: Yoshida et al. (2005) [12].*

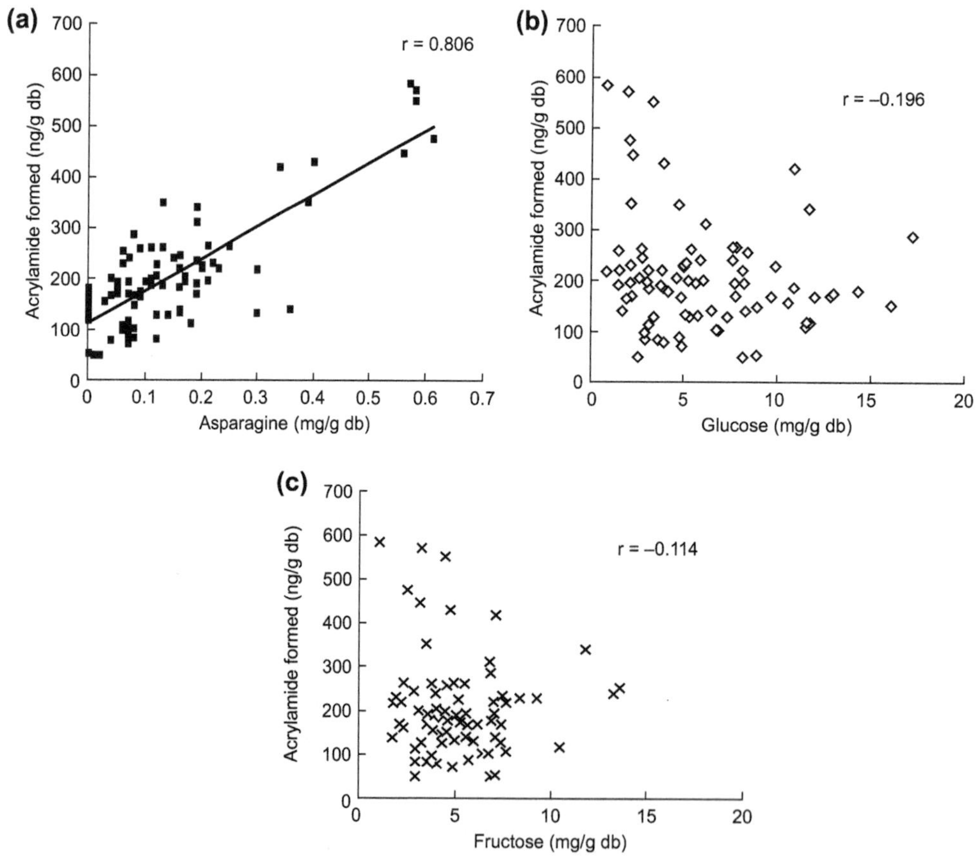

Figure 5 Correlation between asparagine (a), glucose (b), and fructose (c), in tea leaves, and acrylamide formed in the product roasted at 180 °C for 10 min. Each value is presented as the mean ($n = 9$ for acrylamide; $n = 3$ for asparagine and sugar). *Source: Mizukami et al. (2006) [10].*

Additionally, the asparagine of Pu-erh tea samples could not be detected, and there is no significant correlation between the reducing sugars and acrylamide [18]. The formation pathway of acrylamide could not be explained clearly during the fermentation of Pu-erh tea.

SYNOPSIS OF ANALYTICAL TECHNIQUES

The methods for quantification of acrylamide in tea could be generally available for a wide range of food products. As the pretreatment is effective for the complex tea matrix, it also works well in different types of food with minor modification [4–11]. In principle, the analysis follows the same procedure, including extraction, purification, concentration, and instrumental analysis (on the basis of gas chromatography mass spectroscopy (GC–MS) of the brominated derivative and LC–MS of free acrylamide).

The extraction procedure is an important step at the beginning of sample pretreatments. The high water solubility of acrylamide permits its effective extraction from pulverized tea products using water. In order to correct for the loss of analyte during sample preparation and matrix effects, the internal standard acrylamide-d_3 or [$^{13}C_3$]-acrylamide solution was spiked. Shaking was used for the initial extraction steps [10,17]. After extraction, the aqueous phase was homogenized and centrifuged. Chemical purification and purification with solid-phase extraction (SPE) columns are the two principles used in further clean-up steps.

GC-based methods usually demand derivatization of acrylamide with bromination reagent (KBr, HBr, and bromine water) to improve the GC properties [10]. To confirm the SPE eluate used for bromination contained no tea catechins that inhibited acrylamide bromination, and ion-exchange resin was used to remove tea catechins from the extract or tea infusion and the first SPE eluate (0.5 ml) was discarded [21]. The SPE eluate was mixed with a bromination reagent in an ice bath until the reddish color of bromine remained. The excess of bromine was removed after 1 h reaction by titration with sodium thiosulfate solution until the solution became colorless. The brominated acrylamide was less polar compared with the original compound and, therefore, it could be extracted with nonpolar organic solvent (ethyl acetate) from the aqueous phase. The extract was dried over anhydrous sodium sulfate, and most of the solvent was removed by centrifugal evaporation under reduced pressure at 30 °C, the residue was transferred to a micrograss tube to be kept at 4 °C till GC–MS analysis. Fragment ion peaks of 2,3-dibromopropionamide were derived from acrylamide at m/z 150 $[C_3H_5NO^{79}Br]^+$ and 152 $[C_3H_5NO^{81}Br]^+$, and those from acrylamide-d_3 or [$^{13}C_3$]-acrylamide at m/z 153 and 155 were detected by selected ion monitoring.

For LC–MS/MS analysis, co-eluting and undetected matrix components that often reduce or enhance the ion intensity of the analytes and affect the reproducibility and accuracy of the assay, should be removed as much as possible by purification; 10 ml acetonitrile, 4 g of anhydrous magnesium sulfate, and 0.5 g of sodium chloride were added to the extract. The mixture was shaken vigorously for 1 min immediately and then centrifuged. The addition of NaCl decreases the polarity of the acetonitrile layer and a combination of anhydrous $MgSO_4$ and NaCl can induce a distinct phase separation between water and acetonitrile. Hence, the majority of acrylamide was forced into the acetonitrile layer. Although water was necessary to facilitate acrylamide extraction, acetonitrile offered appropriate extraction selectivity and was compatible with reversed-phase LC analysis. Three layers were obtained as follows, an acetonitrile layer containing acrylamide at the top, the matrix layer in the middle, and the water layer with the excessive salts in the bottom (Figure 6). As the acrylamide content was generally low except roasted green tea, enrichment was necessary for tea samples, especially for tea infusion. Thus, the acetonitrile solution (9 ml) was transferred into a glass test tube and evaporated to dryness under a stream of nitrogen in a water bath at 40 °C. The residue on the wall of glass tube was re-dissolved in 0.5 ml of water under vortex, so that some highly lipophilic co-extractives were excluded again. Aqueous extract was filtered through a 0.45 μm syringe filter for further clean-up. The sorbent of Oasis MCX SPE cartridge is effective for extraction of basic compounds from complex matrices, while acrylamide exhibits weak basicity and does not interact strongly with the sorbent. The pass through strategy for the SPE clean-up was applied to retain the matrix interferences and reduce matrix effect in MS analysis. SPE cartridge was conditioned consecutively with 2 ml of methanol and 2 ml water. Re-dissolved extract (0.5 ml) passed through the SPE cartridge and effluent was collected. Subsequently, a wash step was carried out to elute other acrylamide retained on the SPE column. The final test solutions were filtered through a 0.22 μm syringe filter for quantification using LC–MS/MS.

For the chromatographic separation of acrylamide, reversed-phase columns were often used with aqueous mobile phase adding 0.1% formic acid that enhanced the ESI

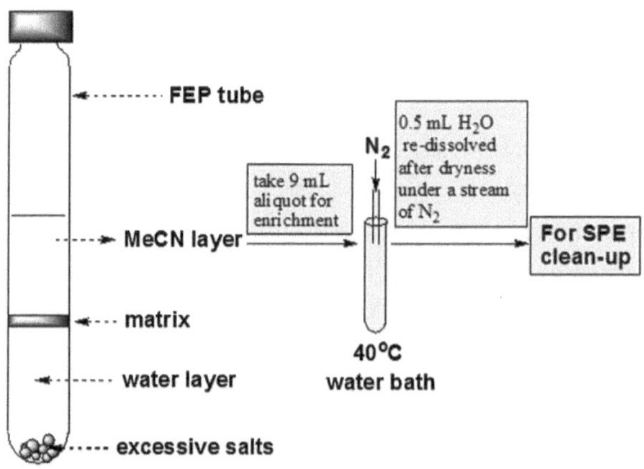

Figure 6 Schematic of sample preparation procedure for LC–MS/MS analysis (solvent layer arrangement in an FEP (fluorinated ethylene propylene) tube after the centrifugation of an extract, enrichment, and purification).

response in MS analysis. For detection of acrylamide after LC separation, tandem mass spectrometry working in multiple reaction monitoring (MRM) is the most often method of choice because of its high selectivity. The MRM mode of the degradation patterns (m/z 72→55) for acrylamide and (m/z 75→58) for isotopically labeled acrylamide were used for quantification. The transition from a precursor ion, which is separated in the first quadruple, to a product ion, generated by collision with argon in the second quadruple, is monitored in the third quadruple.

Besides MS detection methods, HPLC-UV detection [20] and GC-electron capture detection [11] have also been introduced to analyze tea product and tea infusion.

KEY FACTS

Tea was discovered over 4700 years ago by Emperor Shen Nung in Chinese legend. It is a kind of popular cosmopolitan beverage and about 3 billion kilograms of teas were produced and consumed all over the world yearly. The plant of tea was originated from Southeast Asia and is now being cultivated in more than 30 countries. There are in fact four main types of Camellia tea in a multiplicity of blends. They are nonfermented green tea, partially fermented oolong tea, fully fermented black tea, and postfermented Pu-erh tea. Of the tea produced worldwide, 78% is black tea, which is usually consumed in the Western countries, 20% is green tea, which is commonly consumed in Asian countries, and 2% is oolong tea which is produced mainly in southern China [27–30].

MINI DICTIONARY

Houjicha It is a Japanese green tea that is roasted in a porcelain pot over charcoal. The tea is fired at a high temperature, altering the leaf color tints from green to reddish-brown. The infusions have a light-to reddish-brown appearance and are less astringent.

Oolong tea It is a traditional Chinese tea produced through withering the plant under the strong sun and oxidation before curling and twisting. The taste of oolong can be sweet and fruity with honey aromas, or woody and thick with roasted aromas, or green and fresh with bouquet aromas.

Polyphenols These are a class of mainly natural, synthetic, or semisynthetic, organic chemicals with large multiples of phenol structural units.

Pu-erh tea It is a kind of fermented dark tea produced in Yunnan province, China. The tea leaves undergo microbial fermentation and oxidation after they are dried and rolled.

Wet-piling It is a kind of tea processing method in which bacterial and fungal fermentation is under-controlled in a warm humid environment. It involves piling, dampening, and turning the tea leaves in a manner much akin to composting.

SUMMARY POINTS

The acrylamide levels in the tea samples, except roasted green tea, were generally lower than those in other heat-processed foods, especially in tea infusion. Conventional amounts of consumed tea may not negatively affect the human health. GC–MS and

LC–MS/MS appeared to be acknowledged as the most useful and authoritative methods for the quantification of acrylamide. Processing methods, processing conditions, and components of tea leaves are three important factors influencing on acrylamide formation in tea. The processing methods used by different regions dramatically affect the acrylamide content in tea. The acrylamide levels in tested Chinese tea products were much lower than that in Japanese tea, especially in roasted green tea such as Houjicha. The acrylamide content in green tea are less than that in Pu-erh tea. For the roasted green tea, roasting temperature and time play important roles in acrylamide formation. The maximum acrylamide formation was observed when the tea leaves were roasted at 180 °C for 10 min. For the Pu-erh tea, fermentation has the significant impact on the acrylamide formation. In addition, the level of acrylamide in roasted green tea strongly correlated with the level of asparagine in the presence of reducing sugars in the tea leaves, whereas no correlation between reducing sugars or asparagines and acrylamide formation were observed in Pu-erh tea [18].

REFERENCES

[1] Mizukami Y, Sawai Y, Yamaguchi Y. Simultaneous analysis of catechins, gallic acid, strictinin, and purine alkaloids in green tea by using catechol as an internal standard. J Agric Food Chem 2007;55(13):4957–64.

[2] Hernández Figueroa TT, Rodríguez-Rodríguez E, Sánchez-Muniz FJ. The green tea, a good choice for cardiovascular disease prevention? Arch Latinoam Nutr 2004;54(4):380–94.

[3] Yang CS, Hong J. Prevention of chronic diseases by tea: possible mechanisms and human relevance. Annu Rev Nutr 2013;33:161–81.

[4] Chen F, Yuan Y, Liu J, et al. Survey of acrylamide levels in Chinese foods. Food Addit Contam Part B Surveill 2008;1(2):85–92.

[5] Zhuang H, Zhang T, Liu J. Detection of acrylamide content in traditional Chinese food by high-performance liquid chromatography tandem mass spectrometry method. CyTA-J Food 2012; 10:36–41.

[6] Cheng W, Kao Y, Shih D. Validation of an improved LC/MS/MS method for acrylamide analysis in foods. J Food Drug Anal 2009;17:190–7.

[7] Lee S, Yoo M, Koo M, et al. In-house-validated liquid chromatography-tandem mass spectrometry (LC–MS/MS) method for survey of acrylamide in various processed foods from Korean market. Food Sci Nutr 2013;1(5):402–7.

[8] Nemoto S, Takatsuki S, Sasaki K, et al. Determination of acrylamide in foods by GC/MS using 13C-labeled acrylamide as an internal standard. Shokuhin Eiseigaku Zasshi 2002;43(6):371–6.

[9] Takatsuki S, Nemoto S, Sasaki K, et al. Determination of acrylamide in processed foods by LC/MS using column switching. Shokuhin Eiseigaku Zasshi 2003;44(2):89–95.

[10] Mizukami Y, Kohata K, Yamaguchi Y, et al. Analysis of acrylamide in green tea by gas chromatography-mass spectrometry. J Agric Food Chem 2006;54(19):7370–7.

[11] Qu Y, Liu C, Luo F, et al. Electropolymerization of single-walled carbon nanotubes composited with polypyrrole as a solid-phase microextraction fiber for the detection of acrylamide in food samples using GC with electron-capture detection. J Sep Sci 2013;36(24):3889–95.

[12] Yoshida M, Ono H, Chuda Y, et al. Acrylamide in Japanese processed foods and factors affecting acrylamide level in potato chips and tea. Adv Exp Med Biol 2005;561:405–13.

[13] MAFFJ. The ministry of agriculture, forestry, and fisheries of Japan; 2003.

[14] Mottram DS, Wedzicha BL, Dodson AT. Acrylamide is formed in the Maillard reaction. Nature 2002;419(6906):448–9.

[15] Yaylayan VA, Wnorowski A, Perez Locas C. Why asparagine needs carbohydrates to generate acrylamide. J Agric Food Chem 2003;51(6):1753–7.
[16] Taubert D, Harlfinger S, Henkes L, et al. Influence of processing parameters on acrylamide formation during frying of potatoes. J Agric Food Chem 2004;52(9):2735–9.
[17] Liu J, Zhao G, Yuan Y, et al. Quantitative analysis of acrylamide in tea by liquid chromatography coupled with electrospray ionization tandem mass spectrometry. Food Chem 2008;108:760–7.
[18] Li L, Zhang T, Sun Z, et al. Survey of acrylamide contents in Pu-erh tea and study on its formation mechanism. Trans Chin Soc Agric Mach 2012;43:123–9 [in Chinese].
[19] Zhang Y, Ying T, Zhang Y. Reduction of acrylamide and its kinetics by addition of antioxidant of bamboo leaves (AOB) and extract of green tea (EGT) in asparagine-glucose microwave heating system. J Food Sci 2008;73(2):C60–6.
[20] Wakaizumi M, Yamamoto H, Fujimoto N, et al. Acrylamide degradation by filamentous fungi used in food and beverage industries. J Biosci Bioeng 2009;108(5):391–3.
[21] Mizukami Y, Sawai Y, Yamaguchi Y. Changes in the concentrations of acrylamide, selected odorants, and catechins caused by roasting of green tea. J Agric Food Chem 2008;56(6):2154–9.
[22] Hatano T, Hori M, Kusuda M, et al. Characterization of the oxidation products of (-)-epigallocatechin gallate, a bioactive tea polyphenol, on incubation in neutral solution. Heterocycles 2004;63:1547–54.
[23] Amrein TM, Lukac H, Andres L, et al. Acrylamide in roasted almonds and hazelnuts. J Agric Food Chem 2005;53(20):7819–25.
[24] Amrein TM, Bachmann S, Noti A, et al. Potential of acrylamide formation, sugars and free asparagine in potatoes: a comparison of cultivars and farming systems. J Agric Food Chem 2003;51(18):5556–60.
[25] Becalski A, Lau BP, Lewis D, et al. Acrylamide in French fries: influence of free amino acids and sugars. J Agric Food Chem 2004;52(12):3801–6.
[26] Ohara-Takada A, Matsuura-Endo C, Chuda Y, et al. Change in content of sugars and free amino acids in potato tubers under short-term storage at low temperature and the effect on acrylamide level after frying. Biosci Biotechnol Biochem 2005;69(7):1232–8.
[27] Mukhtar H, Ahmad N. Tea polyphenols: prevention of cancer and optimizing health. Am J Clin Nutr 2000;71(Suppl. 6):1698S–702S.
[28] Cabrera C, Artacho R, Giménez R. Beneficial effects of green tea-a review. J Am Coll Nutr 2006;25(2):79–99.
[29] Khan N, Mukhtar H. Tea polyphenols for health promotion. Life Sci 2007;81(7):519–33.
[30] Butt MS, Sultan MT. Green tea: Nature's defense against malignancies. Crit Rev Food Sci Nutr 2009;49(5):463–73.

CHAPTER 12

Acrylamide in Table Olives

Alfredo Montaño[1], Francisco J. Casado[2], Reinhold Carle[3,4]

[1]Food Biotechnology Department, Instituto de la Grasa (CSIC), Seville, Spain; [2]Hohenheim University, Life Science Center, Stuttgart, Germany; [3]Hohenheim University, Institute of Food Science and Biotechnology, Chair Plant Foodstuff Technology and Analysis, Stuttgart, Germany; [4]Biological Science Department, King Abdulaziz University, Jeddah, Saudi Arabia

INTRODUCTION

Acrylamide has attracted the attention of the scientific community since April 2002, when a Swedish research group from the Stockholm University reported the finding of elevated acrylamide levels in potato products and cereals, such as fried potatoes, baked potatoes, bread, breakfast cereals, and biscuits [1]. Particular attention has been paid to these products because of their high acrylamide contents and their important contribution to human nutrition as a staple food. The formation of acrylamide was believed to be generally associated with the application of harsh heat treatment, for example, roasting, baking, and deep-frying, with temperatures exceeding 150 °C. Surprisingly, although such conditions do not apply to any type of table olives, high levels of acrylamide have been reported in canned black ripe olives during the last decade. Roach et al. [2] were the first to report high acrylamide levels around 2000 μg/kg in samples of black ripe olives. This type of table olives was then ranked among the top 20 and top 8 foods by mean acrylamide intake (μg/kg of body weight/day) and per portion, respectively [3]. Subsequently, different surveys on acrylamide levels in different types of table olive samples have been published [4–6]. The latter provided evidence that acrylamide is only detected in California-style black ripe and California-style green ripe olives, while it is not detected in other processing types such as Spanish-style and Greek-style olives. Worldwide, California-style black ripe olives are one of the most important commercialized table olives. In 2013, this processing style accounted for 46% of total table olives exported from Spain, the world's main producer and exporter of table olives [7]. In the United States, about 50% of their production is destined for canned California-style black olives; another 46% is processed into olive oil [8]. Compared to California-style black ripe olives, the production of California-style green ripe olives as well as their acrylamide levels are relatively low [6]. Therefore, most investigations about acrylamide in table olives carried out to date have focused on California-style black ripe olives. This chapter summarizes the state of the art in this field.

Acrylamide in Food
http://dx.doi.org/10.1016/B978-0-12-802832-2.00012-7

CALIFORNIA-STYLE BLACK RIPE OLIVE PROCESSING

The flowchart presented in Figure 1 outlines the production scheme of this type of table olives.

Fruits are harvested prior to full maturity (green or turning color). Once in the production plant, olives are selected and may be immediately processed or (more commonly) preserved before oxidation. Preservation is usually performed by their immersion into brine containing 4–6% NaCl for 2–6 months, depending on the production requirements. The brine may be adjusted to attain pH 4 by the addition of acetic acid, and kept under aerobic conditions to prevent spoilage. Alternatively, the use of acidified water (2.4% acetic acid) under anaerobic conditions has been proposed [9]. By this method, the discharge of sodium chloride into wastewater streams may be significantly reduced. In the oxidation step, fruits are successively treated with sodium hydroxide solutions (lye) for varying periods of time to achieve a progressive penetration of the lye into the flesh. After each alkaline treatment, the olives are washed with water and oxidized by injecting pressurized air into the water. This treatment permits complete blackening of the fruit skin due to oxidation and polymerization of *o*-diphenols, mainly hydroxytyrosol and caffeic acid [10]. The lye treatments are generally repeated between three and five times, although some processors only

Harvesting
↓
Transportation
↓
Previous Handlings
↓
Washing
↓
(Preservation in brine or acidified water)
↓
Lye treatment and air oxidation
↓
Washing (alkali neutralization)
↓
Color fixation
↓
Sorting and grading
↓
(Pitting, Slicing, etc.)
↓
Packing
↓
Sterilization

Figure 1 Process steps for black ripe olives. Operations in brackets are optional.

apply two or even one single lye treatment. Penetration into the fruit is controlled using phenolphthalein as an indicator. After the first treatment, sodium hydroxide merely permeates through the skin (exocarp). Subsequent treatments result in progressive penetration of the lye into the pulp, until the pit is reached. The concentration of sodium hydroxide depends on the ripeness of the fruit, its variety, the environmental temperature, and the desired penetration speed. It varies between 1% and 2% NaOH (w/v). After completion of the lye and oxidation treatments, the olives are washed several times with water to remove excessive lye, resulting in a final pH of around 8 in the flesh (mesocarp). The number of washes may be diminished by adding food grade HCl to the liquid or by injecting CO_2 into the containers [11]. To stabilize color, bivalent iron salts such as ferrous gluconate, sulfate, and lactate are used. Generally, 0.1% (w/v) of ferrous gluconate or 0.06% (w/v) ferrous lactate is added to the final round of washing water. Iron diffusion into the flesh is completed within about 10 h. Olives are then packed in cans or glass containers with brine containing 2–4% of NaCl and 10–40 ppm of iron (added as ferrous gluconate) to prevent deterioration of their black color. The pH value of the final product ranges between 5.8 and 7.9, and the NaCl content is between 1% and 3%. Due to pH > 4.5 and high water activity, which do not ensure microbial safety, ripe olives must be sterilized to prevent growth of foodborne pathogens. Heat sterilization is generally performed between 121 and 126 °C, and to assure proper preservation of the product, it shall reach an accumulated minimum value of lethality of $15F_0$ [12]. In the United States, sterilization at 115.6 and 121.1 °C for 60 and 50 min, respectively, is obligatory [13].

For green ripe olives, the process is mostly identical, except for exposure to air (oxidized), because their green color is retained. Consequently, the addition of ferrous gluconate treatment is unnecessary. Moreover, only fresh olives are processed into green ripe olives, whereas black ripe olives can be made from fresh or preserved olives [14].

FORMATION OF ACRYLAMIDE IN BLACK RIPE OLIVE PROCESSING

Most acrylamide in black ripe olives is formed during heat sterilization. Without heating, the formation of acrylamide in olive pulp is negligible (≈50 µg/kg) [6] or even undetectable [5]. The influence of sterilization time and temperature on acrylamide formation has been studied in green ripe olives prepared on a laboratory scale [15]. At temperatures of 110, 116, 121, or 125 °C, the acrylamide content increases with progressing sterilization time (Figure 2).

Interestingly, for equivalent heat treatments, that is, processes with the same F_0 value, the amount of acrylamide formed is lower at higher temperatures (e.g., longer heat exposure—75 min—at 116 °C produces higher acrylamide levels than does sterilization at 121 °C for 27 min, and 125 °C for 17 min, respectively). These findings are in agreement with the UHT-treatment of foods.

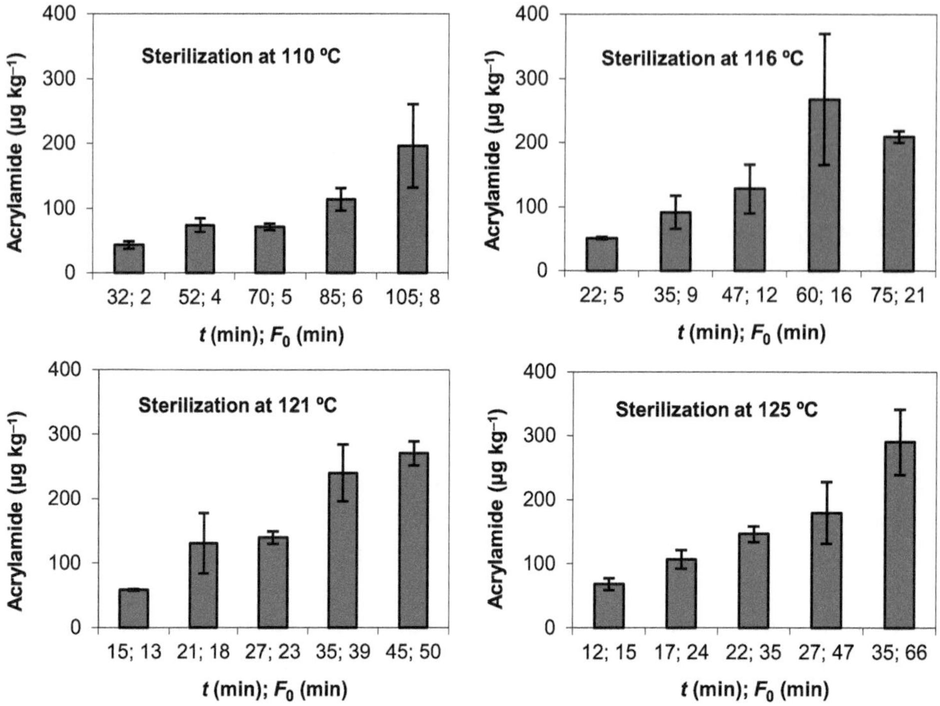

Figure 2 Acrylamide content in Hojiblanca green ripe olives sterilized at 110, 116, 121, and 125 °C for different times. Values of accumulated lethality F_0 together with each time are included. Error bars represent 95% confidence intervals. *Adapted from Montaño et al. [15].*

The kinetic profile of acrylamide formation has been studied using black ripe olive juice in a model system at 121 °C [16]. Experimental data are well fitted using a simple logistic model (Figure 3), characterized by increasing acrylamide contents with progressing heating time until an equilibrium value is reached:

$$C(t) = \frac{a(T)}{1 + \exp\{k_1(T)[t_{c1}(T) - t]\}}$$

where $C(t)$ is the concentration of acrylamide, $t_{c1}(T)$ is a temperature-dependent time characteristic for the inflection point in the formation (t_{c1}) of acrylamide, $k_1(T)$ is a temperature-dependent steepness parameter around the inflection point for the formation (k_1) of acrylamide, and $a(T)$ serves as a temperature-dependent "scale factor" for the acrylamide concentration.

The logistic model (or empirical models in general) as mentioned above only describe the acrylamide formation kinetics, but they do not offer much insight into the mechanism behind the acrylamide formation.

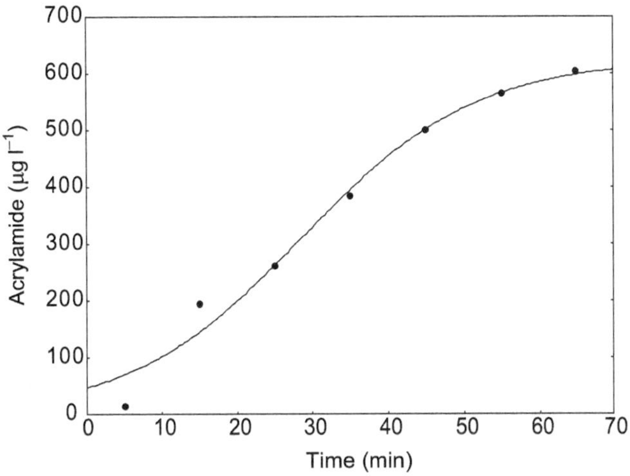

Figure 3 Kinetic profile of acrylamide formation as a function of heating time at 121 °C in black ripe olive model system. Points represent experimental data (mean values) and the line is acrylamide predicted by the kinetic model (logistic function). *Adapted from López-López et al. [16].*

INFLUENCE OF OLIVE CULTIVAR AND PROCESSING CONDITIONS ON ACRYLAMIDE CONTENTS

Acrylamide levels in black ripe olives are affected by olive cultivar (cv.) and variations in processing conditions. In Spain, the most important cultivars used in the production of black ripe olives are Manzanilla, Hojiblanca, and Cacereña. In the US, Manzanilla is the main cultivar destined for black ripe olive production. All studies carried out to date show that acrylamide levels in ripe olives from Manzanilla cultivar are higher than those in ripe olives from cvs. Hojiblanca and Cacereña. For example, in a systematic study carried out to determine the effects of different factors, including olive cultivar, on acrylamide levels during industrial processing of black ripe olives, the mean value of acrylamide content in ripe olives of cv. Manzanilla was 925 μg/kg FW, that being significantly higher than in ripe olives of cv. Hojiblanca (656 μg/kg FW) [5]. When three Spanish cultivars (Manzanilla, Hojiblanca, and Cacereña) were compared after processing under identical conditions at laboratory scale, acrylamide contents (mean ± SD, $n=3$) of 1243 ± 154, 546 ± 68, and 132 ± 13 μg/kg FW were determined (Montaño et al., 2013, data not published). The considerable cultivar-dependent variation of acrylamide levels may be attributed to different concentrations of precursors in the following order: Manzanilla > Hojiblanca > Cacereña.

Regarding the influence of processing conditions on acrylamide contents, the effect of different industrial methods of lye treatment and water washings has been studied [5]. Technological impact on acrylamide formation was found to be even more pronounced than that of olive cultivars. Specifically, three methods (M1–M3) were

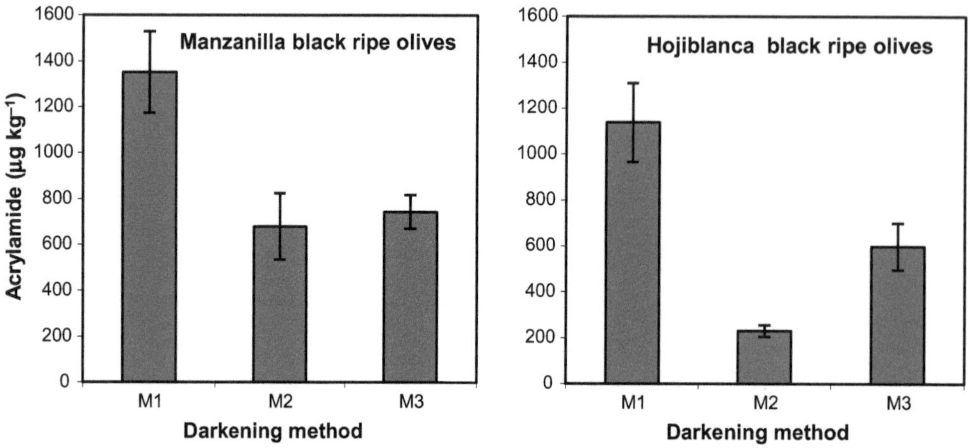

Figure 4 Effect of darkening method on acrylamide content in canned black ripe olives of Manzanilla and Hojiblanca cultivars. Data points represent means ±SD of duplicate cans, each analyzed in duplicate (n=4). *Adapted from Casado and Montaño [5].*

compared: M1 applying a 3% NaOH solution, where lye penetration was completed after 4.5 h followed by one water washing (24 h duration); M2 consisting of one single lye application (3% NaOH for 4.5 h) with two subsequent washing steps, both for 24 h; and M3 including two lye applications plus three water washings, according to the following sequence: (1) lye treatment with 3% NaOH for 2 h just to penetrate the skin; (2) water washing for 18 h; (3) lye treatment with 2.5% NaOH for 3 h to reach the stone; and (4) two water washings, each of 24 h duration. Olives, both cv. Manzanilla and Hojiblanca, treated by M1 showed higher acrylamide values than those treated by M2 and M3 (Figure 4). This might be due to the lower number of washings resulting in less leaching of soluble solids, and thus, higher concentration of acrylamide precursor(s) in olives treated by M1. Despite different number of washings, the leaching effect appeared to be similar in samples of Manzanilla processed by M2 and M3. However, contrary to the expectations, Hojiblanca olives treated by M3 behaved differently, showing higher acrylamide contents than those treated by M2.

The influence of the individual processing steps involved in the darkening operation (lye treatment, air oxidation, alkali neutralization, and addition of ferrous gluconate) has been evaluated (Figure 5) [6]. Ferrous gluconate had no effect on acrylamide levels. Olives processed without air oxidation had significantly lower acrylamide levels than those processed under air exposure (310 vs 569 µg/kg). This observation suggests that acrylamide precursors are mainly formed via oxidation during lye treatment, the latter being converted into acrylamide during sterilization. This may also explain the lower acrylamide content in California-style green ripe olives compared to that in California-style black ripe olives. In addition to air oxidation, pH conditions had a significant effect on acrylamide level. While acrylamide level was 339 µg/kg FW in sterilized olives that

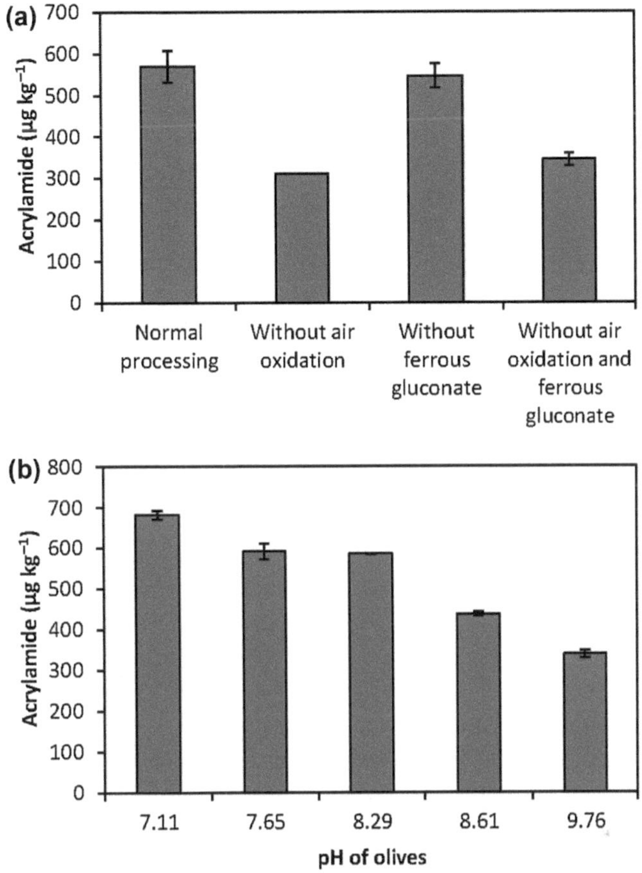

Figure 5 Effect of the individual treatments involved in the darkening step on acrylamide levels in black ripe olives: (a) influence of air oxidation and ferrous gluconate treatment; (b) influence of pH (neutralization with carbon dioxide). *Adapted from Charoenprasert and Mitchell [6].*

were not previously neutralized, 682 μg/kg FW were determined in sterilized olives that had been neutralized, indicating that the formation of acrylamide in olives favored at neutral or low acidic pH. The influence of pH on acrylamide formation has been studied in a model system using olive juice heated at 121 °C for 30 min [17]. Acrylamide levels exhibited a maximum pH value between 5.5 and 6.0, with the content being significantly lowered at pH values above 8 and below 5.

In addition, the preservation method significantly affected acrylamide contents, although this effect is less pronounced than that of darkening method and cultivar. Brine preservation, which is the traditional method, has been compared to acid preservation in water containing 2.4% acetic acid for olives subjected to the black ripe olive processing at industrial scale [5]. On an average, the acrylamide content in brine-preserved ripe olives amounted to 656 μg/kg FW, which was significantly higher than

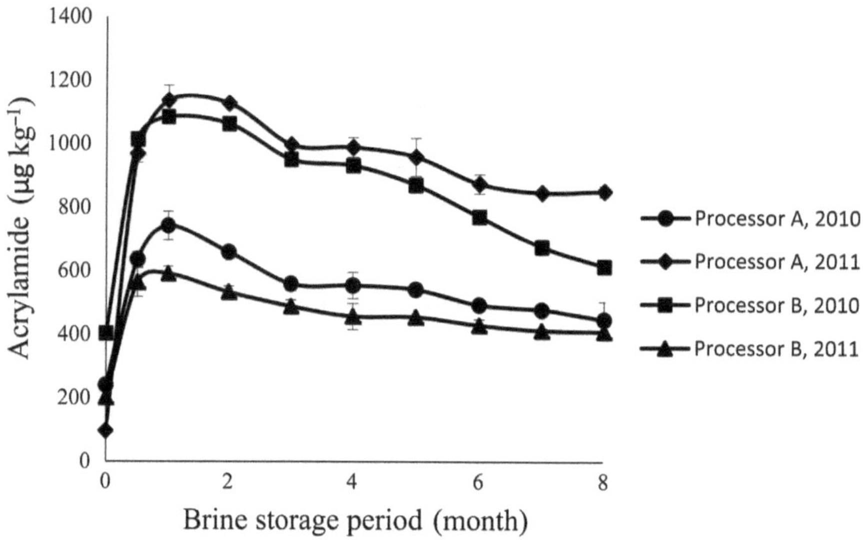

Figure 6 Effect of preservation time (0–8 months) in brine on acrylamide formation. *Reprinted with permission from Charoenprasert and Mitchell [6], Copyright 2014 American Chemical Society.*

that of acid-preserved ripe olives (439 μg/kg FW). This discrepancy may be due to different contents of acrylamide precursors, which may be a result of microbial metabolism during preservation. In this context, de Castro et al. [9] stated that, in addition to yeasts, acetic acid bacteria grew in olives preserved by the addition of 2.4% acetic acid, but the latter were not detected in olives preserved in brine.

Preservation time prior to processing was shown to influence the formation of acrylamide in a time-dependent manner. Charoenprasert and Mitchell [6] monitored acrylamide formation in olives stored in brine over 8 months and then processed into California-style black ripe olives. The brine was composed of calcium chloride (0.2–0.4%), sodium benzoate (0.2–0.3%), and acetic acid to maintain the pH at 3.7–4.0. Acrylamide levels increased during the first month, and then declined with increasing time (Figure 6), thus indicating that acrylamide precursors are either degraded or they diffuse into the surrounding brine during prolonged storage.

The effect of optional operations on acrylamide contents prior to packing, such as pitting or slicing, is of minor importance compared to the above-mentioned factors. For example, in a previous study [5], sliced Hojiblanca ripe olives were found to contain a slightly lower acrylamide level than pitted Hojiblanca ripe olives (523 vs 571 μg/kg FW); however, the differences, although being significant, were negligible.

Acrylamide formed in black ripe olives was found to be stable during storage at room temperature for 6 months [5], but an acrylamide reduction of 25% has been determined after 8 months compared to its content after 20 days of storage [18]. Acrylamide concentrations significantly decreased during storage have been found in other foods like coffee, cacao, and roasted almonds, which was attributed to reactions of acrylamide with food components and/or reaction products [19,20].

STRATEGIES TO MITIGATE THE FORMATION OF ACRYLAMIDE IN RIPE OLIVES

Although the relevance of dietary exposure to acrylamide to human health is unclear, World Health Organization (WHO) continues to encourage food manufacturers to take measures to reduce acrylamide levels in processed foods [21]. The above-mentioned observations about the influence of olive cultivar and processing steps on acrylamide formation in black ripe olives suggest some measures to decrease the amount of acrylamide in the final product: (1) by selecting the most suitable olive cultivar (i.e., favored processing of cvs. Hojiblanca and Cacereña over cv. Manzanilla), (2) by increasing soaking time of olives in brine prior processing, (3) by using a darkening method comprising two water washing steps instead of a single one, and (4) by strict adherence to the recommended sterilization conditions, that is, the minimum value of accumulated lethality should be achieved by applying the highest sterilization temperature; however, the recommended value ($15F_0$) should not be exceeded. Nevertheless, all measures, even when taken altogether, do not resolve the acrylamide problem in black ripe olives.

The use of additives is another strategy to reduce acrylamide content in black ripe olives. A model system based on juice from alkali-treated green olives, which approximated the chemical composition of green ripe olives before sterilization, was used to test the suitability of different salts, protein amino acids, and antioxidant polyphenols for the inhibition of acrylamide formation in ripe olives (Table 1) [17]. Sodium bisulfite was found to be the most efficient inhibitor of acrylamide formation showing a reduction rate of 85% at 50mM followed by L-cysteine, which had the second most reducing effect with a reduction rate of 50% at 50mM level. This result may be attributed to Michael-addition reactions between the sulfydryl group of cysteine and acrylamide. As already assumed by Claeys et al. [22], by this reaction, cysteinyl-S-β-propionamide may be formed. The formation of this reaction product has been confirmed using gas chromatography–mass spectrometry (GC–MS) methods [23]. However, L-cysteine generated an unpleasant odor after heating, thus being unsuitable as a potential additive for acrylamide inhibition. Other amino acids such as glycine, L-lysine, and L-arginine gave lower reduction rates ranging between 24% and 27%. Further proteinogenic amino acids tested (L-aspartic acid, L-glutamic acid, L-glutamine, L-serine, and L-tyrosine) as well as the phenolic compounds hydroxytyrosol (3,4-dihydroxyphenyl ethanol) and 3,4-dihydroxyphenyl glycol, the latter two being genuine constituents of olives exerting high antioxidant activities, did not show any significant effect on acrylamide levels.

Taking into account the previously described results, López-López et al. [16] used an olive model system based on juice from black ripe olives to evaluate the inhibiting effects of selected amino acids and water-soluble vitamins on acrylamide formation (Table 2). Sodium sulfite was also tested for comparison. A good correlation ($R^2 = 0.941$) between

Table 1 Effect of salts, amino acids, and antioxidant polyphenols on the content of acrylamide in ripe olive model system (alkali-treated olive juice heated at 121 °C for 30 min)

Added component	Concentration (mM)	Acrylamide content (μg/L)[a]	% Change in acrylamide content[b]
Sodium bisulfite	1.5	322 ± 27	−15
Sodium bisulfite	25	241 ± 20	−37*
Sodium bisulfite	50	56 ± 5	−85*
Sodium chloride	500	267 ± 22	−30*
Sodium dihydrogen phosphate	100	278 ± 29	−27*
Sodium borate	100	324 ± 22	−15*
Calcium chloride	50	472 ± 39	24*
Magnesium chloride	50	327 ± 27	−14
Ferrous gluconate	2.5	335 ± 27	−12
Ferrous gluconate	25	331 ± 26	−13
Glycine	50	291 ± 24	−24*
L-lysine	50	279 ± 23	−27*
L-cysteine	50	191 ± 16	−50*
L-arginine	50	285 ± 24	−25*
L-aspartic acid	50	347 ± 29	−9
L-glutamine	50	375 ± 31	−2
L-glutamic acid	50	324 ± 27	−15
L-serine	50	322 ± 27	−16
L-alanine	50	310 ± 26	−19*
L-tyrosine	50	349 ± 29	−8
L-valine	50	320 ± 27	−16
Hydroxytyrosol	12	405 ± 34	12
Hydroxytyrosol	25	387 ± 32	7
Hydroxytyrosol	50	373 ± 31	3
3,4-Dihydroxyphenyl glycol	12	424 ± 35	17
3,4-Dihydroxyphenyl glycol	25	396 ± 33	10
3,4-Dihydroxyphenyl glycol	50	409 ± 34	13

An asterisk indicates significant difference from the control ($p < 0.05$).
[a]Data represent mean values ($n = 2$) ± 95% confidence intervals.
[b]Percentage of change in acrylamide content compared to control (olive juice without additive).
Adapted from Casado et al. [17].

the acrylamide contents obtained in the model system and those obtained in the real product was stated, indicating that the ripe olive model system is a simple and reliable tool for screening compounds acting as potential inhibitors of acrylamide formation in ripe olive processing. In addition, the above-mentioned authors observed that (1) acrylamide formation was completely inhibited after the addition of 20 mM of sodium sulfite, which confirmed its efficacy as a strong inhibitor of acrylamide formation as previously reported [17]; (2) of the tested proteinogenic amino acids, L-proline was the most potent inhibitor followed by glycine (reduction rates of 91% and 64%, respectively, at 200 mM

Table 2 Effect of sodium bisulfite, protein and non-protein amino acids, and water-soluble vitamins on the content of acrylamide in black ripe olive model system (black ripe olive juice heated at 121 °C for 30 min)

Added component	Concentration (mM)	Acrylamide content (μg/L)[a]	% Change in acrylamide content[b]
Sodium sulfite	10	29 ± 11	−91*
Sodium sulfite	20	nd	−100*
L-proline	50	125 ± 15	−61*
L-proline	100	71 ± 3	−78*
L-proline	200	27 ± 2	−91*
Glycine	50	225 ± 7	−30*
Glycine	100	146 ± 7	−54*
Glycine	200	115 ± 3	−64*
L-tryptophan	50	212 ± 27	−34*
L-tryptophan	100	205 ± 26	−36*
L-tryptophan	200	198 ± 9	−38*
L-arginine	50	374 ± 10	17
L-arginine	100	242 ± 5	−24*
L-arginine	200	186 ± 15	−42*
L-histidine	50	280 ± 30	−13
L-histidine	100	267 ± 12	−17
L-histidine	200	194 ± 18	−39*
Sarcosine	50	126 ± 12	−61*
Sarcosine	100	80 ± 14	−75*
Sarcosine	200	55 ± 9	−83*
DL-ornithine	50	142 ± 6	−55*
DL-ornithine	100	117 ± 17	−63*
DL-ornithine	200	104 ± 14	−68*
Taurine	50	161 ± 20	−50*
Taurine	100	144 ± 50	−55*
Taurine	200	132 ± 45	−59*
γ-Amino-*n*-butyric acid	50	182 ± 10	−43*
γ-Amino-*n*-butyric acid	100	163 ± 35	−49*
γ-Amino-*n*-butyric acid	200	116 ± 12	−64*
Methiin	25	282 ± 19	−12
Methiin	50	232 ± 11	−27*
Methiin	100	203 ± 27	−36*
Alliin	25	337 ± 39	5
Alliin	50	270 ± 26	−16
Alliin	100	322 ± 37	1
Thiamine (VB1)	25	204 ± 14	−36*
Thiamine (VB1)	50	183 ± 7	−43*
Thiamine (VB1)	100	166 ± 33	−48*
Nicotinic acid (VB3)	25	303 ± 25	−5
Nicotinic acid (VB3)	50	286 ± 45	−11
Nicotinic acid (VB3)	100	247 ± 26	−23*

Continued

Table 2 Effect of sodium bisulfite, protein and non-protein amino acids, and water-soluble vitamins on the content of acrylamide in black ripe olive model system (black ripe olive juice heated at 121 °C for 30 min)—cont'd

Added component	Concentration (mM)	Acrylamide content (µg/L)[a]	% Change in acrylamide content[b]
Pyridoxine (VB6)	25	308 ± 29	−4
Pyridoxine (VB6)	50	295 ± 28	−8
Pyridoxine (VB6)	100	290 ± 20	−9
Biotin (VB7)	25	272 ± 17	−15
Biotin (VB7)	50	272 ± 28	−15
Biotin (VB7)	100	292 ± 43	−9
Sodium-L-ascorbate (VC)	25	306 ± 22	−4
Sodium-L-ascorbate (VC)	50	274 ± 42	−14
Sodium-L-ascorbate (VC)	100	263 ± 40	−18

nd = not detected; An asterisk indicates significant difference from the control ($p < 0.05$).
[a]Data represent mean values ($n = 3$) ± 95% confidence intervals.
[b]Percentage of change in acrylamide content compared to control (olive juice without additive).
Adapted from López-López et al. [16].

level); (3) L-tryptophan inhibited acrylamide formation (~36% reduction rate), and reduction was independent of dose; (4) L-arginine lowered the formation of acrylamide at concentrations of 100 and 200 mM, but it was ineffective at the 50 mM level; (5) L-histidine was only effective at the highest concentration assayed showing 39% reduction at 200 mM; (6) of the tested non-proteinogenic amino acids, sarcosine (*N*-methyl glycine) showed the highest inhibition rates, similar to those of L-proline, although ornithine, taurine, and γ-amino-*n*- butyric acid also yielded high inhibition rates (50–65% at 100 mM); (7) the non-proteinogenic sulfur-containing amino acids alliin and methiin did not show promising results as acrylamide inhibitors. In addition, after heat treatment, they imparted a distinct "garlic odor"; (8) of the tested water-soluble vitamins, only vitamin B1 significantly reduced the acrylamide content compared to the control juice, but this vitamin imparted a noticeable off-odor to the heated juice; and (9) vitamins B3, B6, B7, and C had insignificant acrylamide-reducing effects in the olive model system in general, which is in contrast to the significant acrylamide reductions reported in other model systems [24,25].

Finally, based on the results obtained in model systems, the most efficient additives, along with other compounds (*N*-acetyl-L-cysteine, reduced glutathione, methionine, calcium chloride, and propyl gallate) and several natural products with well-documented antioxidant activities (green tea, oregano, rosemary, and garlic) have been tested in black ripe olives (Table 3) [17,18]. In accordance with the results obtained in ripe model systems, sodium bisulfite at 1.5 mM level, corresponding to about 100 mg/L expressed as SO_2, did not have any significant ($p < 0.05$) effect on acrylamide levels in black ripe olives compared to control. However, at 25 mM, acrylamide levels were reduced by 100%.

Table 3 Effect of various additives on the acrylamide content as well as on the sensory characteristics of black ripe olives (sterilization at 121 °C for 15 min). Results from two different seasons

Added component	Concentration (mM)[a]	Acrylamide (µg/kg pulp)	% Change in acrylamide[b]	ΔF[c]	ΔR_{700}[d]	Flavor test[e]
First season						
Sodium bisulfite	1.5	244 ± 10	−11	−0.9	+0.2	
Sodium bisulfite	25	nd	−100*	+0.4	+1.1*	*
Calcium chloride	50	233 ± 4	−15	+10.8*	+0.1	
L-cysteine	50	nd	−100*	−0.1	+2.0*	*
N-acetyl-L-cysteine	50	nd	−100*	−5.3*	+0.5*	*
Reduced glutathione	25	nd	−100*	−1.9	+0.5*	*
L-methionine	50	261 ± 3	−4	+1.3	+0.1	
L-arginine	50	156 ± 9	−43*	−3.9*	+0.3	*
Propyl gallate	0.9	280 ± 5	3	+1.2	−0.3	
Green tea	3	275 ± 11	1	+5.1*	+0.2	
Rosemary	3	265 ± 3	−3	+0.8	−0.4*	
Oregano	3	264 ± 2	−3	+2.4	−0.1	
Raw garlic	15	303 ± 19	11	+0.2	+0.2	
Blanched garlic	15	209 ± 2	−23*	+0.1	+0.2	*
Second season						
L-arginine	100	433 ± 32	−16*	+1.0	+0.2	ns
L-arginine	200	349 ± 24	−32*	−0.1	+0.5	ns
L-proline	100	227 ± 21	−62*	−0.4	+0.3	*
L-proline	200	115 ± 18	−81*	−0.6	−0.3	*
Sarcosine	100	216 ± 27	−58*	0.7	−1.2*	ns
Sarcosine	200	104 ± 27	−80*	1.3*	−0.2	*
Glycine	100	337 ± 25	−35*	0.8	−0.2	*
Glycine	200	248 ± 30	−52*	1.2	0.9*	*

nd = not detected; ns = not significantly different from control; An asterisk indicates significant difference from the control ($p < 0.05$).
[a]Concentration for compounds are in mM, while natural products are in g/kg net weight.
[b]Percentage of change in acrylamide content compared to control.
[c]Firmness difference with respect to control.
[d]Color difference with respect to control.
[e]Paired comparison test compared with control (only for samples with significant reduction of acrylamide).
Adapted from Casado et al. [17] and Sánchez et al. [18].

From a sensory standpoint, olive color and flavor did only slightly differ compared to control. Moreover, at the above mentioned sodium bisulfite level, the panelists did not detect any abnormal or unpleasant taste. This result is extremely important, because by using this additive, black (or green) ripe olives of good sensory quality may be obtained without acrylamide formation. In an asparagine/sugar model system, the effect of bisulfite has been attributed to the addition reaction of bisulfite with the carbonyl group of the reducing sugar, which condenses with the amino compound, thus inhibiting the Maillard reaction [25]. Since this reaction has been assumed to be circumstantial for the acrylamide formation in ripe olives [5,26], the reducing effect of bisulfite could be due to the addition reaction of bisulfite with acrylamide to form 3-sulfopropionamide [27]. Unfortunately, health risks have been attributed to sulfites in foods [28] (e.g., asthmatic reactions in sensitive individuals) and, in accordance with European regulations (Commission Regulation (EU) No. 1129/2011), sulfites are currently not permitted as additives to table olives.

Acrylamide reductions of 100% were also obtained for L-cysteine, *N*-acetyl-L-cysteine, and reduced glutathione at the 50 mM level. Unfortunately, these sulfur-containing compounds negatively affected the sensory quality of black ripe olives: compared to control, retention of black color was poor, and an unpleasant odor was noted. In contrast, the addition of L-methionine did neither affect sensory quality of the product, nor show any acrylamide-reducing effect.

Calcium chloride at a 50 mM level (corresponding to 0.55%, w/v) did not significantly ($p < 0.05$) reduce the acrylamide content of black ripe olives (Table 3). This is in contrast to the acrylamide-reducing effect of $CaCl_2$ at similar concentrations in model systems and potato products [29–31]. Thus, previous assumptions suggesting a different pathway of acrylamide formation in olives are supported. $CaCl_2$ is used as firming agent, which confirms the positive effect on olive firmness [11].

L-arginine has been proposed as a good candidate to mitigate acrylamide contents in black ripe olives [17]. An acrylamide reduction of 43% was obtained when adding L-arginine at 50 mM level, whereas the impact on sensory quality was acceptable. However, when this amino acid was tested in a later olive harvest season, its efficacy as acrylamide reductor was considerably weaker (32% reduction at 200 mM) [16].

Polyphenols appeared to be ineffective as acrylamide-reducing agents in black ripe olives. When propyl gallate was added, acrylamide levels were not affected (Table 3). As previously mentioned in the case of model systems, other phenolic compounds such as hydroxytyrosol and 3,4-dihydroxyphenyl glycol did not show any effect on acrylamide levels. Moreover, the addition of natural products such as oregano, rosemary, tea, or garlic, with high contents in phenolic compounds, showed no significant reduction of acrylamide content, with the exception of a small reduction (23%) in the case of blanched garlic.

L-proline, sarcosine, and glycine are the most effective amino acids in acrylamide elimination using the ripe olive model system. They showed high acrylamide inhibition rates in black ripe olives ranging from 52% to 81% at 200 mM level (Table 3), slightly lower than those obtained in ripe olive model system (Table 2). The latter amino acids may form amino acid–acrylamide adducts through the Michael addition-type reactions, thereby reducing the acrylamide contents [18,32,33]. Since L-proline and sarcosine showed comparable acrylamide-reducing effects, nucleophilic reactivities of these amino acids are suggested to be highly similar. In general, the impact of these amino acids on firmness and surface color of black ripe olives is negligible, but there are clear differences in flavor compared to control (with the exception of sarcosine at 100 mM). It is known that the three amino acids mentioned above impart sweetness which was clearly noted by panelists at a 200-mM level. Based on these findings, sarcosine at a concentration of 100 mM has been suggested to be a promising candidate for the mitigation of acrylamide formation in black ripe olives without affecting their sensory quality. Consequently, its use has been the subject of a recent patent application [34].

The acrylamide-reducing effect of the above-mentioned amino acids may be extended during storage as demonstrated by Figure 7. Consequently, after prolonged storage (e.g., 8 months or longer) the acrylamide levels may fall below the detection limit, as has been observed for the products containing sarcosine or proline at 200 mM [18].

ACRYLAMIDE PRECURSORS IN RIPE OLIVES

The first mechanistic insight into the formation of acrylamide in heated food was provided by Stadler et al. [35] and Mottram et al. [36], who hypothesized acrylamide to be formed in the Maillard reaction from asparagine and carbonyl sources, such as reducing sugars. Although the exact formation mechanism of acrylamide in the Maillard reaction is still not fully elucidated, its formation from amino acids is well accepted, mainly asparagine, and the reducing sugars. Consequently, the Maillard reaction represents the main route of acrylamide formation in potato products [37,38], roasted almonds [20], and roasted tea [39]. However, in ripe olives, no correlation between their acrylamide content and the levels of reducing sugars and free amino acids prior to sterilization, has been established, suggesting that these compounds are irrelevant as acrylamide precursors in this commodity [5]. Recent investigations, using olive juice as a model system, indicated that small peptides (<10,000 Da) bound to polyphenols may act as precursors of acrylamide formation in ripe olives [40,41]. Since acrylamide was only formed from model peptides and from isolated gluten at elevated temperatures without the involvement of carbonyl compounds [42–44], it has been suggested that peptides need to be activated. In olives, peptides or proteins could form adducts with reactive dialdehydes and diketones (i.e., oleuropein aglycone or *o*-quinones) to form acrylamide following sterilization. The involvement of free asparagine and

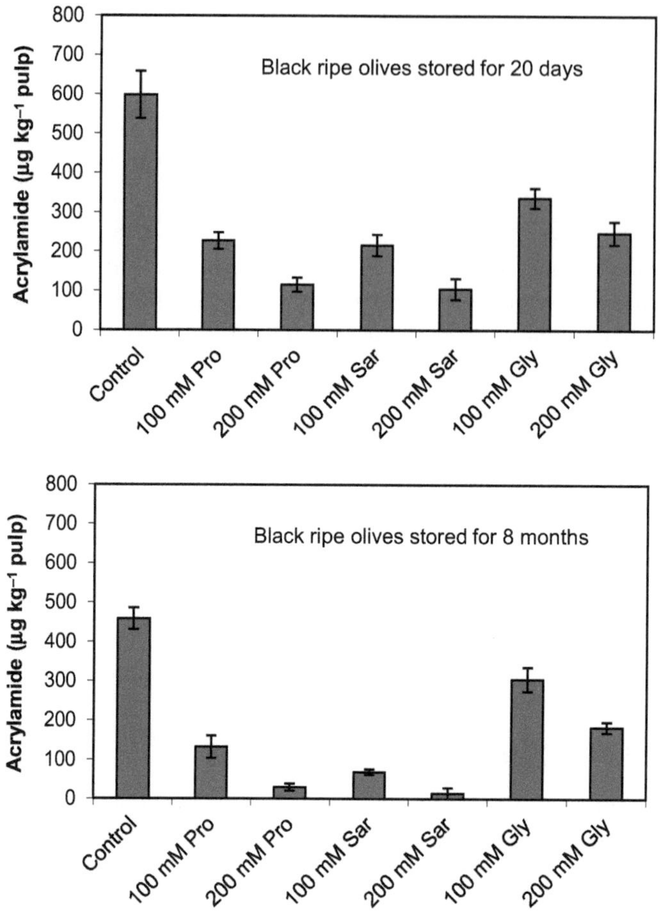

Figure 7 Effect of storage time on acrylamide content in black ripe olives in the presence of selected amino acids at two concentration levels. *Adapted from Sánchez et al. [18].*

reducing sugars has been excluded by previous investigations [5]. Briante et al. [45] and Antunes et al. [46] have confirmed the strong affinity of the oleuropein aglycone, a naturally-occurring secoiridoid polyphenol in olives, for nucleophilic binding sites as amino and thiol groups of proteins, forming protein–polyphenols adducts, thus being used for vegetable leather tanning. Structures of the above mentioned and other reactive compounds produced from transformations of oleuropein during olive processing are depicted in Figure 8.

Ongoing studies are focused on the elucidation of the exact mechanism of acrylamide formation in table olives. Novel findings will not only contribute to the progress of our knowledge in the area of food science and technology, but will also create the basis for developing effective strategies to reduce acrylamide levels by eliminating precursors or intermediates.

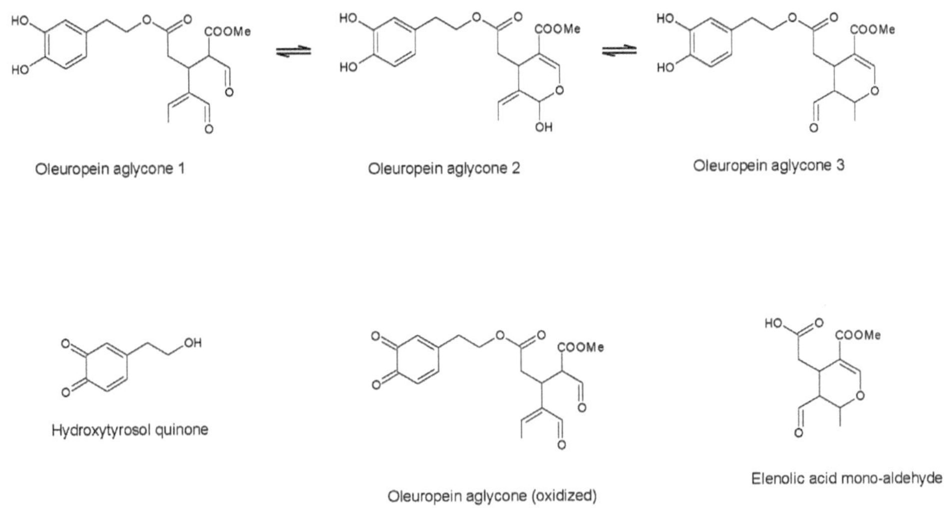

Figure 8 Structures of some reactive compounds occurring in black ripe olive processing.

SYNOPSIS OF ANALYTICAL TECHNIQUES

The main analytical methods for the determination and quantitation of acrylamide in table olives adopt a similar approach: addition of internal standard to the sample, extraction with water, purification of extract using a solid-phase extraction cartridge, and subsequent determination using either GC/MS after bromination, or liquid chromatography with tandem mass spectrometric detection (LC-MS/MS).

GC–MS with Bromination

Although Amrein et al. [26] determined acrylamide in table olives by GC–MS without derivatization, bromination of acrylamide is advantageous, because a more volatile compound is produced, thereby increasing the selectivity of the determination. Consequently, Casado and Montaño [5] applied GC–MS with bromination to quantitate acrylamide in black ripe olives. Validation data showed good precision and accuracy of the method. Therefore, this method was also used in further investigations of this group [16–18]. After pitting and homogenization, olives were spiked with ($^{13}C_3$)acrylamide as the internal standard, and blended with water. The mixture was homogenized and centrifuged at 21,000*g* (4 °C) for 30 min. Subsequently, the aqueous supernatant was applied to an SPE cartridge. The bromination reagent was added to the SPE eluate, and the mixture was left overnight at 4 °C in a refrigerator. Then, the excess bromine was reduced to a colorless solution by adding sodium thiosulfate solution, and the mixture was extracted twice with ethyl acetate. The solvent was evaporated, and the residue was re-dissolved in 100 μL of ethyl acetate. Finally, 10 μL of triethylamine was added to convert 2,3-dibromopropionamide to

2-bromopropenamide. The extract was injected into the GC–MS system, using the splitless injection method. Separation was carried out on a BPX70 capillary column (30 m × 0.22 mm i.d., 0.25 μm film thickness; SGE, Ringwood, Australia) using helium as the carrier gas and a gradient oven temperature. Ions monitored were *m/z* 70, 149, and 151 for 2-bromopropenamide, and *m/z* 110 and 154 for 2-bromo($^{13}C_3$) propenamide. Acrylamide in samples was quantitated using the ion at *m/z* 151 for 2-bromopropenamide, and the ion at *m/z* 154 for 2-bromo($^{13}C_3$) propenamide. Further ions at *m/z* 70, 110, and 149 were only considered for confirmation purposes.

Liquid chromatography with tandem mass spectrometry

The first data about acrylamide in table olives were obtained using the LC–MS/MS technique of Roach et al. [2] After addition of ($^{13}C_3$)acrylamide (internal standard) and extraction with water, the filtered extract is passed through an SPE cartridge (Oasis HLB; Waters, Milford, MA, USA) and then through another SPE cartridge (Bond Elut-Accucat; Varian, Chicago, IL, USA). The eluate may then be analyzed by LC-MS/MS using a Synergi 4 μ Hydro-RP 80A column (250 × 2.0 mm; Phenomenex, Torrance, CA, USA) maintained at 26 °C. The mobile phase consists of 0.5% methanol/0.1% acetic acid in water. The transitions monitored for acrylamide are 72→72 at 5V, 72→55 at 10V, and 72→27 at 19V. For labeled acrylamide, the following transitions are monitored: 75→75 at 5V, 75→58 at 10V, and 75→29 at 19V.

Another LC-MS/MS method for analysis of acrylamide in table olives applied by Casado et al. [40] was adapted from Claus et al. [47] After sample extraction and clean-up using multimode SPE cartridges, further analyte enrichment was accomplished by solid phase-supported liquid–liquid extraction with ethyl acetate prior to LC-MS/MS analysis. Chromatographic separation of acrylamide was carried out on a Hypercarb column (100 × 2.1 mm; 5 μm; Thermo Hypersil, Dreieich, Germany) at 30 °C. The sample was separated with a mobile phase composed of water/acetonitrile/formic acid (99/1/0.05 v/v/v) at a flow rate of 0.2 mL/min isocratically. The mass spectrometer is operated in the positive electrospray ionization mode (ESI+). The signals at *m/z* 55.5 and *m/z* 58.5 are used for the quantitation of acrylamide and D_3-acrylamide (internal standard), respectively, while the signals at *m/z* 44.5 and *m/z* 45.5 are used for qualification.

Charoenprasert and Mitchell [6] applied a similar procedure for the analysis of acrylamide in table olives. Olives are extracted with water, and D_3-acrylamide was added as the internal standard. Lipids and other nonpolar compounds are removed with hexane, and the aqueous layer is applied to an SPE cartridge (Strata-X-C; Phenomenex, Torrance, CA, USA). Acrylamide is eluted with 1 mL of 0.1% formic acid in water/methanol, (90:10, v/v), and the eluted solution is injected into the LC-MS/MS. The olive extracts are separated on a Polaris C18-A column (2.0 × 150 mm, 3 μm, Agilent

Technologies), and the column temperature is set to 30 °C. The mobile phase consists of 0.1% formic acid in water (A) and methanol (B) as follows: 0% B, 0–1 min; 0–20% B, 1–2 min; 20% B, 2–4 min, at a flow rate of 0.2 mL/min. To quantitate acrylamide levels in olives, transition ions of m/z 72→55 for acrylamide and m/z 75→58 for D3-labeled acrylamide are monitored.

KEY FACTS

- Brown to black color during the oxidation process of black ripe olive preparation is mainly due to the polymerization of hydroxytyrosol and caffeic acid after their oxidation to *o*-quinones by exposure of the fruits to air or their aeration in aqueous suspension. The former compound results from the acid hydrolysis of oleuropein (a heterosidic ester of elenolic acid and hydroxytyrosol) during fruit preservation in brine at a pH value below 4.2. Caffeic acid is obtained from the degradation of verbascoside, a rhamnoglucoside being esterified with caffeic acid and hydroxytyrosol.
- Hydroxytyrosol is largely responsible for the manifold health benefits associated with virgin olive oil and table olives. The method applied to process table olives may have a major impact on the phenolic profile of table olives. Due to oxidative degradation, processing of California-style black ripe olive results in the lowest levels of total phenolics and considerably lower levels or even loss of hydroxytyrosol compared to other processed table olives.
- The sterilizing effect, being equivalent to lethality or death rate, indicates the effect of a heat treatment expressed as the number of decimal reductions in the counts of microorganisms. The F value of a process is the number of minutes required to kill a known population of microorganisms in a given food under specified conditions. This F value is usually set at 12D values to ensure a 12 log cycle reduction of the most heat-resistant species of mesophilic spores in a food container. Since *Clostridium botulinum*, affecting low-acid foods under anaerobic conditions, and constituting the most challenging hazard to public health; the destruction of its spores is taken as the minimum requirement for safety of heat-processed food. It is stated that the decimal reduction time is 0.21 min at 121.1 °C for *C. botulinum*. Consequently, 12 decimal reductions are equivalent to an F_0 value of $12 \times 0.21 = 2.52$ min. The minimal process lethality (F_0) required is thus 2.52 min. Due to slow conductive and convective heat transfers, the real processing time of canned food in a retort is always greater than the F value.
- The Michael addition involves the reaction of a nucleophile, a so-called "Michael donor," with an activated electrophilic olefin, the "Michael acceptor," resulting in a "Michael adduct." Although, the Michael addition generally refers to the addition of enolate nucleophiles to activated olefins, a wide range of functional groups possesses enough nucleophilicity to act as Michael donors. Reactions involving non-enolate

nucleophiles such as amines, thiols, and phosphines are typically referred to as "Michael-type additions." For example, the amino acid sarcosine (Michael donor) can react with acrylamide (Michael acceptor) at temperatures exceeding 150 °C according to the following reaction scheme:

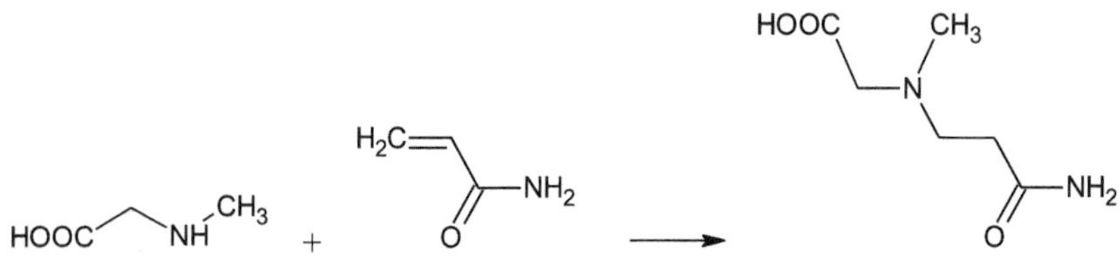

MINI DICTIONARY

Canning It is a heat sterilization process applied to foods to ensure that they are commercially sterile (i.e., the products are free from microorganisms capable of growing in the food at normal non-refrigerated temperatures). Properly sealed and heated canned foods should remain stable and indefinitely unspoiled in the absence of refrigeration.

Color It is a visual sensation (e.g., green or yellow) resulting from the cerebral interpretation of the wavelengths of the perceived light.

Cultivar An assemblage of plants that have been selected for a particular attribute or combination of attributes and that is clearly distinct, uniform, and stable in these characteristics and that when propagated by appropriate means retains these characteristics.

Deep fried Cooked in deep oil or fat.

Firmness Force required for the compression and shearing of a food in an instrument mimicking mastication of a food by the molars.

Flavor Complex combination of the olfactory, gustatory, and trigeminal sensations perceived during tasting.

Food additive Any substance not normally consumed as a food in itself and not normally used as a characteristic ingredient of food, whether or not it has nutritive value, the intentional addition of which to food for a technological purpose in the manufacture, processing, preparation, treatment, packaging, transport or storage of such food results, or may be reasonably expected to result, in it or its by-products becoming directly or indirectly a component of such foods.

Food quality The totality of features and characteristics of a product that bear on its ability to satisfy the stated or the implied needs.

Heat sterilization An act of destroying all forms of life on and in bacteriological media, foods, hospital supplies, and other materials by means of moist or dry heat.

Sensory Relating to the use of the senses, that is, to the experience of a person.

SUMMARY POINTS

- Acrylamide is only detected in California-style black ripe and California-style green ripe olives, while it is not detected in other types of processed table olives such as Spanish-style and Greek-style olives.
- Since California-style green ripe olives are processed without air oxidation, their acrylamide content is significantly lower than that of California-style black ripe olives.

- Acrylamide in black (or green) ripe olives is formed during heat sterilization.
- When applying identical F_0 value for heat sterilization, the amount of acrylamide formed is lower at higher temperatures and has shorter sterilization time.
- Acrylamide levels in processed ripe olives of cv. Manzanilla are higher than those in the processed ripe olives from cvs. Hojiblanca and Cacereña.
- Technological measures related to the darkening step of black ripe olives (i.e., lye treatment, aeration, and washing) have a significant influence on acrylamide content, with the exception of ferrous gluconate addition.
- Acrylamide content in brined-preserved ripe olives is significantly higher than that in acid-preserved ripe olives.
- Preservation time prior to processing has an impact on the formation of acrylamide.
- Acrylamide content in black ripe olives remains stable during storage time (at least within 6 months of storage).
- When using sulfiting agents (e.g., sodium bisulfite or sodium sulfite) as additives, black (or green) ripe olives being devoid of acrylamide may be obtained without affecting the sensory quality. However, these additives are currently not permitted in the EU.
- Sarcosine has been proposed as an additive for the mitigation of acrylamide formation in black ripe olives having almost no impact on their sensory quality.
- Recent investigations suggest small peptides binding to oxidized polyphenols and further carbonyl donators such as elenolic acid mono-aldehyde acting as precursors of acrylamide formation in ripe olives following sterilization without the involvement of free asparagine and reducing sugars.

REFERENCES

[1] Tareke E, Rydberg P, Karlsson P, et al. Analysis of acrylamide, a carcinogen formed in heated foodstuffs. J Agric Food Chem 2002;50:4998–5006.
[2] Roach JAG, Andrzejewski D, Gay ML, et al. Rugged LC-MS/MS survey analysis for acrylamide in foods. J Agric Food Chem 2003;51:7547–54.
[3] DiNovi M, Howard D. The updated exposure assessment for acrylamide. In: Acrylamide in food workshop: update–scientific issues, uncertainties, and research strategies. JIFSAN; 2004. http://www.jifsan.umd.edu/acrylamide2004.htm.
[4] FDA (US Food and Drug Administration). Survey data on acrylamide in food: Individual food products. US Department of Health and Human Services, Center for Food Safety and Nutrition; 2006. http://www.cfsan.fda.gov/~dms/acrydata.html.
[5] Casado FJ, Montaño A. Influence of processing conditions on acrylamide content in black ripe olives. J Agric Food Chem 2008;56:2021–7.
[6] Charoenprasert S, Mitchel A. Influence of California-style black ripe olive processing on the formation of acrylamide. J Agric Food Chem 2014;62:8716–21.
[7] ASEMESA (Asociación de Exportadores e Industriales de Aceitunas de Mesa). Datos generales del sector. 2014. http://www.asemesa.es/content/datos_generales_del_sector.
[8] Hundrods D. Olive profile. Agricultural Marketing Resource Center, Iowa State University; 2013. http://www.agmrc.org/commodities__products/fruits/olive-profile/.

[9] de Castro A, García P, Romero C, et al. Industrial implementation of black ripe olive storage under acid conditions. J Food Eng 2007;80:1206–12.
[10] Brenes-Balbuena M, García-García P, Garrido-Fernández A. Phenolic compounds related to the black color formed during the processing of ripe olives. J Agric Food Chem 1992;40:1192–6.
[11] Sánchez AH, García P, Rejano L. Elaboration of table olives. Grasas Aceites 2006;57:86–94.
[12] IOOC (International Olive Oil Council). Trade standard applying to table olives. Res-2/91-IV/04. Madrid: IOOC; 2004.
[13] Garrido-Fernández A, Fernández-Díez MJ, Adams MR. Table olives, production and processing. London: Chapman and Hall; 1997.
[14] Charoenprasert S, Mitchell A. Factors influencing phenolic compounds in table olives (Olea europaea). J Agric Food Chem 2012;60:7081–95.
[15] Montaño A, Casado FJ, Rejano L, et al. Formación de acrilamida durante el tratamiento de esterilización de aceitunas tratadas con álcali. In: CYTALIA XIII, Congreso Anual en Ciencia y Tecnología de Alimentos. Madrid. 2008.
[16] López-López A, Beato VM, Sánchez AH, et al. Effects of selected amino acids and water-soluble vitamins on acrylamide formation in a ripe olive model system. J Food Eng 2014;120:9–16.
[17] Casado FJ, Sánchez AH, Montaño A. Reduction of acrylamide content of ripe olives by selected additives. Food Chem 2010;119:161–6.
[18] Sánchez AH, Beato VM, López-López A, et al. Comparative study of the use of sarcosine, proline and glycine as acrylamide inhibitors in ripe olive processing. Food Addit Cont A 2014;31:242–9.
[19] Hoenicke K, Gatermann R. Studies on the stability of acrylamide in food during storage. J AOAC Int 2005;88:268–73.
[20] Amrein TM, Lukac H, Andres L, et al. Acrylamide in roasted almonds and hazelnuts. J Agric Food Chem 2005;53:7819–25.
[21] JECFA (Joint FAO/WHO Expert Committee on Food Additives). Summary Report of 64th Meeting. Rome, Italy. 2005.
[22] Claeys WL, De Vleehouwer K, Hendrickx ME. Effect of amino acids on acrylamide formation and elimination kinetics. Biotech Prog 2005;21:1525–30.
[23] Claus A, Mongili M, Weisz G, et al. Impact of formulation and technological factors on the acrylamide content of wheat bread and bread rolls. J Cereal Sci 2008;47:546–54.
[24] Zeng X, Cheng KW, Jiang Y, et al. Inhibition of acrylamide formation by vitamins in model reactions and fried potato strips. Food Chem 2009;116:34–9.
[25] Yuan Y, Shu C, Zhou B, et al. Impact of selected additives on acrylamide formation in asparagine/sugar Maillard model systems. Food Res Int 2011;44:449–55.
[26] Amrein TM, Andres L, Escher F, et al. Occurrence of acrylamide in selected foods and mitigation options. Food Addit Cont A 2007;24:13–25.
[27] Lehmann J, Benson AA. The plant sulfolipid. IX. Sulfosugar syntheses from methyl hexoseenides. J Am Chem Soc 1964;86:4469–72.
[28] Taylor SL, Higley NA, Bush RK. Sulfites in foods: uses, analytical methods, residues, fate, exposure assessment, metabolism, toxicity, and hypersensitivity. Adv Food Res 1986;30:1–76.
[29] Gökmen V, Şenyuva HZ. Acrylamide formation is prevented by divalent cations during the Maillard reaction. Food Chem 2007;103:196–203.
[30] Ou S, Lin Q, Zhang Y, et al. Reduction of acrylamide formation by selected agents in fried potato crisps on industrial scale. Innov Food Sci Emerg Technol 2008;9:116–21.
[31] Mestdagh F, Maertens J, Cucu T, et al. Impact of additives to lower the formation of acrylamide in a potato model system through pH reduction and other mechanisms. Food Chem 2008;107:26–31.
[32] Koutsidis G, Simons SPJ, Thong YH, et al. Investigations on the effect of amino acids on acrylamide, pyrazines, and Michael addition products in model systems. J Agric Food Chem 2009;57:9011–5.
[33] Liu J, Chen F, Man Y, et al. The pathways for the removal of acrylamide in model systems using glycine based on the identification of reaction products. Food Chem 2011;128:442–9.
[34] Montaño A, Sánchez AH, López A, et al. Composición de envasado para aceitunas esterilizadas que comprende sarcosina y sus aplicaciones. Application P201330451 (Spain). 2013.
[35] Stadler RH, Blank I, Varga N, et al. Acrylamide from Maillard reaction products. Nature 2002; 419:449–50.

[36] Mottram DS, Wedzicha BL, Dodson AT. Acrylamide is formed in the Maillard reaction. Nature 2002;419:448–9.
[37] Amrein TM, Schiónbochler B, Rohner F, et al. Potential for acrylamide formation in potatoes: data from the 2003 harvest. Eur Food Res Technol 2004;219:572–8.
[38] Taubert D, Halfinger S, Henkes L, et al. Influence of processing parameters on acrylamide formation during frying of potatoes. J Agric Food Chem 2004;52:2735–9.
[39] Mizukami Y, Kohata K, Yamaguchi Y, et al. Analysis of acrylamide in green tea by gas chromatography–mass spectrometry. J Agric Food Chem 2006;54:7370–7.
[40] Casado FJ, Montaño A, Spitzner D, et al. Investigations into acrylamide precursors in sterilized table olives: evidence of a peptic fraction being responsible for acrylamide formation. Food Chem 2013;141:1158–65.
[41] Casado FJ, Montaño A, Carle R. Contribution of peptides and polyphenols from olive water to acrylamide formation in sterilized table olives. LWT-Food Sci Technol 2014;59:376–82.
[42] Claus A, Weisz GM, Schieber A, et al. Pyrolytic acrylamide formation from purified wheat gluten and gluten supplemented wheat bread rolls. Mol Nutr Food Res 2006;50:87–93.
[43] Buhlert J, Carle R, Majer Z, et al. Thermal degradation of peptides and formation of acrylamide. Lett Org Chem 2006;3:356–7.
[44] Buhlert J, Carle R, Majer Z, et al. Thermal degradation of peptides and formation of acrylamide, Part 2. Lett Org Chem 2007;4:329–31.
[45] Briante R, La Cara F, Febbraio F, et al. Hydrolysis of oleuropein by recombinant b-glycosidase from hyperthermophilic archaeon Sulfolobus solfataricus immobilised on chitosan matrix. J Biotechnol 2000;77:275–86.
[46] Antunes APM, Attenburrow G, Covington AD, et al. Utilisation of oleuropein as crosslinking agent in collagenic films. J Leather Sci 2013;2:17–23.
[47] Claus A, Weisz GM, Kammerer DR, et al. A method for the determination of acrylamide in bakery products using ion trap LC-ESI-MS/MS. Mol Nutr Food Res 2005;49:918–25.

CHAPTER 13

Acrylamide in Battered Products

Eduardo J. Guerra-Hernández
Departamento de Nutrición y Bromatología, Universidad de Granada, Granada, Spain

INTRODUCTION

Battering is a culinary technique that is widely used in the home preparation of fried foods. Its use in industrial food production has increased, in line with changing lifestyles and consumption habits in recent years, as people express greater preferences for ready-to-eat products. Commercial battered foods are usually frozen, ready-to-eat products, containing one or various foodstuffs of animal or vegetable origin, with the addition of flour and/or breadcrumbs, milk, egg, and other authorized substances, seasoned as appropriate, generally subjected to brief frying (par-frying) and then freezing (−20 °C) or deep freezing (−40 °C). Once thawed, or taken directly from the frozen state, they can be eaten after a short frying or baking/frying process. The battering/breading technique is routinely used with meats (especially chicken), fish (anchovies, calamari, hake, cod, etc.), and vegetables (aubergines, onions, etc.) and can be applied in many different ways, according to the local common practice in different countries, using only flour or mixtures of flour–water, flour–milk, or flour–egg, or only breadcrumbs, or a breadcrumbs–egg mixture, and so on. But in every case, the first step is to coat the raw food with batter and the breading mixture, and the second step is to deep fry it.

Fried batters not only enhance the sensory quality of the coated food but may also produce undesirable effects. They normally contain high amounts of oil and contribute to oil-related health problems such as obesity and heart disease. Moreover, the evidence of high levels of acrylamide in battered/breaded foods that are rich in carbohydrates has added to concerns about the excessive consumption of fats, because this substance—possibly carcinogenic—is formed in large quantities in such foods, because the ingredients (flour/bread) and temperatures used are ideal for its formation.

In April 2002, the Swedish National Food Administration and researchers at Stockholm University [1] announced research findings that acrylamide, a toxic and potentially cancer-causing chemical, is formed in many types of food that are prepared/cooked at high temperatures. According to initial data, the acrylamide content in breaded/battered products from Norway, Sweden, Switzerland, the United Kingdom and the United States is 30–39 μg/kg in fish and seafood products, 39–64 μg/kg in poultry [2], about 30 μg/kg in battered fried vegetables (tempura) in Japan [3] and <3–170 μg/kg in deep-fried meat Chinese/Indian style (wheat-flour based) and in deep-fried wheat flour [4].

Acrylamide in Food
http://dx.doi.org/10.1016/B978-0-12-802832-2.00013-9

Since this announcement, many further studies have been conducted to better understand the mechanism, ingredients, and processes that favor the formation of acrylamide and ways to limit its generation. The flour/bread used contains sugars (glucose, fructose) and asparagine, which are the main precursors of acrylamide [5], while the content of free amino acid is the limiting factor in cereals [6].

ACRYLAMIDE CONTENT

Acrylamide in Fried Flour Products

Wheat Flours

Leung et al. [4] studied the influence of temperature on the formation of acrylamide during the preparation of "Chinese fried fritters." In this study, dough was prepared with wheat flour, water, yeast, baking soda, and salt, and fried in vegetable oil at 170, 190, and 210 °C. In this deep-frying process, the oil penetrated into the food matrix; the degree of this penetration depended on the frying time, and was relatively independent of the oil temperature. At 170 and 190 °C, the decrease in moisture content was largely compensated by the increase in fat content.

The rate of acrylamide formation increased steadily with frying time, for all three oil temperatures, from below detection levels in the dough to a leveling value of about 350 μg/kg in the overcooked product. At an oil temperature of 170 °C, the increase in acrylamide was gentle but steady through out the 18-min frying period. At 210 °C, the acrylamide level rose very rapidly initially, but converged after about 5 min to gradually approach the leveling value. At 190 °C, the increase in acrylamide was intermediate between the previous two cases: there was a gentle initial rise for the first 5 min, followed by a rapid increase in the rate, which then converged to the leveling value at about 10 min (Figure 1).

These authors found that the frying time for an acceptable quality product at 170 °C was 12–14 min, which fell to 10–11 min at 190 °C, while at 210 °C a frying time of 5 min produced an acceptable product. The acrylamide content in the acceptable fried fritters prepared at 170 °C was 150–200 μg/kg, which rose to over 250 μg/kg for the products prepared at 190 and 210 °C, respectively.

The amount of acrylamide in the fried fritter was inversely proportional to the moisture content. The lower the temperature, the higher the dependence on moisture loss for acrylamide formation, due to the necessary diffusion of hot oil into the dough. At 190 °C, a moisture content of 20% of the product was associated with an acrylamide content exceeding 200 μg/kg.

Morales and Arribas-Lorenzo [7] studied the level of acrylamide in *churros,* a Spanish fried-dough pastry, when dough prepared with wheat flour, water, and salt was deep-fried at 180, 190, and 200 °C for 2, 3, 5 and 7 min in high-oleic sunflower oil. High temperatures and short periods of time are necessary to avoid excessive oil penetration

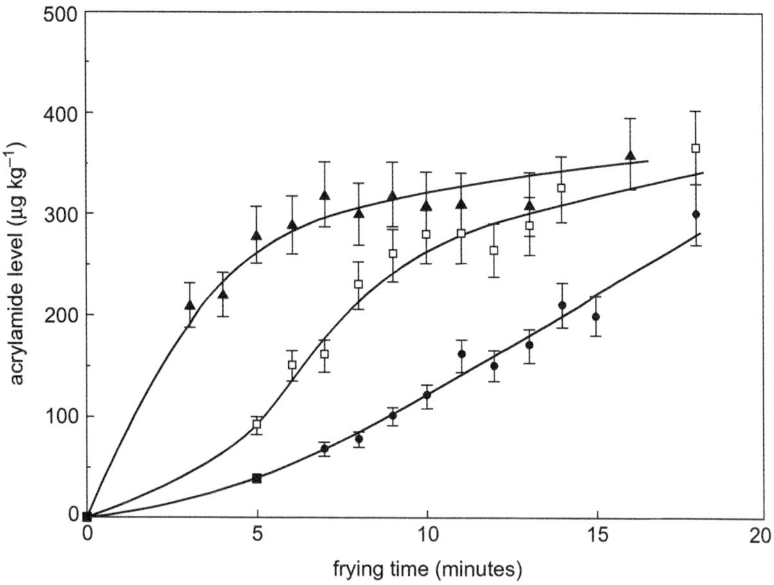

Figure 1 Acrylamide level (µgkg^{-1}) as function of frying time at various oil temperatures: ●, 170 °C; □, 190 °C; ▲, 210 °C. *Leung et al. [4], Figure 1, with permission from Taylor & Francis.*

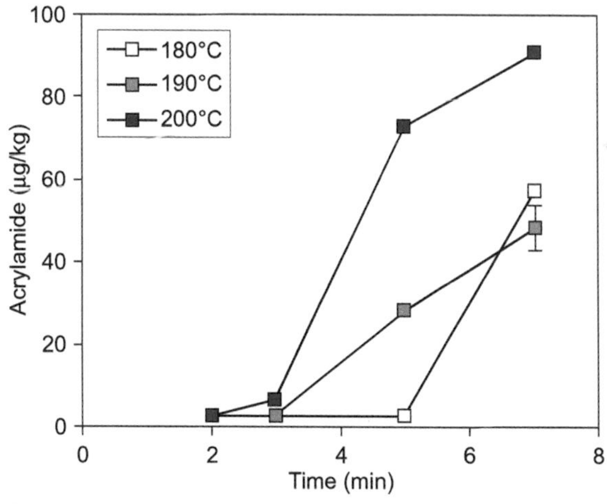

Figure 2 Acrylamide formation in *churros* during deep frying at different times and temperatures. *Morales and Arribas-Lorenzo [7], Figure 4, with permission from Elsevier.*

into the product, which should be served within a few minutes of frying to avoid subsequent hydration. Acrylamide was only formed under the most severe frying conditions, after an induction period of 3–5 min. After 7 min the acrylamide level was around 60 µg/kg for frying temperatures of 180 and 190 °C, rising to 90 µg/kg for a temperature of 200 °C (Figure 2). However, at this temperature, frying times of over 3 min caused excessive moisture loss and browning, resulting in a loss of product quality. Deep frying at temperatures of 185–200 °C for 3–4 min is usually applied by local producers, based on their practical experience. The process has not been automated and remains artisanal. At these frying times, an increase of just 10 °C produces a twofold increase in the level of acrylamide; therefore, the frying temperature has a critical effect on the formation of acrylamide. These authors concluded it was necessary to determine the end point of the frying process in order to minimize the formation of potentially harmful compounds in *churros* and thus reduce their presence in the dietary intake of the Spanish population.

In commercial samples, the average acrylamide content found by Morales and Arribas-Lorenzo [7] was 46 ± 24.5 μg/kg (n = 10), although other authors have reported average levels of 400 μg/kg [8,9].

Other Flours

Shih et al. [10] studied the generation of acrylamide and the absorption of oil in flours commonly used in battering. The flours used in this study were obtained from long-grain rice, waxy rice, wheat, or corn, to which sodium chloride, sodium bicarbonate, disodium pyrophosphate, and water were added to create the appropriate viscosity (1380–1560 cP). The batters were introduced by spoonfuls into Wesson vegetable oil and fried for 6 min at 177 °C. The results for oil uptake and acrylamide content are shown in Table 1. The highest level of acrylamide content was found in corn flour, followed by wheat flour, while the lowest levels were produced in rice flour. The formation of acrylamide does not seem to correlate with the oil uptake of the fried batters, as the highest level of oil was obtained with wheat flour, probably due to its gluten content.

These authors also studied the effects of adding pre-gelatinized rice flour, which is used to obtain textural characteristics comparable to those of traditional wheat batter, and of adding milk. When pre-gelatinized rice flour at 5% and 10% was incorporated, acrylamide levels rose from 180 to 245 ng/g for the low-level incorporation and rose to 319 ng/g for the high-level incorporation, while oil uptake increased only slightly, from

Table 1 Oil uptake and acrylamide content of batters during frying

Flour	Oil uptake (%)	Acrylamide content (μg/kg)
Long-grain (LG) rice	21.4	180
Waxy rice	31.9	194
Wheat	47.3	298
Corn	27.9	378
LG + 5% pre-gelatinized rice flour (LG5P)	26.5	245
LG + 10% P	25.2	319
LG5P + 1.5% milk	30.4	331
LG5P + 3% milk	28.4	336

21.4% to 25.2% for the high-level incorporation and increased to 26.5% for the low-level incorporation. A 5% incorporation to long-grain rice increased the acrylamide content, but the level remained below that obtained with wheat batter, and the oil uptake was about 45% lower. The addition of milk, at 1.5–3.0%, as is commonly done in the food industry, to 95% long-grain rice flour and 5% pre-gelatinized rice flour, raised the acrylamide level to about 330 ng/g, a value that is slightly above that found in wheat batter, while the oil content was about 30%, which is well below the level found in wheat batter (47%). The authors recommended that the batters used should be formulated with 95% long-grain rice flour, 5% pre-gelatinized rice flour and 1.5–3.0% milk in order to minimize the acrylamide content and achieve low-oil frying properties.

Acrylamide in Breaded Meats

Deep Frying

Paleologos and Kontominas [11] evaluated the presence of acrylamide in precooked breaded chicken samples, and measured acrylamide values of 910–970 μg/kg in filleted chicken breast and leg dipped in a batter of beaten eggs, spices, and breadcrumbs; fried in corn oil for 2 min at 180 °C; and then baked for 5 min at 130–150 °C. These values are much higher than those found for the flours, due to the presence of egg and chicken meat (resulting from the interaction between carbohydrates in the breadcrumbs and free amino acids in the egg and chicken meat). When these ingredients were fried separately, the quantity of acrylamide produced was less than 50% than that found when they were heat processed together. Tareke et al. [12] observed a similar phenomenon when they analyzed the acrylamide content in fried potatoes mixed with milk, egg, or flour, compared with fried potatoes without these additives. On similar lines, Wang et al. [13] analyzed the acrylamide content of eight food samples of deep-fried flour-based indigenous Chinese foods commercially available in Hong Kong, and reported that the highest level of acrylamide (198 μg/kg) of the 27 samples corresponded to fried egg pastry, and that the amounts were higher in darker colored products than in lighter ones.

Paleologos and Kontominas [11] referred to an FDA [14] study according to which chicken nuggets, which are similar to the products they evaluated, contained minimal amounts of acrylamide, not exceeding 20–25 μg/kg and suggested that this contradiction may be partly attributed to the amount and composition of the batter, to differences in the heat process used, and to the diet of the chickens. In comparable surveys, Tateo et al. [15] reported an acrylamide content ranging from 50 to 159 μg/kg in Italy, while Eerola et al. [16] measured values <68 μg/kg in Finnish foodstuffs.

Paleologos and Kontominas [11] stored precooked breaded chicken samples in different atmospheres and observed that the acrylamide content increased with storage, especially in aerobically packaged samples. Therefore, acrylamide formation may be accelerated via a non-thermal route. Lipid oxidation and microbial growth during storage may have an impact on the formation of acrylamide through less obvious pathways, but more experimental work is needed before firm conclusions can be drawn. After peaking at about day 15, acrylamide concentrations decreased, perhaps due to subsequent degradation, as would be expected due to the depletion of reactants. The decrease was very small in the aerobically packaged samples, in which the acrylamide concentrations remained high, whereas in the modified atmosphere packing (30% CO_2–70% N_2, 60% CO_2–40% N_2, or 90% CO_2–10% N_2) samples, the acrylamide concentrations tended to return to their initial values. The highest acrylamide values obtained were 1800µg/kg (air/day 15) during storage and 1360µg/kg for samples packaged under a modified atmosphere (60% CO_2–40% N_2).

Microwave Frying

Barutcu et al. [17] studied the effects of microwave frying, in comparison with a commercial bench-top deep fryer, using different flours, on acrylamide formation in the coated part of chicken nuggets. The control flour used was a 50:50 mixture of wheat and corn flour, replacing 30% of this mixture with chickpea flour, rice flour, or soy flour. The batter formulations were composed of a 2:3 (w/w) solid-to-water ratio. Chicken breast portions were individually immersed into the batter formulation, after which the breading was applied and the portions were fried in sunflower oil at 180°C using 365W microwave power level, for four different times (0.5, 1.0, 1.5, and 2.0min) or for 5.0min in a commercial bench-top deep fryer. The breading used was a traditional breadcrumb material from Turkey, made from wheat flour, hydrogenated vegetable oil, salt, coloring, water, and potassium sorbate as a preservative. For the product to be acceptable for consumption, the authors established that a cooking time of 1.5min in the microwave oven or 5min of conventional deep frying was necessary. The conventionally fried samples had a higher acrylamide content and a darker color than the microwave fried samples had, ranging from 88µg/kg for the control flour to 111µg/kg for the rice flour. These values represent 13.8% and 34.5% more acrylamide, respectively, than those obtained with microwave oven cooking (about 76µg/kg). The lower acrylamide content with microwave oven frying was explained by the shorter frying time than with conventional frying. During microwave frying, the moisture content in the coating layers of fried chicken decreased significantly with increasing frying time. Chickpea, rice, and control flours had similar effects on moisture retention, but the use of soy flour significantly increased the moisture content, and an extra 30s of cooking time was needed to achieve the final moisture content of about 20% (Figure 3). After 1.5min of frying, the average acrylamide content ranged from 73µg/kg for the rice flour sample to 79µg/kg for the chickpea flour sample. The color (*L* and *a* values) was similar in the control samples, the rice, and the chickpea flours (about 55 and 17.7, respectively) but very different from the soybean flour (64.9 and 12.4, respectively). This shows that the color parameters may not always be a reliable indicator of acrylamide levels. The cereal flours presented the highest levels of acrylamide, although these values later decreased, with increased frying time, due to the low content of free asparagine: 0.16mg/100g for rice flour, 4.28mg/100g for corn flour and 6.22mg/100g for wheat flour. However, for chickpea and soybean flour, the acrylamide

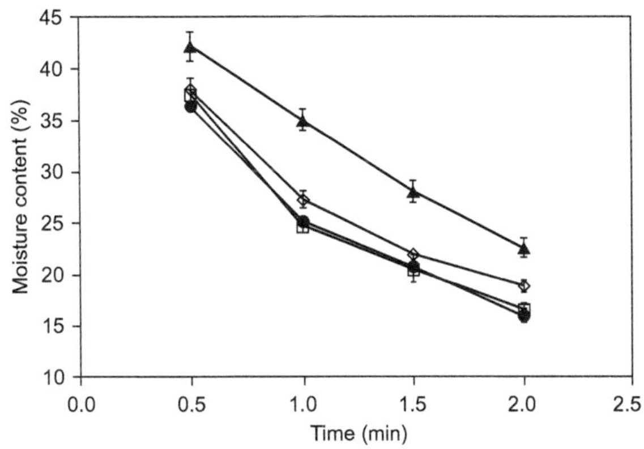

Figure 3 Variation of moisture content in the coating part of chicken having different flours during microwave frying (□) control[a], (◊) chickpea flour[(a)], (●) rice flour[(a)], (▲) soy flour[(b)]. Formulations having different letters are significantly different ($p \leq 0.05$). *Barutcu et al. [17], Figure 1, with permission from Elsevier. Between flours with the same superscript (a) as: Control, chickpea flour and rice there aren't statistical differences, but these samples are statistically different to soy flour with different superscript (b) (p<0.05).*

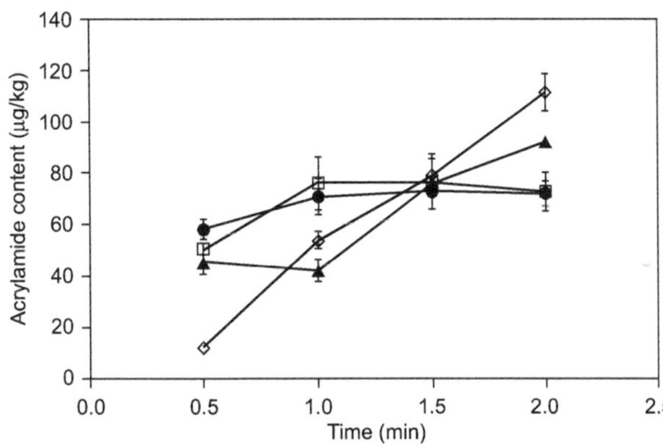

Figure 4 Variation of acrylamide content in the coating part of chicken having different flours during microwave frying (□) control, (◊) chickpea flour, (●) rice flour, (▲) soy flour. *Barutcu et al. [17], Figure 2, with permission from Elsevier.*

content increased continuously during the test, as the asparagine content was higher in these samples, at 22.56 mg/100 g and 42.10 mg/100 g for soy flour and chickpea flour, respectively. At 2 min (30 s longer than the optimal time), the highest levels of acrylamide were found in the chickpea flour sample (≅110 μg/kg), due to its higher level of free asparagine and lower moisture content (about 19%), compared to soy flour, which had an acrylamide value of ≅90 μg/kg and a moisture content of ≅23% (Figure 4). Although the acrylamide contents were similar for all the types of flours in the microwave fried samples, for a frying time of 1.5 min, the different levels of free asparagine and moisture meant that this pattern was different from that for other frying times.

Demirok and Kolsarici [18] measured 314 μg/kg and 285 μg/kg of acrylamide in the coating surface of chicken drumsticks (CDs) and chicken wings (CWs), respectively. These samples were first coated with a marinade sauce (basic ingredients: oil, salt, red pepper, black pepper, coriander, ginger, coconut, garlic, citrate, basil, cumin, rosemary, and thyme) and then breaded. The batter was made by dissolving 500 g of breading in 875 mL of water, and the chicken pieces were then deep fried, in a high pressure fryer at 175 °C and 8 psi pressure (CDs, 8.5 min; CWs, 3.5 min). These authors conducted a marketing study to determine acrylamide levels in the coating of fried CDs and CWs purchased from fast food restaurants in Turkey, and recorded an average of 217 and 19 (ng acrylamide/g coating) in CDs and CWs, respectively.

Acrylamide in Vegetable Foods

Guerra-Hernández et al. [19] studied the presence of acrylamide in zucchini-based product, a vegetable product. This food was cut into sticks and battered with flour, flour–water, flour–milk, or flour–egg, and breaded with breadcrumbs–egg or breadcrumbs–egg–flour. The authors used four commercial flour types (wheat, rice, chickpeas, and a mixture of wheat and corn) and two types of commercial breadcrumbs, with different sized particles. In all cases, the packaging of the products stated that they were suitable for breading and/or battering.

The zucchini were deep fried in sunflower oil at 190 °C until acceptable organoleptic characteristics were achieved, according to a preestablished frying time (see Table 2), but always for ≤4 min. The acrylamide content in the coated material was determined by UPLC–MS/MS, using acrylamide $^{13}C_3$ as an internal standard. The samples were defatted with hexane and extracted with methanol, using Carrez solution as a precipitating agent. The methanol was removed using a SpeedVac concentrator and the aqueous acrylamide solution obtained, with added water, was purified with Oasis HLB extraction cartridges.

The mean acrylamide values in batter-fried and breaded zucchini were 213 μg/kg and 177 μg/kg, respectively, and ranged from 50 μg/kg for the fine breadcrumbs–egg coating to 368 μg/kg for the batter-fried zucchini with wheat flour–water coating and 494 μg/kg for batter-fried zucchini with only a wheat flour coating (Table 2). These data are not the same as those previously reported [19], because a problem was observed in the high performance liquid chromatography mass spectrometer (HPLC–MS) equipment and the chromatographic analyses had to be repeated.

Battered Products

In the battered products, the liquid-to-solid ratio (i.e., water, milk, egg: flour) needed to obtain similarly consistent doughs ranged from 1.5 for the water mixtures to 2.5 for the egg-based mixtures. The frying time required to obtain an organoleptically acceptable product was lower for the milk-based samples (2–3 min) and greater for the egg-based ones (4 min). The mean acrylamide content for the three treatments (water/milk/egg) was 228 μg/kg for the wheat flour, 217 μg/kg for the wheat–corn flour mixture, 177 μg/kg for

Table 2 Preparation conditions and the fat, moisture, and acrylamide content in battered and breaded portions of zucchini

Cereal/Legume	Treatment (battered)	Liquid:solid ratio	Frying time (min)	% Batter	% Fat in batter	Moisture content (%)	Acrylamide in batter (μg/kg)	Acrylamide per serving (μg/kg)[a]	g fat per serving[a]
	–	–	–	–	27.4	8.5	494	–	–
Wheat	Water	1.6	3.5	36.28	23.4	15.5	368	16.03	10.17
	Milk	1.7	3.0	37.34	20.8	35.4	196	8.76	9.31
	Egg	2.4	4.0	41.32	26.1	24.4	122	6.02	12.94
Rice	Water	1.7	3.0	27.73	22.9	10.8	235	7.82	7.61
	Milk	1.8	2.0	31.63	25.2	20.3	177	6.72	9.58
	Egg	2.5	4.0	29.98	44.2	29.0	119	4.28	15.92
Wheat-corn	Water	1.5	2.5	21.22	41.6	17.4	280	7.13	10.59
	Milk	1.6	2.5	40.44	42.6	12.5	246	11.92	20.68
	Egg	2.0	4.0	47.60	42.3	27.5	127	7.23	24.17
Chickpea	Water	1.9	2.5	19.18	49.7	17.4	169	3.89	11.45
	Milk	2.0	2.0	18.14	41.6	24.0	140	3.05	9.06
	Egg	2.5	4.0	32.21	48.5	23.1	99	3.83	18.75
	Treatment (breaded)	**Liquid:solid ratio**	**Frying time (min)**	**% Breaded**	**% Fat in breading mix**	**Moisture content (%)**	**Acrylamide in breading mix (μg/kg)**	**Acrylamide per serving (μg/kg)[a]**	**g fat per serving[a]**
Coarse breadcrumbs	Egg + bread	–	2.5	13.41	41.6	11.8	219	3.53	6.69
	Egg + bread + flour	2.5	2.5	36.15	35.0	13.0	303	13.13	15.18
Fine breadcrumbs	Egg + bread	–	3.0	28.57	34.9	28.3	50	1.70	11.97
	Egg + bread + flour	2.5	3.0	34.12	20.0	21.0	138	5.66	8.19

[a]Serving = 120 g.

rice flour, and 136 μg/kg for the chickpea flour. The highest acrylamide content was obtained with the flour–water mixture, followed by flour–milk and flour–egg. There were statistically significant differences between all the treatments for the wheat flour and wheat–corn flour samples ($p < 0.01$) and for the rice flour samples, between flour–water and flour–eggs. Although the asparagine content was different in each of the flours used, as reported elsewhere, the chickpea flour probably had the highest content and the rice flour, the lowest [17]. Nevertheless, we do not believe this to be a limiting factor because these flours are used in ways that vary (i.e., different liquid/solid ratios and different frying times) but which are suitable to obtain products with organoleptically acceptable characteristics.

There was a correlation ($r = 0.6305$) between the liquid/solid ratio and the frying time, and a highly significant inverse relationship ($r = -0.804$, $p < 0.001$) between this relationship and the acrylamide content, which was >0.9 for the rice, wheat, corn by wheat-corn, and chickpea flours. Although a similar consistency was obtained with each coating material, the greater amount of fluid needed for the egg and milk samples meant that the acrylamide content was lower in the fried product.

The average moisture value in the final product was 21.4%, similar to that obtained by other authors [17]. In general, the greater the amount of liquid added, the greater the amount that remained in the coating, regardless of the frying time. An inverse and significant relationship was obtained between the moisture content of the final product and the acrylamide content ($r = -0.573$, $p < 0.05$). However, there was no such relationship with the frying time when no account was taken of the different flours or treatment used, except when the treatment with egg ($r = 0.7490$) was excluded.

When egg was used, the amount of batter obtained was generally higher and that of acrylamide, lower. In this case, the batter-coated layer was spongy (unlike the texture obtained with water and milk), dark colored on the outside but not on the inside, which explains the lower levels of acrylamide.

The quantity of acrylamide per serving (120 μg) in the battered samples ranged from 3 μg for the chickpea flour–milk sample to 16 μg for the flour–water batter (Table 2). There was less acrylamide in the flour–egg batter than when milk or water was used. However, the fact that the flour–egg mixture gave rise to a higher percentage of batter meant that the total acrylamide content did not vary greatly from the other types. With the chickpea flour, the quantity of acrylamide was lower (3.6 μg); moreover, with this mixture the frying time was shorter, the moisture content of the batter was greater and there was a lower percentage of batter in the final product. The mean acrylamide content for the other flours, regardless of the treatment applied, was 6.3 μg for the rice flour, 8.8 μg for the wheat–corn flour mixture, and 10.3 μg for the wheat flour. These values are 3–8 times lower than the daily estimates of acrylamide exposure for the Spanish population (30 μg).

The amount of fat per serving, not surprisingly, was higher in the samples with egg, for which a value of 24.2 g was recorded in the wheat–corn flour samples. The quantity of fat in the egg-based batter was about twice of that measured in the samples prepared with water or milk, except in the samples in which only wheat flour was used. The average fat content was lowest in the rice flour samples (11 g). The chickpea flour samples had a better acrylamide–fat ratio, and the sample prepared with milk and chickpeas presented the best ratio, with 3.05 μg of acrylamide and 9.06 g of fat.

Breading Products

The mean acrylamide values were similar in the breading and in the batter products. However, there was a large difference in acrylamide content between the use of coarse or fine breadcrumbs (262 and 94 μg/kg, respectively). The frying time was lesser with coarse breadcrumbs, but it should also be noted that a different brand of breadcrumbs was used, and moreover, the coarse breadcrumbs also contained gluten. In these products (coarse breadcrumbs), the fat content in the breading mix was higher than that in fine breadcrumbs, Shih et al. [10] reported similar results in flours with a high gluten content. As in the battered products, there was a significant inverse correlation between acrylamide content and the moisture recorded in the breading ($r = -0.839$, $p < 0.01$). The addition of wheat flour raised the levels of acrylamide in the final breading by about 85 μg/kg for both coarse and fine breadcrumbs.

In the breaded food, the intake of acrylamide ranged from 1.7 μg for the fine breadcrumbs with egg to 13.3 μg for the coarse breadcrumbs with egg and flour (Table 2).

As in the battered food, the samples prepared with only egg presented the lowest acrylamide values. When the breading was performed with flour added the quantity of acrylamide increased by about 3.5 times. In the breaded food, the fat content per serving ranged from 6.7 to 15.2 g, in both cases when coarse breadcrumbs were used. The breaded food made with breadcrumbs and egg, but no flour, presented the best acrylamide–fat relationship, with 3.53 μg:6.69 g, respectively, for coarse breadcrumbs and 1.70 μg:11.97 g, respectively, for fine breadcrumbs.

In Spain during 2014 [20], the average daily consumption of relevant foods in this context was as follows: vegetables grown for their fruit (aubergines, courgettes, tomatoes, etc.,) about 70 g; by onions, 20 g; fresh and frozen fish, 40 g; fresh and frozen shellfish, 21 g, and chicken, 48 g. Application of the highest value obtained for acrylamide (368 μg/kg) and the highest percentage of batter/breading (47.6%), and assuming that at most 10% of the products consumed are battered/breaded, shows that the daily consumption of acrylamide would be about 2.34, 0.67, 1.34, 0.70, and 1.6 μg for vegetables grown for their fruit, onions, fish, shellfish, and chicken, respectively. In the worst case, this would be equivalent to 7.8% of daily acrylamide consumption.

EFFECTS OF OILS ON ACRYLAMIDE FORMATION

Chuang et al. [21] analyzed acrylamide formation in chicken legs during frying, in three types of unsaturated oil (palm oil, soybean oil, and lard), at two temperatures, 160 and 180 °C for 10 min. The results obtained were compared with those for French fries (potato chips), prepared under the same conditions, but fried for only 3 min. Acrylamide was found only in the outer, floured part of the fried chicken legs, and not in the inner, meaty portion. The acrylamide content ranged from 22.7 μg/kg in the chicken cooked in lard to a value of 55.1 μg/kg when soybean oil, at a temperature of 160 °C was used. At a temperature of 180 °C, the acrylamide content ranged from 45.2 μg/kg in lard to a value of 65.3 μg/kg in palm oil. For French fries, the highest level of acrylamide (163.7 μg/kg) was obtained with soybean oil at 180 °C. At 160 °C, the acrylamide content in French fries was similar to that recorded for the chicken legs, but levels at 180 °C were much higher, ranging from 75 μg/kg in lard to 163.7 μg/kg in soybean oil. The authors surmised that this was because soybean oil contains more unsaturated fatty acids (mainly linoleic acid) than does lard and that unsaturated fatty acids such as oleic or linoleic acid are more susceptible during heating to the formation of octanal or decanal [22], which on reaction with asparagine leads to the formation of acrylamide [23]. The effect of carbonyl compounds formed by the degradation of hydro-peroxides during lipid oxidation is probably more important in French fries, for which these compounds are the limiting factor, but not in flour, for which asparagine is the limiting factor. At 180 °C, it was surprising to find that both French fries (143.5 μg/kg) and chicken legs (65.3 μg/kg) presented such a high acrylamide content when fried in palm oil, which is not as unsaturated as soybean oil. In this respect, Gertz and Klostermann [24] reported the presence of 6–8% diglycerides in palm oil products, and this factor may enhance the rate of reaction of acrylamide formation, by the emulsification of the components. Moreover, the 0.3–0.5% monoacylglycerol content in this type of oil can decompose during frying to fatty acid and acrolein. These same authors found that if silicone (dimethylpolysiloxane, E−900), which is legally permitted in Europe as additive E900, and used as an anti-foaming agent, is added to a deep frying fat, the acrylamide content in fried foods is increased. This compound forms a monomolecular layer at the surface of the oil, which is subjected to degradation during the frying process, and the evaporation of water as steam, which would remove acrolein or acrylamide, may thus be hindered and/or reduced, and hence the concentration of acrylamide will be increased. Mestdagh et al. [25] observed significantly greater amounts of acrylamide in model systems containing acrolein and asparagine, compared to the control group to which only asparagine was added. However, the contribution of acrolein to the overall formation of acrylamide appears to be negligible in the presence of a reducing sugar, which suggests that in foodstuffs, acrolein and other oil degradation products such as glycerol, monoacyglycerols, and diacylglycerols are probably of minor importance. Totani et al. [26], too, concluded that the oil used for deep frying would not contaminate foodstuffs with acrylamide.

INHIBITING THE GENERATION OF ACRYLAMIDE

Many proposals have been made for minimizing acrylamide levels in foods, and some successful results have been presented [27–31]. In addition, kinetic model studies have been performed to assess acrylamide formation based on the initial concentration of acrylamide precursors [32,33]. In this field, the European Commission [34] established the HEATOX program [35] to investigate health risks from acrylamide and other chemicals formed in heat-treated foods. Additionally, the experiences of the food industry and of scientists working in this respect, in diverse product categories, have been collated in a guidance document, the Acrylamide Toolbox, published by FoodDrinkEurope [6], formerly known as CIAA. Many of the approaches defined in this Toolbox are summarized in the document *Code of Practice for the Reduction of Acrylamide in Foods* [36]. Recommended strategies include reducing frying/baking time and temperature, decreasing the initial amount of acrylamide precursors, using microwave treatment, and lowering the pH and the quantity of asparaginase, antioxidant compound or divalent cations employed [29,37]. Most research in this field has been conducted on potato-based products or bakery foods, and few studies have addressed the question on methods to reduce the levels of acrylamide in battered or breaded products.

As most of the acrylamide in these products is contained in the coated flour of fried food, some of the methods described by the FDA document *Guidance for Industry: Acrylamide in Foods* [38], aimed at reducing the acrylamide content in cereal products, may be useful. Suggestions include using varieties of wheat with a lower content of asparagine (by modifying the content of sulfur and nitrogen fertilization), adding amino acids such as glycine, adding calcium salts or adding asparaginase if the batter is to be made by mixing the flour with another liquid ingredient, and ensuring that the enzyme has sufficient time to have an effect on the batter. Different flours may be used in the manufacture of these products, and the asparagine content can have a decisive influence on the final acrylamide content. With respect to free asparagine levels in different cereals, the highest free asparagine content is found in rye grass (319–791 mg/kg^{-1}), followed by oats (250–400 mg/kg^{-1}), wheat (69–443 mg/kg^{-1}), and corn (43 mg/kg^{-1}). The lowest quantities are found in rice (1.6 mg/kg^{-1}) [17,39]. Moreover, whole grain products (e.g., whole wheat) have higher asparagine concentrations than refined grains have, and produce more acrylamide in the finished products. In addition to these considerations, the frying time should be carefully controlled.

In addition to the methods described above, and taking into account the characteristics of the products tested, other ingredients may be included with the usual flour or breadcrumbs. Thus, Zhang et al. [40] proposed the addition of antioxidants from bamboo leaves (AOB) to reduce the formation of acrylamide in fried CWs. In the method described, the flour paste used for the coated material was prepared by mixing 100 g of flour with beaten egg and a sequence of prepared aqueous solutions (80 ml), containing 0.001%, 0.01%, 0.05%, 0.1%, 0.5%, and 1% (w/w) AOB. After coating the CWs with the prepared flour paste and special flavoring (pepper, sesame, granulated sugar, refined salt, ginseng, Chinese wolfberry, and chicken essence), they were fried in palm oil for 4 min at 170 °C. The acrylamide content in the control sample was around 160 μg/kg, and reductions of 33.4%, 34.4%, 45.5%, 57.8%, 59.0%, and 53.5% in acrylamide content were obtained for the different concentrations of AOB added (Figure 5). These authors reported good sensory acceptability at an addition of 0.1% AOB. On comparing the reduction obtained at this concentration with that for two commercial extracts of bamboo, they measured reductions in acrylamide content proportional to the content of flavonoids, and concluded that these compounds play an important role in lowering the content of acrylamide.

Antioxidant compounds have been found to be effective in minimizing acrylamide formation, due to three effects: (1) radical scavenging activity; (2) a carbonyl trapping effect; and (3) the limitation of sugar degradation through the Maillard reaction [30,41].

More recently, Demirok and Kolsarici [18] have studied the effects of green tea extract (GTE), either alone or combined with microwave precooking (MPC), to reduce acrylamide formation in fried bone-in CDs and CWs. For this purpose, the authors marinated CDs and CWs for one day and then divided the samples into eight groups. Four were coated in batter and breaded, including 0%, 0.5%, 1.5%, and 3% GTE, and deep fried for 8.5 min (CDs) and 3.5 min (CWs) at 175 °C and 8 psi pressure. The remainder of the marinated CDs and CWs were subjected to MPC for 7 and 5 min, respectively, and then coated in batter and breaded, including 0%, 0.5%, 1.5%, and 3% GTE, and deep fried for 5.5 min (CDs) and 2.5 min (CWs) at 175 °C and 8 psi pressure.

In comparison with the control group (0% GTE), the use of 0.5%, 1.5%, and 3% GTE lowered acrylamide concentrations by about 21%, 39%, and 43% in non-microwaved CDs and by 25%, 32%, and 42% in microwaved CDs, respectively. For CWs with 0.5%, 1.5%, and 3% GTE, the reductions were 1%, 19%, and 34% for non-microwaved treatment and 27%, 40%, and 55% for microwaved treatment, respectively (Figure 6). The synergistic effect of GTE and MPC was greater for the CWs. GTE includes high levels of

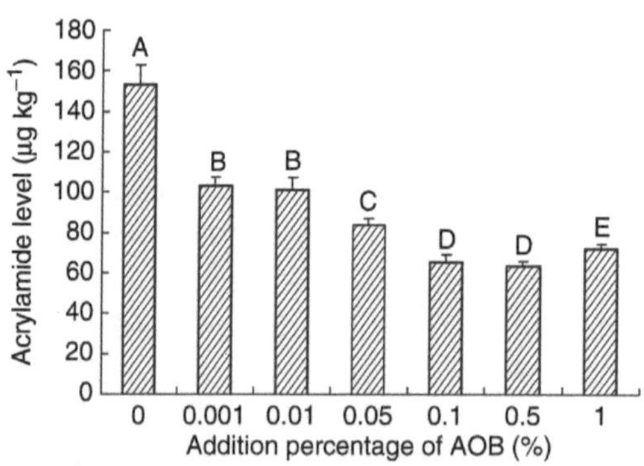

Figure 5 Relationship between acrylamide levels and different conditions of antioxidants from bamboo leaves (AOB) addition treatments in fried chicken wings ($n = 6$). Error bars designate the standard deviation (SD); different letters indicate significant differences via Duncan's multiple comparison ($p < 0.05$). *Zhang et al. [40], Figure 1, with permission from Taylor & Francis.*

polyphenols, predominantly catechins, which are well-known antioxidant compounds [42]. The major classes of catechins are epigallocatechingallate (EGCG), epigallocatechin, epicatechingallate, epicatechin (EC), gallocatechin, and catechin [43]. Previous studies have reported that EC and EGCG inhibit the Maillard reaction by forming covalent bonds with reactive carbonyls (carbonyl trapping effect) during the heating process [28,43,44]. In addition, it has been reported that the inhibiting effect of phenols on the Maillard reaction arises from radical scavenging activity in food model systems [44]. Accordingly, in the current study the effect of GTE on the formation of acrylamide in fried bone-in chicken products could be explained by a carbonyl trapping effect or by the radical scavenging activity of catechin with carbohydrates. Moreover, antioxidants have been reported to reduce acrylamide formation in the presence of large amounts of fat in biscuits [45] and in French fries [46].

From the reductions in acrylamide content achieved, it is clear that this treatment alone was effective for both CDs and CWs, but it was more efficient for CWs when combined with MPC. The addition of GTE to the breading mix is a more practical method for the food industry than the use of MPC, since it does not require additional processing steps. With increasing levels of GTE, the concentration of acrylamide decreased gradually in both CDs and CWs. MPC also decreased acrylamide formation in CWs, due to the shorter frying time. Moreover, MPC was more efficient at reducing acrylamide formation in CWs when it was combined with GTE. Neither GTE nor MPC had any adverse effect on the sensory quality of fried CDs and CWs. Thus it is of crucial importance, as color is a critical parameter of visual quality, while texture, flavor, and odor are equally important parameters of palatability. These authors concluded that the addition of GTE to the coating material is an easy, straight forward method to reduce the formation of acrylamide in fried CDs and CWs.

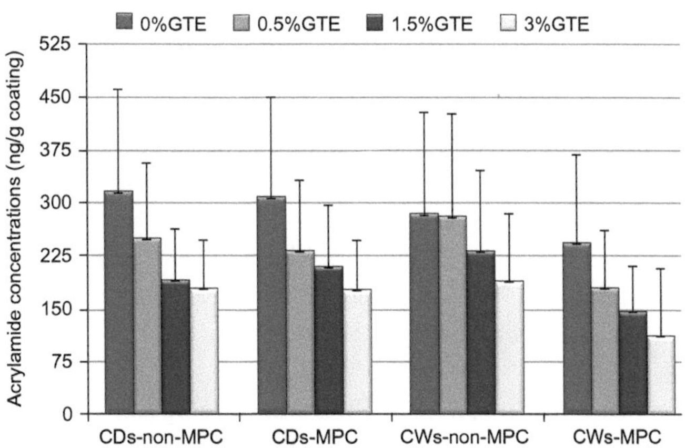

Figure 6 Acrylamide concentration (ng/g) in the coating of fried CDs and CWs ($n=4$). *Demirok and Kolsarici [18], Figure 3, with permission from Elsevier.*

SYNOPSIS OF ANALYTICAL TECHNIQUES

The most commonly applied techniques for acrylamide analysis are gas chromatography–mass spectrometry (GC–MS) and liquid chromatography–mass spectrometry (LC–MS), using an isotope-labeled internal standard [47–50].

LC–MS methods are considered simpler and less time-consuming, but can suffer matrix interferences and ion suppression. Therefore, emphasis is usually placed on the extraction and clean-up stages of the method to ensure that clean sample extracts are obtained.

In diverse studies, the extraction process has been performed using water [3,4,10,11,40], acetic acid 0.2 mM [17], formic acid 10 mM [18], or methanol [19]. Zhang et al. [40] and Guerra-Hernández et al. [19] previously added petroleum ether or hexane, respectively, to defatted samples. An internal standard such as $^{13}C_3$-acryalmide [3,4,19], d_3-acrylamide [10], or methacryalmide [11] is used and Carrez I and II is sometimes added to precipitate the co-extracted colloids [17–19]. After agitation at room temperature or at 70 °C [11], the samples are separated, habitually by cold centrifugation to solidify the oil, and hexane is sometimes added to extract the remaining long-chain fatty acids [11]. The latter authors, and also Barutcu et al. [17], subjected the extract to HPLC without purification, but most authors prefer to conduct a clean-up stage with different solid phase extraction cartridges, type OASIS HLB, with reverse phase sorbent, determining a hydrophilic–lipophilic balance [4,19], using an OASIS MCX mixed-mode sorbent reversed phase and cation exchange [18] or ISOLUTE Multimode with non-polar (C18) strong cation and strong anion exchanges [3,10]. In this respect, Zhang et al. [40], before solid-phase extraction, purified the aqueous extract with ethyl acetate.

C18-reversed phase HPLC columns are commonly employed in the LC method [3,4,10,17–19]; Aminex HPX-87H [11] has also been used, with HPLC or UPLC [18,19]. The mobile phase is aqueous with acetic acid, formic acid, sulfuric acid or methanol, in low concentration.

Most of the LC–MS methods that have been proposed employ a triple quadrupole instrument (MS/MS), together with electrospray ionization [3,4,10,18,19], although Barutcu et al. [17] reported good results with atmospheric pressure chemical ionization. In MS/MS detection, the use of multiple-reaction monitoring significantly enhances the selectivity of the method. Typical transitions are m/z: 72→55 (usually chosen as the quantifier), 72→54, 72→44, 72→27, and for the internal standard ($^{13}C_3$-acrylamide), 75→58 as the quantifier, and 75→30 as the qualifier.

The limit of detection obtained was 3 $\mu g kg^{-1}$ and the limit of quantification was 10 $\mu g kg^{-1}$ [4,18].

UV detection with normal phase-HPLC was developed by Paleologos and Kontominas for the analysis of acrylamide and methacrylamide in precooked breaded chicken products [11]. Acrylamide quantities were confirmed by conversion to acrylic acid after hydrolysis. The quantification limit obtained for both analytes was >10 μg/kg. Wang et al. [13] obtained a quantification limit of 23 μg/kg in their analysis of the acrylamide content of deep-fried flour-based indigenous Chinese foods by HPLC-UV.

In this field, GC–MS has only been used in two studies. Ono et al. [3] analyzed certain battered products processed in Japan (fried vegetables and tempura), and obtained a satisfactory agreement between the results from bromination-GC-MS and underivatized LC–MS. In the column used, the stationary phase contained 50% diphenyl and 50% dimethylsiloxane. Fragment ion peaks of 2,3-dibromopropionamide derived from acrylamide at *m/z* 150 and 152, and those from acrylamide-d_3 at *m/z* 153 and 155 were monitored and used for the calculation of acrylamide content in the food samples. In the second of these studies, Chuang et al. [21] analyzed chicken legs, using ethyl acetate as an extraction solvent and a C-18 cartridge for purification. Triethylamine was added to promote the formation of 2-bromopropenamide, which was used for quantification as an alternative to 2,3-dibromopropionamide, because the latter compound is less stable during chromatographic analysis. In the column used, the stationary phase was composed of 5% diphenyl and 95% dimethylsiloxane. Selected-ion monitoring mode was used to identify and quantify 2-bromopropenamide (*m/z* 106, 108, 149, and 151) and *m/z* 178 and 180 for N,N-dimethylacrylamide as the internal standard. The average recovery obtained was 98% and the detection limit was 5 ppb.

KEY FACTS

- Making frozen breaded products healthier, in a modified, energy-efficient manufacturing process.

In late 2003, researchers at the CSIC Institute of Agrochemistry and Food Technology in Valencia (Spain) patented a new method of preparing a frozen, battered food product to eliminate the most problematic step in standard batter processes, namely par-frying. This method consists, first, in flouring the food product to be battered and then applying a reconstituted batter to the floured food product. This batter contains a cellulose derivative which coagulates in a bath of hot water, in the presence of water vapor. The battered food product is subsequently heated and coagulated using a standard infrared or microwave oven, such that the layer of coagulated batter does not break up when cooled. Finally, the product is cooled and frozen. Cooking, rather than pre-frying, has several advantages: in nutritional terms, the food is less fatty, more nutritious, and easier to store (free from rancidity); furthermore, less energy is consumed in its manufacture, with lower temperatures being used and, in all probability, less acrylamide is produced. This process, which is exclusive to the food company Adin SA, is currently being applied in the manufacture of calamari (squid rings).

- The need for culinary education to achieve optimum frying temperatures with lower levels of fat and acrylamide in products with high nutritional value. The key is to fry the product well.

The market for breaded/battered fish is expanding and this product has been recommended for children to provide an adequate intake of omega-3 long-chain polyunsaturated fatty acids (EPA and DHA). Some processing industries have managed to reduce the added salt in breaded and battered foods by up to 30%. Others have enriched flours with certain nutrients (such as vitamins A and C, phosphorus, and calcium), or developed foods that are rich in dietary fiber with resistant starch, without altering the sensory quality of the product.

- Analyzing the presence of acrylamide in healthier battered/breaded foods prepared with new ingredients, using various, mostly patented, processes.

The use of various flours mixed with different proportions of hydrocolloids (such as high-amylose rice, calcium pectins, methyl cellulose, hydroxypropyl methylcellulose, guar gum, and xanthan gum), immersing the products in a bath after breading them with whey protein, or breading using crumbs made from rice flour and xanthan gum, can achieve reductions in oil absorption of up to 50%.

- Lower nutritional value and higher intake of acrylamide in products that increase the proportion of breading/batter.

The analysis of different brands of the same type of product—frozen squid—reveals proportions ranging from 29% to 40% squid, much lower percentages than are found in home-cooked battered squid. In order to achieve a fluffier batter, some manufacturers add raising agents (sodium bicarbonate and diphosphates, either combined or separately) and thickeners (methylcellulose, guar gum, and locust bean gum). Because the acrylamide content rises in proportion with the amount of batter/breading, and taking into account that the ingredients used in the preparation of the batter can modify the acrylamide content in the product, food quality agencies such as EFSA and FDA should recommend or require the inclusion of the percentage of batter in product labeling and the possible incorporation of messages such as "the type of flour used in the batter/breading of this product may produce higher levels of acrylamide."

- New assessments of acrylamide intake in population groups that consume the largest amounts of ready-to-eat breaded food.

In developed countries, the consumption of prepared meals is increasing significantly. In Spain, between 1987 and 2012 the sales of such foods rose by about 400%, in line with increasing population size and socio–economic status, especially in households without children (or where children are aged six years or older) and among young adults living independently. By contrast, consumption is lower among households where family members are older and numbers fewer. The frequency of consumption of battered/breaded dishes varies significantly, but it is very high for squid and battered hake; in the latter case, only 40% of the population state that they never or hardly ever eat it, while 26% do so more than once a week.

MINI DICTIONARY

Battered Food coated with a layer of flour and then fried.

Breaded Food coated in breadcrumbs and beaten egg and then fried.

Limiting factor Compound present in the flour/bread which determines the amount of acrylamide formed. In this kind of food preparation, the limiting factor is the amino acid asparagine.

Frying fat/oil The oil which is used in frying battered/breaded food and which, in given proportions, is incorporated into the new food. Its degree of unsaturation, number of double bonds present, is related to the health characteristics of the food (the absence of double bonds is associated with increased cardiovascular risk).

Par-frying It is a short frying process to achieve the desired consistency and to enable subsequent handling, freezing and transport, as well as reducing the number of microorganisms present in the food.

Internal standard Compound added, in a known quantity, during the determination of acrylamide content in order to improve the precision and accuracy of the measurement.

Monoglycerides and diglycerides Compounds that have emulsifying properties and may be present in small proportions in a fat (triglyceride), either "per se" or generated during frying.

Acrolein Compound that may be generated in frying by the dehydration of glycerol in the fat/oil used.

Refined grains/flours Flour used in the breading/battering process, from which the outer layers of the grain (which produce most acrylamide) have been removed.

CIE L*, a*, b* color scale. L* = Lightness. a* = Red for positive value and green for negative value. b* = Yellow for positive value and blue for negative value. Reduction in L★ values and increases in a★ values indicate that the flours became darker and get more red.

SUMMARY POINTS

- The acrylamide content in organoleptically acceptable fritters fried at 170 °C is about 150–200 µg/kg. The amount of acrylamide in the fried fritter is inversely proportional to the moisture content. The lower the frying temperature, the greater the dependence of acrylamide formation on moisture loss, due to the necessary diffusion of hot oil into the batter.
- In preparing *churros*, at constant frying times, an increase of just 10 °C in the frying temperature can provoke a twofold increase in the level of acrylamide.
- The formation of acrylamide does not seem to correlate with the oil uptake of the fried batter.
- The replacement of the flour used with another type (wheat, rice, corn, soybean, and chickpea) and the addition of other ingredients (water, milk, and egg) can reduce the generation of acrylamide and the absorption of fat into the batter/breading.
- The highest level of acrylamide observed in these products was 970 µg/kg in filleted chicken breast and leg dipped in a batter of beaten egg, spices, and breadcrumbs, first fried and then baked. In these products, the acrylamide content rose during storage to 1800 µg/kg (in air/15 days) and to 1360 µg/kg (packaged in a modified atmosphere; 60% CO_2–40% N_2).
- Microwave frying generates lower levels of acrylamide in breaded foods than does conventional frying, probably because of the shorter frying time required.

- Similar levels of acrylamide content are found when flours with different contents of precursors (free asparagine) are used, at the optimal preparation times. However, at longer times, this similarity may disappear.
- For coating vegetables, the use of beaten egg with the flour, rather than water or milk, reduces the quantity of acrylamide present, regardless of the type of flour used.
- In breaded food, adding flour to the breadcrumbs and egg mixture increases the acrylamide content more than three-fold.
- The type of oil used in frying is more influential in French fries (potato chips), in which carbonyl compounds are the limiting factor. When palm oil is used to fry breaded food, acrylamide levels are higher than when other oils are used, due to the presence of monoglycerides and diglycerides, which possess emulsifying properties and produce acrolein.
- The use of antioxidants and microwave frying reduces the acrylamide content in these foods.
- Maximum advisable levels of acrylamide have not been established for these products, but in any case the quantity of acrylamide present is small; the amount consumed is not great and the coated material is only a small proportion of the final food.

REFERENCES

[1] Swedish National Food Administration. Information about acrylamide in food. April 24, 2002. http://www.slv.se.

[2] FAO/WHO. Health implications of acrylamide in food. Report of a Joint FAO/WHO Consultation. (Geneva: WHO); 2002. p. 8–9.

[3] Ono H, Chuda Y, Ohnishi-Kameyama M, et al. Analysis of acrylamide by LC-MS/MS and GC-MS in processed Japanese food. Food Addit Contam 2003;20(3):215–20.

[4] Leung KS, Lin A, Tsang CK, et al. Acrylamide in Asian foods in Hong Kong. Food Addit Contam 2003;20(12):1105–13.

[5] Yaylayan VA, Wnorowski A, Perez Locas C. Why asparagine needs carbohydrate to generate acrylamide. J Agric Food Chem 2003;51(6):1753–7.

[6] FDE. Food drink Europe acrylamide toolbox. 2011. http://fooddrinkeurope.eu/uploads/publications_documents/Toolboxfinal260911.pdf.

[7] Morales FJ, Arribas-Lorenzo G. The formation of potentially harmful compounds in churros, a Spanish fried-dough pastry, as influenced by deep frying conditions. Food Chem 2008;109(2):421–5.

[8] Bermudo E, Moyano E, Puignou L, et al. Determination of acrylamide in foodstuffs by liquid chromatography ion-trap tandem mass-spectrometry using an improved clean-up procedure. Anal Chim Acta 2006;559(2):207–14.

[9] Yusa V, Quintas G, Pardo O, et al. Determination of acrylamide in foods by pressurized fluid extraction and liquid chromatography-tandem mass spectrometry used for a survey of Spanish cereal-based foods. Food Addit Contam 2006;23(3):237–44.

[10] Shih FF, Boue SM, Daigle KW, et al. Effects of flour sources on acrylamide formation and oil uptake in fried batters. J Am Oil Chem Soc 2004;81(3):265–8.

[11] Paleologos EK, Kontominas MG. Effect of processing and storage conditions on the generation of acrylamide in precooked breaded chicken products. J Food Prot 2007;70(2):466–70.

[12] Tareke E, Rydberg P, Karlsson P, et al. Analysis of acrylamide, a carcinogen formed in heated foodstuffs. J Agric Food Chem 2002;50(17):4998–5006.

[13] Wang H, Lee AWM, Shuang S, et al. SPE/HPLC/UV studies on acrylamide in deep-fried flour-based indigenous Chinese foods. Microchem J 2008;89(2):90–7.

[14] FDA. Exploratory data on acrylamide in food: total diet study results. 2004. Available at: http://www.cfsan.fda.gov/dms/acrydat2.html.
[15] Tateo F, Bononi M, Andreoli G. Acrylamide levels in cooked rice, tomato sauces and some fast food on the Italian market. J Food Compos Anal 2007;20(3–4):232–5.
[16] Eerola S, Hollebekkers K, Hallikainen A, et al. Acrylamide levels in Finnish foodstuffs analysed with liquid chromatography tandem mass spectrometry. Mol Nutr Food Res 2007;51(2):239–47.
[17] Barutcu I, Sahin S, Sumnu G. Acrylamide formation in different batter formulations during microwave frying. LWT-Food Sci Technol 2009;42(1):17–22.
[18] Demirok E, Kolsarici N. Effect of green tea extract and microwave pre-cooking on the formation of acrylamide in fried chicken drumsticks and chicken wings. Food Res Int 2014;63(Part C):290–8.
[19] Guerra-Hernández E, Plata-Guerrero R, García-Villanova Ruiz B. Acrylamide formation in different batter and breading formulation during deep-frying of zucchini and exposure estimation. Ann Nutr Metab 1841;2013(Suppl. 1):63.
[20] MAGRAMA. La alimentación mes a mes en España. 2014. http://www.magrama.gob.es/es/alimentacion/temas/consumo-y-comercializacion-y-distribucion-alimentaria/panel-de-consumo-alimentario/ultimos-datos/.
[21] Chuang WH, Chiu CP, Chen BH. Analysis and formation of acryalmide in Frech fries and chicken legs during frying. J Food Biochem 2006;30(5):497–507.
[22] Frankel EN, Nett WE, Selke E. Analysis of oxidized fats by gas chromatography–mass spectrometry: VII Volatile thermal decomposition products of pure hydroperoxides from autooxidized and photosensitized oxidized methyl oleate, linoleate and linolenate. Lipids 1981;16(5):279–85.
[23] Becalski A, Lau BP, Lewis D, et al. Acrylamide in foods: occurrence, sources, and modeling. J Agric Food Chem 2003;51(3):802–8.
[24] Gertz C, Klostermann S. Analysis of acrylamide and mechanisms of its formation in deep-fried products. Eur J Lipid Sci Technol 2002;104(11):762–71.
[25] Mestdagh F, Castelein P, Van Peteghem C, et al. Importance of oil degradation components in the formation of acrylamide in fried foodstuffs. J Agric Food Chem 2008;56(15):6141–4.
[26] Totani N, Yawata M, Takada M, et al. Acrylamide content of commercial frying oil. J Oleo Sci 2007;56(2):103–6.
[27] Amrein TM, Schönbachler B, Escher F, et al. Acrylamide in gingerbread: critical factors for formation and possible ways for reduction. J Agric Food Chem 2004;52(13):4282–8.
[28] Capuano E, Ferrigno A, Acampa I, et al. Effect of flour type on Maillard reaction and acrylamide formation during toasting of bread crisp model systems and mitigation strategies. Food Res Int 2009;42(9):1295–302.
[29] Erdoğdu SB, Palazoğlu TK, Gökmen V, et al. Reduction of acrylamide formation of French fries by microwave pre-cooking of potato strips. J Sci Food Agri 2007;87(1):133–7.
[30] Hedegaard RV, Granby K, Frandsen H, et al. Acrylamide in bread. Effect of prooxidants and antioxidants. Eur Food Res Technol 2008;227(2):519–25.
[31] Zhang Y, Zhang Y. Effect of natural antioxidants on kinetic behaviour of acrylamide formation and elimination in low-moisture asparagine–glucose model system. J Food Eng 2008;85(1):105–15.
[32] Knol JJ, Van Loon WAM, Linssen JPH, et al. Toward a kinetic model for acrylamide formation in a glucose–asparagine reaction system. J Agric Food Chem 2005;53(15):6133–9.
[33] Parker JK, Balagiannis DP, Higley J, et al. Kinetic model for the formation of acrylamide during the finish-frying of commercial French fries. J Agric Food Chem 2012;60(36):9321–31.
[34] European Commission (EC). Food contaminants acrylamide. 2006. http://ec.europa.eu/food/food/chemicalsafety/contaminants/acrylamide_en.htm.
[35] HEATOX. The HEATOX Project. 2006. www.heatox.org.
[36] Codex. Code of Practice for the Reduction of Acrylamide in Foods (CAC/RCP 67–2009). 2009. http://www.codexalimentarius.net/download/standards/11258/CXP_067e.pdf.
[37] Sahin S, Sumnu G, Oztop MC. Effect of osmotic pretreatment and microwave frying on acrylamide formation in potato strips. J Sci Food Agri 2007;87(15):2830–6.
[38] FDA. http://www.fda.gov/downloads/Food/GuidanceRegulation/GuidanceDocumentsRegulatoryInformation/ChemicalContaminantsMetalsNaturalToxinsPesticides/UCM374534.pdf; 2013.
[39] Konings EJM, Ashby P, Hamlet CG, et al. Acrylamide in cereal and cereal products: a review on progress in level reduction. Food Addit Contam 2007;24(Suppl. 1):47–59.

[40] Zhang Y, Xu W, Wu X, et al. Addition of antioxidant from bamboo leaves as an effective way to reduce the formation of acrylamide in fried chicken wings. Food Addit Contam 2007;24(3):242–51.
[41] Capuano E, Oliviero T, Açar ÖÇ, et al. Lipid oxidation promotes acrylamide formation in fat-rich model systems. Food Res Int 2010;43(4):1021–6.
[42] NamalSenanayake SPJ. Green tea extract: chemistry, antioxidant properties and food applications—A review. J Funct Foods 2013;5(4):1529–41.
[43] Schamberger GP, Labuza TP. Effect of green tea flavonoids on Maillard browning in UHT milk. LWT-Food Sci Technol 2007;40(8):1410–7.
[44] Totlani VM, Peterson DG. Epicatechin carbonyl-trapping reactions in aqueous maillard systems: Identification and structural elucidation. J Agric Food Chem 2006;54(19):7311–8.
[45] Arribas-Lorenzo G, Fogliano V, Morales FJ. Acrylamide formation in a cookie system as influenced by the oil phenol profile and degree of oxidation. Eur Food Res Technol 2009;229(1):63–72.
[46] Napolitano A, Morales FJ, Sacchi R, et al. Relationship between virgin olive oil phenolic compounds and acrylamide formation in fried crisps. J Agric Food Chem 2008;56(6):2034–40.
[47] Wenzl T, Beatriz de la Calle M, Anklam E. Analytical methods for the determination of acrylamide in food products: a review. Food Addit Contam 2003;20(10):885–902.
[48] Castle L, Eriksson S. Analytical methods used to measure acrylamide concentrations. J AOAC Int 2005;88(1):274–84.
[49] Castle L. Analysis of acrylamide in foods. In: Skog, Alexander, editors. Acrylamide and other hazardous compounds in heat-treated foods. Cambridge: MA WoodheadPubl; 2006. p. 117–29.
[50] Lineback DR, Coughlin JR, Stadler RH. Acrylamide in foods: a review of the Science and Future considerations. Annu Rev Food Sci Technol 2012;3:15–35.

CHAPTER 14

Acrylamide in Surface and Drinking Water

Yalçın Tepe
Biology Department, Faculty of Arts and Science, Giresun University, Giresun, Turkey

INTRODUCTION

Acrylamide and polyacrylamides are industrial chemicals commonly used as flocculants in several industrial fields such as mineral extraction, chemical industry, food-processing industry, and treatment of drinking water and effluents to enhance solid/liquid separation in water that contains suspended matter. Besides this, they are also used in the production of dyes and organic chemicals, pesticides, contact lenses, cosmetics, toiletries, permanent-press fabrics, textiles, pulp and paper products; in ore processing; crude oil production; manufacture of coatings for home appliances, building materials, automotive parts, explosives, adhesives, printing inks, adhesive tapes, latex; sugar refining; and as a chemical grouting agent and soil stabilizer for the construction of tunnels, sewers, wells, reservoirs, and dams [1]. Acrylamide is a solid at room temperature, and is highly soluble (2155 gL^{-1} of water) and mobile in water. It has a low partition to organic matter and low volatilization in water. Acrylamide release in drinking water mainly comes from the residual monomer of polyacrylamide, especially during the treatment process [1–3]. The use of polyacrylamides as a chemical grouting agent and soil stabilizer for the construction of tunnels, sewers, wells, reservoirs, and dams can also be a source of release to drinking water supplies. Plastic and dye industries also release acrylamide into water [4]. Since acrylamide does not bind to soil, but is highly soluble and mobile in water, it will move with seepage rapidly and will thus cause an increased risk of surface or groundwater contamination.

The US Environmental Protection Agency (EPA) set a limit for the monomer level not to be >0.05% at a maximum dosage of 1 mgL^{-1} and the maximum contaminant level goals for acrylamide at zero [5]. After November 5, 2012, the REACH (Regulation on Registration, Evaluation, Authorisation and Restriction of Chemicals) regulation limited the acrylamide content in commercial products to <0.1% by weight for grouting applications in Europe [6]. European Council by the Drinking Water Directive has also set a minimum quality requirement of 0.1 μgL^{-1} for the acrylamide level in potable water [7]. Because of low concentrations and the physicochemical characteristics of this molecule, it is difficult to determine acrylamide levels in natural samples.

Acrylamide in Food
http://dx.doi.org/10.1016/B978-0-12-802832-2.00014-0

ACRYLAMIDE IN SURFACE AND DRINKING WATER

Acrylamide Formation in Water

Water is vital to all people around the world. Thus, the additive effects of contaminants in water are of most particular interest when regulating exposure limits for public health. A detailed knowledge of the acrylamide amount that contaminates drinking water become vital in terms of true risk assessment. Most people are still not aware of its potential presence in drinking water; however, the public's knowledge of acrylamide as a contaminant has grown since 2002. Like many other chemicals, constant monitoring in the treatment process and direct testing of acrylamide also should be performed against mechanical or human error to avoid large temporary or even extended overdosing.

Acrylamide is a highly toxic monomer, and is classified as a carcinogenic, reprotoxic, irritant, and neurotoxin to animals. Although it is soluble in water, methanol, dimethyl ether, and acetone, it is insoluble in benzene and heptane. Acrylamide monomer is the main compound involved in the production of polyacrylamide polymers. Acrylamide and acrylic acid/acrylate together are used to synthesize polyacrylamide polymer, which has a high molecular weight (Figure 1).

The greatest demand for water-soluble polyacrylamides in industries is as flocculants in water treatment processes and as grouting agents in the construction of drinking water reservoir, dams, and wells. Polyacrylamide applications are mainly based on the principles of colloidal suspensions and used to clear up liquids, particularly aqueous media, either for disposal or human consumption. Acrylamide can also get released into the environment in consequence of extended use of polyacrylamide polymers in many fields such as agriculture, cement, cosmetic, soap, skin care, and hair grooming products. A major application of acrylamide in horticulture is to reduce herbicide drift during spray applications and to improve the water and nutrient supply to plants. By using the polymer, the viscosity of the herbicide solution would increase, which allows more uniform spray applications, and increased plant contact time.

Polyacrylamide emulsions with calcium carbonate and clay are applied as a white coating in the manufacture of cardboard cartons. In oil drilling, it is used as additives to mud to reduce friction, give stability to clay, and increase viscosity during the drill process.

Release in Water

The advantages of using flocculants are to increase the recycling rate of processed water and to decrease sludge volumes and thus the surface of settling ponds. Settling

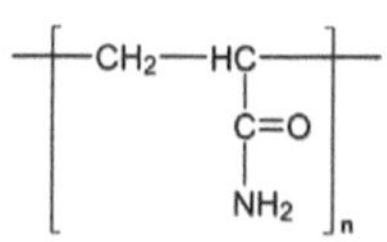

Figure 1 Chemical structure of polyacrylamide.

ponds are open ponds with possible exchanges into the near aquatic environment via water percolation toward soils and aquifers, and water seepages in natural lake, streams, or rivers. Thus, flocculants disperse into surface and groundwaters [8,9]. This dissemination of flocculants causes a new key environmental problem. Polyacrylamide itself does not cause any environmental hazards; it is indeed generally considered as nontoxic [10]. The residual acrylamide, however, is the main hazard for receiver water sources [11]. There is no threshold limit for allowed acrylamide concentrations in residual water discharges in Europe. Residual acrylamide concentrations in 32 polyacrylamide flocculants approved for water treatment plants ranged from 0.5 to 600 ppm [12]. Contamination from acrylonitrile–acrylamide production is another source that releases nearly 1 g of acrylamide per liter of effluent [13]. Acrylamide concentrations have been determined from 0.3 ppb to 5 ppm in terrestrial and aquatic ecosystems around industrial areas that use acrylamide or polyacrylamides [14]. Nearly 800 kg of polyacrylamide was reported to be used yearly by the Kuching Water Board, Malaysia, for drinking water treatment [15]. Polyacrylamide-based flocculant use in France is about 15 kt $year^{-1}$ and 50,000 tons in the European Union (EU) [16]. The United States used 65% of the produced polyacrylamide for water treatment in 1991.

The World Health Organization (WHO) has established a guideline value of 0.5 $\mu g L^{-1}$ for residual concentration of acrylamide in drinking water, while the EU has been stricter and has set a limit value of 0.1 $\mu g L^{-1}$ due to its toxicological properties. The US EPA requires the water supplier to inform them when acrylamide is added to water. The US EPA has established a limit residual concentration of 0.5 $\mu g L^{-1}$ while the EU requires a limit parametric value of 0.1 $\mu g L^{-1}$ of acrylamide in drinking water. The Australian Drinking Water Guidelines [17] impose that the level of acrylamide in drinking water should not be >0.0002 $mg L^{-1}$. This level is equivalent to a daily exposure limit of 5.7×10^{-6} mg kg-bw^{-1} day^{-1}. We can make a simple calculation here; assuming a maximal possible concentration of acrylamide of 0.5 $\mu g L^{-1}$ in drinking water and 2 L consumption per day for a man/woman of 70 kg, it is estimated that the maximum exposure would be 1.42×10^{-5} mg kg-bw^{-1} day^{-1}.

These parametric values are based on the available scientific knowledge and precautionary principles and ensured safe, healthy, and sustainable drinking water. These parametric values for acrylamide refer to residual monomer concentrations in water, and they shall be complied with the following items according to Article 6 of European Council Directive 98/83/EC:

1. In the case of water supplied from a distribution network, at the point, within the premises or an establishment, at which it emerges from the taps that are normally used for human consumption;
2. In the case of water supplied from a tanker, at the point at which it emerges from the tanker;

3. In the case of water put into bottles or containers intended for sale, at the point at which the water is put into the bottles or containers;
4. In the case of water used in a food-production undertaking, at the point where the water is used in the undertaking.

Acrylamide levels in river and tap water in an area where polyacrylamides were used in the treatment of potable water have been reported as being <5 $\mu g L^{-1}$. Results of a study showed that acrylamide levels ranged from 0.024 to 0.041 $\mu g L^{-1}$ from public drinking water supply wells in West Virginia in the United States. Another study reported low microgram levels of acrylamide (0.75–4.5 $\mu g L^{-1}$) from tap water samples in the United Kingdom [18]. Flocculants used in the 1970s contained nearly 50 times more acrylamide than those used currently. Nonetheless, acrylamide can still be transferred into the environment.

In a more recent study, the acrylamide concentrations in surface water were below a detection limit of 0.02 $\mu g L^{-1}$ and in the runoff the concentrations were 0.07–0.16 $\mu g L^{-1}$ in an area where water treatment sludge flocculated with polyacrylamide had been spread on farmland [19]. These values confirm that industrial use of polyacrylamide-based flocculants can release acrylamide into natural waters. Acrylamide release into the environment has received emphasis by the legal authorities of several European countries such as France, Spain, Norway, and Germany as a major environmental issue in the vicinity of quarries.

In 1975, in Japan, 400 $mg L^{-1}$ of acrylamide was reported from drinking water samples taken 2.5 m from a pipeline grouted with acrylamide. In 1997, in the southern part of Sweden, 92 $mg L^{-1}$ of acrylamide was detected in the river water sample taken immediately downstream the construction of an 8.6-km tunnel in the Vadbaken Creek. Water samples were taken from fish ponds that were connected to this creek and 2 $mg L^{-1}$ of acrylamide was determined. The concentration had dropped to <0.1 $mg L^{-1}$ in the same creek and was below the detection limit of 0.05 $mg L^{-1}$ at the furthest point monitored downstream by the end of December 1997. Therefore, the concentration decreased with distance and time.

In a study from Georgia, United States, acrylamide was measured in the processed water of quarries using polyacrylamides at a level of 1.19 $\mu g L^{-1}$. More recently, in a study on the quarry's water circuit acrylamide concentrations were measured to be between 1.01 and 5.66 $\mu g L^{-1}$ in processed water and between 0.08 and 0.41 $\mu g L^{-1}$ in supernatant water. Researchers noticed a contamination in groundwater as well with 0.01–0.02 $\mu g L^{-1}$ [20]. These values also confirm that the presence of acrylamide in processed water can be related to the polyacrylamide-based flocculant use in quarries. Considering that 1.72 billion ton and 348 million ton of aggregates are extracted annually only in the United States and France, respectively, quarries can have impacts on the environment mainly on the biodiversity, landscape, and aquatic life.

While a water sample taken downstream of the effluent of a polyacrylamide producer contained 1500 $\mu g L^{-1}$ of acrylamide, another sample taken from polyacrylamide production and use locations contained <0.8 $\mu g L^{-1}$ and soil from the same areas contained 0.02–0.08 $\mu g g^{-1}$ of acrylamide [21].

Some effluents were analyzed for acrylamide contents, and the following values were reported from the United Kingdom: a clay pit, 16.0 $\mu g L^{-1}$; a tailings lagoon, 39–42 $\mu g L^{-1}$; coal washing lagoon, 1.8 $\mu g L^{-1}$; colliery-cooking plant, 0.74 $\mu g L^{-1}$; treated paper mill, 0.47–14.4 $\mu g L^{-1}$; paper mill process water, 45.4 $\mu g L^{-1}$; and sewage effluent, 17. 4 $\mu g L^{-1}$ [18,22]. More recent studies also have been conducted, and concentrations are presented as follows: 0.082–1 $\mu g L^{-1}$ from four surface water samples [23]; below detection limits to 0.16 $\mu g L^{-1}$ from river and runoff from an agricultural field [19]; 0.28 $\mu g L^{-1}$ from a holding pond after the addition of a flocculent in a sand and gravel mine and 0.047 $\mu g L^{-1}$ in a water-reused process [24].

Agricultural Application of Polyacrylamides

Efficient transfer of precious water resources in arid zones from the water source to the point of use through unlined canals is very important for sustainable water conservation in the future. Water losses via seepage from these unlined canals can be significant with rates 20–30 % of the conveyed volume [25]. Prevention of unwanted seepage loss could be vitally important, if the world's increasing water shortages against increasing demand for water are taken into account. Conventional canal lining methods such as concrete or various types of membranes can effectively reduce seepage losses but are costly. The use of polyacrylamide in unlined canals was evaluated by The United States Bureau of Reclamation to minimize water losses due to seepage and found to reduce losses by >90% in full scale [26,27]. Published data indicate that acrylamide levels 10 times greater than actually measured in canal water would be undetectable within 25 m of the canal due to dilution processes and microbial degradation, with an estimated half life of 30–42 h.

Pollution of surface and groundwater from both human and animal waste has been a threat to public health since these bodies of water are often used for sources of drinking water and/or for recreational activities. Human contact with this water that contains intestinal pathogens can effectively result in disease transmission. Therefore, maintenance of lakes and streams free of intestinal pathogens and excess nutrients is a critical water quality issue throughout the world. Treating irrigation return flows or agricultural or industrial wastewaters with polyacrylamide could work as a filter to remove nutrients and pathogenic microorganisms in both surface and groundwaters.

Polyacrylamide can be notably effective at mitigating off-field transport of suspended sediment, with benefits to nearby surface water with respect to clarification. Anionic polyacrylamide polymers, when added to irrigation water, have been shown to be extremely effective at reducing soil erosion [28]. About 800,000 ha of US irrigated land was reported to use polyacrylamide for erosion and/or infiltration management. The typical dose of polyacrylamide applied by farmers is 3–5 $kg ha^{-1}$ only during irrigation events as required by conservation programs or reducing the high erosion risk. High-molecular-weight polymers, 12–15 Mg per mole, are used in agriculture to inhibit resuspension of sediment, and

Figure 2 Irrigation furrows without polyacrylamide treatment of the water (left) and after polyacrylamide treatment of the water (right).

flocculate those few particles that are resuspended by their long, linear, negatively charged molecules. They can be applied either as a liquid metered into irrigation water at a rate of 1–10 $mg L^{-1}$ or as a solid tablet or granule placed at the bottom of furrows (Figure 2).

The solid forms dissolve in the water as the furrow stream passes over them. Generally, polyacrylamide use results in a 75–95 % reduction in loss of soil through the runoff and increases infiltration by about 15% [29].

Low cost, easy, and quick application, and suitability for use with other pollution reduction techniques are some of the benefits of polyacrylamide treatment of effluents.

Application of polyacrylamide to sprinkler irrigated land can also improve infiltration, which reduces runoff and soil erosion as well. Spraying of 500 ppm polyacrylamide solutions on the soil surface at rates ≥20 $kg ha^{-1}$ increased the final infiltration rate and decreased soil erosion in several laboratory studies [30–32].

Degradation

Since acrylamide is a large molecule, the polymer cannot penetrate the membrane of aquatic organisms and microorganisms [33]. Unlike the polymer, acrylamide is a small, mobile molecule that can penetrate the membrane of aquatic organisms at low levels. Low bioaccumulation of acrylamide in aquatic organisms is because of its high water solubility and degradation by microorganisms [3,34,35]. The microbial degradation of acrylamide begins with the presence of amidase enzyme and also requires enzymes to metabolize intermediate products. Eventually, the biodegradation of acrylamide induces the production of ammonia and acrylic acid acrylate.

Sandy soils have been reported to have a higher mobility and lower rate of degradation than clay soils. Biodegradation in soil depends on the soil type, pH, and temperature. Degradation of acrylamide at pH 7 was maximum; however, a slower and very little degradation was measured at pH 8 and pH 6, respectively [36]. Increasing the temperature from 20 to 60 °C was found to increase the adsorption of anionic polyacrylamide on clay particles. Although, enzyme-catalyzed hydrolysis plays a key role for removal of acrylamide from soil, nonbiological hydrolysis may play an important role in water [37].

Acrylamide readily degrades in water within approximately 8–12 days by biological processes under aerobic and anaerobic conditions, and serves as nitrogen and/or carbon sources [2,3,12,34,35]. Acrylamide monomer, however, was found to remain stable for >2 months in tap water [38]. The same researcher has studied the degradation of acrylamide in different cluster samples of natural and polluted waters, under various conditions such as aeration, light, pH, and sterilization. They showed that acrylamide in surface waters and process water was completely degraded within 4–10 and 0.5 days, respectively. This demonstrates that sterilization of samples inhibited degradation [39].

A more recent study demonstrated that time required for complete acrylamide biodegradation could vary from 480 h to >1100 h depending upon the type of water samples such as tap, river, and lake [40]. Those differences were explained by the author as some species in the water may be interacting with acrylamide, either accelerating or hindering monomer degradation.

Anaerobic biodegradation of acrylamide has been demonstrated by a few studies with nitrate serving as the most favorable anaerobic electron acceptor [11]; however, the degradation kinetics was lower than that observed under aerobic conditions. This suggests that aerobic acrylamide biodegradation is the main biodegradation pathway. Acrylamide may not be completely degraded in domestic sewage and water treatment facilities in the case of a short residence time [41].

According to a study, polyacrylamide does not degrade to acrylamide monomer in the presence of sunlight and glyphosate. On the contrary, glyphosate seems to interact with either the acrylamide monomer or polymer to decrease the rate of monomer degradation [40].

Mechanical degradation of polyacrylamide in aqueous solutions consists of different processes such as stirring, pumping, injection, and movement through porous media. It also occurs due to a shearing hydrodynamic field or by direct mechanical load in the aggregate treatment plant. The main problem associated with mechanical degradation of aqueous solutions of polyacrylamide is the irreversible change of the polymeric material.

Polyacrylamides are susceptible to a wide range of photodegradation processes, depending on the reaction conditions. Similar to mechanical degradation, photodegradation results in irreversible changes in both the physical and chemical properties of polyacrylamides. The release of acrylamide, however, still remains in question and would require further clarification. Another suspicious issue is that studies on photodegradation of polyacrylamide have not been demonstrated in environmental conditions but only in laboratory aqueous solutions.

The degradation of polyacrylamide through an agricultural setting creates an environmental and health hazard. When polyacrylamides were to release acrylamide as a result of the degradation process, acrylamide, because of its chemical properties, could adulterate surface water and/or groundwater systems. Adulteration of these sources would impact both health and the economical welfare of rural communities, since those communities rely upon these types of water sources for livestock and personal use. Acrylamide can remain slightly longer in estuarine or saltwater than in freshwater [22].

Acrylamide, at an initial concentration of 10 ppm, and $8\,\mu g L^{-1}$, was completely degraded in approximately 12 and 9 days from the Hackensack River [42] and the Thames River, England [22], respectively. Studies on degradation of acrylamide in water show that its half life varies from weeks to months but concentrations $>2\,mg L^{-1}$ are not readily and quickly degraded by biotic or abiotic factors [38]; greater concentrations are accounted to be toxic for microorganisms. Bacteria such as *Rhodococcus* sp., *Arthrobacter* [43], *Pseudomonas stutzeri* [44], and *Bacillus cereus* [15] have been isolated from the environment capable of utilizing acrylamide as the sole carbon and/or nitrogen source at concentrations $>1000\,mg L^{-1}$ as high as $3000\,mg L^{-1}$. An optimum pH of 6–8 and a mesophilic temperature of 15–30 °C have been reported for acrylamide biodegradation [45].

Toxic Effects

Human poisoning cases have been documented from contaminated water from sewer grouting. The initial symptoms are skin rash and hands, cramps, sweating, and peeling. Afterward, acrylamide-exposed workers had a headache, breathlessness, muscle weakness, and altered sensitivity in their extremities. Additionally, pain, paresthesias, loss of balance, and truncal ataxia may appear in some cases [46]. In March 1974, there was a case of acrylamide poisoning of a family exposed to subacute acrylamide through ingestion and external use of well water with 400 mg of acrylamide per liter, because of a grouting operation [47]. About a month later, symptoms such as confusion, disorientation, memory disturbances, hallucinations, and truncal ataxia developed because of toxicity. A 23-year-old female survived after intentionally ingesting 18 g of acrylamide crystals. However, hallucinations and hypotension were observed after 5 h and seizures were observed after 9 h. On the third day, gastrointestinal bleeding, adult respiratory distress syndrome, hepatotoxicity, and peripheral neuropathy were observed [48].

Several reproduction studies with mice have been conducted on acrylamide monomer, and they all consistently demonstrated reproductive toxicity at neurotoxic doses. Mice exposed to acrylamide in drinking water at doses of $1.25–24\,mg kg^{-1} day^{-1}$ for 4 weeks had reduced fertility rates and litter sizes, increased incidences of embryo resorption rates, abnormal sperm, and decreased sperm counts [49]. Other effects of acrylamide demonstrated in similar studies include decreased reproductive behavior, testicular atrophy, abnormal spermatogenesis, and reduced sperm quality [50–53]. As demonstrated in animal studies, acrylamide produces reproductive toxicity; however, these effects have not been seen in humans.

In a lifetime oncogenicity study, rats were administered acrylamide in their drinking water at a dosage of 0–2 $mg\,kg^{-1}\,day^{-1}$ to males and 0–3 $mg\,kg^{-1}\,day^{-1}$ to females, respectively. The study demonstrated increases in the incidence of tumors in several organs, especially at higher doses [54]. Thus, the question arises as to whether exposure of humans to acrylamide derived from an environmental source as well as from drinking water constitutes a human health hazard.

There are a wide variety of polyacrylamide formulations available in the market. Polymer flocculants are categorized into three groups based on their charge as anionic, cationic, and neutral. Anionic polyacrylamide is one of the commonly used types in agriculture.

Besides other advantages such as efficacy, product cost, and ease of application, it appears that the use of the solid form and water-based polyacrylamide products are preferable to oil-based products strictly from the standpoint of aquatic toxicity. *Hyalella azteca* and *Ceriodaphnia dubia* were introduced as the most sensitive species to polyacrylamides. A study showed that oil-based polyacrylamides were toxic to the two organisms, but the water-based form was not [29]. Cationic and neutral polyacrylamide polymers, as well as the acrylamide monomer, are all admitted to have significant toxicity issues associated with them; the anionic forms are, however, used in agriculture and typically described as safe for aquatic life. Cationic polyacrylamides have LC_{50} values of 0.3–10 ppm, and they bind to sites rich in hemoglobin, such as fish gills. This poses a barrier to oxygen diffusion rather than causing systemic toxicity.

Toxicity studies have revealed moderate to low effects of acrylamide monomers to aquatic wildlife, especially to fish and crustaceans [55–57]. Environmentally relevant chronic, sublethal doses of anionic polymer products to juvenile rainbow trout resulted in either no damage or very low levels of damage to gill tissue. Mild to moderate effects were found only in fish exposed to the higher concentrations (>100 $mg\,L^{-1}$) of anionic polymer products [58]. Another study has shown minimal toxicity of anionic polyacrylamides to aquatic life, despite concentrations being far in excess of those used in agricultural irrigation systems [29].

Toxicity of cationic acrylamide-based flocculants also has been tested on rainbow trout of two age groups in static bioassay chambers. Severe gill alterations have been noticed in all fish at lethal concentrations. Large-sized fish were more susceptible and died at lower concentrations in the same study [59].

Acrylamide has been shown to cause severe pancreatic tissue lesions in goldfish, which added a most probable effect as a genotoxicant and probable mutagenic agent in fish [57]. Another study with goldfish demonstrated that acrylamide is genotoxic in blood erythrocytes; this leads to cancer in individuals, and also disrupts homeostasis of organisms and introduces genetic diversity [60]. This study consistent with the reported effects of acrylamide in other species [61–63] indicates that acrylamide causes genotoxicity.

Acrylamide did not accumulate significantly in trout when exposed to acrylamide concentrations of 0.338 and 0.710 $mg L^{-1}$ in a bioassay study [64]. In a study, acrylamide bioconcentration factors (BCFs) were reported as 1.44 for the carcass and 1.65 for the viscera of rainbow trout [64]. Another study reported BCFs of 0.77 and 2.53 for carp and Japanese medaka, respectively.

The calculated LC_{50}s were as follows: daphnids, 160 $mg L^{-1}$; midges, 410 $mg L^{-1}$; rainbow trout, 110 $mg L^{-1}$; fathead minnows, 120 $mg L^{-1}$; and bluegill, 100 $mg L^{-1}$. This demonstrates that acrylamide monomer was moderately toxic to the five aquatic organisms studied [56]. Behavioral observations were surfacing and loss of equilibrium of bluegill sunfish and fathead minnow, followed by death. Tested daphnids moved to the bottom of the test chambers with little movement until death. Mysid shrimp were reported to show more sensibility and the calculated LC_{50} was 78 $mg L^{-1}$; mortality increased as the duration of exposure increased.

The 24-, 48-, and 96-h flow through LC_{50} values for harlequin fish (*Rasbora heteromorpha*) were 460, 250, and 130 ppm, respectively. The 7 day LC_{50} value for guppy (*Poecilia reticulate*) was 35 ppm as reported by the US EPA [65].

Algae are the most sensitive of the organism groups studied. The minimum reported EC_{50} value is 34 $mg L^{-1}$ for freshwater algae, and the 72 h EC_{50} value is 90 $mg L^{-1}$ for marine algae.

Elevated LC_{50} for acrylamide in mussels was reported with 411 $mg L^{-1}$; this indicated lower acute toxicity of acrylamide to marine bivalves. According to studies, acrylamide in aquatic ecosystems was reported in micrograms per liter and scaled up to a few milligrams per liter in sewage effluent. Thus, these toxic levels apparently do not easily occur and are a threat to aquatic fauna. Nonetheless, sublethal exposure to acrylamide in goldfish is genotoxic and toxic to both the liver and pancreas [57].

The Food and Agriculture Organization (FAO) and WHO concluded that food makes a significant contribution to total exposure of the general public to acrylamide, with average intake rates in the range of 0.3–0.8 μg of acrylamide intake per kilogram of body weight per day. Acrylamide concentrations in commonly eaten foods are 5–460 times greater than the maximum residual acrylamide concentrations expected in water. The universal human exposure to acrylamide from common foods provides a quantitative contrast for considering risk from exposure to acrylamide from environmental uses of recommended polyacrylamide products at recommended application rates.

SYNOPSIS OF ANALYTICAL TECHNIQUES

Acrylamide (C_3H_5NO) is a polar and small-sized molecule. This makes its analysis at trace levels by direct gas chromatography/mass spectrometry (GC/MS) difficult. Most of the analytical methods for the detection and quantification of acrylamide are targeted toward foods. Environmental aqueous samples are quite different from food samples; therefore, different types of extraction sorbents are required.

Acrylamide can be determined in surface and drinking water samples, by direct injection [4,66–68] or after a preparative step [19,69,70] by GC–MS or GC electron capture detector (ECD), and by liquid chromatography (LC) or MS detection [4,19,67]. Table 1 lists different extraction and analytical techniques developed for water samples, with their limits of quantification (LOQ) and limit of detection (LOD). Depending on the type of samples and their acrylamide levels, both direct and indirect methods can be used to quantify acrylamide.

With regard to low peak efficiency in chromatographic separation and no significant low mass fragmentation in MS detection, direct GC/MS analysis of acrylamide has never become a routine method. Acrylamide is normally analyzed by GC/MS after derivatization (bromination) to limit strong peak broadening and increase the identification power of ion fragments. These methods required acrylamide derivatization and liquid–liquid extraction before analysis, which are laborious. Although a GC/MS method for the quantification of acrylamide has also been developed without derivatization, it required the extraction of 500 mL of surface water on four solid phase extraction (SPE) cartridges coupled in series [72]. The LC/MS method offers a less laborious alternative to GC/MS for acrylamide determination, and involves simpler extraction procedures and elimination of derivatization processes.

In the literature, some analytical techniques to determine acrylamide in water samples based on high-performance liquid chromatography (HPLC) by direct injection without derivatization steps were introduced. Tandem mass spectrometry is used with HPLC in environmental analyses because of its high selectivity and increased sensitivity.

These methods are also often time consuming, laborious, and results can be variable. The US EPA [65] proposed an analytical method based on LC, and ultraviolet (UV) detection is only applicable to determine acrylamide levels $>10\,\mu g L^{-1}$. Considering guideline values of acrylamide in drinking water, high detection limits ($3–10\,\mu g L^{-1}$) make HPLC inadequate to use in sensitive analyses of acrylamide in water samples.

Liquid chromatography with MS [4] or MS/MS detection [67] and direct injection were used to determine acrylamide at sub-ppb levels. The detection limits of the direct methods were satisfactory when compared to the regulated value. These methods, however, requested important injection volumes. Since acrylamide determination in environmental water samples by direct injection induces a high method detection limit, it is not sufficient for drinking water monitoring. An SPE step has been developed to enhance the sensitivity and simplify the extraction step (compared to liquid–liquid extraction).

Although HPLC analyses of acrylamide with UV detection suffered from high detection limits in the past, recently tandem mass spectrometry is introduced with HPLC in environmental analyses because of its high selectivity and sensitivity.

Table 1 Comparison of different methods for acrylamide analysis in water

Sample preparation	Injected volume	LOQ/LOD ($\mu g L^{-1}$)	Separation/ detection	Solvent	Internal standards	Reference
Direct injection	5 μL	0.032	GC/ECD	Ethyl acetate/diethyl ether/methanol/ Benzene/acetone	Dimethyl phthalate	[66]
Direct injection	500 μL	0.5/0.2	LC/MS	Acetonitrile/formic acid	$[^{13}C_3]$ acrylamide	[4]
Direct injection	750 μL	--/0.1	LC/MS/MS	Water/methanol	$[^{13}C_3]$ acrylamide	[67]
Evaporation	25 μL	0.06/0.02	LC/MS/MS	Acetonitrile/water/ formic acid	$[^{13}C_3]$ acrylamide	[19]
Spherical activated carbon	100 μL	0.02/0.1	LC/MS/MS	Acetonitrile/water/ formic acid	$[^{13}C_3]$ acrylamide	[69]
SPE-activated carbon	30 μL	0.021/-	LC/MS/MS	Water/acetonitrile	$[^{13}C_3]$ acrylamide	[23]
Direct injection	750 μL	0.0024/0.017	HILIC	Water/dichloromethane	$[^{13}C_3]$ acrylamide	[24]
Direct injection	25 μL	1/-	HPLC/MS/ MS	Acetonitrile/water/ formic acid	$[^{13}C_3]$ acrylamide	[68]

The method developed using HPLC/MS/MS by DeArmond and DiGoregorio [23] required the use of SPE before analysis to concentrate acrylamide from aqueous samples. Different sorbents have been used, but SPE may be problematic due to the high polarity of acrylamide. Further, it takes a long time because of the preconcentration step.

Another method developed by Chu and Metcalfe [19] avoided the use of SPE and used evaporation with water using a rotary evaporator at 90°C under a vacuum for concentrating acrylamide. Since the sample preparation took 90 min and only one sample at a time could be processed per rotary evaporator, evaporation resulted in a low throughput.

A rapid, simple, and sensitive HPLC/MS/MS method, developed by Mnif et al. [68], can analyze acrylamide in aqueous samples with direct injection of aqueous samples at ppb levels. In this method, the limit of quantification was 1 $\mu g L^{-1}$, and acrylamide was eluted from the chromatographic column within 3 min.

A robust, sensitive, and accurate ultraperformance liquid chromatography–tandem mass spectrometer (UPLC–MS/MS) method combined with SPE was developed with a limit of quantification (0.002 $\mu g L^{-1}$) compatible with toxicity threshold values [70]. This method is suitable to determine trace levels of acrylamide in groundwater, surface water, and processed water. It has been applied to monitor the occurrence of this compound and its potential impacts in the aquatic environment [20] and in laboratory experiments to study the fate and behavior of acrylamide [71].

The hydrophilic interaction liquid chromatography (HILIC) method is designed especially for the retention of small polar molecules and has become popular among analysts to separate small polar analytes. A method using HILIC is developed for the analysis of acrylamide in environmental and drinking waters [24]. Using acetonitrile as an intermediate, this method consisted of solvent exchanging acrylamide from 2 mL of water into 2 mL of dichloromethane, then directly analyzed by large-volume injection HILIC/MS.

Recently, a novel detection method for poly(acrylamide) flocculants was developed using interpolymer complexation between flocculants and a probe (poly(acrylic acid-*co*-acenaphthylene)). It was tested in freshwater situations and found to be robust against a range of contaminants. The method was capable of detecting levels of flocculant levels at <1 $mg L^{-1}$. It shows potential as a clear indicator of correct dosing levels for polymer addition to effluent treatment processes. The main advantage of the method is that it is fast, appears unaffected by the presence of dissolved organic matter, and requires no modification of the sample prior to analysis besides setting the pH and addition of the probe polymer [73].

KEY FACTS

Acrylamide is used as a chemical intermediate or as a monomer in the production of polyacrylamide. Both acrylamide and polyacrylamide are used mainly in the production of flocculants for the clarification of potable water and in the treatment of municipal and industrial effluents. They are also used as grouting agents in the construction of drinking water reservoirs and wells [74].

The level of acrylamide in drinking water should not exceed the established guideline value of 0.5 $\mu g L^{-1}$ by the WHO and US EPA, 0.1 $\mu g L^{-1}$ by the EU, and 0.0002 $mg L^{-1}$ by the Australian Drinking Water Guidelines.

Polyacrylamides refer to a broad class of compounds, and there are hundreds of specific polyacrylamide formulations. Anionic polyacrylamides are preferred for environmental applications because of their extremely low aquatic toxicity compared to nonionic or cationic forms. Sublethal doses of anionic polymer products to juvenile rainbow trout resulted in either no damage or very low levels of damage to gill tissue. Severe gill alterations have been noticed in all fish at lethal concentrations. Large-sized fish were more susceptible and died at lower concentrations compared to small-sized fish [59].

Acrylamide is highly mobile in aqueous environments and readily leachable in soil. As it has a higher mobility and lower rate of degradation in sandy soils than in clay soils [75], it may contaminate groundwater. Anionic polyacrylamide polymers are extremely effective at reducing soil erosion when added to irrigation water. Some of the benefits of polyacrylamide treatment of effluents and irrigation water are low cost, easy and quick application, and suitability for use with other pollution reduction techniques.

Acrylamide can readily biodegrade in both soil and surface water. Acrylamide concentrations decreased from 20 to 1 μg/L in 24 h in the effluent from a sludge dewatering process [76]. One of the most important mechanisms for the removal of acrylamide from soils is enzyme-catalyzed hydrolysis; nonbiological hydrolysis may be important in natural water. Volatilization is not an important removal process. As acrylamide is both highly soluble in water and degraded by microorganisms, it is not likely to bioconcentrate significantly [77].

Human poisoning cases have been reported from contaminated water from sewer grouting. Initial symptoms are skin rash and hands, cramps, sweating, and peeling. Afterward, the acrylamide-exposed workers had a headache, breathlessness, muscle weakness, and altered sensitivity of their extremities. Additionally, pain, paresthesias, loss of balance, and truncal ataxia may appear in some cases [46].

The methods used for measuring acrylamide include polarography, electron capture GC, and HPLC. An HPLC/UV absorption detection procedure for the determination of acrylamide in water has a detection range of 0.2–100 μg/L [18]. GS–MS detection has a detection limit of 0.008 μg/L and can be used with concentrations typically up to 0.5 μg/L. A newly developed method, that is, HILIC is used for the analysis of acrylamide in environmental and drinking waters with a sensitive LOD values [24].

MINI-DICTIONARY

Flocculation This is a process wherein colloids come out of suspension in the form of floc or flake, either spontaneously or due to the addition of clarifying agent. According to The International Union of Pure and Applied Chemistry definition, a flocculant is "a process of contact and adhesion whereby the particles of a dispersion form larger size cluster." Flocculation is synonymous with agglomeration and coagulation/coalescence. The action differs from precipitation in that, prior to flocculation, colloids are merely suspended in a liquid and not actually dissolved in a solution. In the flocculated system, there is no formation of a cake, since all the flocs are in the suspension.

Polyacrylamide This is a water-soluble polymer formed from acrylamide (a compound with a molecular formula C_3H_5NO) subunits. Polyacrylamide is most often used to increase the viscosity of water (creating a thicker solution) or to encourage flocculation of particles present in water. Polyacrylamide is a synthetic chemical that can be tailored to fit a broad range of applications. There are a wide variety of polyacrylamide formulations available in the market. Polymer flocculants are categorized into three groups based on their charge as anionic, cationic, and neutral. Anionic polyacrylamide is one of the commonly used types in agriculture.

Bioconcentration Factor This is the ratio of the steady-state chemical concentrations in an aquatic water-respiring organism and the water determined in a controlled laboratory experiment in which the test organisms are exposed to a chemical in the water but not in the diet. The BCF is usually determined at steady state as follows:

$$\mathrm{BCF} = \frac{\text{Concentration in fish at steady state } \left(\text{mg kg}^{-1} \text{ wet fish}\right)}{\text{Concentration in water at steady state } \left(\text{mg L}^{-1}\right)}$$

Photodegradation Exposure to UV radiation may cause significant degradation of many materials. UV radiation causes photooxidative degradation, which results in breaking of the polymer chains, produces radicals, and reduces the molecular weight. This causes deterioration of mechanical properties and leads to useless materials, after an unpredictable time.

Biodegradation This is nature's way of recycling wastes, or breaking down organic matter into nutrients that can be used by other organisms. "Degradation" means decay, and the "bio" prefix means that the decay is carried out by a huge assortment of bacteria, fungi, insects, worms, and other organisms that eat dead material and recycle it into new forms.

SUMMARY POINTS

Acrylamide is an unsaturated amide with diverse applications, mainly the production of polymers used as additives in water treatment, agricultural processes, textile manufacturing, and oil drilling. Acrylamide is reported to be a neurotoxin to humans and also classified as a carcinogenic, mutagenic, and reprotoxic compound. The main concerns toward human health are mostly related to the presence of acrylamide in food stuff; however, the environmental fate of acrylamide monomers after degradation of polyacrylamide polymers has recently aroused interest. Acrylamide may remain in water after treatment due to its high solubility and is not readily absorbed by sediment. Acrylamide in raw water cannot be removed by conventional water treatment processes, treatment with ozone, or potassium permanganate; however, it may decrease in concentration. Acrylamide is likely to partially biodegrade in water within approximately 8–12 days. Bacterial use of acrylamide as a source of nitrogen and carbon is the main cause acrylamide degradation in water.

REFERENCES

[1] WHO. Acrylamide in drinking-water. Background document for development of WHO guidelines for drinking-water quality. 2011. Vol WHO/SDE/WSH/03.04/71/rev1.

[2] Abdelmagid HM, Tabatabai MA. Decomposition of acrylamide in soils. J Environ Qual 1982;11:701–4.

[3] EPA US. Toxicological review of acrylamide in support of summary information on the integrated risk information systems (IRIS). US Environmental Protection Agency; 2010.

[4] Cavalli S, Polesello S, Saccani G. Determination of acrylamide in drinking water by large-volume direct injection and ion-exclusion chromatography–mass spectrometry. J Chromatogr A 2004;1039:155–9.

[5] U.S. Environmental Protection Agency. List of drinking water contaminants and MCLs. Cincinnati, OH: EPA 816-F-02-013; 2002.

[6] European Parliament, European Council. Registration, Evaluation, Authorisation and Restriction of Chemicals (REACH), establishing a European Chemicals Agency; 2006. No 907/2006.

[7] European Council Directive 98/83/EC of 3 November 1998 on the quality of water intended for human consumption, European Union, Brussels. Off J Eur Commun 1998;L330:32.

[8] de Rosemond SJC, Liber K. Wastewater treatment polymers identified as the toxic component of a diamond mine effluent. Environ Toxicol Chem 2004;23:2234–42.

[9] Liber K, Weber L, Levesque C. Sublethal toxicity of two wastewater treatment polymers to lake trout fry (*Salvelinus namaycush*). Chemosphere 2005;61:1123–33.

[10] Andersen FA. Amended final report on the safety assessment of polyacrylamide and acrylamide residues in cosmetics. Int J Toxicol 2005;24:21–50.

[11] Labahn SK, Fisher JC, Robleto EA, Young MH, Moser DP. Microbially mediated aerobic and anaerobic degradation of acrylamide in a western United States irrigation canal. J Environ. Qual 2010;39:1563–9.

[12] Howard PH. Large production and priority pollutants. Handbook of environmental fate and exposure data for organic chemicals. Chelsea: Lewis; 1989.

[13] Rogacheva SM, Ignatov OV. The respiratory activity of *Rhodococcus rhodochrous* M8 cells producing nitrile-hydrolyzing enzymes. Appl Biochem Microbiol 2001;37:282.

[14] Weideborg M, Källqvist T, Ødegård KE, Sverdrup LE, Vik EA. Environmental risk assessment of acrylamide and methyloacrylamide from a grouting agent used in the tunnel construction of Romeriksporten, Norway. Water Res 2001;35:2645.

[15] Shukor MY, Gusmanizar N, Azmi NA, Hamid M, Ramli J, Shamaan NA, et al. Isolation and characterization of an acrylamide degrading *Bacillus cereus*. J Environ Biol 2009;30(1):57–64.

[16] European Union. Risk assessment report on acrylamide. CAS no: 79-06-1; 2002. http://europa.eu.int.

[17] NHMRC. Australian drinking water guidelines 6. National Health and Medical Research Council; 2004. www.nhmrc.gov.au/publications/synopses/_files/adwg_11_06.pdf.

[18] Brown L, Rhead MM. Liquid chromatographic determination of acrylamide monomer in natural and polluted aqueous environments. Analyst 1979;104:391–9.

[19] Chu S, Metcalfe CD. Analysis of acrylamide in water using a coevaporation preparative step and isotope dilution liquid chromatography tandem mass spectrometry. Anal Chem 2007;79(13):5093–6.

[20] Touzé S, Guerin V, Guezennec AG, Binet S, Togola A. Dissemination of acrylamide monomer from polyacrylamide based flocculant use—sand and gravel quarry case study. Environ Sci Pollut Res 2014:1–8.

[21] EPA. Environmental monitoring near industrial sites acrylamide. Washington, DC: U.S. Environmental Protection Agency; 1978. EPA560678001. PB281879.

[22] Croll BT, Arkell GH, Hodge RPJ. Residues of acrylamide in water. Water Resour 1974;8(11):989–93.

[23] DeArmond PD, DiGoregorio AL. Characterization of liquid chromatography–tandem mass spectrometry method for the determination of acrylamide in complex environmental samples. Anal Bioanal Chem 2013;405:4159–66.

[24] Backe WJ, Yingling V, Johnson T. The determination of acrylamide in environmental and drinking waters by large-volume injection—hydrophilic-interaction liquid chromatography and tandem mass spectrometry. J Chromatogr A 2014;1334:72–8.
[25] Tanji KK, Kielen NC. Agricultural drainage water management in arid and semi-arid areas. FAO Irrigation and Drainage Paper No. 61. Rome: Food and Agriculture Organization, FAO; 2002.
[26] Young MH, Tappen JJ, Miller GC, Caarroll RB, Susfalk RB. Initial risk characterization: using linear anionic polyacrylamide (LAPAM) to reduce water seepage from unlined water delivery canal systems. Publication no. 41226. Las Vegas, NV: Desert Research Institute; 2007.
[27] Susfalk RB, Young MH, Smith DM. Application guidelines: use of granular linear anionic polyacrylamide (LA-PAM) in water delivery canals for seepage control. Publication no. 41239. Reno, NV: Desert Research Institute; 2007.
[28] Sojka RE, Bjorneberg DL, Entry JA, Lentz RD, Orts WJ. Polyacrylamide in agriculture and environmental land management. Adv Agron 2007;92:75–162.
[29] Weston DP, Lentz RD, Cahn MD, Ogle RS, Rothert AK, Lyde MJ. Toxicity of anionic polyacrylamide formulations when used for erosion control in agriculture. J Environ Qual 2009;38:238–47.
[30] Levy GJ, Agassi M. Polymer molecular-weight and degree of drying effects on infiltration and erosion of three different soils. Soil Res 1995;33(6):1007–18.
[31] Ben-Hur M, Keren R. Polymer effects on water infiltration and soil aggregation. Soil Sci Soc Am J 1997;61(2):565–70.
[32] Bjorneberg DL, Santos FL, Castanheira NS, Martins OC, Reis JL, Aase JK, et al. Using polyacrylamide with sprinkler irrigation to improve infiltration. J Soil Water Cons 2003;58(5):283–9.
[33] Berndt WO, Bergfeld WF, Boutwell RK, Carlton WW, Hoffmann DK, Schroeter AL, et al. Final report on the safety assessment of polyacrylamide. J Am Coll Toxicol 1991;10(1):193–203.
[34] Habermannn CE. Acrylamide. In: Kirk-Othmer encylcopedia of chemical technology. John Wiley & Sons; 2002. p. 1–22.
[35] WHO. 2003. Acrylamide in drinking-water. World Health Organization; August 3, 2011. WHO/SDE/WSH/03.04/71. http://www.who.int/water_sanitation_health/dwq/chemicals/acrylamide.pdf.
[36] Nawaz MS, Billedeau SM, Cerniglia CE. Influence of selected physical parameters on the biodegradation of acrylamide by immobilized cells of *Rhodococcus* sp. Biodegradation 1998;9:381–7.
[37] NICNAS (National Industrial Chemicals Notification and Assessment Scheme). Acrylamide. Priority existing chemical. Assessment report no. 23. Comonwealth of Australia. May 2002.
[38] Brown L, Rhead MM, Bancroft KCC, Allen N. Case studies of acrylamide pollution resulting from industrial use of acrylamides. Water Pollut Control 1980a;79(4):507–10.
[39] Brown L, Rhead MM, Bancroft KCC, Allen N. Model studies of the degradation of acrylamide monomer. Water Res 1980b;14:775–9.
[40] Vers Ver LM. Determination of acrylamide monomer in polyacrylamide degradation studies by high-performance liquid chromatography. J Chromatogr Sci 1999;37:486–94.
[41] Brown L, Rhead MM, Hill D, Bancroft KCC. Rapid screening technique utilizing high-performance liquid chromatography for assessing acrylamide contamination in effluents. Analyst 1982;107(1276):749–54.
[42] Cherry AB, Gabaccia AJ, Senn HW. The assimilation behavior of certain toxic organic compounds in natural water. Sew Ind Waste 1956;28(9):1137–46.
[43] Yamada H, Asano Y, Hino T, Tani Y. Microbial utilisation of acrylonitrile. J Ferment Technol 1979;57:8–14.
[44] Wang CC, Lee CM. Denitrification with acrylamide in pure culture of bacteria isolated from acrylonitrile–butadiene–styrene resin manufactured for wastewater treatment. Chemosphere 2001;44:1047–53.
[45] Charoenpanich J. Removal of acrylamide by microorganisms. In: Patil Y, editor. Applied bioremediation—active and passive approaches. InTech; 2013. ISBN: 978-953-51-1200-6. http://dx.doi.org/10.5772/56150. Available from: http://www.intechopen.com/books/applied-bioremediation-active-and-passive-approaches/removal-of-acrylamide-by-microorganisms.
[46] Pennisi M, Malaguarnera G, Puglisi V, Vinciguerra L, Vacante M, Malaguarnera M. Neurotoxicity of acrylamide in exposed worker. Int J Environ Res Pub Health 2013;10:3843–54.
[47] Igisu H, Goto I, Kawamura Y, Kato M, Izumi K. Acrylamide encephaloneuropathy due to well water pollution. J Neurol Neurosurg Psychiatry 1975;38:581–4.

[48] Donovan JW, Pearson T. Ingestion of acrylamide with severe encephalopathy, neurotoxicity and hepatotoxicity. Veterinary Hum Toxicol 1987;29:462. [abstract].
[49] Sakamoto J, Hashimoto K. Reproductive toxicity of acrylamide and related compounds in mice: effects on fertility and sperm morphology. Arch Toxicol 1986;59:201–5.
[50] Hashimoto K, Sakamoto J, Tanii H. Neurotoxicity of acrylamide and related compounds and their effects on male gonads in mice. Arch Toxicol 1981;47:179–89.
[51] Sublet VH, Zenick H, Smith MK. Factors associated with reduced fertility and implantation rates in females mated to acrylamide-treated rats. Toxicology 1989;55:53–67.
[52] Wise LD, Gordon LR, Soper KA, Duchai DM, Morrissey RE. Developmental neurotoxicity evaluation of acrylamide in Sprague–Dawley rats. Neurotoxicol Teratol 1995;17(2):189–98.
[53] Yang HJ, Lee SH, Jin Y, Choi JH, Han DU, Chae C, et al. Toxicological effects of acrylamide on rat testicular gene expression profile. Reprod Toxicol 2005;19(4):527–34.
[54] Friedman MA, Duak LH, Stedham MA. A lifetime oncogenicity study in rats with acrylamide. Fundam Appl Toxicol 1995;27:95–105.
[55] Bridié AL, Wolff CJM, Winter M. The acute toxicity of some petrochemicals to goldfish. Water Res 1979;13.
[56] Krautter GR, Mast RW, Alexander HC, Wolf CH, Friedman MA. Acute aquatic toxicity tests with acrylamide monomer and macroinvertebrates and fish. Environ Toxicol Chem ETOCDK 1986;5(4): 373–7.
[57] Larguinho M, Costa PM, Sousa G, Costa MH, Diniz MS, Baptista PV. Histopathological findings on *Carassius auratus* hepatopancreas upon exposure to acrylamide: correlation with genotoxicity and metabolic alterations. J Appl Toxicol 2013;34:1293–302.
[58] Kerr JL, Lumsden JS, Russell SK, Jasinska EJ, Goss GG. Effects of anionic polyacrylamide products on gill histopathology in juvenile rainbow trout. Environmental Toxicol Chem 2014;33(7):1552–62.
[59] Albassam M, Moore J, Sharma A. Ultrastructural and clinicopathological studies on the toxicity of cationic acrylamide-based flocculant to rainbow trout. Veterinary Pathol 1987;24:34–43.
[60] Tan D, Li L, Wang S, Wei B, Zhang X, Sun B, et al. The cytogenetic effects of acrylamide on *Carassius auratus* peripheral blood cells. Food Chem Toxicol 2013;62:318–22.
[61] Taubert D, Glockner R, Muller D, Schomig E. The garlic ingredient diallyl sulfide inhibits cytochrome P450 2E1 dependent bioactivation of acrylamide to glycidamide. Toxicol Lett 2006;164:1–5.
[62] Von Tungeln LS, Churchwell MI, Doerge DR, Shaddock JG, McGarrity LJ, Heflich RH, et al. DNA adduct formation and induction of micronuclei and mutations in B6C3F1/Tk mice treated neonatally with acrylamide or glycidamide. Int J Cancer 2009;124:2006–15.
[63] Yener Y, Dikmenli M. Increased micronucleus frequency in rat bone marrow after acrylamide treatment. Food Chem Toxicol 2009;47:2120–3.
[64] Petersen DW, Kleinow KM, Kraska RC, Lech JJ. Uptake disposition and elimination of acrylamide in rainbow trout *Salmo-gairdneri*. Toxicol Appl Pharmacol 1985;80(1):58–65.
[65] USEPA. Chemical summary for acrylamide. Prepared by Office of Pollution Prevention and Toxics. EPA 749-F-94–005a. U.S. Environmental Protection Agency; 1994. Available on line at: http://www.epa.gov/chemfact/s_acryla.txt.
[66] US Environmental Protection Agency. Method 8032A: acrylamide by gas spectrometry. 1996. Cincinnati, OH. http://www.epa.gov/SW-846/pdfs/8032a.pdf.
[67] Marin JM, Pozo OJ, Sancho JV, Pitarch E, López FJ, Hernández F. Study of different atmospheric-pressure interfaces for LC–MS/MS determination of acrylamide in water at sub-ppb levels. J Mass Spectrom 2006;41:1041–8.
[68] Mnif I, Hurel C, Guezennec AG, Marmier N. Interaction of polyacrylamide flocculants and acrylamide with clays, soil and sediments. International conference on interfaces against pollution. Netherlands: De Harmonie, Leeuwarden; May 2014.
[69] Lucentini L, Ferretti E, Veschetti E, Achene L, Turrio-Baldassarri L, Ottaviani M, et al. Determination of low-level acrylamide in drinking water by liquid chromatography/tandem mass spectrometry. J AOAC Int 2009;92:263–70.
[70] Togola A, Coureau C, Guezennec AG, Touzé S. A sensitive analytical procedure for monitoring acrylamide in environmental water samples by offline SPE–UPLC/MS/MS. Environ Sci Pollut Res 2014:1–7.

[71] Guézennec AG, Michel C, Desroche N, Togola A, Touzé S. Microbial aerobic and anaerobic degradation of acrylamide in sludges and industrial and natural waters—sand and gravel quarry case study. Environ Sci Pollut Res 2014:1–12.
[72] Kawata K, Ibaraki T, Tanabe A, Yagoh H, Shinoda A, Suzuki H, et al. Gas chromatographic-mass spectrometric determination of hydrophilic compounds in environmental water by solid-phase extraction with activated carbon fiber felt. J Chromatogr A 2001;911(1):75–83.
[73] Swift T, Swanson T, Bretherick A, Rimmer S. Measuring poly (acrylamide) flocculants in fresh water using inter-polymer complex formation. Environmental Science: Water Research & Technology; 2015. http://dx.doi.org/10.1039/c4ew00092g.
[74] IPCS (International Programme on Chemical Safety). Acrylamide. International programme on 8178 chemical Safety. Poisons information monograph 652. 1999. Available at: http://www.inchem.org/documents/pims/chemical/pim652.htm.
[75] Lande SS, Bosch SJ, Howard PH. Degradation and leaching of acrylamide in soil. J Environ Qual 1979;8:133–7.
[76] Arimitu H, Ikebukuro H, Seto I. The biological degradability of acrylamide monomer. J Jpn Water Works Assoc 1975;487:31–9.
[77] Neely WB, Branson DR, Blau GE. Partition coefficient to measure bioconcentration potential of organic chemicals in fish. Environ Sci Technol 1974;8:1113–5.

SECTION C

Interactions and Reductions

CHAPTER 15

Use of Nucleophilic Compounds, and Their Combination, for Acrylamide Removal

Rosario Zamora, Rosa M. Delgado, Francisco J. Hidalgo
Instituto de la Grasa, Consejo Superior de Investigaciones Científicas, Seville, Spain

INTRODUCTION

Because of its toxicity, acrylamide mitigation has been a major objective for industry, academic, and research centers since the discovery of the formation of this toxicant in food products more than one decade ago [1]. These studies have employed different strategies: to decrease reactants content, to modify reaction conditions to inhibit acrylamide formation, to favor alternative pathways, and to provide compounds that can react with acrylamide and produce its elimination once it is formed [2–18]. Among the different strategies and additives employed, compounds having nucleophilic groups have provided promising results because they can contribute to both the inhibition of acrylamide formation and the scavenging of the acrylamide formed.

The formation of acrylamide in foods is produced as a consequence of the thermal degradation of asparagine in the presence of carbonyl compounds [19–21]. As can be observed in Figure 1 (*pathway A*), in the presence of reactive carbonyls, asparagine is decarboxylated to 3-alkylaminopropionamides or 3-aminopropionamide [22]. These compounds lately suffer a Hofmann elimination to produce acrylamide, a step in which the presence of carbonyl compounds reduces the required activation energy [23]. When other nucleophiles such as amino acids are present, alternative routes appear. First, there is a rivalry between the amino acids and asparagine for the carbonyl compounds and the amount of the acrylamide formed through *pathway A* decreases. However, the reaction between amino acids and the carbonyl compounds produces new amino acid-derived carbonyl compounds: mainly α-keto acids and Strecker aldehydes [24]. These last compounds also produce asparagine degradation according to *pathways B* and *C* [16]. In fact, some of these carbonyl compounds are more active than original carbonyls for asparagine degradation. For that reason, increases in the formation of acrylamide have been observed when some amino acids are added to asparagine/carbonyl compound reaction mixtures [25].

Acrylamide in Food
http://dx.doi.org/10.1016/B978-0-12-802832-2.00015-2

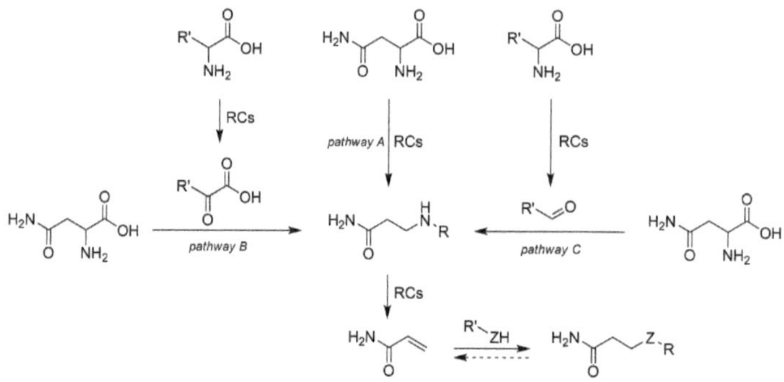

Figure 1 Effect of amino acids and other nucleophiles on acrylamide formation. In the presence of reactive carbonyls (RCs), asparagine is decarboxylated to aminopropionamide or alkylaminopropionamides (*pathway A*). When other amino acids are present, they are also degraded by reactive carbonyls to produce α-keto acids and Strecker aldehydes, among other compounds. These carbonyls also decarboxylate asparagine (*pathways B* and *C*). An elimination reaction in aminopropionamide or alkylaminopropionamides, favored by reactive carbonyls, is responsible for acrylamide formation. Finally, acrylamide reacts reversibly or irreversibly with amino acids and other nucleophiles to produce the corresponding Michael adduct. Z is a sulfur atom or an NH group. R and R′ are alkyl groups.

Once produced, acrylamide is not stable and can react with nucleophiles as a consequence of its chemical structure. Acrylamide is characterized by a conjugated structure in which an electron-withdrawing group (the amide group) is linked to a carbon–carbon double bond. This conjugated α,β-unsaturated carbonyl structure is an electrophile that forms Michael-type adducts with nucleophiles. In biological and food systems, sulfur atoms are the principal nucleophilic targets [26]. Therefore, the formation of adducts between acrylamide and the thiol group of cysteine residues of proteins should be expected. In addition, the free amino group of lysine or the imidazolic NH group of histidine residues are also nucleophilic and, therefore, they should also produce Michael adducts with acrylamide. Furthermore, amino groups of free amino acids can also be added to acrylamide. This chapter reviews our present knowledge of the potential use of nucleophilic compounds, and their combination, for the removal of acrylamide in food products.

MICHAEL ADDITION OF THIOLS TO ACRYLAMIDE

Many studies have dealt with the role of thiols for acrylamide removal. However, most of these studies were carried out with the amino acid cysteine or the peptide glutathione. In both cases, these molecules have two nucleophilic groups and the observed effects are likely a combination of the simultaneous presence of thiol and amino groups in the molecule. This combined action of both nucleophilic groups will be described in other section (see below).

One of the few studies carried out to determine the role of thiol groups in the absence of other groups was carried out by Hidalgo et al. [27]. These authors found that,

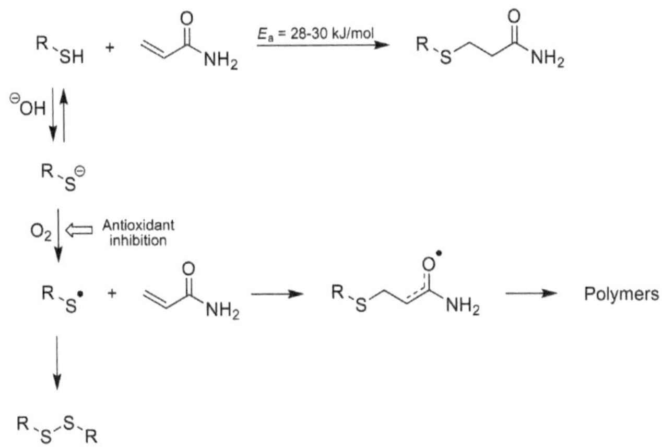

Figure 2 Addition of thiols to acrylamide. In the absence of oxygen the Michael adduct is irreversibly produced. In the presence of oxygen, the corresponding thiolyl radical is produced that can polymerize and induce the polymerization of acrylamide. This second pathway is inhibited in the presence of antioxidants. R is an alkyl group.

in the absence of other nucleophilic groups, thiols add rapidly to the carbon–carbon double bond of acrylamide producing the corresponding adduct (Figure 2). The activation energy of the reaction between acrylamide and benzylthiol or *N*-acetylcysteine is quite low (28–30 kJ). Therefore, it should be expected to be produced to a high extent at relatively low temperatures. Although this reaction has been found to be reversible by different authors when using cysteine [28], Hidalgo et al. [27] found that it was not reversible under their reaction conditions. In fact, they found that the heating of the Michael adduct produced in the reaction of benzylthiol with acrylamide [3-(benzylthio) propanamide] under nitrogen for 35 min at 180 °C only produced trace amounts of acrylamide.

Reaction changed in the presence of oxygen (Figure 2). Under oxidizing conditions, thiols were rapidly converted into the corresponding thiolyl radicals that both polymerized themselves and induced the polymerization of acrylamide [27]. Thus, when the reaction was carried out under air, the addition product between the thiol and acrylamide was not observed, although most of the acrylamide disappeared. However, addition of gallic acid—a well-known antioxidant—prevented thiol oxidation and identical losses of acrylamide were observed in the presence of nitrogen or in the presence of air/gallic acid. Therefore, antioxidants can also contribute to acrylamide preservation in addition to their described function for preventing acrylamide production [29].

Although it cannot be considered a thiol, sodium bisulfite has been shown to reduce acrylamide formation in food products [30,31]. Thus, for example, sodium bisulfite was the best additive among the assayed compounds to reduce the acrylamide level in black ripe olives without a negative repercussion on sensory quality, which is mostly observed

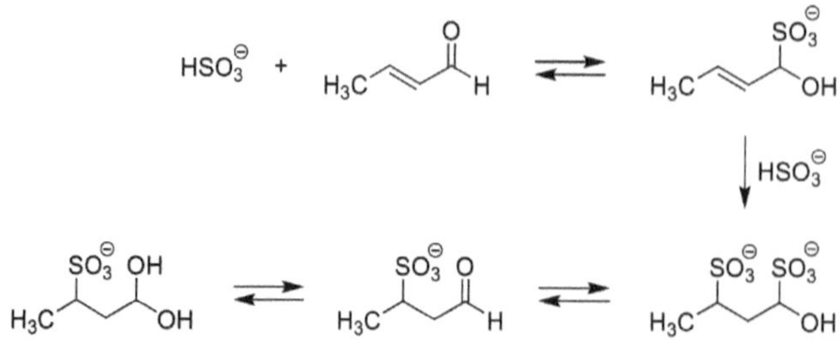

Figure 3 Addition of sodium bisulfite to 2-butenal. The first step is a reversible addition of bisulfite to the carbonyl group to produce a sulfonate. The latter irreversible addition of a second molecule of bisulfite produces a disulfonate from which the carbobyl group can be recovered to form the corresponding Michael adduct. A similar pathway can be hypothesized to be produced for the addition of sodium bisulfite to acrylamide.

when thiols are used for controlling acrylamide formation. The mechanism for acrylamide removal by sodium bisulfite is not known, but it can be supposed to be analogous to the reaction between sodium sulfite and α,β-unsaturated carbonyl compounds. This reaction is widely used in brewing industry for reducing the flavor impact of a variety of carbonyl compounds [32].

Adduct formation between 2-butenal and sodium bisulfite was studied by Dufour et al. [33]. They proposed a two-step mechanism that ultimately yields a sulfonate as the thermodynamic-stable product (Figure 3). The reaction starts with a reversible binding between the carbonyl group and bisulfite. Then, a second molecule of sulfite is added to produce irreversibly a disulfonate, which is in equilibrium with a monosulfonate and its acetal. A similar mechanism was also suggested by Youngblood [34] for the irreversible addition of sulfite to *p*-benzoquinone. Therefore, the reaction between sodium bisulfite and acrylamide can be hypothesized to result in the irreversible formation of a sulfonate derivative.

MICHAEL ADDITION OF AMINO COMPOUNDS TO ACRYLAMIDE

Analogously to thiols, amino compounds are also able to react with acrylamide to form adducts, although thiols are 100–300 times more reactive with α,β-unsaturated carbonyl compounds than amino groups [35]. In addition, the produced reaction is not a simple Michael addition of the amino compound to the carbon–carbon double bond of the acrylamide [36]. As shown in Figure 4, acrylamide reacts very easily and rapidly with amino compounds to produce the corresponding Michael adduct. However, because this Michael adduct still has one nucleophilic group (NH), it may react with another molecule of acrylamide to produce a new adduct. Nevertheless, the formation of this second adduct is not usually produced in food products because acrylamide is always present to a much lower extent than amino compounds. Although the single Michael adduct is produced very easily, it is not stable, and its thermal decomposition is also

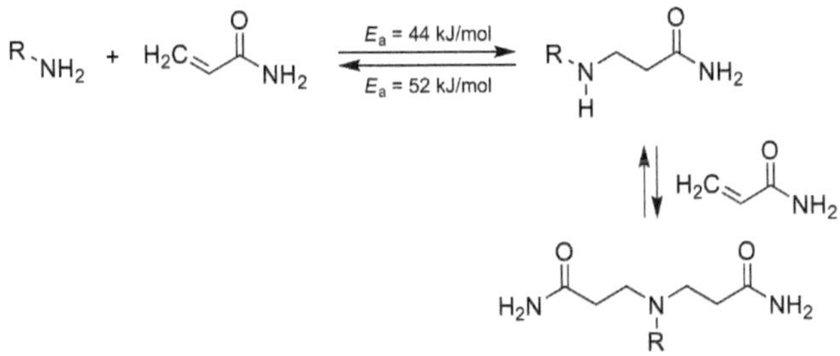

Figure 4 Addition of amino compounds to acrylamide. The reaction produces reversibly the corresponding Michael adduct in a first step. However, because the adduct still has an NH group, this group can be added reversibly to a second molecule of acrylamide forming the corresponding Michael adduct with two molecules of acrylamide. This second adduct is not usually formed in food products because acrylamide is present to a much lower amount than amino compounds. R is an alkyl group.

produced very easily by an elimination mechanism [23]. This elimination can also be produced from the adduct with two molecules of acrylamide.

Therefore, and differently to the reaction between acrylamide and thiols described above, the reaction between acrylamide and amino compounds is reversible. However, activation energies (E_a) of one reaction and its reverse are not the same. Thus, for example, the addition of butylamine to acrylamide had an E_a of 44 kJ/mol and the E_a of the cleavage of 3-(butylamino)propionamide to produce acrylamide was 52 kJ/mol. This means that the equilibrium is shifted to the formation of the Michael adduct, but acrylamide is only hidden in this adduct and can appear again if reaction conditions change.

The scavenging ability of amino compounds for acrylamide will depend on the number of nucleophilic groups present in relation to the molecular weight of the compound. Thus, for example, butylamine and lysine exhibited a similar reactivity for acrylamide at short reaction times. However, at longer reaction times, lysine was considerably more reactive, which is the consequence of the presence of two amino groups in lysine [36]. Furthermore, when polylysines having different molecular weights were assayed, the lower the molecular weight of the polylysine, the higher the acrylamide-scavenging ability. This is not only a consequence of the number of amino groups (the higher the molecular weight, the smaller the number of terminal amino groups), but also of the accessibility of amino groups (the higher molecular the weight of the polypeptide, the lesser the accessibility of amino groups to acrylamide). A similar conclusion was also achieved by Adams et al. [28] while comparing different food-relevant nucleophiles. These authors found that all assayed nucleophiles yielded comparable condensation products, but reactivity depended on the number and kind of nucleophilic groups involved.

A lesser known group of amino compounds with possibilities of being employed for scavenging acrylamide are aminophospholipids. Differently to other additives, lecithins are recognized as GRAS substances and are widely employed in foods as emulsifiers and also because of their antioxidant properties. Zamora et al. [37] found that phosphatidylethanolamine (PE), but not phosphatidylcholine (PC), decreased the acrylamide produced in mixtures of asparagine/glucose and asparagine/2,4-decadienal, an effect that was also observed for soybean and egg lecithins. PE had a very similar behavior to that of the amino acid glycine, therefore suggesting that its mitigating action should be related to the presence of a primary amino group in the phospholipid. This nucleophilic group can be added to the carbon–carbon double bond of the formed acrylamide as described previously. In addition, the amino group can also react with carbonyl compounds, therefore protecting asparagine from degradation to produce acrylamide.

Both PE and ethanolamine were good acrylamide scavengers. However, ethanolamine was more protective than PE in the asparagine/glucose system and less protective than PE in the asparagine/2,4-decadienal system, which is likely related to the different hydrophilic nature of both compounds [37]. Therefore, PE and lecithins may be better acrylamide-scavenger additives in lipophilic environments than other lipophilic additives having analogous nucleophilic groups.

COMBINED ACTION OF THIOL AND AMINO GROUPS FOR ACRYLAMIDE REMOVAL

Most studies regarding the use of nucleophilic compounds to decrease acrylamide content use the combination of several nucleophilic groups to potentiate their action. However, there are very few studies that try to determine if this combination of nucleophilic groups is positive for acrylamide removal. In this sense, Hidalgo et al. [38] studied the reaction of acrylamide with glutamic acid (as an amino acid having only an amino group), *N*-acetylcysteine (as an amino acid having only a thiol group), the mixture of glutamic acid and *N*-acetylcysteine (where both amino and thiol groups are present but in different molecules), glutathione (where both amino and thiol groups are present in the same molecule but not close together), and cysteine (where amino and thiol groups are closer than in glutathione). They found that E_a for acrylamide disappearance in the presence of glutathione was 27 kJ/mol, which was lower than acrylamide disappearance in the presence of *N*-acetylcysteine (30 kJ/mol), and much lower than acrylamine disappearance in the presence of an amino acid having only one amino group (52 kJ/mol). However, E_a for acrylamide disappearance in the presence of cysteine was reduced to 24 kJ/mol. These differences in E_a have not been yet explained, but Hidalgo et al. [38] found that the reactivities of these nucleophiles were inversely correlated ($r=-0.975$, $p=0.025$) to the E_a of the reaction (the lower E_a, the higher decrease of acrylamide). These authors concluded that the E_a of acrylamide/nucleophile reactions, and therefore the carbonyl-scavenging ability of the compounds having these nucleophilic groups, is determined not only by the simultaneous presence of amino and thiol groups in the same molecule but also by the distance among them (the closer the groups, the higher the interaction between them which facilitates acrylamide removal).

The E_a of these reactions are so low that they are produced spontaneously during digestion. Hamzalioglu and Gokmen [39] have recently found that acrylamide levels gradually decreased through gastric, duodenal, and colon phases during in vitro digestion of biscuits as a consequence of the reactions between acrylamide and the amino acids released as a consequence of the digestive process. They were able to identify the adduct produced between cysteine and two molecules of acrylamide. The structure of this adduct had two Michael adducts: the addition of the thiol group to one of the molecules of acrylamide and the addition of the amino group to the other molecule of acrylamide. Analogous acrylamide disappearance was observed by Hoenicke and Gatermann [40] in coffee and cacao power upon storage. They also suggested that this decrease was a consequence of the acrylamide-scavenging ability of thiol-containing substances, although no attempts were carried out to isolate and characterize the compounds responsible for this action.

PRACTICAL USES OF NUCLEOPHILIC COMPOUNDS FOR ACRYLAMIDE REMOVAL IN FOOD PRODUCTS

The use of amino acids from the group of cysteine, lysine, glycine, histidine, alanine, methionine, glutamic acid, aspartic acid, proline, phenylalanine, valine, and arginine for reducing acrylamide formation in thermally processed foods has been patented [41]. In addition, the addition of amino acids for acrylamide reduction in food products has been employed, for example, by Brathen et al. in cereal and potato products [42], Kim et al. in fried snacks [43], Fink et al. in yeast-leavened bread [44], Levine and Smith in crackers [31], Elder in potato dough [45], and Flueckiger and Erdjan in crisp-bread [46], among others.

In addition to amino acids, proteins can also be employed to reduce acrylamide content in food. However, the effectivity of all proteins is not the same because it depends on the amino acid composition. To this respect, amaranth protein has been suggested to be especially adequate for acrylamide mitigation because it is rich in lysine (4.8–6.4 g/100 g of protein), and its sulfur amino acid concentration (3.7–5.5 g/100 g of protein) is higher than that of the most important legumes (~1.4 g/100 g of protein) [47]. When amaranth protein isolate was added to cookies, fried tortilla chips, and baked tortilla chips, the acrylamide content decreased by 89%, 51%, and 62%, respectively, without relevant changes in food properties. Furthermore, amaranth protein has been suggested to increase nutritional properties of these foods because of its antioxidative activity [48], cholesterol-lowering effect [49], and the antihypertensive activity of their hydrolysates [50]. Other proteins suggested for acrylamide removal have been legume proteins [51] or casein [31]. In addition, protein hydrolysates have also been observed to reduce acrylamide content [52]. However, addition of proteins can cause allergies. Therefore, their use for acrylamide reduction should be carefully assessed [3].

The formation of acrylamide–protein adducts has also consequences for acrylamide uptake. Schabacker et al. [53] found that, by using a human intestine model (Caco-2 cells), acrylamide monomers were highly bioavailable and passed the cell monolayer via passive diffusion. However, acrylamide bonded to dietary proteins such as chicken egg albumin under intestinal and cooking conditions. This binding reduced the concentration of acrylamide monomers and led to a reduced uptake by Caco-2 cells.

SYNOPSIS OF ANALYTICAL TECHNIQUES

Different analytical techniques have been employed in the study of acrylamide removal by nucleophiles. Isolation and characterization of Michael adducts between acrylamide and nucleophiles were carried out in model systems employing chromatographic techniques for the isolation (either column chromatography or preparative high-performance liquid chromatography) and spectroscopic and spectrometric techniques for the characterization (mainly nuclear magnetic resonance spectroscopy and high-resolution mass spectrometry). Detection of Michael adducts in food systems has been carried out mainly by high-resolution mass spectrometry. A description of the method employed for this detection can be found in the article of Hamzalioglu and Gokmen [39].

KEY FACTS

- Nucleophiles are added either reversibly or irreversibly to acrylamide, which produces acrylamide disappearance.
- The most potent nucleophiles for this reaction are compounds having a thiol group.
- Cysteine and derivatives are the most important food-relevant nucleophiles having a thiol group that have been employed for acrylamide scavenging.
- Sodium bisulfite reacts similarly to cysteine and derivatives.
- Although less reactive than thiols, amino compounds (amino acids, proteins, and aminophospholipids) are also an option for acrylamide scavenging.
- Reactivity of amino compounds depends on the number of amino groups/molecular weight ratio.
- Simultaneous presence of thiol and amino groups in the same molecule produces an increased acrylamide-scavenging potential.
- These reactions are produced during digestion and are relevant for the bioavailability of the ingested acrylamide.

MINI DICTIONARY

α,β-unsaturated carbonyl compound It is a compound in which the carbonyl group is conjugated with a carbon–carbon double bond. These compounds are susceptible to the attack by nucleophiles at the β carbon.

Acrylamide Potential occupational carcinogen frequently used in industry which was found to be produced in foods in 2002.

Amaranth Common name for the plants in the genus *Amaranthus. Amaranthus hypochondriacus* is an ornamental plant originally endemic to Mexico. It is valued as source of food.

Caco-2 cells They are cells that, when cultivated under specific conditions, become differentiated and polarized such that their phenotype resembles the enterocytes lining the small intestine. They are widely used as an in vitro model of the human small intestinal mucosa to predict the absorption of chemical compounds.

Electrophile It is a compound attracted to electrons or to an electron rich center.

Michael reaction It is the addition reaction of a nucleophile to an α,β-unsaturated carbonyl compound.

Nucleophile It is a chemical compound that donates an electron pair to an electrophile to form a chemical bond. Amino acids are food-relevant nucleophiles.

Sodium bisulfite Chemical substance employed in food industry for preventing oxidation and preserving flavor.

SUMMARY POINTS

Acrylamide is an electrophile that reacts easily with nucleophiles. Thiols are better nucleophiles than amino compounds for acrylamide removal. However, the best nucleophiles for scavenging acrylamide are those having both thiol and amino groups. The reaction between nucleophiles and acrylamide is a Michael reaction. This reaction is produced very easily under soft conditions. It also plays a major role in the disappearance of acrylamide during the digestive process and the decrease of its bioavailability.

REFERENCES

[1] Tareke E, Rydberg P, Karlsson P, Eriksson S, Tornqvist M. Analysis of acrylamide, a carcinogen formed in heated foodstuffs. J Agric Food Chem 2002;50:4998–5006.
[2] Capuano E, Fogliano V. Acrylamide and 5-hydroxymethylfurfural (HMF): a review on metabolism, toxicity, occurrence in food and mitigation strategies. LWT—Food Sci Technol 2011;44:793–810.
[3] Claus A, Carle R, Schieber A. Acrylamide in cereal products: a review. J Cereal Sci 2008;47:118–33.
[4] Foot RJ, Haase NU, Grob K, Gondé P. Acrylamide in fried and roasted potato products: a review on progress in mitigation. Food Addit Contam 2007;24:37–46.
[5] Friedman M. Chemistry, biochemistry, and safety of acrylamide. A review. J Agric Food Chem 2003;51:4504–26.
[6] Friedman M, Levin CE. Review of methods for the reduction of dietary content and toxicity of acrylamide. J Agric Food Chem 2008;56:6113–40.
[7] Guenther H, Anklam E, Wenzl T, Stadler RH. Acrylamide in coffee: review of progress in analysis, formation and level reduction. Food Addit Contam 2007;24:60–70.
[8] Keramat J, LeBail A, Prost C, Soltanizadeh N. Acrylamide in foods: chemistry and analysis. A review. Food Bioprocess Technol 2011;4:340–63.
[9] Konings EJM, Ashby P, Hamlet CG, Thompson GAK. Acrylamide in cereal and cereal products: a review on progress of level reduction. Food Adit Contam 2007;24:47–59.
[10] Lineback DR, Coughlin JR, Stadler RH. Acrylamide in foods: a review of the science and future considerations. Ann Rev Food Sci Technol 2012;3:15–35.
[11] Matthaus B, Haase NU. Acrylamide—still a mattern of concern for fried potato food? Eur J Lipid Sci Technol 2014;116:675–87.
[12] Pedreschi P, Mariotti MS, Granby K. Current issues in dietary acrylamide: formation, mitigation and risk assessment. J Sci Food Agric 2014;94:9–20.
[13] Rydberg P, Eriksson S, Tareke E, Karlsson P, Ehrenberg L, Tornqvist M. Investigations of factors that influence the acrylamide content of heated foodstuffs. J Agric Food Chem 2003;51:7012–8.

[14] Vinci RM, Mestdagh F, De Meulenaer B. Acrylamide formation in fried potato products—present and future, a critical review on mitigation strategies. Food Chem 2012;133:1138–54.
[15] Wenzl T, de la Calle MB, Anklam E. Analytical methods for the determination of acrylamide in food products: a review. Food Addit Contam 2003;20:885–902.
[16] Zamora R, Delgado RM, Hidalgo FJ. Strecker aldehydes and α-keto acids, produced by carbonyl–amine reactions, contribute to the formation of acrylamide. Food Chem 2011;128:465–70.
[17] Zhang Y, Zhang G, Zhang Y. Occurrence and analytical methods of acrylamide in heat-treated foods: review and recent developments. J Chromatogr A 2005;1075:1–21.
[18] Zhang Y, Ren Y, Zhang Y. New research developments on acrylamide: analytical chemistry, formation mechanism, and mitigation recipes. Chem Rev 2009;109:4375–97.
[19] Mottram DS, Wedzicha BL, Dobson AT. Acrylamide is formed in the Maillard reaction. Nature 2002;419:448–9.
[20] Stadler RH, Blank I, Varga N, Robert F, Hau J, Guy PA, et al. Acrylamide from Maillard reaction products. Nature 2002;419:449–50.
[21] Zamora R, Hidalgo FJ. Contribution of lipid oxidation products to acrylamide formation in model systems. J Agric Food Chem 2008;56:6075–80.
[22] Hidalgo FJ, Delgado RM, Navarro JL, Zamora R. Asparagine decarboxylation by lipid oxidation products in model systems. J Agric Food Chem 2010;58:10512–7.
[23] Zamora R, Delgado RM, Hidalgo FJ. Conversion of 3-aminopropionamide and 3-alkylpropionamide into acrylamide in model systems. Mol Nutr Food Res 2009;53:1512–20.
[24] Hidalgo FJ, Zamora R. Amino acid degradations produced by lipid oxidation products. Crit Rev Food Sci Nutr, http://dx.doi.org/10.1080/10408398.2012.761173, in press.
[25] Koutsidis G, Simons SPJ, Thong YH, Haldou-Pis Y, Mojica-Lazaro J, Wedzicha BL, et al. Investigations on the effect of amino acids on acrylamide, pyrazines, and Michael addition products in model systems. J Agric Food Chem 2009;57:9011–5.
[26] LoPachin RM, Barber DS, Gavin T. Molecular mechanisms of the conjugated α,β-unsaturated carbonyl derivatives: relevance to neurotoxicity and neurodegenerative diseases. Toxicol Sci 2008;104:235–49.
[27] Hidalgo FJ, Delgado RM, Zamora R. Role of mercaptans on acrylamide elimination. Food Chem 2010;122:596–601.
[28] Adams A, Hamdani S, Van Lancker F, Méjri S, De Kimpe N. Stability of acrylamide in model systems and its reactivity with selected nucleophiles. Food Res Int 2010;43:1517–22.
[29] Jin C, Wu X, Zhang Y. Relationship between antioxidants and acrylamide formation: a review. Food Res Int 2013;51:611–20.
[30] Casado FJ, Sanchez AH, Montano A. Reduction of acrylamide content of ripe olives by selected additives. Food Chem 2010;119:161–6.
[31] Levine RA, Smith RE. Sources of variability of acrylamide levels in a cracker model. J Agric Food Chem 2005;53:4410–6.
[32] Guido LF. How do sulfites help to control beer ageing? Cerevisia 2005;30(2). article 2.
[33] Dufour JP, Leus M, Baxter AJ. Characterization of the reaction of bisulfite with unsaturated aldehydes in a beer model system using nuclear magnetic resonance spectroscopy. J Am Soc Brew Chem 1999;57:138–44.
[34] Youngblood MP. Kinetics and mechanism of the addition of sulfite to *p*-benzoquinone. J Org Chem 1986;51:1981–5.
[35] Friedman M, Cavins JF, Wall JS. Relative nucleophilic reactivities of amino groups and mercaptide ions in addition reactions with α,β-unsaturated compounds. J Am Chem Soc 1965;87:3672–82.
[36] Zamora R, Delgado RM, Hidalgo FJ. Model reactions of acrylamide with selected amino compounds. J Agric Food Chem 2010;58:1708–13.
[37] Zamora R, Delgado RM, Hidalgo FJ. Amino phospholipids and lecithins as mitigating agents for acrylamide in asparagine/glucose and asparagine/2,4-decadienal model systems. Food Chem 2011;126:104–8.
[38] Hidalgo FJ, Delgado RM, Zamora R. Positive interaction between amino and sulfhydryl groups for acrylamide removal. Food Res Int 2011;44:1083–7.
[39] Hamzalioglu A, Gokmen V. Investigation of the reactions of acrylamide during in vitro multistep enzymatic digestion of thermally processed foods. Food Funct 2015;6:108–13.

[40] Hoenicke K, Gatermann R. Studies on stability of acrylamide in food during storage. J AOAC Int 2005;88:268–73.

[41] Elder VA, Fulcher JG, Leung HKH, Topor MG. Method for reducing acrylamide formation in thermally processed foods. US Patent 7267834; 2007.

[42] Brathen E, Kita A, Knutsen SH, Wicklund T. Addition of glycine reduces the content of acrylamide in cereal and potato products. J Agric Food Chem 2005;53:3259–64.

[43] Kim CT, Hwang ES, Lee HJ. Reducing acrylamide in fried snack products by adding amino acids. J Food Sci 2005;70:C354–8.

[44] Fink M, Andersson R, Rosen J, Aman P. Effect of added asparagine and glycine on acrylamide content in yeast-leavened bread. Cereal Chem 2006;83:218–22.

[45] Elder VA. Deactivating asparagine in an asparagine-containing food ingredient and heating; adding a free thiol compound to a starch-based dough for a thermally processed food and a reducing agent to said starch-based dough. US Patent 20050118322; 2005.

[46] Flueckiger R, Erdjan S. Method to limit acrylamide in heated foods. Patent WO2006017526; 2006.

[47] Salazar R, Arambula-Villa G, Vazquez-Landaverde PA, Hidalgo FJ, Zamora R. Mitigating effect of amaranth (*Amarantus hypochondriacus*) protein on acrylamide formation in foods. Food Chem 2012;135:2293–8.

[48] Pasko P, Barton H, Zagrodzki P, Chlopicka J, Izewska A, Gawlik M, et al. Effect of amaranth seeds in diet on oxidative status in plasma and selected tissues of high fructose-fed rats. Food Chem 2011;126:85–90.

[49] Mendonca S, Saldiva PH, Cruz RJ, Areas JAG. Amaranth protein presents cholesterol-lowering effect. Food Chem 2009;116:738–42.

[50] Fritz M, Vecchi B, Rinaldi G, Anon MC. Amaranth seed protein hydrolysates have in vivo and in vitro antihypertensive activity. Food Chem 2011;126:878–84.

[51] Vattem D, Shetty K. Compositions of legume proteins and methods of use thereof for reducing acrylamide in cooked foods. US Patent 20050048172; 2005.

[52] Cook DJ, Taylor AJ. On-line MS/MS monitoring of acrylamide generation in potato- and cereal-based systems. J Agric Food Chem 2005;53:8926–33.

[53] Schabacker J, Schwend T, Wink M. Reduction of acrylamide uptake by dietary proteins in a Caco-2 gut model. J Agric Food Chem 2004;52:4021–5.

CHAPTER 16

Lipid Oxidation Promotes Acrylamide Formation in Fat-Rich Systems

Edoardo Capuano
Wageningen University and Research Centre, Food Quality and Design Group, The Netherlands

BACKGROUND

After many years of thorough investigations it is nowadays widely accepted that acrylamide (AA) in food mostly formed from degradation of asparagine through the Maillard reaction (MR). Despite asparagine can theoretically form AA directly, the reaction yield is extremely low. Instead, in the presence of reducing sugars AA is produced in a substantially higher rate and yield. Formally, however, reducing sugars are not strictly necessary for the degradation of asparagine to AA. Any carbonyl compound can in principle react with asparagine and yield AA, although with different efficiency. Reducing sugars are key factors in AA generation in plant-derived food because they represent by far the most important source of carbonyl compounds (i.e., reducing sugars themselves and their degradation products, e.g., glyoxal, methylglyoxal, and deoxyosones). However, additional carbonyl compounds (e.g., flavor compounds, vitamin B_6, and products of lipid oxidation (LO)) may be present in food, albeit in small amount. LO is one of the major chemical reactions occurring in foods during processing and cooking. It can drastically affect food sensory quality and plays an important role in the final acceptance of edible fats/oils or lipid-rich foods by the consumers. As a result of LO, LO products (LOPs) are formed which comprise a vast array of relatively unstable compounds of various molecular weight, flavor threshold, and biological significance including aldehydes, ketones, alcohol, epoxides, and hydrocarbons. Since the amount of LOPs which possess a carbonyl moiety may be significant in some cases, the hypothesis was put forward that LO can significantly contribute to AA formation in lipid-rich food. In this respect, it is important to recall that another mechanism for AA formation related to lipids has been postulated, i.e., reaction of acrolein with ammonia. Acrolein is a typical product of lipid degradation, mainly stemming from dehydration of glycerol. This pathway is usually considered minor compared with the one from MR. In any case, in this chapter we solely speculate on the potential contribution of carbonyl compounds from LO on AA formation.

Acrylamide in Food
http://dx.doi.org/10.1016/B978-0-12-802832-2.00016-4

GENERALITIES ON LIPID OXIDATION

Lipid oxidation (LO) (more properly referred to as lipid autoxidation to be distinguished from the enzymatic oxidation) is a complex chemical phenomenon involving several consecutive steps and intermediate. Oxidation proceeds by a sequential free radical chain-reaction mechanism which schematic representation is provided in Figure 1. The first step in the reaction chain is the extraction of one hydrogen atom from the aliphatic chain of the lipid that produces a lipid radical R•. This step is called **initiation step**. The extraction is triggered by radical species that act as initiators. In polyunsaturated lipids, the generation of the lipid radical is followed by its molecular rearrangement that produces a thermodynamically favourable conjugated diene. The next step is the incorporation of an oxygen molecule with formation of a peroxy radical ROO•. This radical is able to extract a hydrogen atom from another lipid molecule producing a hydroperoxide, ROOH and regenerating the lipid radical R•. Hydroperoxides are known as **primary oxidation products**. The bulk of them will accumulate in the system but some may decompose producing RO• and OH• radicals that can further react with other lipid molecules. These steps globally constitutes the **propagation step**. After a while, **termination step** occurs by collision of pair of radicals that wipes out almost all the reactive radicals. The rate of hydroperoxides formation depends, among several other factors, on the lipid chemical structure. The energy required for the homolytic extraction of H• from the lipid molecule is in the order vinylic > methylic > methylenic > allylic > diallylic which means that linolenic acid has a higher number of hydrogens more easily extractable (4 diallylic) than linoleic acid

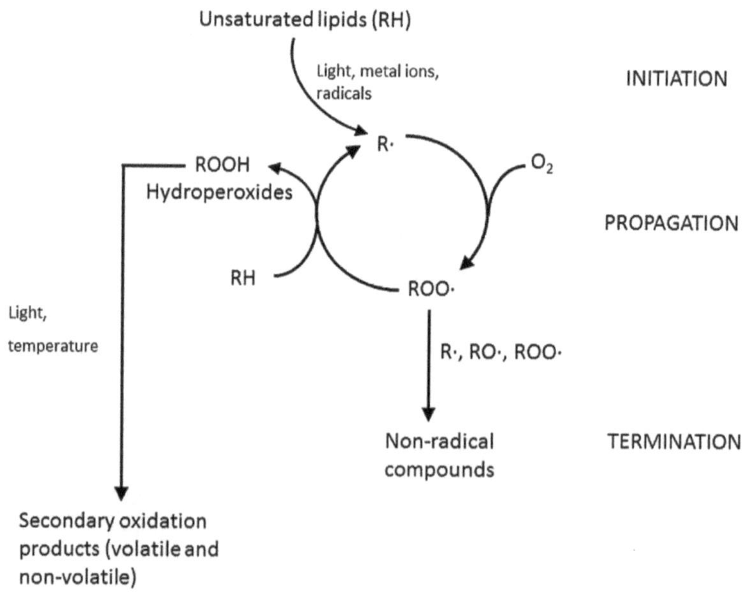

Figure 1 Schematic representation of lipid oxidation.

(2 diallylic) and oleic acid (no diallylic H•). This explains the order in reactivity for different fatty acids and the type and relative distribution of resulting hydroperoxides reported in Table 1. After some time and depending on the system conditions (mainly its temperature) hydroperoxides decompose producing a vast array of **secondary oxidation products** that confer specific (and almost always undesired) taste and flavor to oxidized fats/oils. These products mostly comprise aldehydes but may also include alkanes, alkenes, esters, epoxides, ketones, and malondialdehyde (MDA). Aldehydes are mainly formed from the homolytic scission of the labile O–O bond of the hydroperoxides (catalyzed, e.g., by a radical species or metal ions) followed by β-scission of the ensuing alkyl radicals. The specific secondary oxidation products that form depend on the chemical structure of the lipid being oxidized: that is, the number and the position of the double bonds and the fatty acid chain length (Table 2). As an example, the formation of 2,4-decadienal from linoleic acid 9-hydroperoxide is shown in Figure 2. Despite the type of fat/oil that plays a major role essentially through its level of unsaturation, numerous other factors affect the rate of LO and the pattern and concentration of primary and secondary oxidation products, among which the presence and activity of pro-oxidants (e.g., cations, heme, and preformed radicals) and antioxidants (flavonoids or vitamins C and E), partial pressure of oxygen, the nature of the surface exposed to oxygen and the storage/processing conditions (e.g., temperature, light exposure, and moisture content) of fat/oil (containing food).

Table 1 Induction time, relative rate of oxidation at 25 °C and hydroperoxides (HP) distribution for oleic, linoleic, and linolenic acid

Fatty acid	Induction time (h)	Oxidation rate relative to stearic acid (C18:0)	HOO– group position in HP	Double bonds position in HP	Relative proportion of HP (%)
Oleic acid (C18:1)	82	100	8	9	27
			9	10	23
			10	8	23
			11	9	27
Linoleic acid (C18:2)	19	1200	8	9,12	1.5
			9	10,12	46.5
			10	8,12	0.5
			12	9,13	0.5
			13	9,11	49.5
			14	9,12	1.5
Linolenic acid (C18:3)	1.34	2500	9	10,12,15	31
			10	8,12,15	
			12	9,13,15	11
			13	9,11,15	12
			15	9,12,16	
			16	9,12,14	46

Adapted from Ref. [20].

Table 2 Major volatile compounds formed by autoxidation of oleic, linoleic, and linolenic acid (in descending order of abundance)[a]

Oleic acid	Linoleic acid	Linolenic acid
Nonanal	Hexanal	Propanal
Octanal	(Z)-2-octenal	(E,Z)-2,4-heptadienal
E−(2)-Undecenal	(E)-2-heptenal	(Z)-3-hexenal
Decanal	(E)-2-octenal	2,4,7-decatrienal
E−(2)-Decenal	(E,Z)-2,4-decadienal	(E,E)-2,4-heptadienal
Heptanal	(E,E)-2,4-decadienal	(Z)-2-pentenal

[a] Each fatty acid in amount of 1 g was autoxidized at 20 °C by an uptake of 0.5 mol oxygen/mol fatty acid. Adapted from Ref. [20].

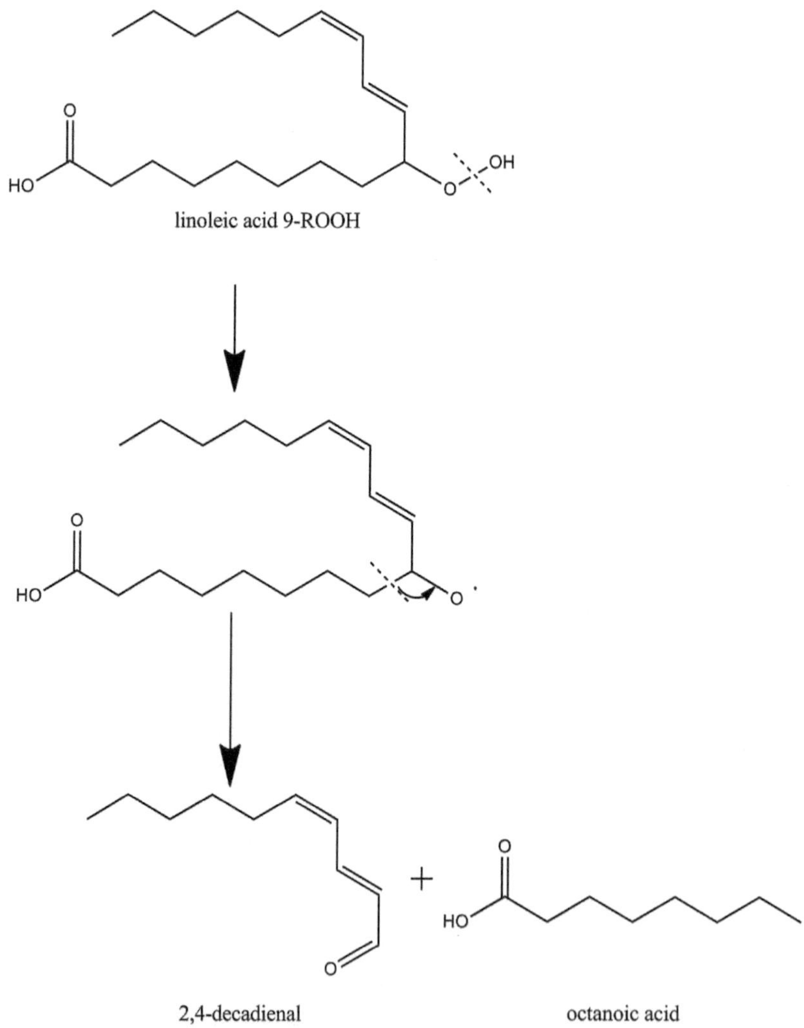

Figure 2 Formation of 2,4-decadienal from linoleic acid 9-hydroperoxide.

Beside autoxidation, lipids in food can also undergo what is called **lipid peroxidation** that is an enzymatic degradation phenomenon catalyzed by the enzyme lipoxygenase (linoleic acid oxygen oxidoreductase, EC 1.13.11.12), which occurs in many plants and also in erythrocytes and leukocytes. Lipoxygenase catalyzes the oxidation of some unsaturated fatty acids, preferably linoleic and linolenic acid, but not oleic acid and monounsaturated fatty acids, to their corresponding hydroperoxides, which are the same primary products of lipid autoxidation. This enzymatic reaction requires much lower temperatures to proceed at significant rate in food. The resulting hydroperoxides can be further degraded by means of enzymatic or nonenzymatic reactions to yield secondary oxidation products that may contribute to the sensorial properties of natural food products. In addition to auto- and peroxidation, hydroperoxides can be formed from direct addition of oxygen to the fatty acid double bond in a process known as **photo-oxidation**. Photo-oxidation requires the ground-state oxygen molecule, that is, the triplet oxygen $^3O_2$ to be activated by light (usually through the action of an intermediate photosensitizer molecule) into a much more reactive singlet oxygen $^1O_2$. Singlet oxygen can add to the double bond without the prior formation of an allyl radical and, for a specific fatty acid, it produces a pattern of hydroperoxides different from that produced by autoxidation. The relative contribution of lipid peroxidation and photo-oxidation to the generation of carbonyl reactive species in food has to be considered marginal compared with the autoxidation and would therefore give a negligible contribution to AA formation in lipid-rich food.

LO CONTRIBUTION TO AA FORMATION IN MODEL SYSTEMS

The potential contribution of LO to AA formation has been thoroughly investigated in model systems and tested as early as soon after AA discovery in heated foods [1]. The potential contribution of LOPs to AA generation has been investigated in aqueous binary or ternary model systems of pure compounds [2–5]. The results of those investigation can be summarized as follows: (1) asparagine degradation to AA only occurs if the LOP possesses a carbonyl group; (2) LOP aldehydes are more reactive than LOP ketones of comparable chain length that is consistent with the higher reactivity of aldehydes in asparagine degradation to AA through carbonyl compounds; (3) short-chain carbonyl compounds (either aldehydes and ketones) are more reactive than long-chain carbonyl compounds presumably because of their higher solubility in water; and (4) α,β,γ,δ-diunsaturated carbonyl compounds have the greatest potential of generating AA when reacted with asparagine when compared with alkenals and alkanals. This reactivity is reduced when the alkedienal is further oxidized to the corresponding epoxide. Asparagine degradation by alkadienals would likely proceed through the decarboxylation of the amino acid to 3-aminopropinamide which is then further converted to AA by deamination but decarboxylation is the key step in AA formation from asparagine/alkadienals model systems due to its higher activation energy compared with the deamination step. Lipid hydroperoxides can also produce AA when reacted with asparagine, likely as a consequence of their thermal

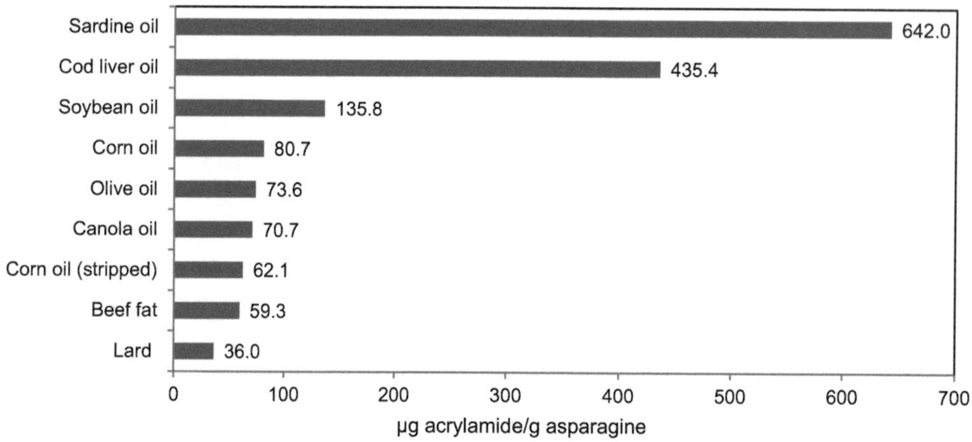

Figure 3 Formation of acrylamide from asparagine (1.76 g) and various lipids (5 g) at 180 °C (30 min). Analysis performed by GC/NPD. Values are the average of two experiments. *From Ref. [6].*

decomposition to alkanals, alkenals, or α,β,γ,δ-diunsaturated alkadienals. A synergic effect of glucose with certain LOPs when reacted with asparagine was also reported, the most notable effect being reported with methyl linolenate but not in methyl stearate/glucose mixtures nor in the presence of antioxidants. This synergism is proposed to be a consequence of the formation of free radicals during the reaction between asparagine and glucose via MR, which would oxidize the lipid into reactive carbonyls. These results suggest that both unoxidized and oxidized lipids are able to contribute to the conversion of asparagine into AA, but unoxidized lipids need to be oxidized as a preliminary step.

The differential contribution of oil/fat unsaturation level and the amount of AA generated in the presence of asparagine was also demonstrated by reacting asparagine with different oils/fats as well as fatty acids. It has been reported, for instance, that the when fats/oils are reacted with asparagine in model systems, the greatest amount of AA is generated in model systems containing sardine oil and cod oil which have a very high degree of unsaturation compared with olive oil or sunflower oil [6] (Figure 3). The least amount of AA was detected in systems containing saturated fat like lard and beef. Interestingly, beef fat produced more AA than lard fat, an outcome that is explained with the fact that beef fat is not refined unlike lard. The refining step is supposed to remove preformed carbonyl LOPs or compounds that can favor LO during the processing/cooking, that is, pro-oxidants. When asparagine was heated in presence of different fatty acids/triglycerides, linoleic acid (more unsaturated and thus more prone to oxidation) is produced in much higher levels than stearic acid (saturated C18) and triolein (triglyceride with three oleic acid), but numerically less AA than oleic acid. As expected triolein produces less AA than free fatty acids which is in agreement with the fact that fatty acids are more prone to oxidation in the free form than when in the triglyceride structure. In a detailed study on the impact of LO on AA formation in model systems, Capuano et al. [7] also demonstrated that replacing palm oil (more saturated, iodine value <20) with sunflower oil (more unsaturated, iodine value ≈130) increases AA formation

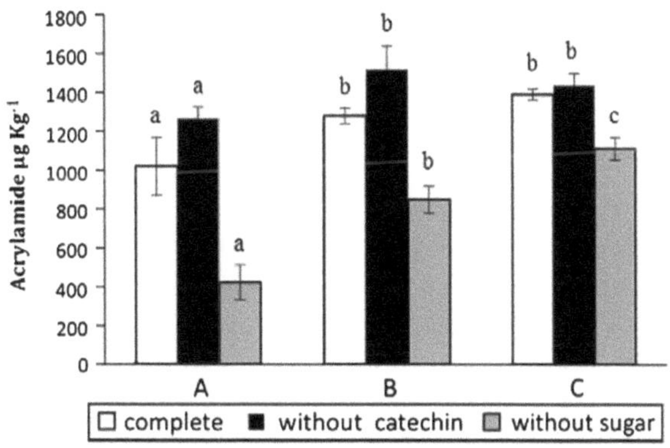

Figure 4 Acrylamide concentration formed in model systems containing oil at different oxidation level. Letter on X-axis indicate the model systems with different oxidized oils. A = model systems with unoxidized oil; B = model systems with sunflower oil heated for 30 min (mildly oxidized); C = model systems with sunflower oil heated for 3 h (severely oxidized). Within each formulation, bars with different letters are significantly different ($p < 0.05$). Results are expressed on dry basis. *From Ref. [7].*

by 35% in solid model system formulated with water and reducing sugars. Moreover, the effect of LO was investigated upon heating of differently formulated solid model system at different water activity (a_w). The systems contained sugars, lipids, and catechin or were formulated without sugars or catechin. Three sunflower oil samples with different oxidation levels were used: untreated sunflower oil with a peroxide value (PV) of $1.5\,m_{eq}\,O_2\,kg\,oil^{-1}$ and a thiobarbituric acid reactive substances (TBARS) value of $0.1\,mmol\,MDA\,kg\,oil^{-1}$; sunflower oil heated for 30 min with a PV of $11.2\,m_{eq}\,O_2\,kg\,oil^{-1}$ and a TBARS value of $8.2\,mmol\,MDA\,kg\,oil^{-1}$, and sunflower oil heated for 3 h with a PV of $21.5\,m_{eq}\,O_2\,kg\,oil^{-1}$ and a TBARS value of $12.5\,mmol\,MDA\,kg\,oil^{-1}$. The results are summarized in Figure 4. The results showed that in the complete formulations, the presence of previously oxidized lipid increases by about 25% AA formation compared with untreated oil, but further oil oxidation did not increase AA levels any more. When lipids are the only potential source of carbonyl compounds, AA progressively increases with the oxidation level. Catechin presence reduced AA formation in systems containing untreated oil and partially oxidized oil (bars A and B), while no significant lowering effect was observed when highly oxidized oil was used (bars C). This was explained by the ability of catechin in preventing further LO, and thus further generation of reactive carbonyls from that source which would explain the lack of effect in model systems formulated with severely oxidized oils. In the latter case, indeed, most of the carbonyl compounds have already formed.

REAL FOOD SYSTEMS

The potential contribution of lipid oxidation to AA formation has also been investigated in real foods processed and/or cooked under real-life relevant conditions. Since the content in lipids and the physical and chemical evolution of the systems are different among different food categories it would be appropriate to discuss the state-of-art for each of the main AA contributing food categories: coffee, bakery products, and potato products.

Coffee

Currently, only one study has investigated the potential of LO during roasting on the final AA content in coffee. Green coffee (before roasting) contains on average almost equal amount of soluble carbohydrates (6–12%) and lipid (8–18%, higher in *arabica* variety). Soluble carbohydrates are mainly constituted of sucrose (8%) and nonreducing di- and trisaccharides, whereas monosaccharides are almost virtually absent (0.5%). Coffee lipids are constituted mainly of linoleic acid followed by palmitic acid. These two fatty acids are still the major fatty acids after coffee roasting. Chlorogenic acid (ester of caffeic acid and quinic acid) is also very abundant in green coffee beans where it can occur at level as high as 10%. Kocadağlı and coworkers [8] identified 2-octenal, 2,4-decadienal, 2,4-heptadienal, 4-hydroxynonenal, and 4,5-epoxy-2-decenal in relatively high quantities in roasted coffee, in concentrations >1 $mg\,g^{-1}$ coffee which would be about 20% of the highest content in reducing monosaccharides. This would suggest that in roasted coffee the contribution of LOPs to total asparagine degradation may not be negligible. However, in roasted coffee sucrose can also be a source of reactive carbonyls. Sucrose decomposition is favored at high temperature and yield reactive monosaccharides and/or hydroxymethylfurfural. These can react with amino compounds including asparagine through the MR. Sucrose and linoleic acid occur in green coffee at almost the same concentration, and they show almost the same reactivity toward AA in model systems with binary systems with asparagine under simulated coffee roasting conditions (Figure 5).

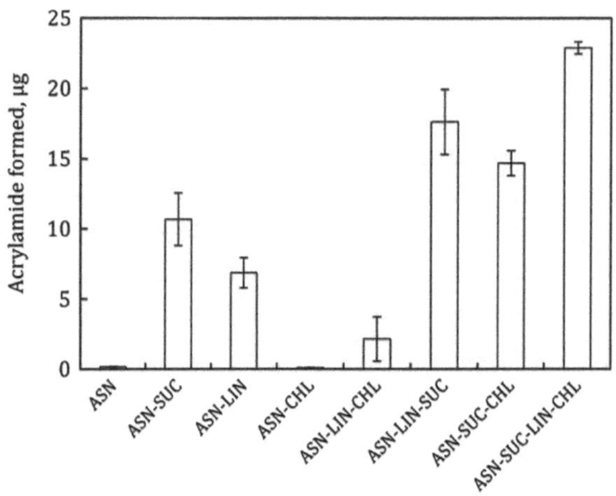

Figure 5 The amount of acrylamide formed in different model systems heated at 180 °C for 10 min. *From Ref. [8].*

Chlorogenic acid can, however, further trigger sucrose decomposition. In the presence of chlorogenic acid, the amount of AA generated in ternary systems with sucrose and asparagine is higher than that generated in binary systems with sucrose/asparagine.

Biscuits and Bakery Products

In nonsugar-added cereal-based products, asparagine is notoriously the limiting factor for AA formation. In such products the lipid content is generally very low (about 1.5–2.0%) and carbonyl compounds from LO are unlikely to play a significant role in asparagine degradation. However, the issue has not been properly investigated. In sweet bakery products such as biscuits or cookies, besides asparagine, reducing sugars are also correlated with AA formation during baking. In biscuits and cookies sugars, mainly sucrose or invert sugar syrup can make up to 20% of the final formulation. Fat/oil is also added to a considerable amount, usually accounting for up to 10% of the final formulation. The effect of LO level or fat unsaturation on the formation of AA in sweet bakery products has been addressed by Arriba-Lorenzo and coworkers [9]. They reported that AA formation dramatically increased (+59%) when cookies were formulated with oxidized sunflower oil (PVs were $33.5\,m_{eq}\,O_2\,kg^{-1}$ and $TBARS = 1.98\,nmol\,MDA\,g^{-1}$) compared with control cookies formulated with untreated sunflower oil ($PV = 1.8\,m_{eq}\,O_2\,kg^{-1}$ and TBARS <the detection limit). In addition, AA content was the lowest in cooked, formulated with high phenolic virgin oil compared with those formulated with olive oils with medium and low phenolic content. While there is no doubt that stressed oils can produce more AA than untreated oils, the practical significance of the reported results is questionable. The PV for the oxidized sunflower oil used in that study was much higher than the limit established by, for instance, the American Oil Chemists Society for edible oils ($10\,m_{eq}\,O_2\,kg\,oil^{-1}$). The oil would have unacceptable sensory attributes and would not be used in any case.

Potato Products

Potato products are the food category where AA formation poses the greatest concern because of the relatively high amount that can form therein, and the contribution of such category to the overall AA dietary intake for the general population. It is widely recognized that reducing sugars content is the limiting factor for AA formation in potato products. Comparatively, asparagine content of potato is considerably of less important. Mitigation strategies aiming at lowering the reducing sugar content in potato slices before the heat treatment such as soaking in properly formulated aqueous solutions, selection of potato varieties lower in reducing sugars, and proper tubers storage have been proposed. For fried potato products, namely French fries and potato chips, the question is whether the oxidation of frying oil during frying generates enough amount of LOP carbonyls to significantly affect the rate and the extent of AA formation. In addition to this chemical effect, LO can affect the rate and the extent of AA formation in fried foods and also through a physical effect: that is, the modification of oil thermal properties due to the accumulation of polar oxidation products, such as LOPs, glycerol, mono- and diglycerides, as well as dimers, trimers, and other polymerization products at the interface between the oil/fat and the food. The accumulation of these compounds generally not only increases oil viscosity and foaming capacity (due to their emulsifying ability), but also changes the rate of heat transfer from oil to the food being fried. On one hand, the increased viscosity decreases the heat transfer rate from oil to food. On the other hand, the polar degradation/oxidation products would change the surface tension at the interface oil/food thus allowing more heat transferred from the oil to the food in a fixed frying time.

The results from the scientific literature are conflicting. Gertz and Klostermann [10], for instance, found a substantial variability in AA content in potato slices fried with different oils (from 772 to 1271 $\mu g\,kg^{-1}$ of potato). However, the AA content did not always correlate with the oil unsaturation level. While the lower amount of AA was found in potato slices fried with palm olein which is the most saturated oil used in the experiment, the highest amount was found in high-oleic sunflower oil, even higher than the amount found in sunflower oil and rapeseed oil that are more unsaturated oils. Even though the iodine value of the used oils was not reported nor was it the level of polar compounds (the fraction putatively containing reactive LOP carbonyls), it could be argued from the data that the amount of AA produced in the chips could not be explained solely on the extent of the LO level. Similarly, Mestdagh and coworkers [11] investigated the impact of oil oxidation and hydrolysis on AA formation in potato homogenates and French fries. They failed in finding a statistically significant effect of soybean oxidation level on AA content of French fries and potato homogenates heated at different temperatures. Soybean oil was selected because of the high linoleic and linolenic acid content. Potato slices were fried in progressively abused oils but the level of AA was not significantly different when the most abused soybean oil (*p*-anisidine value (PAV) > 500) was used compared with fresh oil (PAV ≈ 1.3). Interestingly, the same authors could not find any effect of the oil hydrolysis level (measured by the content in free fatty acids and mono- and diglycerides) on AA formation either. In a companion paper, Mestdagh and coworkers [12] investigated the effect of different frying oils and fats on AA formation in French fries and potato homogenates. Fats and oils exhibited low levels of oxidation at the starting of the frying step. Again the authors could not find significant differences in AA content in different frying media nor could they find a significant correlation with the oil/fat unsaturation level. The same conclusions were drawn by Matthaus and coworkers [13] and by Williams [14]. In the former investigation, the amount of AA was numerically but not significantly higher in potato slices fried with semisolid or solid deep-frying fats compared with several frying oils. Moreover, no effect could be found that the frying oil age (measured by the level of triglycerides oligomers) increases the AA concentration upon frying. On the other hand, Lim et al. [15] reported that AA content in potato slices does significantly depends on the frying cycle and the type of oil used for deep frying. The AA content after the 10th frying cycle showed from a 3-fold (coconut oil) to an 8-fold (sunflower oil) increase compared with the content after the first cycle. Potato slices deep fried in palm oil showed the lowest AA content after the first cycle (298 $\mu g\,kg^{-1}$) and an average across the 10 cycles (1443 $\mu g\,kg^{-1}$) whereas potato slices fried in sunflower oil showed the highest value at first cycle (1060 $\mu g\,kg^{-1}$) and on average across the 10 cycles (2019 $\mu g\,kg^{-1}$). Coconut oil produced more AA than canola oil after the first frying, but the same average level of AA was maintained across the 10 cycles. However, no significant correlation could be found between the level of AA after the first cycle or the average level of AA across the 10 cycles and the PV, PAV, iodine value, free FA concentration, and levels of polyunsaturated fatty acids of the tested oils. During frying the PV and the PAV of the frying oils significantly increased, especially PAV which were from 5 to 12 times as high as in fresh oils.

In lipid-rich, nonfried potato-based products the effect of oil type and oxidation level is likely to be very limited or negligible. In potato dough containing 10% oil, potato powder, flour, and water, shaped into thin discs and baked at 180 °C for different heating times neither the type of oil used (olive, sunflower, or soybean) nor the oil oxidation level (low oxidation: 5.51 $m_{eq}\,O_2\,kg\,oil^{-1}$, MDA content: not detected; medium oxidation: 7.18 $m_{eq}\,O_2\,kg\,oil^{-1}$, MDA content: 23.11 $\mu mol\,kg^{-1}$; high oxidation: 8.84 $m_{eq}\,O_2\,kg\,oil^{-1}$, MDA content: 51.64 $\mu mol\,kg^{-1}$) significantly affected AA formation [16].

THE ROLE OF ANTIOXIDANT COMPOUNDS IN PREVENTING AA FORMATION

Since there is convincing evidence that carbonyl LOPs may contribute to AA formation, **antioxidants** have been proposed as mitigation strategy for AA formation in lipid-rich food. Antioxidant compounds are defined as compounds that are able to inhibit or delay the oxidation of other molecules. A number of antioxidant compounds can be naturally found in foods or formulated into food as additives (vitamins C and E, phenolic acids, and polyphenols, carotenoids, butylated hydroxyanisol, butylated hydroxytoluene, and so on) or can be formed during heat treatment of carbohydrate-rich food through MR, for example, melanoidins (brown anionic nitrogenous polymers) in coffee, bread, and bakery products. The addition of antioxidant compounds or natural extracts rich in antioxidants provides conflicting effects with respect to AA mitigation. A comprehensive review on the subject can be found in Jin et al. [17]. Some (mixture of) antioxidants are effective in reducing AA formation in several real food formulation whereas other compounds or combinations are ineffective or even enhance AA formation. This contradictory results stem from the fact that antioxidant compounds are a class of chemically heterogeneous compounds that may modulate AA formation in different steps and with diverse mechanisms and from the fact that model systems compositions and/or reaction conditions may be very different across scientific studies.

Specific antioxidants can trigger ingredients decomposition (e.g., sucrose), which can generate reactive carbonyls, trap specific MR products or intermediates, prevent lipids from oxidation, react with asparagine directly to form AA, cause precipitation of asparagine, directly react with AA. In the perspective of what has been discussed in this chapter, antioxidant compounds may mitigate AA formation by limiting formation of LOPs through their mere antioxidant capacity, that is, by reducing or delaying LO and thus accumulation of additional carbonyls (i.e., LOPs). However, this does not result necessarily in the mitigation of AA formation because of the multiple ways antioxidants can interact with other food components. A notable example is provided by chlorogenic acid in coffee. Chlorogenic acid reduces LOPs formation from linoleic acid less than it triggers formation of reactive carbonyls from sucrose degradation, the net effect being an increase in AA formation when chlorogenic acid is present (see Figure 5).

CONCLUSIONS

As proven by several studies in model systems there is convincing evidence that secondary oxidation carbonyl compounds generated during advanced LO can react with asparagine and contribute to its degradation to AA. Secondary oxidation carbonyl compounds can not only react directly with asparagine but also react with other amino acids generating reactive intermediates that can contribute even further to asparagine degradation. What is not clear at the present is the practical significance of this additional source of reactive carbonyls to AA formation under industrial processing or domestic cooking real-life conditions. The fraction of secondary oxidation carbonyl compounds to reducing sugars is usually minor in most food products, unless severely oxidized lipids are used in food formulations or severe processing/cooking conditions are applied which would be the exception rather than the norm in almost all the practical cases. All in all, AA formation from LO is considered minor in most of the scientific literature and the evidence for a significant effect in real food can be considered at best conflicting, if not very thin. At the present, no mitigation strategy or suggested practice referring to, or based on, formation of AA from LO is mentioned in the most updated version of the CIAA acrylamide toolbox [18] or in the Codex Alimentarius Code of Practice for the reduction of AA in foods [19]. Overall, the evidence available point to the fact that LO may contribute to AA formation in lipid-rich food but its actual contribution is likely negligible or minor compared with other sources of AA for most of its potential dietary source, the only notable exception potentially being roasted coffee.

SYNOPSIS OF ANALYTICAL TECHNIQUES

Lipid Oxidation

Lipid oxidation level can be measured in different ways: by measuring (1) the amount of oxygen taken up by the oil/fat, (2) the reactants change, (3) the amount of primary oxidation products, (4) the amount of secondary oxidation products, (5) free radicals, and (6) by other methods. The amount of oxygen taken up by the lipid sample can be assessed by measuring the weight gain of the sample or the consumption of headspace oxygen. The measurement of reactant change involves mainly measuring the change in specific fatty acid concentration. Fatty acids can be easily measured after trans-esterification to corresponding methyl esters by gas chromatography with flame ionization or mass spectrometry detector. Primary oxidation products can be expressed by the PV which represents the total hydroperoxides content of the test fat. PV can be measured by iodometric titration, ferric ion complex measurement spectrophotometry, and Fourier Transform Infrared spectroscopy. Detailed information on the structure and the amount of specific hydroperoxides can be obtained by direct or reverse-phase HPLC. As an alternative, dienes and trienes (produced during formation of hydroperoxides from unsaturated fatty acids) can be assayed by measuring the increase of absorbance of the test sample at 234 nm (dienes) and 268 nm (trienes). To avoid interferences from substances absorbing at the same wavelength (e.g., carotenoids), hydroperoxides can be converted to conjugated chromophores followed by measurement of absorbance at 268 and 301 nm (tetraenes). Secondary oxidation products are expressed as TBARS assay. With this assay, secondary oxidation products (mainly MDA) are conjugated to thiobarbituric acid and the concentration of the conjugates measured spectrophotometrically at 530–535 nm. As an alternative, PAV can be used. PAV measures the content of aldehydes (principally 2-alkenals and 2,4-alkadienals) generated during the decomposition of hydroperoxides. It is based on the color reaction of *p*-methoxyaniline (anisidine) and the aldehydic compounds under acidic conditions that produces a derivative that absorbs at 350 nm. Total content in carbonyl compounds (aldehydes and ketones) can be quantified by reacting the carbonyls with 2,4-dinitrophenylhydrazine followed by the reaction of the resulting hydrazones with alkali. The resulting derivative is measured spectrophotometrically at a given wavelength. Volatile secondary oxidation products can also be measured by gas chromatography/mass spectrometry (GC–MS) or by the oxidative stability instrument or Rancimat. The latter instrument quantifies the volatile compounds by monitoring the change in electrical conductivity when effluent from oxidizing oils is passed through water. Another approach to measuring lipid oxidation level is represented by the measurement of free radicals. This is possible by electron spin resonance, also referred to as electron paramagnetic resonance spectroscopy which relies on the paramagnetic properties of the unpaired electrons in radicals. Finally, the evolution of lipid oxidation can be followed by differential scanning calorimetry which monitors the thermally induced transitions occurring during oxidation such as transfer of oxygen molecules to unsaturated fatty acids or by ^{1}H nuclear magnetic resonance spectroscopy. Theoretically, the extent of lipid oxidation can also be quantified by means of the sensory evaluation of the fat sample by expert and trained sensory panels. The sensory evaluation consists in the descriptive analysis of the test samples which includes detection and description of the qualitative and quantitative sensory aspects of the product.

Lipid Unsaturation Level

Lipid unsaturation level can be conveniently measured by the iodine value that gives a measure of the average *degree of unsaturation* of a lipid: the higher the iodine value, the greater the number of C=C double bonds. By definition, the iodine value is expressed as the grams of iodine absorbed per 100g of lipid. One of the most commonly used methods for determining the iodine value of lipids is the "Wijs method" which consists in adding an excess of iodine chloride that reacts with C=C double bonds and titrate the excess of iodine chloride (which did not react with double bonds) with sodium thiosulfate after quantitative liberation of iodine from iodine chloride after addition of excess potassium iodine. Lipid unsaturation level can also be determined by profiling fatty acid composition that consists in the quantitative saponification of triglycerides and simultaneous trans-esterification of FA in the corresponding fatty acid methyl ester (FAME) followed by FAME determination by GC-FID or GC-MS.

KEY FACTS

- Carbonyl compounds generated through lipid oxidation can contribute to asparagine degradation to AA.
- The amount of carbonyl generated through lipid oxidation depends on the lipid properties (fatty acids composition, level of unsaturation, content in antioxidants, and pro-oxidants) as well as storage and processing conditions (such as temperature, time, and level of oxygen).
- A significant contribution to AA formation from lipid oxidation has been unambiguously demonstrated in model systems.
- In potato and cereal-based products the contribution to AA formation from lipid oxidation may be of limited significance whereas it may be significant in coffee.
- Antioxidant compound can theoretically reduce AA formation by limiting the rate and the extent of lipid oxidation but their actual effect is difficult to predict because of the multiple ways antioxidants can interact with other precursors/intermediate of AA formation and with AA itself.

MINI DICTIONARY

Lipid oxidation (LO) It is the degradation of fat/oil through the incorporation of oxygen. In the final stage, LO produces volatile compounds that are detrimental for the oil/fat sensory quality.

Radical Chemical species (atom, molecule, or ion) that has unpaired valence electrons or an open electron shell.

Primary oxidation products Products of incorporation of oxygen in the fat/oil molecule. They are hydroperoxides.

Secondary oxidation products Products of chemical decomposition of primary oxidation products. They are responsible for the rancidity of oxidized oils/fats and may contribute to AA formation.

Homolysis It is the dissociation of a chemical bond of a neutral molecule in two free radicals.

Lipid unsaturation It is the number of double bonds in the aliphatic chain of fatty acids in an oil/fat.

Iodine value It is the grams of iodine incorporated by 100 g of test substance. It is an indicator of oil/fat unsaturation level.

Peroxide value It is the amount of peroxide oxygen per 1 kg of fat or oil. It is expressed in milliequivalent of oxygen. It is an indicator of the content of primary oxidation products (hydroperoxides) in fat/oil.

TBARS Thiobarbituric acid reactive substances are by-product of lipid peroxidation, which can be detected by the TBARS assay which uses thiobarbituric acid as a reagent. It is essentially, if not exclusively, malondialdehyde (MDA).

Decarboxylation It is a reaction where a carboxyl group is removed from a molecule and released as carbon dioxide.

Deamination It is a reaction where an amino group is removed from a molecule and released as ammonia.

Antioxidant compounds These are substances that inhibit or delay the oxidation of other molecules.

Oxidation It is a chemical reaction where electrons or hydrogen are transferred from a substance (which is oxidized in the process) to an oxidizing agent (which is reduced in the process).

Catechin It is a plant secondary metabolite belonging to the family of flavan-3-ols, part of the chemical family of flavonoids. It possesses antioxidant activity.

SUMMARY POINTS

Lipid oxidation can contribute to AA formation in lipid-rich model systems. The contribution depends on the generation of reactive carbonyl oxidation products. Level of unsaturation and oxidative status of lipid affect AA formation. Level and nature of pro-oxidants and antioxidants occurring in lipid also can affect AA formation. Significance of lipid oxidation contribution in real food is questionable because typically level of lipid oxidation products is much lower than reducing sugars and their carbonyl degradation products. In coffee lipid oxidation during roasting may significantly contribute to AA formation.

REFERENCES

[1] Becalski A, Lau BPY, Lewis D, Seaman SW. Acrylamide in foods: occurrence, sources, and modeling. J Agric Food Chem 2003;51:802–8.

[2] Hidalgo FJ, Delgado RM, Zamora R. Degradation of asparagine to acrylamide by carbonyl-amine reactions initiated by alkadienals. Food Chem 2009;116:779–84.

[3] Hidalgo FJ, Delgado RM, Navarro JL, Zamora R. Asparagine decarboxylation by lipid oxidation products in model systems. J Agric Food Chem 2010;58:10512–7.

[4] Zamora R, Hidalgo FJ. Contribution of lipid oxidation products to acrylamide formation in model systems. J Agric Food Chem 2008;56:6075–80.

[5] Zamora R, Delgado RM, Hidalgo FJ. Conversion of 3-aminopropionamide and 3-alkylpropionamide into acrylamide in model systems. Mol Nutr Food Res 2009;53:1512–20.

[6] Ehling S, Hengel M, Shibamoto T. Formation of acrylamide from lipids. In: Friedman M, Mottram D, editors. Chemistry and safety of acrylamide in food. New York, US: Springer; 2005.

[7] Capuano E, Oliviero T, Açar ÖÇ, Gökmen V, Fogliano V. Lipid oxidation promotes acrylamide formation in fat-rich model systems. Food Res Int 2010;43:1021–6.

[8] Kocadağlı T, Göncüoğlu N, Hamzalioğlu A, Gökmen V. In depth study of acrylamide formation in coffee during roasting: role of sucrose decomposition and lipid oxidation. Food Funct 2012;3:970–5.

[9] Arribas-Lorenzo G, Fogliano V, Morales FJ. Acrylamide formation in a cookie system as influenced by the oil phenol profile and degree of oxidation. Eur Food Res Technol 2009;229:63–72.

[10] Gertz C, Klostermann S. Analysis of acrylamide and mechanisms of its formation in deep-fried products. Eur J Lipid Sci Technol 2002;104:762–71.

[11] Mestdagh F, De Meulenaer B, Van Peteghem V. Influence of oil degradation on acrylamide formation in a model system and in French fries. Food Chem 2007;100:1153–9.

[12] Mestdagh FJ, De Meulenaer B, Van Poucke C, Detavernier C, Cromphout C, Van Peteghem C. Influence of oil type on the amounts of acrylamide generated in a model system and in French fries. J Agric Food Chem 2005;53:6170–4.

[13] Matthäus B, Haase NU, Vosmann K. Factors affecting the concentration of acrylamide during deep-fat frying of potatoes. Eur J Lipid Sci Technol 2004;106:793–801.

[14] Williams JSE. Influence of variety and processing conditions on acrylamide levels in fried potato crisps. Food Chem 2005;90:875–91.

[15] Lim PK, Jinap S, Sanny M, Tan CP, Khatib A. The influence of deep frying using various vegetable oils on acrylamide formation in sweet potato (*Ipomoea batatas* L. Lam) chips. J Food Sci 2013;79:115–21.

[16] Kotsiou K, Tasioula-Margari M, Capuano E, Fogliano V. Effect of standard phenolic compounds and olive oil phenolic extracts on acrylamide formation in an emulsion system. Food Chem 2010;124:242–7.

[17] Jin C, Wu X, Zhang Y. Relationship between antioxidants and acrylamide formation: a review. Food Res Int 2013;51:611–20.

[18] CIAA. Acrylamide toolbox. 2011. http://www.fooddrinkeurope.eu/uploads/publications_documents/Toolboxfinal260911.pdf.

[19] Codex Alimentarius. Code of practice for the reduction of acrylamide in foods. 2009. Freely available at: http://www.codexalimentarius.net/input/download/standards/11258/CXP_067e.pdf.

[20] Belizt HD, Grosch W, Schieberle P. Food Chemistry, 4th revised and extended edition. Berlin Heidelberg: Springer Verlag; 2009.

CHAPTER 17

Relationship between Antioxidants and Acrylamide Formation

Ying Zhang[1,2], Cheng Jin[1]
[1]Department of Food Science and Nutrition, College of Biosystems Engineering and Food Science, Zhejiang University, Hangzhou, PR China; [2]Zhejiang Key Laboratory for Agro-Food Processing, Zhejiang University, Hangzhou, PR China

INTRODUCTION

As a reaction widely applied to the preparation of heat-processed foods in the food industry, Maillard reaction can produce hundreds of flavored compounds as well as hazardous chemical contaminants, among which acrylamide is one of the most famous and studied ones.

Acrylamide has been proven to be neurotoxic in animal models [58–60,67] and human–work-related exposure cases [5,15,32,33,38,45,46,64,65]. It can be metabolized to a chemically reactive epoxide, glycidamide, in vivo. Glycidamide may further react with proteins, including hemoglobin, or with deoxyribonucleic acid (DNA) to exhibit genotoxicity, as a mutagen and clastogen [67]. Also, acrylamide is classified by the International Agency for Research on Cancer (IARC) as "probably carcinogenic to humans (Group 2A)" for there is "sufficient evidence in experimental animals for the carcinogenicity of acrylamide" [43]. As for epidemiological studies regarding human cases, the research results are discordant. Both positive [11,42,86] and negative [54,71] results were reported or meta-analyzed. More specific methods to estimate the acrylamide exposure by measuring acrylamide biomarkers directly in the blood of humans are needed.

For the wide application of polyacrylamide in water purification, the European Union [26] set the parametric value in drinking water at 0.1 μg/L. However, in 2002, acrylamide was detected in heat-processed carbohydrate-rich foods at a level of mg/kg [81], which shocked the world and immediately attracted intensive research efforts.

Former studies have revealed the formation mechanism of acrylamide [63,77,78,99], the accurate analytical methods of acrylamide [9,73,93], as well as possible strategies to reduce the content in heat-processed foods [21,24,31]. The Confederation of the Food and Drink Industries of the EU (French: Confédération des Industries Agro-Alimentaires de l'UE; CIAA) created the "CIAA Toolbox" to help advise manufacturers and consumers what to do to minimize acrylamide hazard, which included agronomical factors (sugars and asparagine), recipe factors (fermentation, dilution and piece size, pH, other minor ingredients such as amino acids and calcium salts), processing factors (asparaginase,

Acrylamide in Food
http://dx.doi.org/10.1016/B978-0-12-802832-2.00017-6

thermal input and moisture, pretreatment, and finished product color), and final preparation factors (consumer preparation) for potato-based products [22].

Among all the reported approaches to control the acrylamide content in foods, the effect of antioxidants has not yet been elucidated satisfactorily. As a large group of substance in nature, the relationship between antioxidants and acrylamide formation deserves to be discussed. A broad overview and evidence-based understanding regarding the relationship between antioxidants and acrylamide formation was given by us in 2013 [44], however, with the publishing of newly conducted researches in the past two years, some points of view and hypotheses are updated in this chapter based on observations and results published recently.

DISCORDANT RESULTS AND QUESTIONS RAISED

In recent years, there has been an increasing interest of applying antioxidants to reduce acrylamide content in different matrices. However, both positive and negative results were obtained. Interesting phenomena where the same antioxidant exhibits opposite effects, or the antioxidative extract and its representative components exhibit opposite effects can be found in different researches.

Different Effects of Various Antioxidants on Acrylamide Formation

Rosemary, a spice with a well-known antioxidant property, had been proposed as one effective inhibitor of acrylamide. Acrylamide contents were lowered in fried potato slices when rosemary herb was added to either corn or olive oil [7]. The above finding was confirmed by Hedgaard et al. [39] who added aqueous rosemary extract, rosemary oil, and dried rosemary leaves to a bread model, which thus reduced the contents of acrylamide by 62%, 67%, and 57%, respectively, compared with bread without rosemary. Zhu et al. [98] applied 35 kinds of crude aqueous extracts of plants and 11 phenolic compounds to mitigate acrylamide in an asparagine–glucose (Asn–Glc) model system, finding that 34 of 35 plant extracts exerted reduction effect while nine phenolic compounds except for ferulic acid and hesperetin reduced acrylamide formation.

On the other hand, there was also a study which reported that grape seed extract had no effect on acrylamide formation in bakery products using a crust-like model [1]. Tareke [80] also found out that the addition of commonly used antioxidants like BHT, sesamol, and vitamin E to meat before heating even enhanced the formation of acrylamide.

Cheng et al. [19] tested six fruit extracts (from apple, blueberry, mangosteen, longan, and dragon fruit with white or red flesh) for their activities against acrylamide formation, and found that different fruit extracts have different effects. Apple extract demonstrated potent inhibitory effect. Extracts of blueberry, mangosteen, and longan did not show significant impact, whereas dragon fruit extracts enhanced acrylamide formation.

Discordant Effects of the Same Kind of Antioxidant on Acrylamide Formation

Even the same kind of antioxidant may exert opposite effects in different researches. One typical example is about the commonly consumed antioxidant ascorbic acid and its derivatives. Biedermann et al. [10] found a relatively weak reduction of acrylamide formation by the addition of ascorbic acid to a potato model. Similar results were obtained by [51] when using ascorbic acid and ascorbate in a cracker model. However, by investigating the effect of ascorbyl palmitate and sodium ascorbate, Rydberg et al. [74], found no effect on the net amount of acrylamide in a potato model.

Another example was about the use of two kinds of phenolic acids (caffeic and gallic acids) in two independent experiments [6,48]. Results obtained from the two experiments were completely contradicting both phenolic acids were noneffective in reducing acrylamide in the former one, but effective in the latter one.

Discordant Effects between Antioxidative Extract and Its Representative Components

We showed that addition of two natural antioxidants, extract of bamboo leaves (AOB) and extract of green tea (EGT), and also their representative components homoorientin and gallocatechin gallate (EGCG) to an Asn–Glc model system both significantly reduced acrylamide formation [97]. On the contrary, statistics showed that applying virgin olive oil (VOO) extracts to a potato model system and fresh potatoes led to an enhanced level of acrylamide, while commonly occurring phenols in VOO were proven to be effective in inhibiting acrylamide formation [49].

When the researches are comparatively studied together, the results seem confusing to some extent. Also, questions like "why such phenomena occur and what is the role of antioxidant capacity in acrylamide formation" are raised. In this chapter, we intend to summarize the related researches and their results, try to give comprehensive explanations for the discordant results and answer the question why antioxidants may influence the acrylamide content in different ways.

WHY DO DIFFERENT ANTIOXIDANTS HAVE DISCORDANT EFFECTS ON ACRYLAMIDE?

To answer this question, a quick review of the formation mechanism of acrylamide is needed. Shortly after the discovery of acrylamide in foods, it has been established that the major pathway for acrylamide formation in foods is Maillard reaction with free asparagine as the main precursor [63,77,78]. Asparagine was proven to provide the backbone chain of acrylamide formation by mass spectral studies demonstrating that all the three carbon atoms and the nitrogen atom are from asparagine [99]. However, when asparagine is heated alone, a very limited amount of acrylamide is formed [34], so it needs a carbonyl compound to accelerate its conversion to acrylamide. Classically, the carbonyl compound is an α-hydroxycarbonyl compound (e.g., reducing sugars like glucose or fructose). Subsequent researches showed that the carbonyl can come from multiple sources, including sugar fragments such as glyoxal, hydroxyethanal, and glyceraldehydes [3], some Maillard reaction products like 5-hydroxymethylfurfural [30] as well as specific lipid oxidized products like alkadienals and their analogous ketodienes and hydroperoxides [40]. Different antioxidants may participate in different reactions during the process of Maillard reaction, thus causing different effects toward acrylamide formation and elimination.

Reaction/Interaction with Asparagine

Structure–function analysis shows that the functional group in the α-position and γ-position to the carbonyl group might be the key factor to decide if the carbonyl is a reactive carbonyl for asparagine or not. Hydroxyl group in the α-position favors the rearrangement from azomethine ylide to the decarboxylated Amadori product to afford acrylamide. $\alpha, \beta, \gamma, \delta$-diunsaturated carbonyl compounds and α-dicarbonyl compounds are also preferred. So antioxidants that bear such functional groups may react with asparagines to yield acrylamide.

For example, as a natural antioxidant and colorant, curcumin is widely used in food industry. However, for the $\alpha, \beta, \gamma, \delta$-diunsaturated carbonyl groups in the structure, curcumin is expected to produce acrylamide after a reaction with asparagine. A study was conducted by Hamzalıoğlu et al. [37] in which curcumin was heated in binary and ternary model systems of asparagine–curcumin (Asn–Cur) and asparagine–curcumin–fructose (Asn–Cur–Fru). High-resolution mass spectrometry (HRMS) was used to identify the structure of the reaction intermediates and products, thus presenting the exact mechanism (Figure 1). The results showed that curcumin significantly increased acrylamide formation, though to a lesser extent compared with fructose, and acrylamide formation was also enhanced in the ternary model system to binary system of Asn–Fru. Curcumin can contribute to acrylamide formation under long-term heating as long as asparagine is present, since fructose reacts more rapidly with asparagine than curcumin.

For further understanding, Hamzalıoğlu and Gökmen [36] designed an experiment where five reactive carbonyl compounds (vanillin, dehydroascorbic acid, ascorbic acid, silymarin, and curcumin) bearing either $\alpha, \beta, \gamma, \delta$-diunsaturated carbonyl group, α-dicarbonyl group, or α-hydroxycarbonyl group were heated with asparagine, respectively. Results showed that all the five carbonyls could react with asparagine to form acrylamide. Comparing with fructose, the conversion efficiency of the carbonyl compounds was in the following order: vanillin > fructose > curcumin > silymarin > ascorbic acid > dehydroascorbic acid. Interestingly, the order is accordant with the order of melting points of the five compounds (low to high). Through this experiment, the theory that α-hydroxycarbonyl compounds, α-dicarbonyl compounds, and $\alpha, \beta, \gamma, \delta$-diunsaturated

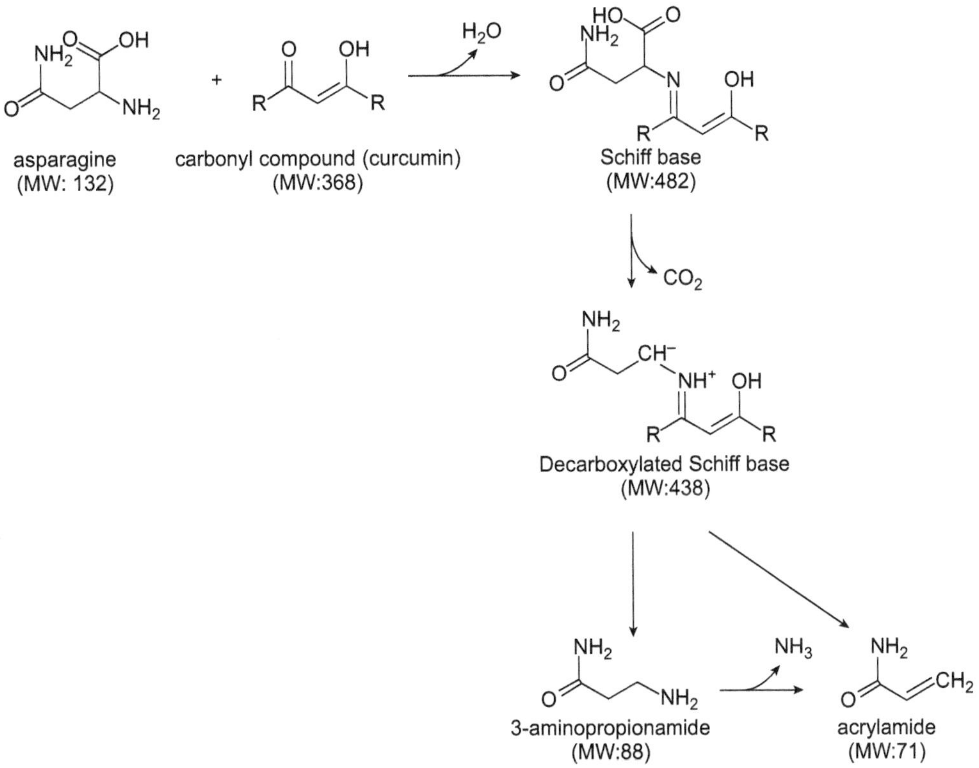

Figure 1 Proposed reaction pathway of curcumin and Asn leading to acrylamide formation confirmed by HRMS. *Adapted from Hamzalıoğlu et al. [37].*

carbonyls can convert asparagine to acrylamide is confirmed and proven. Also, an explanation that lower melting point leads to better contact of the antioxidant with the asparagine precursor, thus promoting acrylamide formation can be interpreted.

Otherwise, several types of phenolic antioxidants like tannins can precipitate amino acids through complexation [76]. Hence, an observed reduction effect of proanthocyanidins (condensed tannins) on acrylamide was assumed for this reason [98]. Many phyto-extracts containing tannins may reduce acrylamide in this way.

Reaction with Sucrose to Trigger Its Decomposition

5-Hydroxymethylfurfural [5-(hydroxymethyl)furan-2-carbalde-hyde, HMF] is a furanic compound which forms as an intermediate in the Maillard reaction [2] or from direct dehydration of sugars under acidic conditions (caramelization) during thermal treatments applied to foods [50]. Figure 2 depicts the main pathways for HMF formation in foods, which includes key intermediate 3-deoxyosone formation via 1,2-enolization and dehydration of glucose or fructose, and further dehydration and cyclization of 3-deoxyosone to yield HMF. Under dry and pyrolytic conditions, an alternative pathway to HMF formation

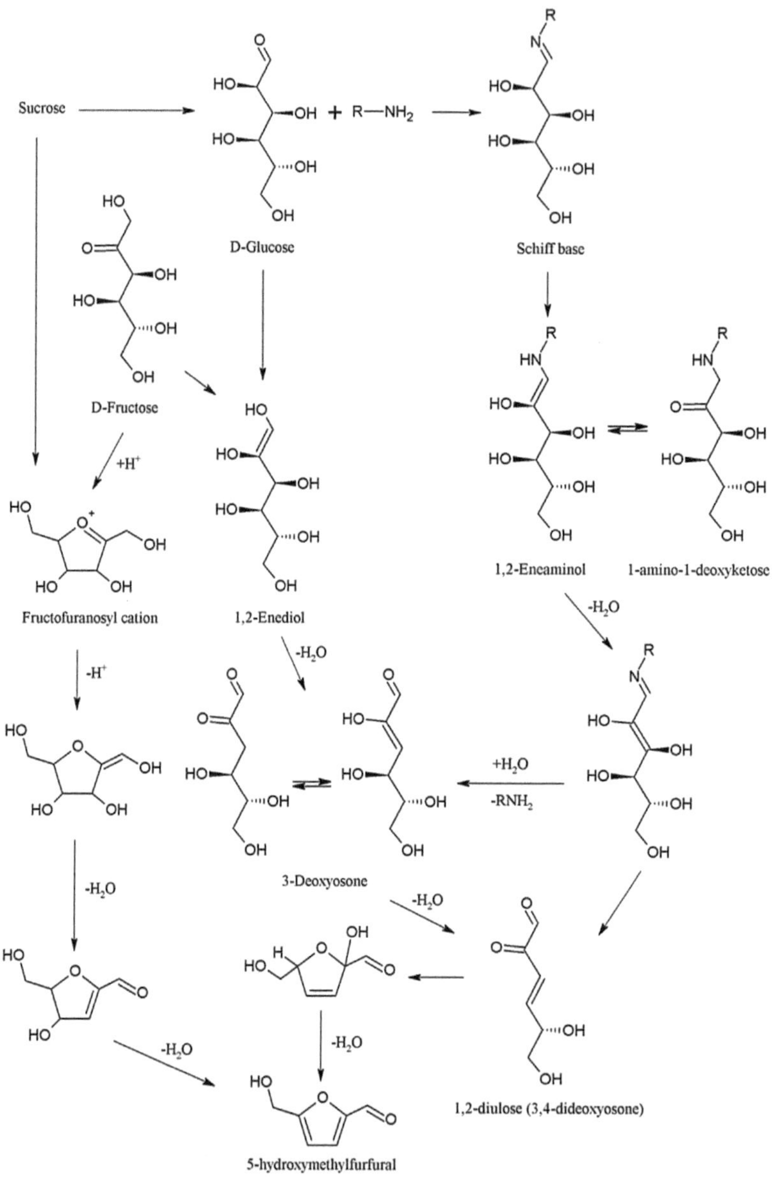

Figure 2 Proposed reaction scheme for the formation of 5-hydroxymethylfurfural. *Adapted from Capuano and Fogliano [16].*

from fructose and sucrose has been proposed. It involves the formation of a highly reactive fructo-furanosyl cation which can be effectively and directly converted to HMF [72].

Recently, it has been shown that HMF, which contains an $\alpha, \beta, \gamma, \delta$-diunsaturated carbonyl group, can convert free asparagine to acrylamide very efficiently during heating. The Asn–HMF model system generated more acrylamide than did the Asn–Glc model system, though the reaction rate was higher in the Asn–Glc system. HRMS analysis confirmed its role as a potent carbonyl accelerating acrylamide formation [30].

A corresponding research conducted by Kocadağlı et al. [47] claimed an observation of chlorogenic acid (CGA), a main phenolic constituent in green coffee, promoting sucrose decomposition to form HMF and thus increasing the acrylamide formation. Statistics obtained showed that the amount of acrylamide formed in the Asn–Suc–CGA model system was found to be 1.38 times higher than that in the Asn–Suc model system. Mass spectrometry confirmed the presence of 3,4-dideoxyosone and HMF as sucrose decomposition products, and conversion of CGA into caffeic acid and quinic acid during roasting of coffee. The authors believed that CGA can trigger sucrose decomposition to form HMF and 3,4-dideoxyosone, which both can react with asparagine to yield acrylamide. The result was confirmed in a recent research where CGA was found to increase acrylamide formation through promoting HMF formation [14].

Trapping of Maillard Reaction Intermediates

As stated above, reactive carbonyls heating with asparagine can form a series of intermediates, and finally leads to yield of acrylamide. During the process, several antioxidants, especially some specific flavonoids, were found to be able to trap Maillard reaction intermediates by C6 or C8 position of the A-ring.

Cheng et al. [20] tested the effect of a citrus flavonoid, naringenin, on the formation of acrylamide in chemical model systems under mild heating conditions. Results showed that naringenin significantly and dose-dependently inhibited the formation of acrylamide by about 20%–50% relative to the control. Structural information obtained by LC–MS/MS and nuclear magnetic resonance (NMR) showed that naringenin reacted with Maillard intermediates, forming two new derivatives 8-C-(E-propenamide)-naringenin and 6-C-(E-propenamide)-naringenin (Figure 3). In other words, naringenin scavenged asparagine-derived intermediates (also amide source of the reaction) at positions 6 and 8 of the A-ring, diverted them from the pathways leading to acrylamide formation, thus reducing the acrylamide content.

Totlani and Peterson [82] reported that epicatechin (EC) in aqueous glucose–glycine (Glc–Gly) model systems functioned as a carbonyl trapping agent of C2, C3, and C4 sugar fragments or key transient precursors of the Maillard reaction (e.g., glyoxal, methylglyoxal, acetol, and erythrose). The authors suggested that EC underwent electrophilic aromatic substitution reactions with these carbonyl compounds. This mechanism was further confirmed by NMR detection of an EC-methylglyoxal adduct product, which indicated that the C6 or C8 position of the flavanol (A-ring) and the carbonyl-bearing carbon of the aldehyde compound were the bonding sites [83]. EC exerts a trapping effect on sugar fragments or 3-deoxy-2-hexosulose intermediates and consequently inhibits Maillard products (including acrylamide) formation. It can be inferred that

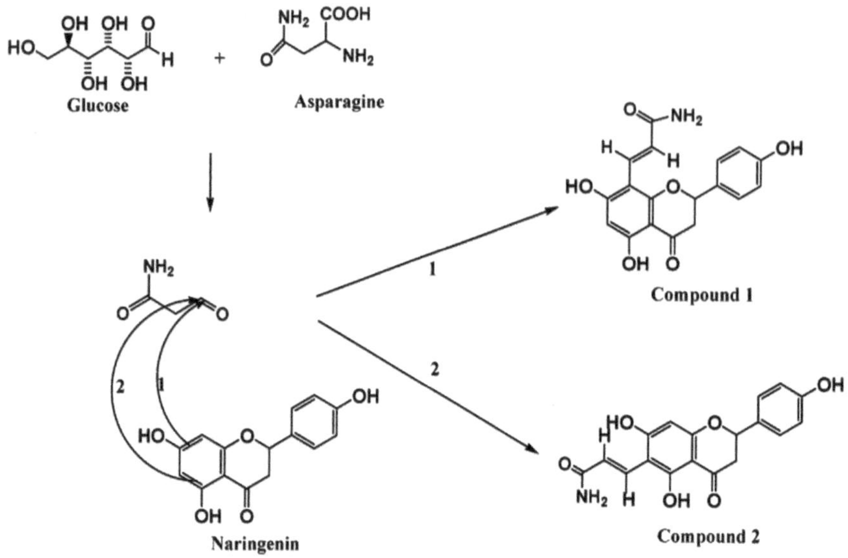

Figure 3 Proposed pathways for the scavenging of amide source to inhibit acrylamide formation by naringenin. *Adapted from Cheng et al. [20].*

flavonoids with similar structure may also scavenge carbonyl Maillard intermediates at the C6 or C8 position of the A-ring to inhibit acrylamide formation.

Destructive and Protective Reactions with Acrylamide

Once formed, the acrylamide content is not always constant. After it reaches the highest concentration, the acrylamide content decreases again due to exhaustion of one of the reactants and/or by the elimination of the acrylamide itself. It is obvious that acrylamide possesses two functional groups, an amino group and an electron-deficient vinylic double bond, making it available for a wide range of reactions, including nucleophilic and Diels–Alder addition and radical reactions [16].

Acrylamide may undergo Michael addition type reactions to the vinylic double bond with nucleophiles, including amino and thiol groups of amino acids and proteins. On the other hand, the amide group can undergo many reactions including hydrolysis, dehydration, alcoholysis, and condensation with aldehydes [28].

For example, Zamora et al. [88] proved that acrylamide can react with amino compounds by means of a Michael addition to form corresponding 3-(alkylamino)propionamides, which may reversibly produce acrylamide again by heating. Liu et al. [55] also revealed that glycine removes acrylamide by forming four kinds of Michael addition products with acrylamide before or after initial oxidation of glycine itself.

For further understanding, Hidalgo et al. [41] tested the removing effect of amino and sulfhydryl groups toward acrylamide using glutamic acid (Glu, with an amino group),

N-acetylcysteine (AcCys, with a sulfhydryl group), glutamic acid/*N*-acetylcysteine mixture, glutathione (GSH, with both amino and sulfhydryl groups), and cysteine (Cys, with both amino and sulfhydryl groups). Results showed that the acrylamide elimination decreased in the following order: Cys > GSH > AcCys ≫ Glu. It can be inferred that the sulfhydryl group can scavenge acrylamide, and the presence of amino group in the same molecule can encourage the effect of the sulfhydryl group. Also, the closer the position of the two groups, the stronger the positive interaction between them. When the two groups are present in different molecules, the positive interaction can only be observed when small concentrations are employed. Yuan et al. [87] also reported an inhibitory effect of allicin, a natural sulfhydryl-containing antioxidant, toward acrylamide.

Alternatively, another hypothesis of interaction between antioxidants and acrylamide was raised by Ou et al. [69] who related it to possible radical-induced elimination of acrylamide. Six kinds of antioxidants, three of which are relatively stable (not easily oxidized) and the rest of which are unstable were tested in both the acrylamide model system and Asn–Glc model system. Results showed that stable antioxidants (BHA, BHT, and TBHQ) had no effect or even promoted acrylamide formation, while unstable antioxidants (ferulic acid, EGCG, and vitamin C) slightly inhibited acrylamide. And when H_2O_2 was added to the acrylamide model system, or oxidized antioxidants were used in the Asn–Glc system, acrylamide formation was significantly reduced. The authors assumed that acrylamide suffered from radical-induced elimination, so antioxidants could help protect acrylamide against radicals by scavenging them. But if the antioxidant was unstable or could be easily oxidized, it might turn into its oxidized form to attack the acrylamide, resulting in reduced acrylamide content.

WHAT IS THE ROLE OF ANTIOXIDANT CAPACITY IN ACRYLAMIDE FORMATION?

It has been found that in lipid-rich systems, lipid oxidation products are able to convert asparagine into acrylamide. This conversion is favored by oxidized lipids having an $\alpha, \beta, \gamma, \delta$-diunsaturated carbonyl group, which comes from oxidation of unsaturated fatty acids and their corresponding esters. For example, lipid oxidation products were prepared via oxidation of methyl linoleate, the structures of which are shown in Figure 4 and numbered as compounds 2–9. Benzaldehyde (compound 10) was also included for comparison purposes. Heating experiments showed that compounds 2, 4, 7, and 9 were reactive to asparagine, and 2,4-decadienal (compound 7) showed the highest reactivity. While heating together with asparagine and glucose, compound 1 promoted the reaction by a synergism factor of 1.6, which may be due to the oxidation of compound 1 to form compound 2 and other primary and secondary oxidized products [89]. Capuano et al. [17] confirmed that lipid oxidation positively influenced the formation of acrylamide, especially in sugar-free systems where lipids became the main sources of carbonyls. Under this frame,

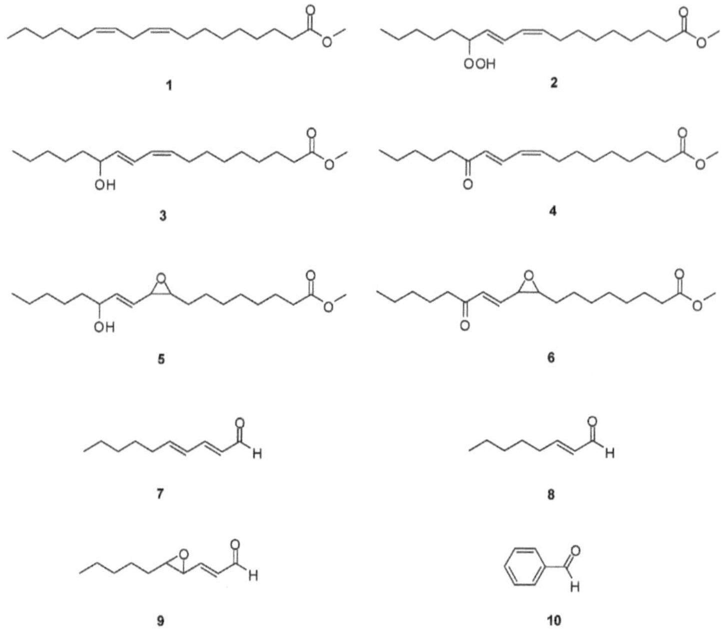

Figure 4 Chemical structures of the fatty acid derivatives employed. They were derived from the ω6 linoleic acid. **1**, Methyl linoleate; **2**, methyl 13-hydroperoxyoctadeca-9,11-dienoate; **3**, methyl 13-hydroxyoctadeca-9,11-dienoate; **4**, methyl 13-oxooctadeca-9,11-dienoate; **5**, methyl 9,10-epoxy-13-hydroxy-11-octadecenoate; **6**, methyl 9,10-epoxy-13-oxo-11-octadecenoate; **7**, 2,4-decadienal; **8**, 2-octenal; **9**, 4,5-epoxy-2-decenal. Also included is the chemical structure of benzaldehyde; **10**, which was employed for comparison purposes. *Adapted from Zamora and Hidalgo [89].*

antioxidants can limit the accumulation of carbonyls by preventing lipid oxidation, thus inhibiting acrylamide formation. This theory was proven by data from studies by Kocadağlı et al. [47], which showed that chlorogenic acid lowered both lipid oxidation product accumulation as well as acrylamide formation, comparing data from an asparagine–linoleic acid system and a system in which chlorogenic acid was added.

When it comes to antioxidant capacity (AOC) of the system, the results are controversial. Serpen and Gökmen [75] found a close linear correlation ($r=0.8322$, $n=36$) between acrylamide and total AOC in a potato matrix. It can be concluded that antioxidative compounds and acrylamide were formed in the similar stages of the Maillard reaction, and at similar rates. Similarly, a close relationship between the formation of acrylamide and AOC of Maillard reaction products [79] was demonstrated. Ciesarová et al. [23] also reported a close relationship between acrylamide content and AOC of spice extract before adding to a potato matrix. However, Bassama et al. [6] in their article stated that the AOC of the model system was not positively correlated to the mitigation of acrylamide formation. After adding caffeic acid at different levels, the AOC changed but no significant effect on the kinetics of acrylamide formation/elimination was observed Arribas-Lorenzo et al. [4] used three kinds of oil which were at high, intermediate, and low levels of phenolic compounds, individually. These different oil samples were used to make cookies, and the acrylamide content and AOC were determined. Results showed that as time passed, the AOC was always not significantly different among the oils, but the acrylamide content gradually showed significant difference.

Our recent research has shown that the correlation between AOC of the system (measured by multiple assays, and in form of ΔTEAC) and acrylamide inhibitory rate compared with the control is significant. What is more, a prediction model to estimate the inhibitory effect of certain flavonoids can be built based on the AOC of the food matrix [18,92]. Actually, the AOC of the final food matrix after heating consists of the following parts: antioxidant capacity caused by residue of the antioxidant added (if not all reacted or decomposed) and the antioxidant formed through Maillard reaction, mainly melanoidins. Melanoidins are brown anionic nitrogenous polymers formed during the final stage of the Maillard reaction [12]. It was known to us, long before the discovery of acrylamide as a product formed during the Maillard reaction, that melanoidins have strong antioxidant properties comparable to those of commonly used food antioxidants [53]. ΔTEAC, which is calculated by AOC of one specific sample minus AOC of the control group, can be interpreted as "additional AOC brought by adding antioxidant minus AOC removed by inhibition of the overall Maillard reaction caused by adding antioxidants (which correlates with inhibitory rate of the overall Maillard reaction)." On the other hand, acrylamide inhibitory rate may also correlate well with inhibitory rate of the overall Maillard reaction (given that acrylamide and melanoidins were formed in the similar stages of the Maillard reaction, and at similar rates). So it makes sense that ΔTEAC correlates well with the acrylamide inhibitory rate. This finding indicated that for regular antioxidants which do not react directly with asparagine, measuring ΔTEAC and acrylamide content in the control group might be a way to predict the acrylamide content, faster than regular determination.

In the same article [92], the structure–activity effect of certain flavonoids toward acrylamide inhibition was also discussed. Four flavone C-glycosides, that is, homoorientin (luteolin-6-C-glucoside), orientin (luteolin-8-C- glucoside), isovitexin (apigenin-6-C-glucoside), vitexin (apigenin-8-C-glucoside), and their corresponding flavone O-glycosides (apigenin-7-O-glucoside, luteolin-7-O-glucoside and luteolin-4′-O-glucoside) were tested in a microwave model system (Figure 5). Structure–activity analysis showed that compared with the effect of flavone O-glucosides, the inhibitory effect of flavone C-glucosides was obviously better, although they share the same aglycone structure (e.g., homoorientin > luteolin-7-O-glucoside), indicating the important role of phenol hydroxyl groups rather than alcoholic hydroxyl groups in the chemical structure of flavonoids for the reduction of acrylamide formation. The insertion of C-glucoside maintains the original number of phenol hydroxyls, while the insertion of O-glucoside removes one phenol hydroxyl. Also, the theory was confirmed by the fact that the reduction effect

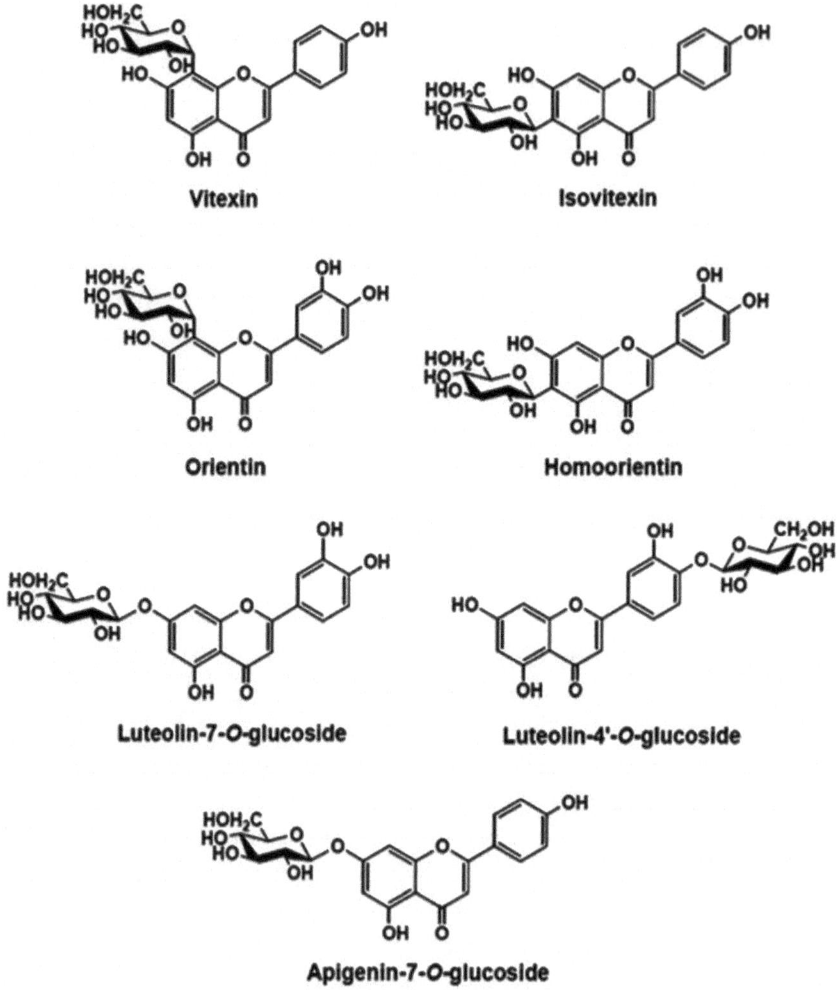

Figure 5 Chemical structures of the flavone C- and their corresponding O-glycosides used. *Adapted from Zhang et al. [92].*

of homoorientin and orientin, which bears one more phenol hydroxyl, was better than the effect of isovitexin and vitexin when their addition levels were equal. Furthermore, by comparing the effect of luteolin-7-O-glucoside and luteolin-4′-O-glucoside, we can see that although they have the same number of phenol hydroxyls, their effectiveness are not the same. Luteolin-7-O-glucoside, which contains an *o*-dihydroxyl (3′,4′-diOH) structure in the B ring, was more effective in inhibiting acrylamide. These results showed that both the number and position of the phenol hydroxyl functional groups play an important role in the ability of flavone glycosides to inhibit the formation of acrylamide.

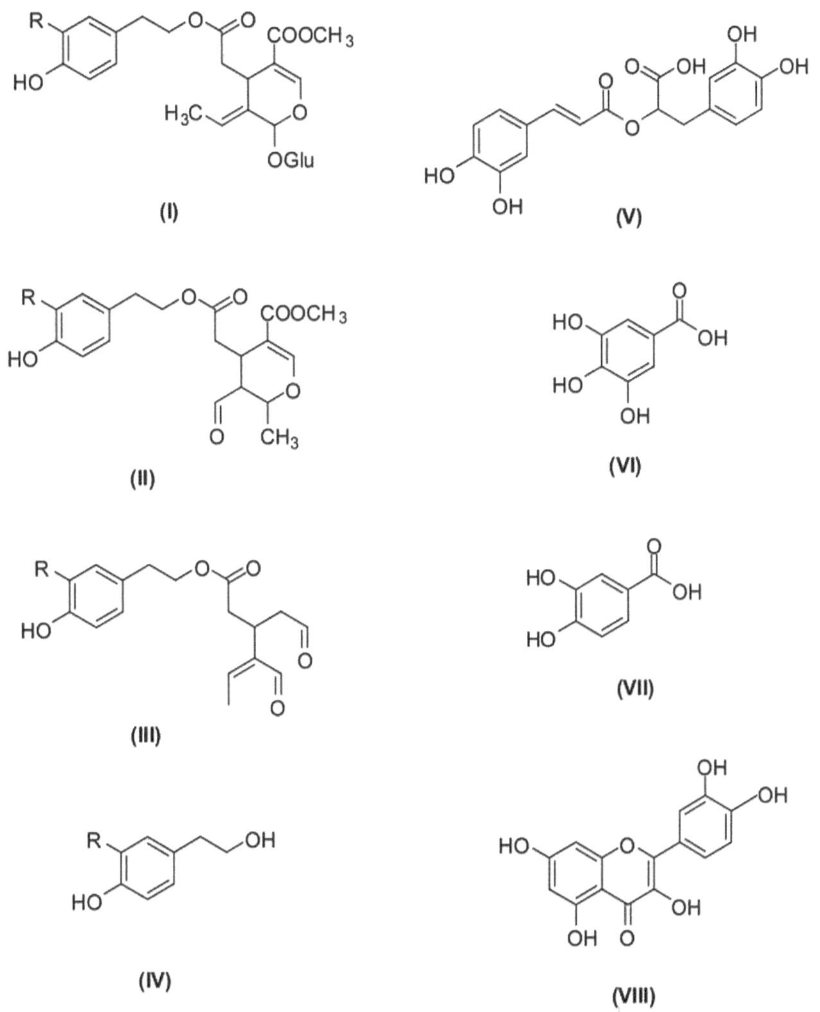

Figure 6 Structure of major polar phenolics occurring in virgin olive oil and oregano. Virgin olive oil secoiridoids: R=OH: (I) oleuropein, (II) aldehydic form of oleuropein aglycone (AFOA) (III) dialdehydic form of decarboxymethyl oleuropein aglycone (DAFOA), and (IV) hydroxytyrosol, R = H: (I) ligstroside, (II) aldehydic form of ligstroside aglycone (AFOA) (III) dialdehydic form of decarboxymethyl ligstroside aglycone (DAFOA), and (IV) tyrosol. Oregano phenolics: (V) rosmarinic acid, (VI) gallic acid, (VII) protocatechuic acid, and (VIII) quercetin. Adapted from [49].

It was reported that applying VOO extracts to a potato model system and fresh potatoes led to an enhanced level of acrylamide, while applying oregano extracts to the same systems led to a reduced level [49]. Now that the reaction conditions were the same, the authors attributed it to the different compositions of phenolic compounds in VOO and oregano extracts. As shown in Figure 6, I–IV are major phenolics occurring in VOO, V–VIII are major phenolics occurring in oregano. It was interpreted that phenolic compounds occurring in VOO differ from oregano phenols as they contain more aldehyde groups in the side chain, which may undergo reactions with asparagine to form acrylamide. Given that phenol hydroxyls especially *o*-dihydroxyls are important structures, we can also interpret the results in the way that compounds V–VIII inhibit acrylamide because they contain more phenol hydroxyls and a lot of them are ortho-diphenolic compounds. Such hypothesis can be confirmed by Napolitano et al. [66], who used four different kinds of VOO with different phenolic profiles to prepare fried potato crisps. Ravece olive oil, which contains the highest content of ortho-diphenolics (3,4-DHPEA-EDA and 3,4-DHPEA-EA), produced the lowest amount of acrylamide.

Since the *o*-dihydroxyphenolic structure is highly linked to antioxidant capacity, correlation between AOC (ΔTEAC) and acrylamide inhibition in systems where the added antioxidant does not react directly with asparagine can be expected. However, it may be influenced greatly by the reaction condition, and the exact mechanism still needs more in-depth studies.

WHY DOES THE SAME KIND OF ANTIOXIDANT EITHER INHIBIT OR ENHANCE ACRYLAMIDE IN DIFFERENT RESEARCHES?

The answer to this question will be discussed in aspects of reaction conditions, polarity of the antioxidant/matrix, and concentration of the antioxidant.

Reaction Conditions

The reaction condition here in the case of Maillard reaction is a multiple concept which includes the reaction matrix, the pH value, moisture, and the overall heat treatment. All these factors have to be taken into consideration discussing the relationship between antioxidants and acrylamide formation, because they are able to affect either the final acrylamide content, or the ways how antioxidants interact with the reactants.

The influence of temperature and heating time on acrylamide formation has been repeatedly demonstrated and results indicated that the acrylamide amount increased with processing temperatures [74]. But acrylamide levels were reduced under high temperature combined with prolonged heating time, due to the elimination process of acrylamide [25]. It was also clarified that acrylamide elimination did not occur at low temperatures, even after prolonged heating times [13].

The heat treatment may also affect the status of antioxidants since some antioxidants with poor heat stability might have decomposed under high temperature, thus being unable to protect lipids from oxidation, or take part in the Maillard reaction. What is more, some heat-sensitive or easily oxidized antioxidants may transform into other forms to exert different effect toward acrylamide formation. For example, Cai et al. [14] heated chlorogenic acid together with polyphenol oxidase to obtain quinone derivative of chlorogenic acid (which may also be formed with sufficient oxygen present), and found a totally opposite effect compared with original chlorogenic acid.

Researches also indicated that the initial pH value and the moisture content of the reaction system may also remarkably influence acrylamide formation. For example, Weisshaar [85] reported that in an Asn–Glc model system, the acrylamide content significantly increased when the pH was beyond 7.0, and the highest acrylamide content was found when the pH was at 8.0. Mestdagh et al. [61] found that acrylamide content reached the highest point between pH 7.0 and 7.5 in homogeneous potato powder mixture systems with different initial pH values. Similar phenomenon was also observed by Yuan et al. [87]. When the initial pH values of the aqueous model Asn–Fru and Asn–Glc systems varied from 5.0 to 7.5, the acrylamide content increased, and when the pH passed 7.5, the acrylamide content decreased with the increase of the pH values from 7.5 to 9.0. It can be concluded that the pH values between 7.0 and 8.0 are the most suitable range for acrylamide to form. As a result, we can modify and maintain the pH value in this range to obtain higher acrylamide response when we are setting up model systems, and we can adjust the pH to a lower value to reduce the formation of acrylamide in industrial applications. What is worth noting is that the stability of a specific antioxidant may vary under different pH values, and that the tolerance of different antioxidants toward severe pH conditions is also different. For instance, [29] revealed that caffeic, chlorogenic, and gallic acids are not stable to relatively high pH (the contents were reduced even under pH of 4.0) and that the pH- and time-dependent spectral transformations are not reversible. On the other hand, (−)-Catechin, (−)-epigallocatechin, ferulic acid, rutin, and trans-cinnamic acid resisted major pH-induced degradation. For unstable phenolic acids like caffeic, chlorogenic, and gallic acids, the situation is a little complicated. We know that acidic condition is good for phenolic acids, but acidic condition itself may inhibit the formation of acrylamide. So there is a dilemma in the situation when investigating phenolic acids. Such findings may help explain why certain phenolic acids may behave differently among researches where the environmental pH values are different.

As for moisture content, Capuano et al. [17] found that there was a large difference in acrylamide contents in two different food systems with 4% and 16% water, respectively. They attributed it to the different time–temperature profiles of the heating procedure caused by the different water contents. In the model system with 4% of water, both on the surface and in the core, the temperature reached the oven temperature more rapidly than that with 16% of water. As a consequence, the more the initial water content, the more the surface and the core time taken to reach the optimum acrylamide formation condition; thus, acrylamide formation will be suppressed. Also, water content may affect the distribution and the status of the antioxidant applied, as well as its combination and reaction with acrylamide or its intermediates.

In order to study the scientific problems under a simplified condition, scientists have set up different model systems, for example, dry systems (oven conditions), oily systems (frying conditions), semidry systems (baking conditions), emulsion systems (multiphase system), and aqueous systems. Mainly, matrix and moisture content have become the key factors to distinguish different systems. Also, there are reaction systems made of food stuffs, such as potato, cookies, bread, and crust. As far as we are concerned, model systems might be more suitable for mechanism analyzing since food systems are much more complicated, while food systems are better for application experiments. Choosing either the model system or the food system may lead to different results. Kotsiou et al. [48] compared the effect of olive oil extracts on acrylamide formation in both emulsion and potato systems, and results showed that the potato system had a significant higher level of acrylamide than the emulsion system, but the increase effect of olive oil extracts on acrylamide was, however, remarkably higher in the emulsion system. Oral et al. [68] also found a discordant effect of oleuropein toward acrylamide formation in a Asn–Fru model system and in a biscuit system.

Furthermore, the effect of the same antioxidant may still differ among different model systems. As mentioned above, two kinds of phenolic acids (caffeic and gallic acids) were tested in an aqueous system [6] and an emulsion system [48], respectively. Results showed that both phenolic acids were effective in reducing acrylamide in the emulsion system but were non-effective in the aqueous system. With the understanding that oxidized lipids produce reactive carbonyls to react with asparagine, the key may be oil (lipids) which is absent in the aqueous system but present in the emulsion one. An alternative possibility is that it was the difference in pH value (pH was adjusted to 5.8 in the aqueous system, but not adjusted in the emulsion system) that influenced the stability of phenolic acids.

Polarity of the Solvent/Matrix and the Antioxidant

Zeng et al. [90] investigated the effects of 12 kinds of vitamins on acrylamide formation, water-soluble vitamins in a chemical model system (aqueous model made of asparagine, glucose, and water) and a food system (fried snack model made of 64.5% wheat flour, 1.0% glucose, 0.5% asparagine, 0.5% testing vitamin, and 34.0% water) while fat-soluble ones only in the latter. The results showed that several vitamins, including thiamin (V_{B1}), nicotinic acid (V_{B3}), biotin (V_{B7}), pyridoxamine (V_{B6}, PM), pyridoxine (V_{B6}, PN), and L-ascorbic acid (V_C) strongly reduced the formation of acrylamide. None of the four fat-soluble vitamins (retinol acid, V_A; α-($\pm$)-tocopherol, V_E; ergocalciferol, V_D, and phylloquinone, V_K) was effective in suppressing the formation of acrylamide. V_D even enhanced acrylamide formation by more than 20% compared with the control. However, a relevant research conducted by us [52] showed that V_E in a cookie system (made of 29.2% shortening, 20.3% sugar, 0.6% cake oil, 8.9% egg, 40.5% wheat flour, 0.4% baking soda, and 0.1% V_E) performed a 71.2% mitigation of acrylamide. A probable explanation for the apparent discrepancy might be the difference in the choice of solvent for antioxidants. Water, the solvent in the fried snack model, is only suitable for the water-soluble antioxidants, while eggs in the cookie model are suitable for both water-soluble and fat-soluble antioxidants. As we all know, it is necessary for the antioxidants to combine adequately with the substrate to exert its ability. Thus, fat-soluble antioxidants showed their potential to inhibit acrylamide in the cookie model rather than the fried snack model. These results show the importance of choosing a suitable antioxidant for a specific food system in practice.

Concentration of the Antioxidant

Kotsiou et al. [49] treated potato model system with different concentrations of VOO phenolic extracts, and found out that the lowest concentrations resulted in a reduction of 15% ($p < 0.05$), while at intermediate concentrations an increase up to 21% ($p < 0.05$) on acrylamide formation was observed. Finally, a remarkable increase of 49% was induced by the highest concentration. Also, similar phenomenon occurs in ascorbic acid cases where at low concentrations (0.2%, 0.5%, 0.8%, and 1.2%), ascorbic acid inhibited acrylamide formation, with a highest inhibitory rate of 58%, but the increase of ascorbic acid concentration to 1.5% promoted acrylamide formation by more than 90% in an Asn–Glc model system [87]. Such findings suggest that the effect of antioxidant toward acrylamide is not only determined by the type but also the concentration of the antioxidants, which makes the case more complicated.

We demonstrated in an experiment that two kinds of natural antioxidants "AOB" and "EGT," of which the main functional ingredients were flavonoids and phenolic acids, could significantly reduce the acrylamide formed in an Asn–Glc microwave heating model system and could achieve a maximum reduction rate when the addition levels of both AOB and EGT were at 10^{-6} mg/mL [95]. Regardless of whether the addition level went up or down, the mitigation effect decreased. Similar results were gained when applying the representative flavonoids in AOB [92]. A non-linear concentration-dependent relationship was found, and the maximum inhibitory effect was observed when the addition level of all flavone glycosides was at 10^{-9} mol/L.

These phenomena might be attributed to the fact that certain antioxidants may participate in multiple reactions due to the complexity of the matrix and the functional groups in the compound. The predominant role or reaction undertaken may be determined by the concentration of the antioxidant. Also, for antioxidants like ascorbic acid that contains reactive carbonyls that can react with asparagine to form acrylamide, the concentration of precursors and the antioxidant might be very important. If asparagine is not excessive, certain reactive carbonyls may competitively react with asparagine and reduce the final acrylamide content for its lower conversion efficiency compared with fructose or glucose. But if asparagine is sufficient, all the reactive carbonyls can be consumed to produce products including acrylamide.

WHY DO SOME PHYTO-EXTRACTS AND THEIR REPRESENTATIVE COMPONENTS BEHAVE DIFFERENTLY?

Research results showed that the standard compounds of commonly occurring phenols in VOO (tyrosol, oleuropein, and 4-hydroxyphenyl acetic acid) were effective in inhibiting acrylamide formation, in contrast to VOO phenolic extracts' behavior which enhanced acrylamide formation [49]. It may be ascribed to the preparation procedure of the phenolic extract. Oleuropein is hydrolyzed during VOO extraction to form dialdehydic derivatives. As mentioned before in this article, dicarbonyl groups may increase the formation of acrylamide as reactive carbonyls, thus preparative methods of the antioxidant may influence acrylamide formation by changing the composition of the antioxidative extract during extraction.

CONCLUSIONS AND OUTLOOKS

In recent years, there have been significant advances in the research field of acrylamide including its formation mechanism and mitigation strategies. Antioxidants have been used extensively to mitigate acrylamide in model and actual food systems. However, there are confusing results in the literature, where some studies claim mitigation while others declare no effect or even an increase. This may be attributed to the ability of antioxidants with different structures or functional groups to react with acrylamide precursors, with intermediates of the reaction or with acrylamide itself, leading to either reducing or promoting effects.

Figure 7 summarizes the possible reactions antioxidants may take part in. As shown in the figure and discussed above, a reactive carbonyl compound is needed to accelerate the conversion of asparagine to acrylamide. The carbonyl compound can come from ingredients of the food matrix, from sucrose decomposition, from Maillard reaction, from lipid oxidation and from the antioxidant added. Among these pathways to form the so-called "reactive carbonyl pool," specific antioxidants can trigger sucrose decomposition (marked as (1)) to promote acrylamide formation, trap specific Maillard reaction products (marked as (2)) to inhibit acrylamide formation, prevent lipids from oxidation (marked as (3)) to lower acrylamide content or even react with asparagine directly to form acrylamide. Also, some antioxidants can cause precipitation of asparagine, which clearly mitigates acrylamide formation (marked as (4)). As the Maillard reaction proceeds, antioxidants of some specific structures can react with the intermediates, thus altering the reaction pathway to reduce acrylamide formation (marked as (5)). Acrylamide may undergo Michael addition type reactions with antioxidants bearing nucleophilic groups to be eliminated (marked as (6)). Either protective or destructive reactions can be expected from antioxidants or their oxidized forms against acrylamide (marked as (7)). For better understanding, in Table 1, the specific structures or compounds and their action sites in Figure 7 are summarized. Since reaction (7) has not yet been confirmed, it is not listed in Table 1.

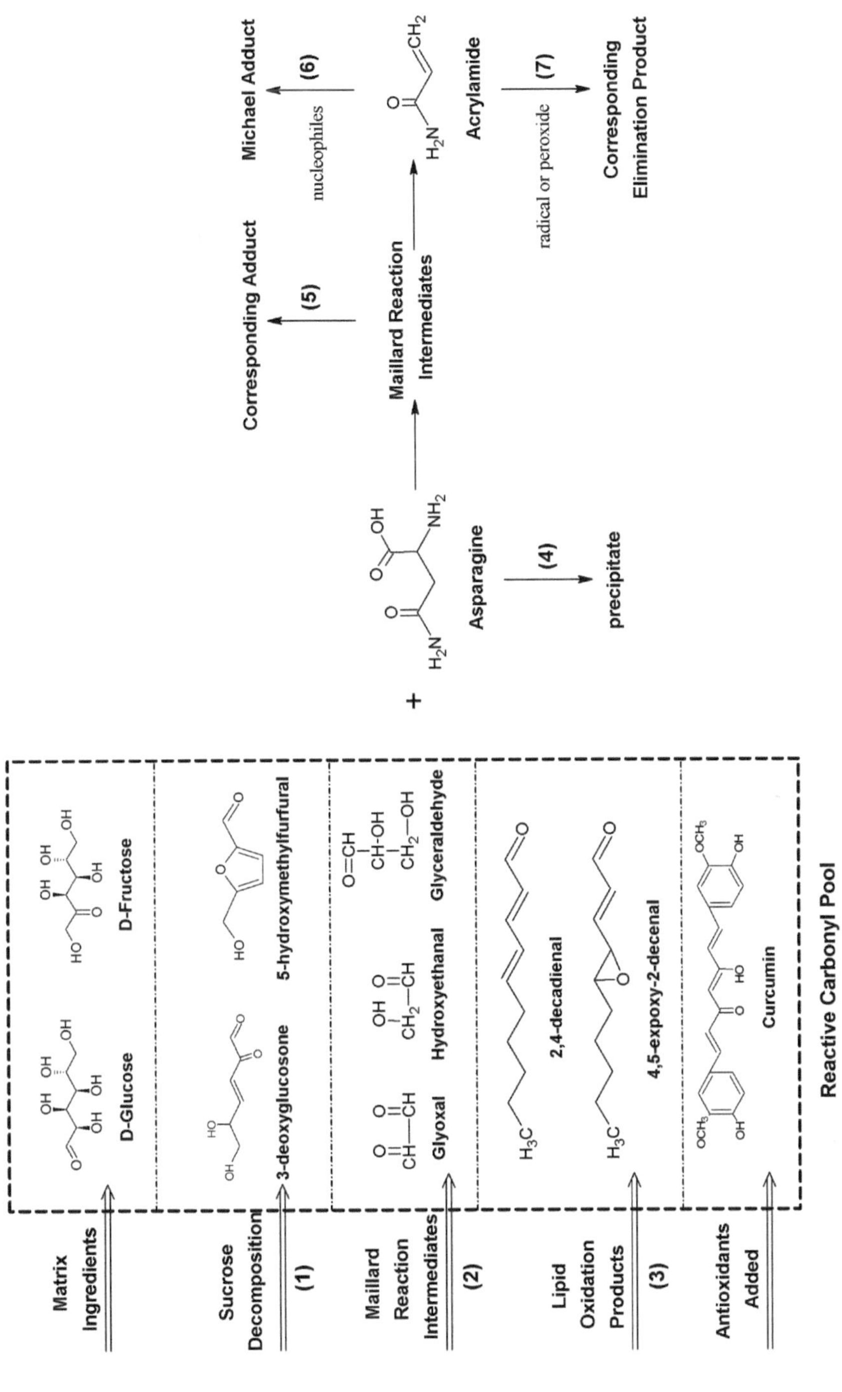

Figure 7 Reactive carbonyl pool from various carbonyl sources and possible reactions' positions for antioxidant (marked as (1)–(7)).

Table 1 Certain structures/compounds and their action sites and outcomes

Structure or specific compound	Possible action site in Figure 7	Outcome
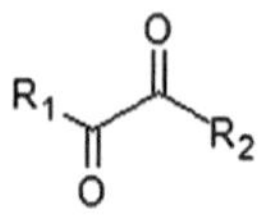 α-hydroxycarbonyl group	React with asparagine	Formation of acrylamide
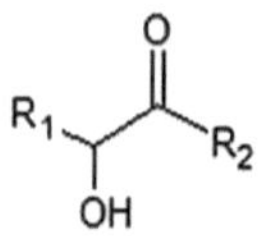 α-dicarbonyl group	React with asparagine	Formation of acrylamide
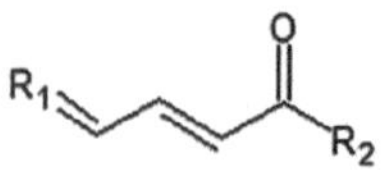 α, β, γ, δ-diunsaturated carbonyl group	React with asparagine	Formation of acrylamide
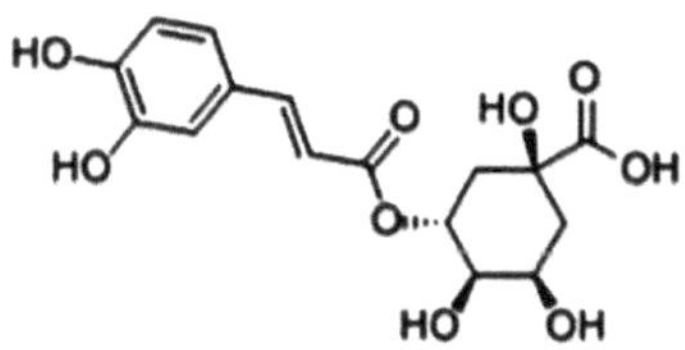 Chlorogenic acid	(1) (3)	Promotion Inhibition
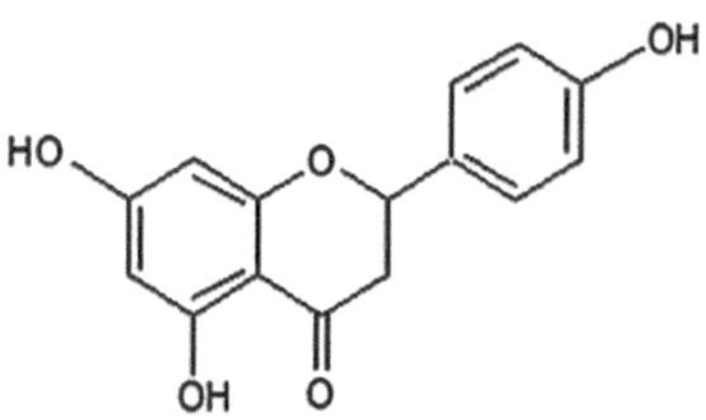	(2) and (5)	Inhibition
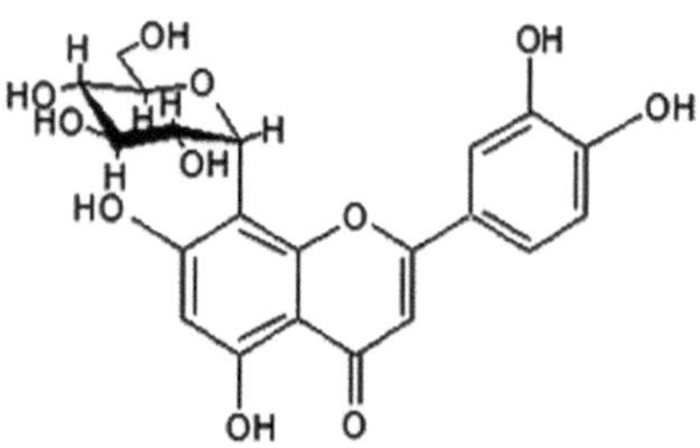 Flavonoids and glucoside	(3)	Inhibition

Table 1 Certain structures/compounds and their action sites and outcomes—cont'd

Structure or specific compound	**Possible action site in Figure 7**	**Outcome**
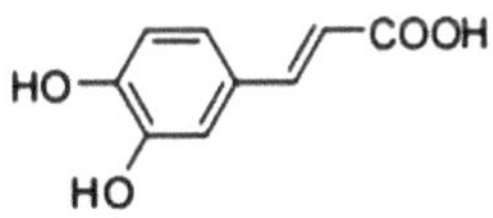 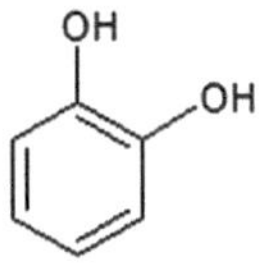 Phenolic acids and polyphenols	(3)	Inhibition
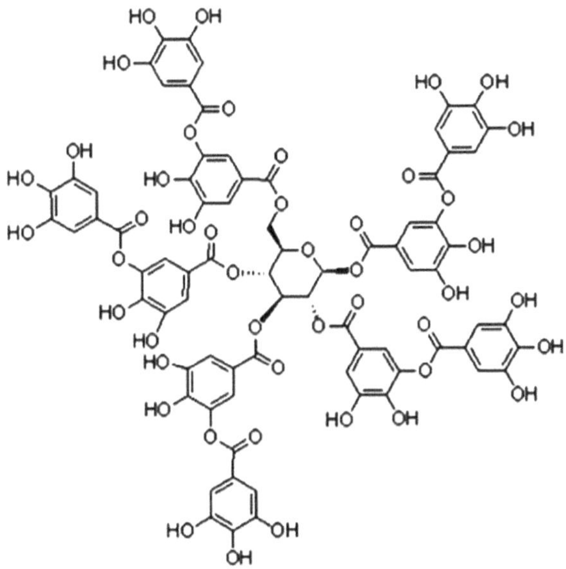 Tannins	(4)	Inhibition
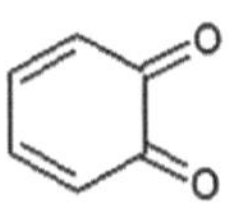 Quinones	(5)	Inhibition
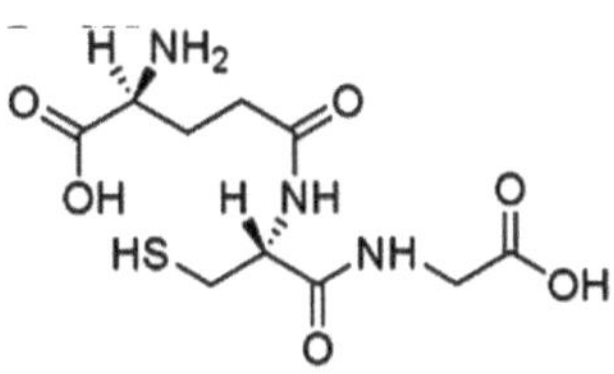 Nucleophilic groups like amino and sulfhydryl groups	(6)	Inhibition

Due to the diversity of functional groups in different antioxidants and the complexity of the food matrix, the antioxidant may participate in several reactions and exert multiple effects at the same time. For example, as a widely occurring and also frequently used natural phenolic acid, chlorogenic acid can influence acrylamide in different ways. It can inhibit acrylamide due to its acidity and its ability to reduce lipid oxidation, also, it can be transformed to its quinone derivative form and reduce acrylamide formation by quinone–amine interactions with key intermediates, including amine groups such as 3-aminopropionamide (3-APA). On the other hand, research showed that chlorogenic acid can promote acrylamide by triggering sugar decomposition to form HMF, decreasing activation energy for conversion of 3-APA to acrylamide and reducing radical-induced acrylamide elimination [14]. As the final result of several positive and negative reactions, acrylamide content depends on the predominant reaction antioxidants and substrates undergo, which may change when the concentration of the antioxidant varies. What is more, the reaction conditions can dramatically influence the final content of acrylamide, making it harder to compare results obtained from different laboratories.

The preparation procedure of antioxidative extract, which often includes heat-assisted extraction or concentration steps, may change the composition of the extract, thus making it different from the raw material or the representative component, especially for heat unstable antioxidants. Also, since the extract is relatively more complicated compared to the pure compounds, the effect of co-extracts and the interaction between representative components and the co-extracts should be taken into consideration.

The relationship between antioxidant capacity and acrylamide remains controversial. Antioxidants can prevent lipid oxidation to limit carbonyl accumulation, by its antioxidant capacity. Recent research showed a close correlation between ΔTEAC and the inhibitory rate of acrylamide, although there are also published papers claiming no correlation between AOC and acrylamide. However, the exact mechanism of how antioxidant capacity influences acrylamide formation still needs more in-depth studies.

In our opinion, future studies should focus on: finding other reactive structure patterns or functional groups in acrylamide formation and reduction; mechanism study regarding antioxidant capacity and acrylamide formation/elimination; foundation and optimization of standard reaction conditions to compare results among laboratories; and pilot-scale or production line-scale experiments using antioxidant as acrylamide inhibitor and simultaneous control of multiple hazards in the Maillard reaction. The application of HRMS and NMR has made it possible to identify the structure of reaction intermediates and products. Therefore, specific pathways and mechanisms can be raised based on these results.

Phenolic compounds have been widely used in recent years as phytochemicals having antioxidant and bioactive properties, to prevent lipid oxidation as well as reduce acrylamide. However, the application has been severely limited due to the low solubility in the fatty phase. Since the usage of synthetic antioxidants such as BHA, BHT, and TBHQ is also limited due to health concerns, scientists have to pay more attention to natural oil-soluble antioxidants. Rosemary extract presented its potential for inhibiting acrylamide and stabilizing the oil [84]. Also, our latest studies showed that oil-soluble AOB (AOB-o) obtained by acylation of water-soluble AOB (AOB-w) is a promising choice for its potent ability to protect oil from oxidation [56,57] as well as to inhibit acrylamide formation (data not published yet). As an antioxidant additive officially approved by the National Health and Family Planning Commission of the People's Republic of China, AOB can be used in 16 categories of foods. So far, AOB was proven to effectively mitigate the potential risk of acrylamide in potato crisps and French fries [91], fried chicken wings [94], cookies [52], and oriental fried bread sticks [96]. Potent acrylamide inhibition and oil stabilizing can be expected when applying the two forms of AOB preparations both in water phase and oil phase simultaneously.

SYNOPSIS OF ANALYTICAL TECHNIQUES

Acrylamide

Acrylamide can be determined by HPLC, GC–MS, or LC-MS/MS methods; the LC-MS/MS method was recommended by the US Food and Drug Administration (FDA) for its sensitivity and accuracy [27]. Briefly, ground food sample is spiked with $^{13}C_3$-labeled acrylamide, extracted with water for 20 min on a rotating shaker. The extract is centrifuged and spin filtered and then cleaned by OASIS HLB SPE cartridge and Varian Bond Elut-Accucat SPE cartridge before LC–MS/MS analysis.

LC–MS/MS analysis is based on collision-induced dissociation (CID) of the target compound and detection of specific ion pairs by a mass analyzer to quantify the target compound.

Antioxidant Capacity

The determination of antioxidant capacity can be achieved by several assays. The most commonly used assays are: the DPPH assay [62], the ABTS assay [70], and the FRAP assay [8].

Substrates

Simultaneous analysis of asparagine, sugars, and acrylamide contents can be achieved according to Zhang et al. [93], by using LC-MS/MS. The two-step simple pretreatment procedures included the addition of isotope internal standards $^{15}N_2$-asparagine, $^{13}C_6$-glucose, and D_3-acrylamide, followed by appropriate dilution with the mobile phase and filtration. Analytes were separated on a Hypercarb column and monitored by MS/MS.

HMF

HMF can be determined according to Kocadağlı et al. [47]. Briefly, ground food sample was extracted with 10 mM formic acid in triple stages (10-, 5-, and 5 mL) by vortexing for 3 min. The co-extracted colloids in the combined extract were precipitated by Carrez clarification. The mixture was centrifuged and 2 mL of the supernatant was passed through a 0.45-μm nylon syringe filter into an auto-sampler vial prior to HPLC-PDA analysis.

KEY FACTS

- There is an interesting term called "Antioxidant Paradox," which is used to describe paradoxical results obtained when studying the function of antioxidants against health threats like cancer in human cases [35]. For example, people with diets rich in fruit and vegetables have a decreased chance of developing cancer and an increase in the concentration of β-carotene in the blood. Supplements of β-carotene, however, do not have an anti-cancer effect, rather it is the opposite in smokers. Fruit and vegetable consumption decreases the amount of free-radical damage to DNA in the human body (a risk factor for cancer development), but supplements of ascorbate, vitamin E, or β-carotene do not decrease DNA damage in most studies. What is more, intake of vitamin C below the recommended daily allowance is associated with increased free-radical damage to DNA but, paradoxically, so is the supplementation with high-dose vitamin C. These cases are similar to what we are facing when it comes to acrylamide: positive and negative effects, discordant results among studies, nondose-dependent functions and so on. Our mission, however, is trying to find a rule and a reasonable explanation for the results. Obviously, the work is not yet complete. As more and research efforts are put into this area, we are sure that the mystery will be solved very soon.
- A wide sequence of spectrophotometric assays has been adopted to estimate the antioxidant capacities in foods, including 2,2′-diphenyl-1-picrylhydrazyl (DPPH), 2,2-azinobis (3-ethyl-benzothiazoline-6-sulphonic acid) (ABTS), ferric reducing antioxidant power (FRAP), the oxygen radical absorption capacity (ORAC), the total radical absorption potentials (TRAP), and the photochemiluminescence (PCL). The principles of most of the above assays use the generation of a redox-active or colored radical coordination compound as well as the detection of scavenging radicals or reducing redox-active compounds by the spectrophotometer. The above assays can mainly be divided into two alternative reaction mechanisms: single electron transfer pathway (e.g., DPPH, ABTS, and FRAP) and hydrogen atom transfer pathway (e.g., ORAC and TRAP). To compare results obtained from different laboratories under different conditions, the antioxidant capacity is often described as "Trolox equivalent antioxidant capacity" (TEAC), using Trolox as the calibration reference. However, the outcomes of antioxidant capacity analysis may vary between assays when applied to food matrices; hence combining the outcomes of multiple assays is recommended.

MINI DICTIONARY

Antioxidants These are substances that slow the rate of oxidation reactions. They exert their antioxidant capacity by either removing oxygen free radicals or chelating metal ions that which catalyze oxidation reactions. There are naturally occurring antioxidants like vitamin C (ascorbic acid), vitamin E (tocopherol), and β-carotene, and also artificial antioxidants like butylated hydroxytoluene (BHT) and butylated hydroxyanisole (BHA).

HMF 5-Hydroxymethylfurfural, a harmful furanic compound forms in the Maillard Reaction or through caramelization of sugars, also proven to be able to react with asparagine to form acrylamide.

Melanoidins These are brown, high molecular weight heterogeneous polymers that are formed through the Maillard reaction at high temperatures and low water activity. Melanoidins are commonly present in foods that have undergone some form of non-enzymatic browning, such as fried potato products, bread crust, bakery products, and coffee.

Trolox 6-hydroxy-2,5,7,8-tetramethylchroman-2-carboxylic acid, a water-soluble analog of vitamin E. Trolox is often used as a calibration reference in antioxidant capacity measurement. The results are often expressed as TEAC (Trolox equivalent antioxidant capacity) to remove environmental error and enable cross-laboratory comparison.

AOB Antioxidant of bamboo leaves (AOB). It is an extract obtained from bamboo leaves, which mainly contains flavonoids and phenolic acids. AOB can be classified as water-soluble (AOB-w) and oil-soluble (AOB-o) types. The AOB-o is obtained by chemical or enzymatic acylation of AOB-w.

SUMMARY POINTS

- Different kinds of antioxidants with different structures or functional groups can react with acrylamide precursors, with intermediates of the reaction, or with acrylamide itself and lead to either reducing or promoting effects.
- Reaction conditions (including matrix, pH, moisture, overall heat treatment, and reaction time) can influence acrylamide content and the status of antioxidants dramatically. The fact that the same kind of antioxidant behaves differently in different studies might be due to the different reaction conditions among the studies.
- Antioxidants may participate in multiple reactions due to the complexity of antioxidant structures and food matrices. Acrylamide content, as the final result of several positive and negative reactions, depends on the predominant reaction antioxidants and substrates undergo, which may change when the concentration of the antioxidant varies.
- The relationship between antioxidant capacity and acrylamide remains controversial. A close correlation between ΔTEAC and inhibitory rate of acrylamide was recently found. A structure–activity relationship between number and position of phenol hydroxyl functional groups in flavonoids and acrylamide inhibition is established. However, the exact mechanism still needs more in-depth studies.

REFERENCES

[1] Açar ÖÇ, Gökmen V. Investigation of acrylamide formation on bakery products using a crust-like model. Mol Nutr Food Res 2009;53(12):1521–5.

[2] Ames JM. The Maillard reaction. In: Biochemistry of food proteins. Springer; 1992. p. 99–153.

[3] Amrein TM, Andres L, Manzardo GGG, Amadò R. Investigations on the promoting effect of ammonium hydrogencarbonate on the formation of acrylamide in model systems. J Agric Food Chem 2006;54(26):10253–61.

[4] Arribas-Lorenzo G, Fogliano V, Morales FJ. Acrylamide formation in a cookie system as influenced by the oil phenol profile and degree of oxidation. Eur Food Res Technol 2009;229(1):63–72.

[5] Auld RB, Bedwell SF. Peripheral neuropathy with sympathetic overactivity from industrial contact with acrylamide. Can Med Assoc J 1967;96(11):652.
[6] Bassama J, Brat P, Bohuon P, Boulanger R, Günata Z. Study of acrylamide mitigation in model system: effect of pure phenolic compounds. Food Chem 2010;123(2):558–62.
[7] Becalski A, Lau BPY, Lewis D, Seaman SW. Acrylamide in foods: occurrence, sources, and modeling. J Agric Food Chem 2003;51(3):802–8.
[8] Benzie IFF, Strain JJ. The ferric reducing ability of plasma (FRAP) as a measure of "antioxidant power": the FRAP assay. Anal Biochem 1996;239(1):70–6.
[9] Biedermann M, Biedermann-Brem S, Noti A, Grob K, Egli P, Mändli H. Two GC–MS methods for the analysis of acrylamide in foods. Mittl aus Leb Hyg 2002;93(6):638–52.
[10] Biedermann M, Noti A, Biedermann-Brem S, Mozzetti V, Grob K. Experiments on acrylamide formation and possibilities to decrease the potential of acrylamide formation in potatoes. Mittl aus Leb Hyg 2002;93(6):668–87.
[11] Bongers ML, Hogervorst JGF, Schouten LJ, Goldbohm RA, Schouten HC, van den Brandt PA. Dietary acrylamide intake and the risk of lymphatic malignancies: the Netherlands Cohort Study on diet and cancer. PloS one 2012;7(6):e38016.
[12] Borrelli RC, Visconti A, Mennella C, Anese M, Fogliano V. Chemical characterization and antioxidant properties of coffee melanoidins. J Agric Food Chem 2002;50(22):6527–33.
[13] Bråthen E, Knutsen SH. Effect of temperature and time on the formation of acrylamide in starch-based and cereal model systems, flat breads and bread. Food Chem 2005;92(4):693–700.
[14] Cai Y, Zhang Z, Jiang S, Yu M, Huang C, Qiu R, et al. Chlorogenic acid increased acrylamide formation through promotion of HMF formation and 3-aminopropionamide deamination. J Hazard Mater 2014;268:1–5.
[15] Calleman CJ, Wu Y, He F, Tian G, Bergmark E, Zhang S, et al. Relationships between biomarkers of exposure and neurological effects in a group of workers exposed to acrylamide. Toxicol Appl Pharmacol 1994;126(2):361–71.
[16] Capuano E, Fogliano V. Acrylamide and 5-hydroxymethylfurfural (HMF): a review on metabolism, toxicity, occurrence in food and mitigation strategies. LWT-Food Sci Technol 2011;44(4):793–810.
[17] Capuano E, Oliviero T, Açar ÖÇ, Gökmen V, Fogliano V. Lipid oxidation promotes acrylamide formation in fat-rich model systems. Food Res Int 2010;43(4):1021–6.
[18] Cheng J, Chen X, Zhao S, Zhang Y. Antioxidant-capacity-based models for the prediction of acrylamide reduction by flavonoids. Food Chem 2015;168:90–9.
[19] Cheng K-W, Shi J-J, Ou S-Y, Wang M, Jiang Y. Effects of fruit extracts on the formation of acrylamide in model reactions and fried potato crisps. J Agric Food Chem 2010;58(1):309–12.
[20] Cheng K-W, Zeng X, Tang YS, Wu J-J, Liu Z, Sze K-H, et al. Inhibitory mechanism of naringenin against carcinogenic acrylamide formation and Nonenzymatic Browning in Maillard model reactions. Chem Res Toxicol 2009;22:1483–9.
[21] Chuda Y, Ono H, Yada H, Ohara-Takada A, Matsuura-Endo C, Mori M. Effects of physiological changes in potato tubers (*Solanum tuberosum* L.) after low temperature storage on the level of acrylamide formed in potato chips. Biosci Biotechnol Biochem 2003;67(5):1188–90.
[22] CIAA. Acrylamide toolbox 2011. 2011. [online], available: http://www.fooddrinkeurope.eu/uploads/publications_documents/Toolboxfinal260911.pdf. [accessed 30.09.11].
[23] Ciesarová Z, Suhaj M, HorváThová J. Correlation between acrylamide contents and antioxidant capacities of spice extracts in a model potato matrix. J Food Nutr Res 2008;47(1):1–5.
[24] Claeys WL, De Vleeschouwer K, Hendrickx ME. Effect of amino acids on acrylamide formation and elimination kinetics. Biotechnol Prog 2005;21(5):1525–30.
[25] Corradini MG, Peleg M. Linear and non-linear kinetics in the synthesis and degradation of acrylamide in foods and model systems. Crit Rev Food Sci Nutr 2006;46(6):489–517.
[26] E.U.Council. Council Directive 98/83/EC of 3 November 1998 on the quality of water intended for human consumption. 1998.
[27] FDA. Detection and quantitation of acrylamide in foods. 2003. [online], available: http://www.fda.gov/Food/FoodborneIllnessContaminants/ChemicalContaminants/ucm053537.htm. [accessed 11.06.14].
[28] Friedman M. Chemistry, biochemistry, and safety of acrylamide. A review. J Agric Food Chem 2003;51(16):4504–26.

[29] Friedman M, Jürgens HS. Effect of pH on the stability of plant phenolic compounds. J Agric Food Chem 2000;48(6):2101–10.
[30] Gökmen V, Kocadağlı T, Göncüoğlu N, Mogol BA. Model studies on the role of 5-hydroxymethyl-2-furfural in acrylamide formation from asparagine. Food Chem 2012;132(1):168–74.
[31] Gökmen V, Şenyuva HZ. Acrylamide formation is prevented by divalent cations during the Maillard reaction. Food Chem 2007;103(1):196–203.
[32] Garland TO, Patterson MW. Six cases of acrylamide poisoning. BMJ-British Med J 1967;4(5572):134–8.
[33] Goffeng LO, Kjuus H, Heier MS, Alvestrand M, Ulvestad B, Skaug V. Colour vision and light sensitivity in tunnel workers previously exposed to acrylamide and *N*-methylolacrylamide containing grouting agents. Neurotoxicology 2008;29(1):31–9.
[34] Granvogl M, Jezussek M, Koehler P, Schieberle P. Quantitation of 3-aminopropionamide in potatoes A minor but potent precursor in acrylamide formation. J Agric Food Chem 2004;52(15):4751–7.
[35] Halliwell B. The antioxidant paradox. Lancet 2000;355(9210):1179–80.
[36] Hamzalıoğlu A, Gökmen V. Role of bioactive carbonyl compounds on the conversion of asparagine into acrylamide during heating. Eur Food Res Technol 2012;235(6):1093–9.
[37] Hamzalıoğlu A, Mogol BA, Lumaga RB, Fogliano V, Gökmen V. Role of curcumin in the conversion of asparagine into acrylamide during heating. Amino Acids 2013;44(6):1419–26.
[38] He F, Zhang S, Wang H, Li G, Zhang Z, Li F, et al. Neurological and electroneuromyographic assessment of the adverse effects of acrylamide on occupationally exposed workers. Scand J Work Environ Health 1989:125–9.
[39] Hedegaard RV, Granby K, Frandsen H, Thygesen J, Skibsted LH. Acrylamide in bread. Effect of prooxidants and antioxidants. Eur Food Res Technol 2008;227(2):519–25.
[40] Hidalgo FJ, Delgado RM, Navarro JL, Zamora R. Asparagine decarboxylation by lipid oxidation products in model systems. J Agric Food Chem 2010;58(19):10512–7.
[41] Hidalgo FJ, Delgado RM, Zamora R. Positive interaction between amino and sulfhydryl groups for acrylamide removal. Food Res Int 2011;44(4):1083–7.
[42] Hogervorst JG, Schouten LJ, Konings EJ, Goldbohm RA, van den Brandt PA. A prospective study of dietary acrylamide intake and the risk of endometrial, ovarian, and breast cancer. Cancer Epidemiol Biomarkers Prev 2007;16(11):2304–13.
[43] IARC. IARC monographs on the evaluation of carcinogenic risks to humans: some industrial chemicals. International Agency for Research on Cancer; 1994.
[44] Jin C, Wu X, Zhang Y. Relationship between antioxidants and acrylamide formation: a review. Food Res Int 2013;51(2):611–20.
[45] Kesson CM, Baird AW, Lawson DH. Acrylamide poisoning. Postgrad Med J 1977;53(615):16–7.
[46] Kjuus H, Goffeng LO, Heier MS, Sjöholm H, Øvrebø S, Skaug V, et al. Effects on the peripheral nervous system of tunnel workers exposed to acrylamide and *N*-methylolacrylamide. Scand J Work Environ health 2004:21–9.
[47] Kocadağlı T, Göncüoğlu N, Hamzalıoğlu A, Gökmen V. In depth study of acrylamide formation in coffee during roasting: role of sucrose decomposition and lipid oxidation. Food Funct 2012;3(9):970–5.
[48] Kotsiou K, Tasioula-Margari M, Capuano E, Fogliano V. Effect of standard phenolic compounds and olive oil phenolic extracts on acrylamide formation in an emulsion system. Food Chem 2011;124(1):242–7.
[49] Kotsiou K, Tasioula-Margari M, Kukurová K, Ciesarová Z. Impact of oregano and virgin olive oil phenolic compounds on acrylamide content in a model system and fresh potatoes. Food Chem 2010;123(4):1149–55.
[50] Kroh LW. Caramelisation in food and beverages. Food Chem 1994;51(4):373–9.
[51] Levine RA, Smith RE. Sources of variability of acrylamide levels in a cracker model. J Agric Food Chem 2005;53(11):4410–6.
[52] Li D, Chen Y, Zhang Y, Lu B, Jin C, Wu X, et al. Study on mitigation of acrylamide formation in cookies by 5 antioxidants. J Food Sci 2012;77(11):C1144–9.
[53] Lingnert H, Waller GR. Stability of antioxidants formed from histidine and glucose by the Maillard reaction. J Agric Food Chem 1983;31(1):27–30.
[54] Lipworth L, Sonderman JS, Tarone RE, McLaughlin JK. Review of epidemiologic studies of dietary acrylamide intake and the risk of cancer. Eur J Cancer Prev 2012;21(4):375–86.

[55] Liu J, Chen F, Man Y, Dong J, Hu X. The pathways for the removal of acrylamide in model systems using glycine based on the identification of reaction products. Food Chem 2011;128(2):442–9.
[56] Liu L, Pang M, Zhang Y. Lipase-catalyzed regioselective synthesis of flavone C-glucosides esters and high-efficiency oil-soluble antioxidant of bamboo leaves (eAOB-o). Eur J Lipid Sci Technol 2015;117.
[57] Liu L, Xia B, Jin C, Zhang Y, Zhang Y. Chemical acylation of water-soluble antioxidant of bamboo leaves (AOB-w) and functional evaluation of oil-soluble AOB (cAOB-o). J Food Sci 2014;79(10):C1886–94.
[58] Lopachin Jr RM, Lehning EJ. Acrylamide-induced distal axon degeneration: a proposed mechanism of action. Neurotoxicology 1994;15(2):247–59.
[59] LoPachin RM, Lehning EJ, Opanashuk LA, Jortner BS. Rate of neurotoxicant exposure determines morphologic manifestations of distal axonopathy. Toxicol Appl Pharmacol 2000;167(2):75–86.
[60] LoPachin RM, Ross JF, Lehning EJ. Nerve terminals as the primary site of acrylamide action: a hypothesis. Neurotoxicology 2002;23(1):43–59.
[61] Mestdagh F, Maertens J, Cucu T, Delporte K, Van Peteghem C, De Meulenaer B. Impact of additives to lower the formation of acrylamide in a potato model system through pH reduction and other mechanisms. Food Chem 2008;107(1):26–31.
[62] Mishra K, Ojha H, Chaudhury NK. Estimation of antiradical properties of antioxidants using DPPH assay: a critical review and results. Food Chem 2012;130(4):1036–43.
[63] Mottram DS, Wedzicha BL, Dodson AT. Food chemistry: acrylamide is formed in the Maillard reaction. Nature 2002;419(6906):448–9.
[64] Mulloy KB. Two case reports of neurological disease in coal mine preparation plant workers. Am J Ind Med 1996;30(1):56–61.
[65] Myers JE, Macun I. Acrylamide neuropathy in a South African factory: an epidemiologic investigation. Am J Ind Med 1991;19(4):487–93.
[66] Napolitano A, Morales F, Sacchi R, Fogliano V. Relationship between virgin olive oil phenolic compounds and acrylamide formation in fried crisps. J Agric Food Chem 2008;56(6):2034–40.
[67] Ng JC. Evaluation of certain contaminants in food: Seventy-second report of the Joint FAO/WHO Expert Committe on Food Additives. World Health Organization; 2011.
[68] Oral RA, Dogan M, Sarioglu K. Effects of certain polyphenols and extracts on furans and acrylamide formation in model system, and total furans during storage. Food Chem 2014;142:423–9.
[69] Ou S, Shi J, Huang C, Zhang G, Teng J, Jiang Y, et al. Effect of antioxidants on elimination and formation of acrylamide in model reaction systems. J Hazard Mater 2010;182(1):863–8.
[70] Ozgen M, Reese RN, Tulio AZ, Scheerens JC, Miller AR. Modified 2, 2-azino-bis-3-ethylbenzothiazoline-6-sulfonic acid (ABTS) method to measure antioxidant capacity of selected small fruits and comparison to ferric reducing antioxidant power (FRAP) and 2, 2′-diphenyl-1-picrylhydrazyl (DPPH) methods. J Agric Food Chem 2006;54(4):1151–7.
[71] Pelucchi C, Galeone C, Talamini R, Negri E, Polesel J, Serraino D, et al. Dietary acrylamide and pancreatic cancer risk in an Italian case–control study. Ann Oncol 2011;22(8):1910–15.
[72] Perez Locas C, Yaylayan VA. Isotope labeling studies on the formation of 5-(hydroxymethyl)-2-furaldehyde (HMF) from sucrose by pyrolysis-GC/MS. J Agric Food Chem 2008;56(15):6717–23.
[73] Rosén J, Hellenäs K-E. Analysis of acrylamide in cooked foods by liquid chromatography tandem mass spectrometry. Analyst 2002;127(7):880–2.
[74] Rydberg P, Eriksson S, Tareke E, Karlsson P, Ehrenberg L, Törnqvist M. Investigations of factors that influence the acrylamide content of heated foodstuffs. J Agric Food Chem 2003;51(24):7012–8.
[75] Serpen A, Gökmen V. Evaluation of the Maillard reaction in potato crisps by acrylamide, antioxidant capacity and color. J Food Compos Analysis 2009;22(6):589–95.
[76] Shahidi F, Ho C. Phenolic compounds in foods and natural health products. American Chemical Society; 2005.
[77] Stadler RH, Blank I, Varga N, Robert F, Hau J, Guy PA, et al. Food chemistry: acrylamide from Maillard reaction products. Nature 2002;419(6906):449–50.
[78] Stadler RH, Robert F, Riediker S, Varga N, Davidek T, Devaud S, et al. In-depth mechanistic study on the formation of acrylamide and other vinylogous compounds by the Maillard reaction. J Agric Food Chem 2004;52(17):5550–8.
[79] Summa C, Wenzl T, Brohee M, De La Calle B, Anklam E. Investigation of the correlation of the acrylamide content and the antioxidant activity of model cookies. J Agric Food Chem 2006;54(3): 853–9.

[80] Tareke E. Identification and origin of potential background carcinogens: endogenous isoprene and oxiranes, dietary acrylamide [unpublished thesis (Doctoral Thesis)]. University Stockholm.
[81] Tareke E, Rydberg P, Karlsson P, Eriksson S, Törnqvist M. Analysis of acrylamide, a carcinogen formed in heated foodstuffs. J Agric Food Chem 2002;50(17):4998–5006.
[82] Totlani VM, Peterson DG. Reactivity of epicatechin in aqueous glycine and glucose Maillard reaction models: Quenching of C2, C3, and C4 sugar fragments. J Agric Food Chem 2005;53(10):4130–5.
[83] Totlani VM, Peterson DG. Epicatechin carbonyl-trapping reactions in aqueous Maillard systems: identification and structural elucidation. J Agric Food Chem 2006;54(19):7311–8.
[84] Urbančič S, Kolar MH, Dimitrijević D, Demšar L, Vidrih R. Stabilisation of sunflower oil and reduction of acrylamide formation of potato with rosemary extract during deep-fat frying. LWT-Food Sci Technol 2014;57(2):671–8.
[85] Weisshaar R. Acrylamide in heated potato products–analytics and formation routes. Eur J Lipid Sci Technol 2004;106(11):786–92.
[86] Wilson KM, Mucci LA, Rosner BA, Willett WC. A prospective study of dietary acrylamide intake and the risk of breast, endometrial, and ovarian cancers. Cancer Epidemiol Biomarkers Prev 2010;19(10): 2503–15.
[87] Yuan Y, Shu C, Zhou B, Qi X, Xiang J. Impact of selected additives on acrylamide formation in asparagine/sugar Maillard model systems. Food Res Int 2011;44(1):449–55.
[88] Zamora R, Delgado RM, Hidalgo FJ. Model reactions of acrylamide with selected amino compounds. J Agric Food Chem 2010;58(3):1708–13.
[89] Zamora R, Hidalgo FJ. Contribution of lipid oxidation products to acrylamide formation in model systems. J Agric Food Chem 2008;56(15):6075–80.
[90] Zeng X, Cheng K-W, Jiang Y, Lin Z-X, Shi J-J, Ou S-Y, et al. Inhibition of acrylamide formation by vitamins in model reactions and fried potato strips. Food Chem 2009;116(1):34–9.
[91] Zhang Y, Chen J, Zhang X, Wu X, Zhang Y. Addition of antioxidant of bamboo leaves (AOB) effectively reduces acrylamide formation in potato crisps and French fries. J Agric Food Chem 2007; 55(2):523–8.
[92] Zhang Y, Chen X, Cheng J, Jin C, Zhang Y. The reduction effect of dietary flavone C-and O-glycosides on the formation of acrylamide and its correlation and prediction with the antioxidant activity of Maillard reaction products. RSC Adv 2014;4(46):24147–55.
[93] Zhang Y, Ren Y, Jiao J, Li D, Zhang Y. Ultra high-performance liquid Chromatography−tandem mass spectrometry for the simultaneous analysis of asparagine, sugars, and acrylamide in Maillard reactions. Anal Chem 2011;83(9):3297–304.
[94] Zhang Y, Xu W, Wu X, Zhang X, Zhang Y. Addition of antioxidant from bamboo leaves as an effective way to reduce the formation of acrylamide in fried chicken wings. Food Addit Contam 2007;24(3):242–51.
[95] Zhang Y, Ying T, Zhang Y. Reduction of acrylamide and its kinetics by addition of antioxidant of bamboo leaves (AOB) and extract of green tea (EGT) in asparagine–glucose microwave heating system. J Food Sci 2008;73(2):C60–6.
[96] Zhang Y, Zhang Y. Study on reduction of acrylamide in fried bread sticks by addition of antioxidant of bamboo leaves and extract of green tea. Asia Pac J Clin Nutr 2007;16(1):131–6.
[97] Zhang Y, Zhang Y. Effect of natural antioxidants on kinetic behavior of acrylamide formation and elimination in low-moisture asparagine–glucose model system. J Food Eng 2008;85(1):105–15.
[98] Zhu F, Cai YZ, Ke J, Corke H. Evaluation of the effect of plant extracts and phenolic compounds on reduction of acrylamide in an asparagine/glucose model system by RP-HPLC-DAD. J Sci Food Agric 2009;89(10):1674–81.
[99] Zyzak DV, Sanders RA, Stojanovic M, Tallmadge DH, Eberhart BL, Ewald DK, et al. Acrylamide formation mechanism in heated foods. J Agric Food Chem 2003;51(16):4782–7.

CHAPTER 18

Interaction between Bioactive Carbonyl Compounds and Asparagine and Impact on Acrylamide

Aytül Hamzalıoğlu, Vural Gökmen
Department of Food Engineering, Hacettepe University, Beytepe, Ankara, Turkey

INTRODUCTION

Bioactive compounds are extra-nutritional constituents that are found in small quantities in foods providing health benefits beyond the basic nutritional value of the product [1]. They are being intensively studied to evaluate their effects on health, and bioactive compounds appear to have beneficial physiological, behavioral, and immunological effects. To date, numerous bioactive compounds have been discovered. These compounds vary widely in chemical structure and function and are grouped accordingly. Some examples of bioactive compounds are carotenoids, flavonoids, carnitine, choline, coenzyme Q, dithiolthiones, phytosterols, phytoestrogens, glucosinolates, polyphenols, and taurine. Since vitamins and minerals elicit pharmacological effects, they can be categorized as bioactive compounds as well.

Bioactive compounds may naturally be found in various foods. Most of the bioactive compounds have antioxidant, anticarcinogenic, antiinflammatory, and antimicrobial properties. Therefore, many epidemiologic studies report that some of them also have protective effects on cardiovascular diseases.

Phenolic compounds, including their subcategory flavonoids, are present in almost all plants and have been found extensively in cereals, legumes, nuts, olive oil, tea, red wine, vegetables, and fruits. They mostly have antioxidant properties and some studies demonstrated favorable effects on cardiovascular diseases risk factors [2].

Various phytoestrogens are present not only in soy, but also in flaxseed oil, whole grains, fruits, and vegetables. Besides their antioxidant properties, their similarities with estrogen at a molecular level allow them to mimic [3].

Since carotenoids are efficient free radical scavengers, they are one of the most potent antioxidant compounds [4]. They are found in most fruits and vegetables, especially apricots, carrots, mangoes, tomatoes, and pumpkins.

The glucosinolates are natural components of many pungent plants such as mustard, cabbage, and horseradish. Since they induce Phase I and Phase II enzymes, inhibiting the enzyme activation, they are under investigation for mitigating cancer [5].

Acrylamide in Food
http://dx.doi.org/10.1016/B978-0-12-802832-2.00018-8

Vitamins are vital nutrients that an organism requires in limited amounts. They cannot be synthesized by human body and must be obtained through diet. They have diverse functions and biological effects on health. Besides their regulatory and catalytic activities in the body, some of them, such as vitamin E and vitamin C, are also functioning as antioxidants [6].

Although bioactive compounds are naturally present in various foods, they are being used as an additive and a processing aid as well. Bioactive compounds are generally added to foods or food products for the enhancement of their health promoting properties. It is a fact that, carotenoids, anthocyanins, and curcumin are the most commonly known coloring bioactive compounds. They are being added to some food products for the purpose of coloring. To prevent oxidation, ascorbic acid is one of the widely used food additive. The most obvious application of cinnamaldehyde and vanillin is as flavoring in sweet foods, chewing gums, and beverages.

A **bioactive carbonyl** is a bioactive compound bearing a carbonyl group. Some of the bioactive carbonyl compounds, their occurrence in foods, and their health promoting properties are listed in Table 1.

Owing to their functional carbonyl group, bioactive carbonyl compounds may also be involved in some reactions during food production processes. Their potential reactivity should be taken into consideration from the point of view of health risks for humans.

ROLE OF CARBONYL COMPOUNDS ON ACRYLAMIDE FORMATION

Several researchers have stated that the main pathway of acrylamide formation in foods is linked to the Maillard reaction [41]. Studies to date clearly show that asparagine is mainly responsible for acrylamide formation in heated foods after condensation with a carbonyl source [42]. Asparagine alone may form only a limited amount of acrylamide through decarboxylation and deamination upon heating at elevated temperatures [43]. However, presence of carbonyl compounds accelerates its conversion to acrylamide.

The first step in the reaction is formation of Schiff base upon condensation of a carbonyl group with the amine group of asparagine. Then, decarboxylation of the Schiff base can occur directly through Schiff betaine or Schiff betaine rearranges to oxazolidine-5-one intermediate which is known to decarboxylate easily to form azomethine ylide (decarboxylated Schiff base) [41,42]. Azomethine ylide rearranges by tautomerization to form decarboxylated Amadori compound, which then forms acrylamide via β-elimination [44]. It was revealed that decarboxylation of Schiff base is the rate limiting step [45]. Carbonyl compounds play an accelerator role in the decarboxylation step making the formation of decarboxylated Schiff base easier.

The carbonyl compounds may come from multiple sources. Besides the significant contribution of mostly known α-hydroxycarbonyl compounds such as fructose and glucose to acrylamide formation [46], glyoxal, methylglyoxal, 5-hydroxymethyl-2-furfural (HMF),

Table 1 Bioactive carbonyl compounds, their occurrence in foods, and health-promoting properties

Bioactive carbonyls	Occurrence in foods	Health-promoting property
Flavonoids		
Flavones		
Luteolin	Broccoli, green pepper, parsley, oregano, carrots, and rosemary	Anticarcinogenic activity [7]
Apigenin	Many fruits and vegetables, parsley, celery, celeriac, and chamomile tea	Anticarcinogenic activity [8]
Tangeritin	Citrus peels	Antihypercholesterolemia effect [9], antioxidant activity [10], and anticarcinogenic activity [11]
Flavonol		
Quercetin	Fruits and vegetables, especially onions, citrus, apples, dark berries, grapes, olive oil, green tea, and red wine	Encouraging blood flow and promoting cardiovascular health properties [2]
Kaempferol	Apples, grapes, tomatoes, green tea, potatoes, onions, broccoli, brussels sprouts, squash, cucumbers, lettuce, green beans, peaches, blackberries, raspberries, and spinach	Anticarcinogenic activity and antioxidant activity [12]
Myricetin	Vegetables, fruits, nuts, berries, tea, and red wine	Antiinflammatory activity, anti-nonenzymatic glycation, and antihyperlipidemia [13]
Galangin	Propolis	Antimicrobial activity, antioxidant activity, anticarcinogenic activity, and antiinflammatory activity [14]
Flavanone		
Hesperitin	Citrus fruits	Antioxidant activity, anticarcinogenic activity, hypolipidemic activity, and vasoprotective activity [15]
Naringenin	Grapefruits, oranges, and tomatoes	Antioxidant activity [16], antiinflammatory activity [17], and immune system modulator [18]
Isoflavones		
Genistein	Soybeans, legumes, and chickpeas	Antioxidant activity and estrogen mimicking property [3]
Daidzein	Soybeans and soy products like tofu	Antioxidant activity and estrogen mimicking property [13]

Continued

Table 1 Bioactive carbonyl compounds, their occurrence in foods, and health-promoting properties—cont'd

Bioactive carbonyls	Occurrence in foods	Health-promoting property
Glycitein	Soy products	Antioxidant activity and estrogen mimicking property [3]
Vitamins		
Pyridoxal	Meats, whole-grain products, vegetables, nuts, and bananas	Reducing cardiovascular disease risk property [19] and anticarcinogenic activity [20,21]
Niacin	Liver, chicken, beef, fish, cereal, peanuts, legumes, avocados, tomatoes, leaf vegetables, carrots, sweet potatoes, asparagus, nuts, whole grain products, and legumes	Reducing cardiovascular disease risk property [22,23]
Riboflavin	Milk, cheese, leafy vegetables, liver, kidneys, legumes, yeast, mushrooms, and almonds	Antioxidant property [24] and protection from migraine [25]
Nicotinamide	Meat, fish, nuts, mushrooms, and some vegetables	Antiinflammatory activity and antianxiety activity [26]
Ascorbic acid	Citrus fruits, tomatoes, tomato juice, potatoes, red and green peppers, kiwifruit, broccoli, strawberries, brussels sprouts, and cantaloupe	Antioxidant activity [27], anticarcinogenic activity [28], and reducing cardiovascular disease risk property [29]
Others		
Silymarin	Milk thistle	Anticarcinogenic activity [30], preventing liver from alcoholic cirrhosis [31,32]
Vanillin	Vanilla, paraguay orchid, and Chinese red pine	Antimicrobial activity [33] and antioxidant activity [34]
Cinnamaldehyde	Cinnamon bark	Antimicrobial activity [35] and anticarcinogenic activity [36]
Punicalagin	Pomegranates	Antioxidant activity [37]
Curcumin	Turmeric	Antibacterial activity [38], antiinflammatory activity [39], and antioxidant activity [40]

and other sugar degradation products bearing carbonyl groups have also been reported to trigger acrylamide formation during heating [47–49]. In addition to all these carbonyl compounds, bioactive carbonyl compounds also participate in Maillard reaction yielding to acrylamide formation [50,51].

Among the bioactive carbonyl compounds, it is known that curcumin, ascorbic acid, dehydroascorbic acid, silymarin, and vanillin react with asparagine for the formation of acrylamide [50]. Pyridoxal and some carbonyls of virgin olive oil phenolic extracts also act like precursor of acrylamide [52,53].

ROLE OF BIOACTIVE CARBONYL COMPOUNDS ON ACRYLAMIDE FORMATION

Role of curcumin on acrylamide formation during heating was studied in different model systems [51]. It was reported that asparagine alone was capable in forming acrylamide to a very lesser extent. But the rate of acrylamide formation increased significantly in the presence of equimolar amounts of fructose (FRU) or curcumin (CUR), besides asparagine, in the model reaction system (Figure 1(a)). This indicated that curcumin also participates in Maillard reaction and converts asparagine into acrylamide, although fructose is much more effective in converting asparagine to acrylamide than curcumin.

As shown in Figure 1(a), acrylamide concentration reached a maximum within 10 min followed by a rapid decrease afterward in the fructose–asparagine (FRU-ASN) model system. On the other hand acrylamide concentration increased linearly within 30 min in the curcumin–asparagine (CUR-ASN) model system. It is the net result of lower reactivity of curcumin compared to that of fructose.

When the concentrations of reactants were monitored in both CUR-ASN and FRU-ASN model systems to understand the efficiency of curcumin in reacting with asparagine, it was clearly seen that ASN and FRU rapidly decreased, and only traces of these reactants were detected after 5 min of reaction at 180 °C (Figure 1(b)). Asparagine and curcumin in CUR-ASN model system disappeared at a slower extent than that in FRU-ASN model system. Since approximately 30% of ASN and CUR still remained in the reaction mixture after 30 min of heating at 180 °C, it was considered that curcumin

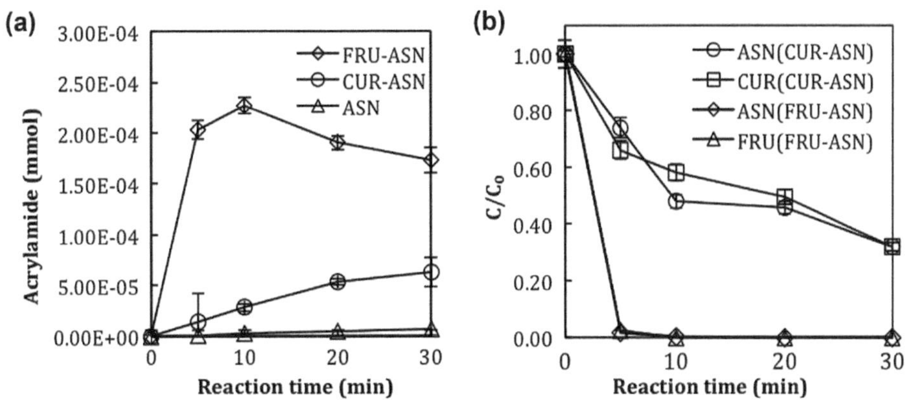

Figure 1 (a) Formation of acrylamide in asparagine (ASN), fructose–asparagine (FRU-ASN) and curcumin–asparagine (CUR-ASN) model system during heating at 180 °C. (b) Change of reactant concentrations in CUR-ASN and FRU-ASN model systems with time during heating at 180 °C [51].

could potentially contribute to acrylamide formation under long-term heating conditions as long as asparagine is present in the medium.

In the other study where the roles of some bioactive carbonyls including vanillin (VAN), ascorbic acid (ASC), dehydroascorbic acid (DHA), silymarin (SIL), and curcumin (CUR) on the conversion of asparagine into acrylamide were investigated, all bioactive tests generated significant amounts of acrylamide from asparagine at 180 °C (Table 1).

Taking the acrylamide contents of different binary model systems given in Table 2 into account, vanillin was found to be the most efficient bioactive carbonyl in acrylamide formation. The efficiency of carbonyls in converting asparagine to acrylamide was found in the following order: VAN > SIL > ASC > CUR > DHA.

Kotsiou et al. [53] reported that the addition of virgin olive oil extracts to a potato model system containing both potato starch and equimolar mixture of asparagine–glucose caused an increase in acrylamide levels whereas addition of oregano extracts caused a decrease. According to the phenolic composition analysis, it was confirmed that dialdehydic form of decarboxylated aglycone of oleuropein and dialdehydic form of decarboxylated aglycone of ligstroside are the dominant phenolic compounds in virgin olive oil extracts. They are the phenolic compounds containing both phenyl and carbonyl groups; thus could also be categorized as bioactive carbonyl compounds. Oleuropein is a glycoside found in olives and is hydrolyzed during virgin olive oil extraction to form dialdehyde or aldehyde derivatives. In contrast to oleuropein, virgin olive oil phenols contain aldehyde groups. Results of the studies carried out with the standard oleuropein (without any aldehyde group in structure) confirmed the reduction in acrylamide formation whereas virgin olive oil extract containing aldehyde groups in its structure enhanced the acrylamide formation up to 20% [53].

Different concentration levels, ranging between 0 and 230 mg/kg, were also applied to potato model systems and it was found that there was a concentration-dependent relationship in acrylamide formation [53]. They found a remarkable increase of

Table 2 The amount of acrylamide formed in different binary and ternary model systems during heating at 180 °C for 10 and 20 min [50]

	Acrylamide formed (nmol)	
Model system	**180 °C × 10 min**	**180 °C × 20 min**
ASN-FUR	227.02 ± 20.52^{d}	190.36 ± 6.86^{j}
ASN-ASC	$29.78 \pm 3.47^{b,c}$	56.21 ± 0.65^{i}
ASN-DHA	22.49 ± 1.45^{b}	$43.04 \pm 1.66^{c,i}$
ASN-SIL	39.39 ± 2.34^{c}	56.68 ± 0.96^{i}
ASN-CUR	$27.98 \pm 2.82^{b,c}$	56.46 ± 2.95^{i}
ASN-VAN	227.60 ± 7.96^{d}	520.43 ± 49.18^{k}

Different letters in the columns indicate statistically significant differences ($p < 0.05$).

approximately 50% with the highest virgin olive oil phenolic extract concentration. This confirmed that, besides the carbonyl groups found in virgin olive oil phenolic compounds structure, concentration of those carbonyl compounds are also important.

Kotsiou et al. [54] also studied the effect of virgin olive oil phenolic extracts on acrylamide formation in an emulsion system containing glucose and asparagine besides sunflower oil with emulsifier. They reported that they observed a remarkable increase in acrylamide formation with the addition of different virgin olive oil phenolic extracts (VOO 1, VOO 2, and VOO 3) at different concentrations (Figure 2). Increase in acrylamide content of model systems where the virgin olive oil extract 1 (VOO 1) added was found to be the highest. This was attributed to the high content of carbonyl groups in VOO 1 extracts. Compared to the effect of virgin olive oil extracts on acrylamide formation in potato model systems, increase in acrylamide content in emulsion systems was remarkably higher. This is the net result of higher efficiency of carbonyl compounds in an emulsion system than in a bulk system.

Zeng et al. [52] reported a study about the effects of different vitamins on acrylamide formation in both model and actual food systems. They prepared aqueous chemical model systems containing 1 mmol of asparagine, glucose, and 30 mg of several water-soluble vitamins, including vitamin B_1, B_3, B_7, C, pyridoxine, pyridoxal, and pyridoxamine. Among the vitamins tested, most of them strongly reduced the formation of acrylamide. Pyridoxine and vitamin B_3 were the most effective inhibitors. On the other hand, pyridoxal, one of the mostly consumed B_6 vitamins was not that effective in inhibiting acrylamide formation. That was attributed to the presence of carbonyl groups in its structure, too.

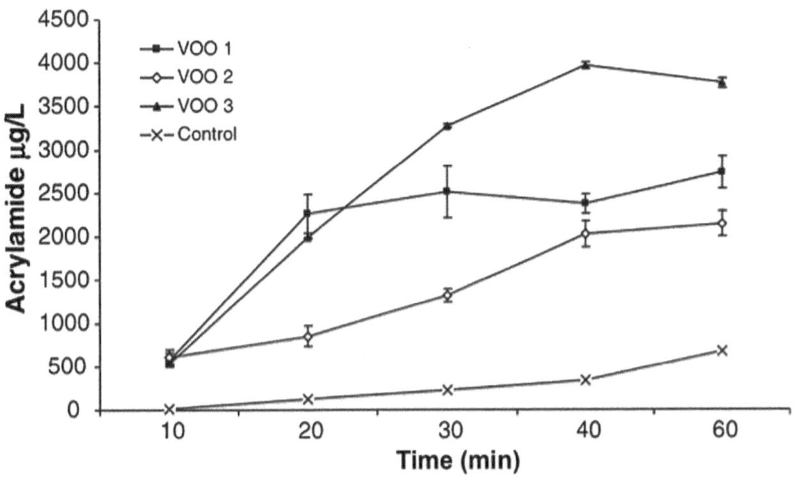

Figure 2 Acrylamide content in the emulsion system, heated at 125 °C as affected by VOO phenolic extracts addition and time [54].

Unlike in asparagine-bioactive carbonyl model systems, there was a significant disappearance in acrylamide content of ASN-FRU model system during heating. It was attributed to rapid consumption of asparagine thus reaching the maximum level in a short while (Table 1). As it is known, Schiff base of asparagine and reducing carbonyls rearranges to form Amadori compounds that cannot decarboxylate. Flavored and colored compounds are also formed via decomposition of these intermediates. Thus, when an α-hydroxycarbonyl compound is present instead of a bioactive carbonyl compound in the reaction medium, reaction not only tends to form acrylamide but also to produce flavor and color compounds. But then, Schiff base formed between asparagine and bioactive carbonyl compounds directly decarboxylates to form acrylamide if bioactive carbonyl is present in the reaction medium. Contrary to α-hydroxycarbonyl compounds, asparagine is the "limiting reactant" in bioactive carbonyl compound-based model systems.

COMPETITION WITH α-HYDROXYCARBONYL COMPOUNDS

It is a fact that foods may contain a variety of carbonyl compounds that may compete with one other during the Maillard reaction. For a specific case of acrylamide formation from asparagine, the competition may occur between different carbonyls that may be present in food or formed during heating. Since both bioactive and other carbonyl compounds are present together in some foods, competition between carbonyl compounds in reacting with asparagine and the determination of the dominating carbonyl is also of importance.

Competition between curcumin and fructose was studied in ternary model systems composed of curcumin–fructose–asparagine (CUR-FRU-ASN) in a study by Hamzalıoğlu et al. [51]. The results show that more acrylamide is formed when curcumin besides fructose is present in the reaction medium (Figure 3(a)). Similar kinetic trends observed in both binary and ternary model systems during heating are the clues about the dominating carbonyl compound. As fructose easily reacts with asparagine and is more efficient under short-term heating, it gives a direction to the reaction. Thus, even the acrylamide content of CUR-FRU-ASN ternary model system is higher; fructose leads the reaction to decrease.

The amounts of acrylamide formed in different ternary model systems composed of fructose and asparagine besides ascorbic acid, dehydroascorbic acid, vanillin, or silymarin as a result of heating at 180 °C for 20 min were shown in Figure 3(b).

It seems like ascorbic acid and dehydroascorbic acid caused a significant decrease in acrylamide content compared to FRU-ASN binary model system heated at the same conditions. On the contrary, presence of vanillin or silymarin significantly increased the conversion of asparagine into acrylamide in the ternary model system of asparagine–fructose–vanillin (ASN-FRU-VAN) or asparagine–fructose–silymarin (ASN-FRU-SIL). As it has been mentioned before, silymarin and vanillin are the most reactive ones among the bioactive carbonyls tested. This evidence clearly confirms that, bioactive carbonyls

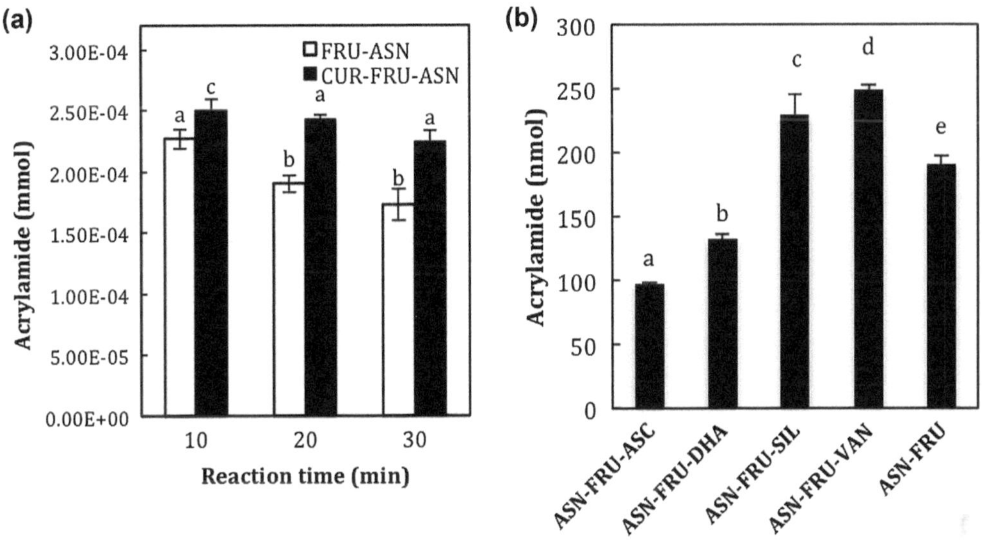

Figure 3 (a) Formation of acrylamide in FRU-ASN and CUR-FRU-ASN model systems during heating at 180 °C [51]. (b) Formation of acrylamide in different ternary model systems during heating at 180 °C for 20 min [51].

compete with fructose in reacting with asparagine and their reactivity plays an important role here. When bioactive carbonyl compounds having lower reactivity compared to that of fructose are added to FRU-ASN system, they cause a "pseudo-decrease" in acrylamide formation. However, they may dominate when the elimination of acrylamide is higher than its formation under long-term heating conditions.

Because of the antioxidant and nutritional properties, ascorbic acid is one of the most studied bioactive carbonyls in both model and actual food systems where it is together with reducing sugars. In a study by Levine and Smith, with the addition of ascorbic acid solution, acrylamide content of a cracker recipe baked for 30 min at 180 °C was reduced from 1850 to 490 μg/kg [55].

In a similar study where the effects of antioxidants on the acrylamide content of potato samples heated in the oven were examined, ascorbic acid at different concentrations ranging between 0.30% and 8.3% was added to the potato and after homogenization it was immediately heated in a microwave oven, for 3 min at 750W. In this case, with 0.30% ascorbic acid addition the acrylamide level was reduced from 3100 to 420 μg/kg [56].

Zeng et al. [52] reported that ascorbic acid addition, caused 50% and 42% reduction in acrylamide content of the chemical and food model systems, respectively. On the other hand, no significant inhibition was found when it was applied to potato samples prior to frying.

In a study by Chen and Gu [57], the effect of antioxidant ascorbic acid on acrylamide formation in ammonia caramel was studied. A mixture of glucose and ammonia was allowed to react at 120 °C for 60 min in the presence of ascorbic acid at varying concentrations ranging from 0 to 0.08 M. It was reported that the acrylamide content of ammonia caramels showed the opposite concentration-dependent relationships at different levels of ascorbic acid. When the concentration of ascorbic acid added reached 0.04 M, the content of acrylamide in the ammonia caramel was approximately 44% lower than that without ascorbic acid. On the contrary, disappearance in acrylamide content of ammonia caramels was significantly lower when the ascorbic acid concentration was above 0.04 M.

Ou et al. [58] reported that addition of some antioxidants caused less reduction or even production of acrylamide formation in asparagine–glucose model system as a result of heating at 80 °C for 10 min. According to the results obtained, acrylamide formation was inhibited 75.72% with the addition of ascorbic acid. They concluded that, carbonyl groups from ascorbic acid reacted with asparagine and thus prevented it from further reaction with glucose; as some carbonyl groups can rapidly react with α-amino groups of the amino acids.

Yuan et al. [59] studied the effect of selected additives, namely NaCl, $NaHSO_3$, ascorbic acid, on acrylamide formation in asparagine–fructose and asparagine–glucose model systems by microwave processing. In asparagine–fructose model system, the reduction rates were 37.24%, 69.53%, 80.91%, and 93.13%, after the addition of ascorbic acid solution at concentrations of 0.2%, 0.5%, 0.8%, and 1.2%, respectively (Figure 4).

After being treated with 1.5% of ascorbic acid, no acrylamide could be detected. Interestingly, in asparagine–glucose model system, increasing the ascorbic acid concentration to 1.5% promoted the acrylamide formation while it caused a 57.76% reduction with the addition of 0.5% ascorbic acid [59]. It was attributed to the formation of

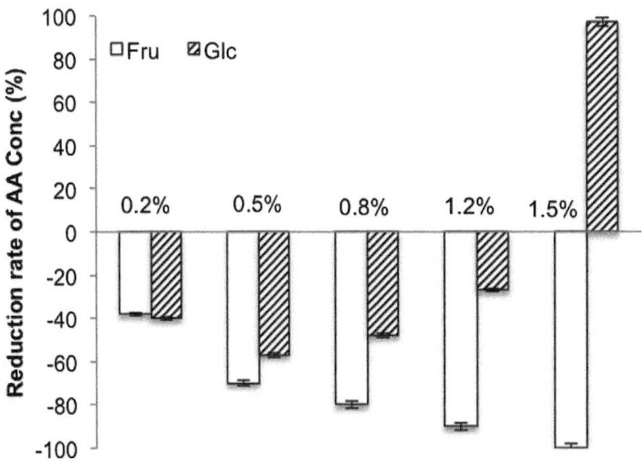

Figure 4 Percentage reduction rates of acrylamide concentration in both asparagine–fructose and asparagine–glucose model systems with the addition of ascorbic acid solutions at different concentrations [59].

thermal degradation products of ascorbic acid that may participate in Maillard reaction. In this study, they concluded that ascorbic acid showed effective reduction on acrylamide formation in asparagine–fructose model system compared to asparagine–glucose model system.

Zhu et al. [60] reported the effects of 34 different plant extracts on acrylamide formation in asparagine–glucose model systems. Addition of different extracts significantly reduced acrylamide formation to varying extents. Curcumin was one of the tested compounds and addition of curcumin extract to the asparagine–glucose model system caused a 42.2% reduction in acrylamide content after it was heated at 180 °C for 15 min.

Zhu et al. [61] confirmed the inhibitory effect of curcumin in cookies and potato starch-based model systems. They tested different raw powders and the results showed that the phenolic compounds could reduce acrylamide content in both cookies and potato starch-based model systems to various extents. Addition of 4% curcumin to cookies reduced acrylamide content by 40.7%, whereas it caused a reduction of 46.8% with the addition of 2% curcumin.

Hamzalıoğlu et al. [51] also determined the contribution of curcumin to acrylamide formation in food model system. A bread crust model system having 1.0% of curcumin and 1.0% of glucose was prepared and baked at 180 °C for 10, 20, and 30 min. Control sample having no carbonyl compound contained low amounts of acrylamide after it was baked for 10, 20, and 30 min at 180 °C, respectively (Figure 5). Acrylamide concentration of crust models having 1.0% of glucose (GLC) was significantly higher than that of samples having 1.0% of curcumin (CUR) just after 10 min of baking at 180 °C. On the other hand, the acrylamide concentration of crust models containing 1.0% of CUR was found to be slightly higher than those containing 1.0% of GLC when baking time was extended.

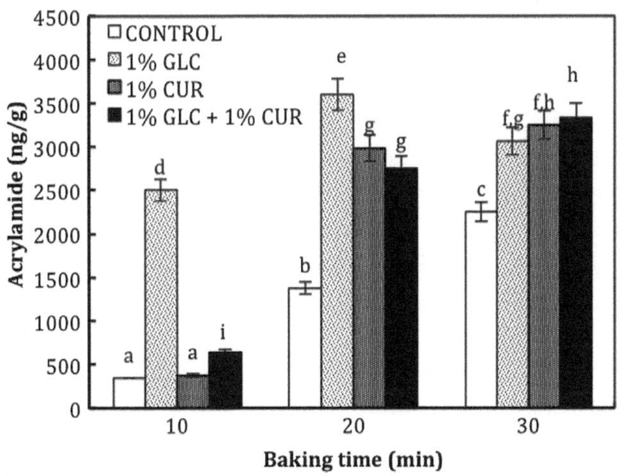

Figure 5 Amounts of acrylamide formed in control, 1% CUR, 1% GLC, and 1% CUR+1% GLC crust models during heating at 180 °C for different times [51].

Owing to its high reactivity, glucose consumed free asparagine at a faster extent. The acrylamide content of crust models reached maximum levels in a short time, while acrylamide formation proceeded with a slower rate in crust models containing CUR. In the crust model systems where CUR was together with GLC, it was clearly seen that the addition of CUR along with GLC reduced the amount of acrylamide generated initially. But it was clearly observed that CUR along with GLC in crust models promoted acrylamide formation with further heating.

Because of the remarkable efficiency of vanillin on acrylamide formation in model system, Hamzalıoğlu and Gökmen [50] also tested its role on acrylamide formation in a crust-resembling food system. Concentration of acrylamide in control crust sample was found to be 245 ng/g after it was baked at 180 °C for 60 min, whereas its concentration was found to be 519 ng/g in 1% vanillin added crust samples baked at the same conditions.

Oral et al. [62] reported a study about the effect of some polyphenols including punicalagin, a bioactive carbonyl compound found in pomegranate husk, on acrylamide formation using asparagine–fructose model system and model biscuits. They observed slight decreases in both the model systems and biscuits where punicalagin was present in the reaction medium besides fructose.

FACTORS AFFECTING REACTIVITY OF CARBONYL COMPOUNDS AND THE ROLE OF MELTING POINT

As it is shown in Figure 6, a carbonyl is any compound that has a carbon double-bonded to oxygen. Polar double bond creates a partial positive charge (δ^+) on the carbonyl carbon and a partial negative charge (δ^-) on oxygen. Carbonyl compounds can behave as a nucleophile or an electrophile because of having both nucleophilic oxygen and electrophilic carbon. δ^- and δ^+ charges arise from resonance because the oxygen atom is more electronegative compared to carbon.

Reactivity of carbonyl compounds depends on their resonance, leaving group (if there is), magnitude of δ^+ charge on carbon atom, and steric effects. When there is no or poor resonance, reaction will be faster. Loss of resonance will spike up the transition state energy for nucleophilic attack. On the contrary, if there is good resonance, carbonyl compounds are less likely to react due to resonance stabilization that reduces the partial δ^+/δ^- charges. The reaction will also be faster in the presence of good leaving groups in the Y position of carbonyl compound structure (Figure 6). Aldehydes and ketones are

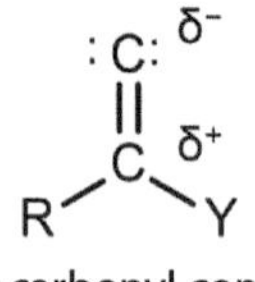

Figure 6 Generic carbonyl compound structure.

less reactive than carbonyl compounds if there is more electronegative atom in the Y position.

One of the factors affecting reactivity of carbonyl compounds is the magnitude of δ^+ charge on carbon atom. Larger δ^+ charge on the carbon atom will make it more electrophilic thus making it easily prone to nucleophiles. The presence of neighboring groups having high electronegativity renders carbonyl more electrophilic. Also, the presence of the neighboring electron-donating groups (such as alkyl groups) reduces the magnitude of δ^+ charge on carbon making it less reactive. Bulky substituents hinder nucleophilic attack as well.

In the light of this information, reactivity of a bioactive carbonyl compound may be estimated. Besides the chemical reactivity of carbonyl sources, their physical states have a crucial role under low moisture conditions. It is a fact that aldose sugars tend to be more reactive compared to ketoses due to their highly reactive aldehyde groups in liquid reaction systems. However, water rapidly evaporates on heating making the reactants concentrated in dry conditions, as on the surface of foods, during heating. When the water is evaporated, the reaction conditions become unfavorable because of the loss of molecular mobility and thus physical states of reactants become more important for a reaction under low moisture conditions. Since the melting point of fructose is lower compared to that of glucose, it converts asparagine to acrylamide more efficiently [63]. Similarly, reactivity of the bioactive carbonyls is found to be directly related with their melting points in acrylamide formation [50].

Among the carbonyl compounds tested, vanillin having the lowest melting point (84 °C) caused the highest conversion of asparagine into acrylamide (Figure 7). The conversion efficiency of the carbonyl compounds was found to be in the following

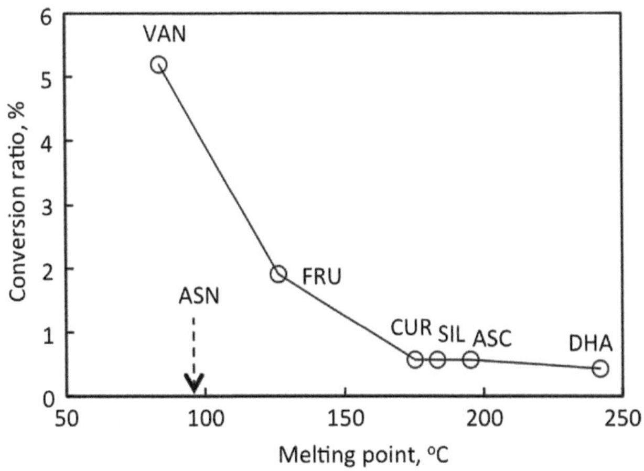

Figure 7 Relationship between the melting point of carbonyl compounds and the conversion rate of asparagine into acrylamide during heating at 180 °C. *Adapted from the data of Hamzalıoğlu et al. [50].*

order: VAN < CUR < SIL < ASC < DHA. Besides, melting points of bioactive carbonyl compounds except for vanillin were lower compared to that of fructose. Since they need more energy to melt under low moisture conditions, fructose becomes more favorable in reacting with asparagine. Melting point is also indicative in food systems where the bioactive carbonyl compounds are found besides α-hydroxycarbonyl compounds.

As discussed above, there are contradictory reports published about the effects of some antioxidants as well as bioactive carbonyls on acrylamide formation: they work as mitigation agents under certain conditions whereas they promote its formation under the same conditions. However, the reactivity of carbonyl compounds plays an important role for the conversion of asparagine into acrylamide. Melting point of carbonyl compounds should also be taken into consideration. Melting points of some bioactive carbonyl compounds are given in Table 3.

Table 3 Melting point of some bioactive carbonyl compounds

Bioactive carbonyls	Melting point (°C)
Luteolin	329.5
Apigenin	345–350
Tangeritin	155–156
Quercetin	216
Kaempferol	267–278
Myricetin	357
Galangin	214–215
Hesperetin	226–228
Naringenin	247–250
Genistein	297–298
Daidzein	315–323
Glycitein	337–339
Umbelliferone	230
Pyridoxal	165
Niacin	237
Riboflavin	282.2
Nicotinamide	129.5
Cyannocobalamine	>300
Ascorbic acid	195
Dehydroascorbic acid	242
Phylloquinone	105–107
Silymarin	183
Curcumin	175
Punicalagin	232
Cinnamaldaheyde	−7.5
Vanillin	84

FORMATION MECHANISM OF ACRYLAMIDE BETWEEN ASPARAGINE AND BIOACTIVE CARBONYL COMPOUNDS

As mentioned before, a reactive carbonyl compound is a prerequisite to accelerate the conversion of asparagine to acrylamide. The carbonyl compound may come from food or produced during a thermal treatment. Hamzalıoğlu et al. [51] have performed high-resolution mass spectrometry analysis to identify the structure of the reaction intermediates and products and propose the formation mechanism in asparagine–curcumin model system. According to the results of the study, a proposed reaction mechanism between curcumin and asparagine at elevated temperatures should proceed according to the reaction scheme shown in Figure 8.

At first, it was expected that the carbonyl group of curcumin would react with the α-amino group of asparagine forming the corresponding Schiff base during the heat treatment of the asparagine–curcumin model system.

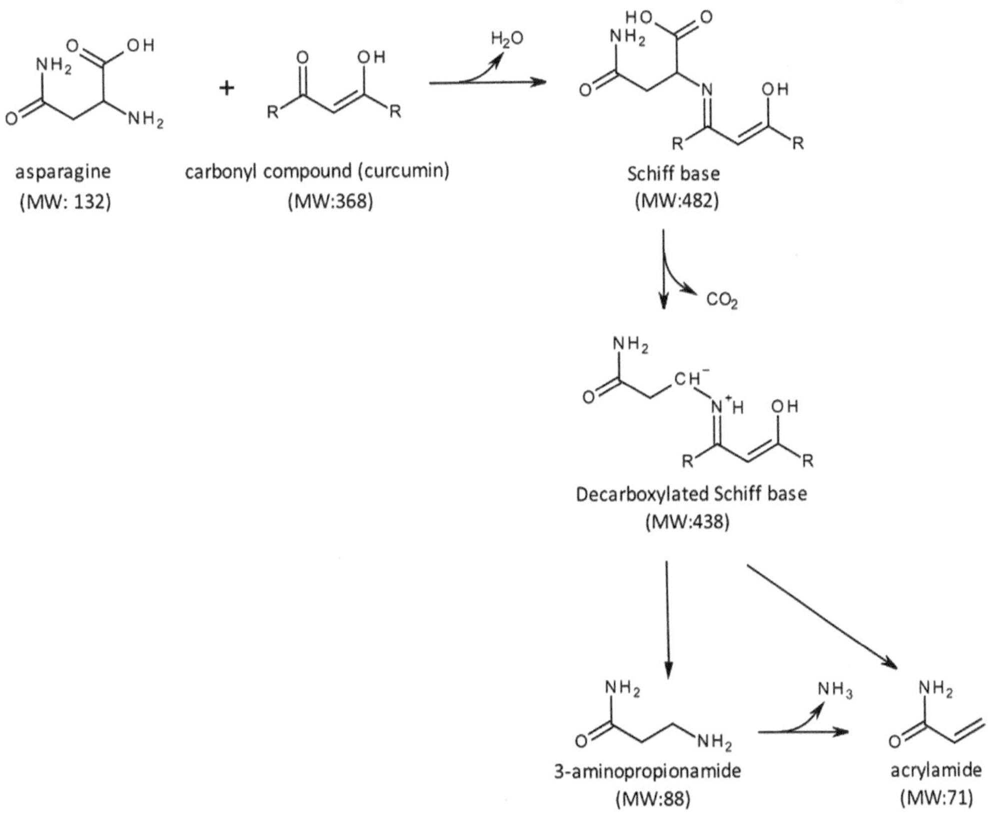

Figure 8 Proposed mechanism for the contribution of curcumin on acrylamide formation from asparagine [51].

It is a fact that acrylamide may be produced via the decarboxylation of Schiff base through different paths. Acrylamide may directly be formed from decarboxylated Schiff base via β-elimination [46,64] as well as it may be formed via Hofmann elimination [65]. After decarboxylation step, there are two possible ways pointing to acrylamide. Decarboxylated Schiff base may also be a direct precursor of acrylamide or hydrolyze to 3-aminopropionamide (3-APA) that is capable of yielding acrylamide through deamination [42].

In the asparagine–curcumin model system, the formation of reaction intermediates and 3-APA during heating was confirmed. There was a significant decrease in 3-APA content when the heating time is prolonged, while the amount of acrylamide increased.

With this confirmation, it can be concluded that the formation of acrylamide in bioactive carbonyl compound-based model systems through 3-APA formation is the dominating way.

ENCAPSULATION TO LIMIT THE REACTIVITY OF BIOACTIVE CARBONYL COMPOUNDS

Encapsulation is one of the solutions to limit the reactivity of reactants. As discussed above, curcumin participates in Maillard reaction in converting asparagine into acrylamide. To confirm the limitation of the reactivity of curcumin in converting asparagine into acrylamide, Hamzalıoğlu [66] baked some crust samples containing 1% free or encapsulated curcumin. According to the results, they observed an 8% decrease with the addition of encapsulated curcumin (Figure 9). It was confirmed that the reactive carbonyl group in curcumin structure was blocked properly with the encapsulation.

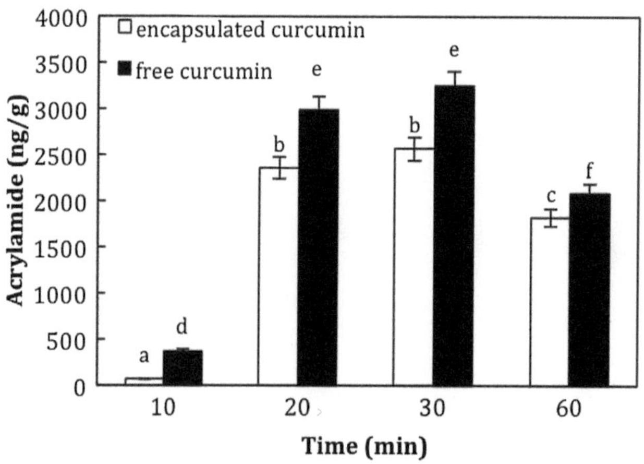

Figure 9 Acrylamide contents of baked crust samples containing 1% encapsulated or free curcumin at 180 °C for different times [66].

Therefore, it is concluded that proper encapsulation techniques may be a good way to limit the reactivity of bioactive carbonyl compounds when they are desired to be used in heat-treated foods as functional ingredients.

SYNOPSIS OF ANALYTICAL TECHNIQUES

Model systems are one of the good ways to evaluate the mechanism behind the reactions. To date, so many studies based on model systems helped for the evaluation of the pathway of acrylamide formation. Food resembling model systems are mostly composed of main reactants found in those foods. To determine the role of bioactive carbonyl compounds in acrylamide formation, various binary and ternary model systems are formulated based on the main reactant asparagine and bioactive carbonyl compounds or/and α-hydroxycarbonyl compounds.

The binary model systems based on bioactive carbonyl compounds are composed of equimolar amounts of asparagine and a bioactive carbonyl compound. In the ternary model system, an equimolar amount of α-hydroxycarbonyl compound is added to the reaction medium together with asparagine and a given bioactive carbonyl compound.

Certain amount of asparagine and the given carbonyl compound are dissolved in deionized water and transferred to glass test tubes as their aqueous solutions. Total volume in the reaction medium is adjusted to a small volume with distilled water to promote the physical interaction between asparagine and the carbonyl compound. Some amount of silica gel is added to cover the reaction mixture to create a porous layer above the reaction mixture in order to prevent any evaporative loss of the reactants during heating. Finally, the tube is tightly closed with a screw cap and dipped into hot oil to initiate the reaction. The reactions should be performed in an oil bath with dipping the whole tube in to compensate the heat transfer. Also, the outer wall of the glass tube should reach the temperature of the oil bath within a short while.

Model system mixtures are extracted with water–methanol mixtures prior to analysis. In general, water is preferred to extract the acrylamide owing to its high solubility in water. For the reaction intermediates and products formed in bioactive carbonyl containing model systems, water–methanol mixtures are effective in extraction since some of them are more hydrophobic. Obtained extracts are usually frozen–thawed before the filtration step to avoid the further turbidity that may result.

Heating the mixture of asparagine and carbonyl compounds forms a pool of reaction intermediates and products as a result of the Maillard reaction. The reaction mixtures are then usually analyzed by high-resolution mass spectrometry (HRMS) in order to identify the chemical structures of key intermediates leading to acrylamide.

The chromatographic separations are performed in reversed-phase chromatography, in which C_{18} columns are used. Atlantis T3 column is the most useful one providing balanced retention of polar and non-polar molecules. Elution is performed by using a gradient mixture of methanol and slightly acidified water with formic acid. As hydrophilicity of both bioactive carbonyl compounds and its reaction intermediates with asparagine are different, gradient mixture provides a better separation. Besides, acidification of water is also a sort of protonation, which is crucial for ionization.

The HRMS system is operated in positive electrospray ionization mode, scanning the ions in m/z range of 50 and 600 at ultra-high resolving power (R = 100,000). Ultra high resolution is a prerequisite to eliminate the isobaric interferences and increase mass accuracy providing more confident analysis. Orbitrap mass spectrometry technology, one of the bench-top HRMS systems, allows achieving a mass level and maintaining excellent mass accuracy. It also allows detecting the calculated theoretical mass with $\Delta < 5$ ppm. Corresponding ions of main intermediates such as Schiff bases, decarboxylated Schiff bases of asparagine with given carbonyls are extracted from the total ion chromatograms.

KEY FACTS

- β-Elimination reactions are chemical reactions in which a nucleophilic pair of electrons heads into a new pi bond when a leaving group departs. It is called β-elimination because the bond β to the nucleophilic pair of electrons breaks. In acrylamide formation mechanism, azomethine ylides are formed after Strecker-type degradation of Schiff base. Acrylamide is then released, via a β-elimination reaction and cleavage of the carbon–nitrogen covalent bond. OH group in β position to the N atom is a prerequisite for this type of elimination. Since α-hydroxycarbonyl compounds are main precursors to yield such azomethine ylides, in contrast to α-dicarbonyl compounds, reactants such as fructose or glucose take part in β-elimination.
- The Strecker degradation is a chemical reaction involving the conversion of an α-amino acid to an aldehyde containing the side chain, by way of an imine intermediate. It is named after its discoverer Adolph Strecker, a German chemist.
- Encapsulation is a process involving the entrapment of bioactive agents within carrier materials. It is usually applied for the production of functional foods with higher nutritional value, lower dose of synthetic preservatives, and better organoleptic features. It has so many applications because of the enhancement of the stability of sensitive compounds during production, storage, and ingestion, masking unpleasant tastes or limiting exposure to oxygen, water, or light.
- Curcumin, one of the bioactive carbonyl compounds, is a traditional Chinese medicine used to treat diseases associated with abdominal pain. It was first used thousands of years ago by the Chinese and the Indians. It was a principal healing agent in Ayurveda, the traditional Indian system of medicine, and recognized as a valuable ingredient. In India, it is still considered to be highly auspicious and it is still used extensively in many occasions from religious ceremonies to wedding ceremonies.
- Vanilla, source of bioactive carbonyl compound vanillin, was derived from orchids in the genus Vanilla native to Mexico. Totonaca Indians of Mexico were cultivating

vanillin as a flavoring agent. Totonac mythology has it that the tropical orchid was born when her father executed Princess Xanat and the vine of the tropical orchid grew where her blood touched the ground. Vanilla beans were so valuable at that time that Totonacs sent vanilla beans to Aztec Kings as tribute when most tribes paid their tributes in the form of maize or gold. They consumed vanillin with chocolate beverages. Introduction of both chocolate and vanilla in Europe happened after Aztecs were defeated by Spain around the mid-1500s.

MINI DICTIONARY

3-Aminopropionamide (3-APA) It is a biogenic amine formed from azomethine ylide of asparagine and easily deaminates to form acrylamide.

5-Hydroxymethyl-2-furfural (HMF) It is a furanic compound that is formed via loss of three molecules of water from hexose sugars during caramelization or Maillard reaction.

α-hydroxycarbonyl compound A class of organic compounds in organic chemistry consisting of a hydroxyl group placed on a neighboring carbon to carbonyl carbon atom.

Azomethine ylide An iminium ion that occurs via decarboxylation of Schiff base during Maillard reaction.

Electrophile A positively charged compound attracted to an electron rich center, thus participating in a chemical reaction by accepting an electron pair to bond.

Encapsulation It is an inclusion of one entity within another to make the included entity not apparent for different purposes.

Nucleophile It is a chemical species having free electron pairs which donates an electron pair to an electrophile to form a chemical bond.

Resonance A method of bonding that describes the delocalization of electrons within molecules, comparing and contrasting two or more possible Lewis structures that can represent a particular molecule.

SUMMARY POINTS

- This chapter focuses on the interaction between bioactive carbonyl compounds and asparagine and how it affects the acrylamide formation.
- Bioactive compounds are extra-nutritional constituents that provide health benefits beyond their nutritional values.
- A bioactive carbonyl is a bioactive compound bearing a carbonyl group and is naturally present in various foods.
- Bioactive carbonyl compounds participate in Maillard reaction for the conversion of asparagine into acrylamide.
- Among the bioactive carbonyl compounds, curcumin, silymarin, ascorbic acid, vanillin, dehydroascorbic acid, pyridoxal, and some virgin olive oil phenolic compounds were found to be efficient in converting asparagine into acrylamide.
- Bioactive carbonyl compounds are more likely to convert asparagine into acrylamide directly.
- Some α-hydroxycarbonyl compounds, like fructose or glucose, are not only effective in forming acrylamide but also in producing flavor and color compounds.

- Contrary to α-hydroxycarbonyl compounds, asparagine is the "limiting reactant" in bioactive carbonyl compound-based systems.
- Bioactive carbonyl compounds compete with α-hydroxycarbonyl compounds in reacting with asparagine and their reactivity plays an important role here.
- When bioactive carbonyl compounds having lower reactivity compared to α-hydroxycarbonyl compounds are together with them, they cause a "pseudo-decrease" in acrylamide formation.
- Besides the chemical reactivity of carbonyl compounds, melting point is also indicative in food systems where the bioactive carbonyl and α-hydroxycarbonyl compounds are together.
- It is possible to limit the reactivity of bioactive carbonyl compounds by encapsulation.

REFERENCES

[1] Kitts DD. Bioactive substances in food: identification and potential uses. Can J Physiol Pharmacol 1994;72:423–4.
[2] Perez-Vizcaino F, Duarte J. Flavonols and cardiovascular disease. Mol Asp Med 2010;31(6):478–94.
[3] Kaufman PB, Duke JA, Brielmann H, et al. A comparative survey of leguminous plants as sources of the isoflavones, genistein and daidzein: implications for human nutrition and health. J Altern Complement Med 1997;3(1):7–12.
[4] Re R, Pellegrini N, Proteggente A, et al. Antioxidant activity applying an improved ABTS radical cation decolorization assay. Free Radic Biol Med 1999;26:1231–7.
[5] Srinibas D, Amrish KT, Harjit K. Cancer modulation by glucosinolates: a review. Curr Sci 2000;79(12):166.
[6] Bender DA. Nutritional biochemistry of the vitamins. Cambridge, UK: Cambridge University Press; 2003.
[7] Lin Y, Shi R, Wang X, et al. Luteolin, a flavonoid with potential for cancer prevention and therapy. Curr Cancer Drug Targets 2008;8(7):634–46.
[8] Patel D, Shukla S, Gupta S. Apigenin and cancer chemoprevention: progress, potential and promise. Int J Oncol 2007;30(1):233–45.
[9] Kurowska EM, Manthey JA. Hypolipidemic effects and absorption of citrus polymethoxylated flavones in hamsters with diet-induced hypercholesterolemia. J Agric Food Chem 2004;52(10):2879–86.
[10] Stuetz W, Prapamontol T, Hongsibong S, et al. Polymethoxylated flavones, flavanone glycosides, carotenoids, and antioxidants in different cultivation types of tangerines (Citrus reticulata Blanco cv. Sainampueng) from Northern Thailand. J Agric Food Chem 2010;58(10):6069–74.
[11] Hirano T, Abe K, Gotoh M, Oka K. Citrus flavone tangeretin inhibits leukaemic HL-60 cell growth partially through induction of apoptosis with less cytotoxicity on normal lymphocytes. Br J Cancer 1995;72(6):1380–8.
[12] Chen AY, Chen YC. A review of the dietary flavonoid, kaempferol on human health and cancer chemoprevention. Food Chem 2013;138(4):2099–107.
[13] Li Y, Ding Y. Minireview: Therapeutic potential of myricetin in diabetes mellitus. Food Sci Hum Well 2012;1(1):19–25.
[14] Patel DK, Patel K, Gadewar M, et al. Pharmacological and bioanalytical aspects of galangin-a concise report. Asian Pac J Trop Biomed 2012;2(1):449–55.
[15] Chakraborty S, Chaudhuri S, Pahari B, et al. A critical study on the interactions of hesperitin with human hemoglobin: fluorescence spectroscopic and molecular modeling approach. J Lumin 2012;132(6):1522–8.
[16] Cavia-Saiz M, Busto MD, Pilar-Izquierdo MC, et al. Antioxidant properties, radical scavenging activity and biomolecule protection capacity of flavonoid naringenin and its glycoside naringin: a comparative study. J Sci Food Agric 2010;90(7):1238–44.

[17] Hämäläinen M, Nieminen R, Vuorela P, et al. Anti-inflammatory effects of flavonoids: genistein, kaempferol, quercetin, and daidzein inhibit STAT-1 and NF-kappaB activations, whereas flavone, isorhamnetin, naringenin, and pelargonidin inhibit only NF-kappaB activation along with their inhibitory effect on iNOS expression and NO production in activated macrophages. Mediators Inflamm 2007;2007:456–73.

[18] Fang F, Tang Y, Gao Z. A novel regulatory mechanism of naringenin through inhibition of T lymphocyte function in contact hypersensitivity suppression. Biochem Biophys Res Commun 2010;397(2):163–9.

[19] Graham IM, Daly LE, Refsum HM, et al. Plasma homocysteine as a risk factor for vascular disease. The European Concerted Action Project. JAMA 1997;277(22):1775–81.

[20] Folsom AR, Nieto FJ, McGovern PG, et al. Prospective study of coronary heart disease incidence in relation to fasting total homocysteine, related genetic polymorphisms, and B vitamins: the Atherosclerosis Risk in Communities (ARIC) study. Circulation 1998;98(3):204–10.

[21] Ramaswamy PG, Natarajan R. Vitamin B6 status in patients with cancer of the uterine cervix. Nutr Cancer 1984;6(3):176–80.

[22] Bruckert E, Labreuche J, Amarenco P. Meta-analysis of the effect of nicotinic acid alone or in combination on cardiovascular events and atherosclerosis. Atherosclerosis 2010;210(2):353–61.

[23] Duggal JK, Singh M, Attri N, et al. Effect of niacin therapy on cardiovascular outcomes in patients with coronary artery disease. J Cardio Pharmac Ther 2010;15(2):158–66.

[24] Ashoori M, Saedisomeolia A. Riboflavin (vitamin B2) and oxidative stress: a review. Br J Nutr 2014;20:1–7.

[25] Boehnke C, Reuter U, Flach U, et al. High-dose riboflavin treatment is efficacious in migraine prophylaxis: an open study in a tertiary care centre. Eur J Neurol 2004;11(7):475–7.

[26] Niren NM. Pharmacologic doses of nicotinamide in the treatment of inflammatory skin conditions: a review. Cutis 2006;77(Suppl. 1):11–6.

[27] Cameron E, Campbell A. The orthomolecular treatment of cancer. II. Clinical trial of high-dose ascorbic acid supplements in advanced human cancer. Chem Biol Interact 1974;9(4):285–315.

[28] Cameron E, Pauling L. Supplemental ascorbate in the supportive treatment of cancer: prolongation of survival times in terminal human cancer. Proc Natl Acad Sci USA 1976;73:3685–9.

[29] Ye Z, Song H. Antioxidant vitamins intake and the risk of coronary heart disease: meta-analysis of cohort studies. Eur J Cardiovasc Prev Rehabil 2008;15:26–34.

[30] Mokhtari MJ, Motamed N, Shokrgozar MA. Evaluation of silibinin on the viability, migration and adhesion of the human prostate adenocarcinoma (PC-3) cell line. Cell Biol Int 2008;32(8):888–92.

[31] Sharma G, Singh RP, Chan DC, et al. Silibinin induces growth inhibition and apoptotic cell death in human lung carcinoma cells. Anticancer Res 2003;23(3B):2649–55.

[32] Saller R, Brignoli R, Melzer J, et al. An updated systematic review with meta-analysis for the clinical evidence of silymarin. Forsch Komplementärmedizin 2008;15(1):9–20.

[33] Fitzgerald DJ, Stratford M, Gasson MJ, et al. Mode of antimicrobial action of vanillin against *Escherichia coli, Lactobacillus plantarum and Listeria innocua*. J Appl Microbiol 2004;97:104–13.

[34] Kumar SS, Priyadarsini KI, Sainis KB. Free radical scavenging activity of vanillin and o-vanillin using 1,1-diphenyl-2-picrylhydrazyl (DPPH) radical. Redox Rep 2002;7:35–40.

[35] Yossa N, Patel J, Macarisin D, et al. Antibacterial activity of cinnamaldehyde and sporan against *Escherichia coli* O157:H7 and Salmonella. J Food Process Preserv 2014;38(3):749–57.

[36] Cabello CM, Bair WB, Lamore SD, et al. The cinnamon derived Michael acceptor cinnamic aldehyde impairs melanoma cell proliferation, invasiveness and tumor growth. Free Radic Biol Med 2009;46(2):220–31.

[37] Gil MI, Tomás-Barberán FA, Hess-Pierce B, et al. Antioxidant activity of pomegranate juice and its relationship with phenolic composition and processing. J Agric Food Chem 2000;48(10):4581–9.

[38] Negi PS, Jayaprakasha GK, Jaganmohan Rao L. Antimicrobial activity of turmeric oil: a by-product from curcumin manufacturer. Agric Food Chem 1999;47:4297–300.

[39] Arora RB, Basu N, Kapoor V, et al. Anti-inflammatory studies on *Curcuma longa*. Indian J Med Res 1971;59:1289–95.

[40] Toda S, Myase T, Arichi H, et al. Natural antioxidants III. Antioxidative components isolated from rhizome of *Curcuma longa* L. Chem Pharmac Bullet 2003;33:1725–8.

[41] Yaylayan VA, Wnorowski A, Locas CP. Why asparagine needs carbohydrates to generate acrylamide. J Agric Food Chem 2003;51:1753–7.

[42] Zyzak DV, Sanders RA, Stojanovic M, et al. Acrylamide formation mechanism in heated foods. J Agric Food Chem 2003;51:4782–7.
[43] Granvogl M, Jezussek M, Koehler P, et al. Quantitation of 3-aminopropionamide in potatoes—a minor but potent precursor in acrylamide formation. J Agric Food Chem 2004;52:4751–7.
[44] Stadler RH, Robert F, Riediker S, et al. In-depth mechanistic study on the formation of acrylamide and other vinylogous compounds by the Maillard reaction. J Agric Food Chem 2004;52:5550–8.
[45] Blank I. Current status of acrylamide research in food: measurement, safety assessment, and formation. Ann NY Acad Sci 2005;1043:30–40.
[46] Yaylayan VA, Stadler RH. Acrylamide formation in food: a mechanistic perspective. J AOAC Int 2005;88(1):262–7.
[47] Amrein TM, Andres L, Manzardo GGG, et al. Investigations on the promoting effect of ammonium hydrogen carbonate on the formation of acrylamide in model systems. J Agric Food Chem 2006;54(26):10253–61.
[48] Amrein TM, Schonbachler B, Escher F, et al. Acrylamide in gingerbread: critical factors for formation and possible ways for reduction. J Agric Food Chem 2004;52:4282–8.
[49] Gokmen V, Kocadagli T, Goncuoglu N, et al. Model studies on the role of 5-hydroxymethyl-2-furfural in acrylamide formation from asparagine. Food Chem 2012;132:168–74.
[50] Hamzalıoğlu A, Gökmen V. Role of bioactive carbonyl compounds on the conversion of asparagine into acrylamide during heating. Eur Food Res Technol 2012;235:1093–9.
[51] Hamzalıoğlu A, Mogol B, Lumaga RB, et al. Role of curcumin on the conversion of asparagine into acrylamide during heating. Amino Acids 2013;44(6):1419–26.
[52] Zeng X, Cheng KW, Jiang Y, et al. Inhibition of acrylamide formation by vitamins in model reactions and fried potato strips. Food Chem 2009;116:34–9.
[53] Kotsiou K, Tasioula-Margari M, Kukurová K, et al. Impact of oregano and virgin olive oil phenolic compounds on acrylamide content in a model system and fresh potatoes. Food Chem 2010;123:1149–55.
[54] Kotsiou K, Tasioula-Margari M, Capuano E, et al. Effect of standard phenolic compounds and olive oil phenolic extracts on acrylamide formation in an emulsion system. Food Chem 2011;124:242–7.
[55] Levine RA, Smith RE. Sources of variability of acrylamide levels in cracker model. J Agric Food Chem 2005;53:4410–6.
[56] Rydberg P, Eriksson S, Tareke E, et al. Investigations of factors that influence the acrylamide content of heated foodstuffs. J Agric Food Chem 2003;51:7012–8.
[57] Chen H, Gu Z. Effect of ascorbic acid on the properties of ammonia caramel colorant additives and acrylamide formation. J Food Sci 2014;79(9):1678–82.
[58] Ou SY, Shi JJ, Huang CH, et al. Effect of antioxidants on elimination and formation of acrylamide in model reaction systems. J Hazard Mat 2010;182:863–8.
[59] Yuan YA, Shu C, Zhou B, et al. Impact of selected additives on acrylamide formation in asparagine/sugar Maillard model systems. Food Res Int 2011;44:449–55.
[60] Zhu F, Cai YZ, Ke J, et al. Evaluation of the effect of plant extracts and phenolic compounds on reduction of acrylamide in an asparagine/glucose model system by RP-HPLC-DAD. J Sci Food Agric 2009;89:1674–81.
[61] Zhu F, Cai YZ, Ke JX, et al. Dietary plant materials reduce acrylamide formation in cookie and starch-based model systems. J Food Sci Agric 2011;91:2477–83.
[62] Oral RA, Dogan M, Sarioglu K. Effects of certain polyphenols and extracts on furans and acrylamide formation in model system, and total furans during storage. Food Chem 2014;142:423–9.
[63] Robert F, Vuataz G, Pollien P, et al. Acrylamide formation from asparagine under low- moisture Maillard reaction conditions. 1. Physical and chemical aspects in crystalline model systems. J Agric Food Chem 2004;52:6837–42.
[64] Stadler RH, Blank I, Varga N, et al. Acrylamide from Maillard reaction products. Nature 2002;419:449–50.
[65] Locas CP, Yaylayan VA. Further insight into thermally and pH-induced generation of acrylamide from glucose/asparagine model systems. J Agric Food Chem 2008;56:6069–74.
[66] Hamzalıoğlu A. Investigation of the effect of bioactive carbonyls on acrylamide formation [Master Thesis]. Hacettepe University, Graduate School of Science and Engineering; 2011.

CHAPTER 19

Effect of Inorganic Salts on Acrylamide Formation in Cereal Matrices

Kristína Kukurová
NPPC VUP National Agricultural and Food Centre, Food Research Institute Bratislava, Priemyselná, Bratislava, Slovak Republic

INTRODUCTION

Inorganic salts represent a common ingredient for various cereal products. They are used most often as table salt or as leavening agents. NaCl is used in almost all recipes for its flavor and flavor-enhancing properties. Its most effective concentration is around 1.0–1.5%, based on the flour weight, but at a level of >2.5%, the taste becomes unpleasant. In dough with significant gluten development, crackers and semisweet types, salt toughens the gluten and gives less sticky dough. Salt is commonly used as a surface dressing and for decoration of savory biscuits [1]. Chemical leavening agents are used to give cookies, cakes, and other baked goods their characteristic textures. They produce gas from the reaction that takes place when a carbon dioxide source and an acid are mixed together and come into contact with water. Baking soda (sodium bicarbonate) is the most common carbon dioxide source. It is low in cost, high in purity, easy to handle, and leaves no after-taste [2]. In the presence of moisture, soda can react with any acidic materials to liberate carbon dioxide and remain as sodium carbonate [1]. Baking powder is a mixture of sodium bicarbonate and leavening acids that dissociate to give an acidic reaction in solution. The purpose of this combination of chemicals is to produce bubbles of carbon dioxide gas either before baking or, more particularly, when the dough piece warms up in the oven [2]. Leavening acids are selected primarily on the basis of reactivity, which depends mostly on solubility; this in turn depends on chemical composition, particle sizes, and special treatments as coating [2]. The most common acidulants are phosphate salts such as monocalcium orthophosphate or sodium acid pyrophosphate (SAPP) [3]. Ammonium bicarbonate is often used in a biscuit manufacture. It is decomposed completely when heated, and it breaks down into carbon dioxide gas, ammonia gas, and water used in biscuit manufacture. Ammonium bicarbonate is not suitable for use in baked products such as cakes or sponges, which have appreciable moisture contents due to residual ammonia [1]. Potassium bicarbonate, known as potash, is used instead of baking soda to minimize the sodium intake from food [2]. Raising agents and inorganic salts are known to significantly influence the formation of acrylamide during

Acrylamide in Food
http://dx.doi.org/10.1016/B978-0-12-802832-2.00019-X

the baking of cereal products in the Maillard reaction process that has been investigated in model systems followed by successful food applications in bakery wares including biscuits, cookies, wafers, bread, snacks, ready-to-eat cereals, confectioneries, crisp breads, breakfast cereals, cakes, pastries, and gingerbread [4].

FROM MODEL SYSTEMS TO FOOD APPLICATION

An inhibiting effect of inorganic salts on acrylamide formation during heating was observed at first in such model systems as an equimolar mixture of asparagine and glucose [5–8], asparagine and fructose [9], or more complex cereal systems with starch addition [10] at temperatures >120 °C, generally in the range from 160 to 230 °C.

The impact of table salt as a very common food additive on the formation of acrylamide was studied in an equimolar mixture of asparagine and glucose at 170 °C/10 min of thermal treatment. It was observed that even 1% of NaCl addition lowered acrylamide content from $267 \pm 30\,\mu g/g$ of the mixture to $181 \pm 9\,\mu g/g$ of the mixture. The highest decrement of acrylamide was determined in the range of 0–1% (w/w) of NaCl. However, further increasing of NaCl addition from 1% up to 10% (w/w) did not lead to acrylamide mitigation in a linear trend, while a maximum of 40% reduction was reached [5,6].

The effects of monovalent (Na^+, K^+) and divalent cations (Ca^{2+}, Mg^{2+}) on the formation of acrylamide were compared in a fructose–asparagine model system at 150 and 180 °C. Although monovalent cations, such as Na^+ present in table salt, prevented the formation of acrylamide to a certain extent, it was less efficient than the divalent cations Ca^{2+} and Mg^{2+}. At amounts equivalent to those of asparagine and fructose, added divalent cations were found to prevent acrylamide completely, whereas monovalent cations almost halved the acrylamide formed in the model system. Also, in the fructose–glucose model system, $CaCl_2$ and $MgCl_2$ decreased or even inhibited acrylamide formation [7].

It is noteworthy that not only cations but also anions play an important role in affecting acrylamide formation during food processing. It was observed that calcium caused an acrylamide reduction if it was added as a carbonate or chloride. However, calcium propionate, which was added as a preservative (up to 0.2%), resulted in more than a 90% increase in acrylamide [9,11]. By comparing different chlorides (Na^+, K^+, Ca^{2+}), it was observed that Ca^{2+} was more efficient than K^+ or Na^+ ions. Phosphates and pyrophosphates were also very effective tools in acrylamide suppression. For this reason, their application in food processing can be promising, particularly in combination with other acceptable salts such as $NaHCO_3$ or $KHCO_3$ [12].

The impact of different salts, such as monovalent and divalent chlorides, hydrogen carbonates, phosphates, and lactate on acrylamide formation was compared systematically in a complex model of a cereal system while heating at 190 °C for 9 min [10].

The cereal model system consisted of wheat starch, fructose, glucose monohydrate, L-asparagine monohydrate, and water simulating a common composition of acrylamide precursors in dough. NaCl, $CaCl_2 \cdot 2H_2O$, $KHCO_3$, $NaH_2PO_4 \cdot 2H_2O$, KH_2PO_4, $NaHCO_3$, NH_4HCO_3, $Ca(C_3H_5O_3)_2 \cdot 5H_2O$ (marked as CaL_2), $Na_2H_2P_2O_7$, and KCl were added at concentrations of 0.01, 0.05, and 0.1 $\mu mol\,g^{-1}$ into the model system. The results of the presented study are summarized in Figures 1 and 2. It was reported that the presence of salts in a reaction mixture resulted in a decrease of acrylamide concentration with an exception of ammonium hydrogen carbonate (NH_4HCO_3), a well-known accelerator of acrylamide formation [13]. Calcium chloride was the most effective with close to 90% acrylamide elimination ability from 603 ± 29 to $70 \pm 4\,\mu g\,kg^{-1}$. Sodium acid pyrophosphate and sodium as well as potassium dihydrogen phosphate were also very effective and brought about a 75% acrylamide content decrease, followed by calcium lactate, sodium chloride, and potassium chloride. This causes 40–45% of acrylamide elimination and finally sodium and potassium hydrogen carbonates that achieved a 30% reduction of acrylamide. It was concluded that $CaCl_2$ addition was the most promising tool for acrylamide mitigation in cereal-based food products followed by pyrophosphates and phosphates such as $Na_2H_2P_2O_7$, $NaH_2P_2O_4$, and $KH_2P_2O_4$ [10].

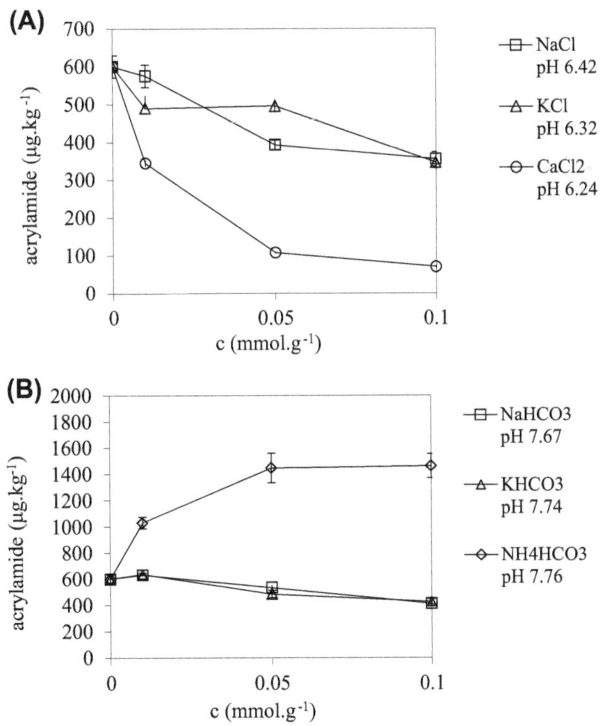

Figure 1 Effect of chloride (A) and hydrogen carbonate anions (B) of sodium, calcium, potassium, and ammonium salts on acrylamide formation in a cereal model system (190 °C/9 min) [10].

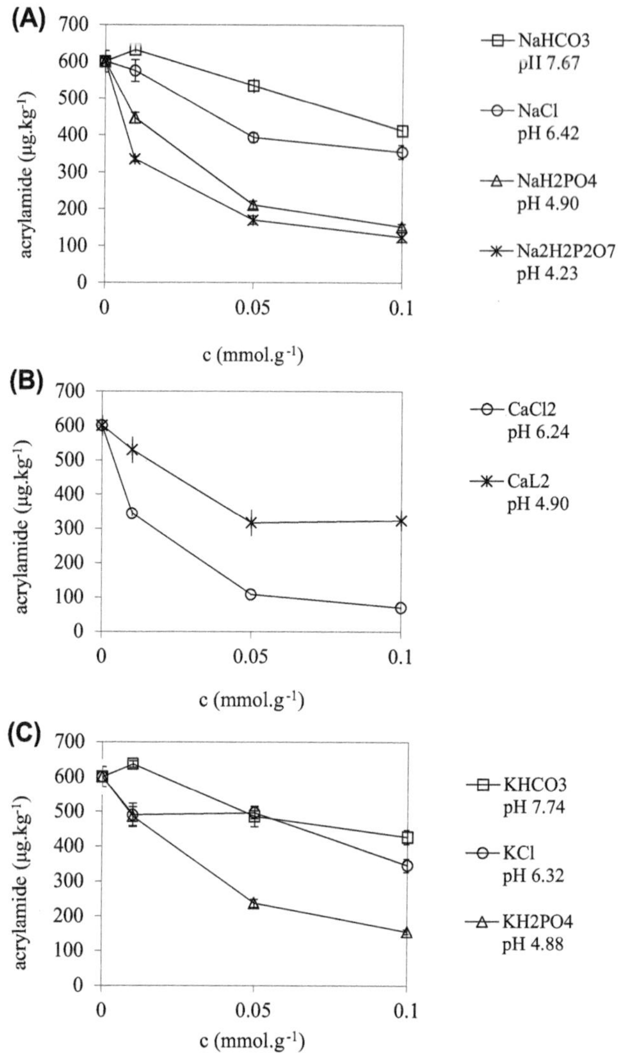

Figure 2 Effect of sodium (A), calcium (B), and potassium cations (C) of different salts on acrylamide formation in a cereal model system (190°C/9 min) [10].

MECHANISM OF ACTION

Acrylamide formation was found to occur during a thermal process, by the Maillard reaction, of reducing sugars with asparagine at temperatures of >120°C [14]. It was proposed that the presence of inorganic salts in the food system influences the formation of acrylamide by the Maillard reaction in several stages from interactions with acrylamide precursors [9,11,15], changing pH of the reaction system [7,10], or the acceleration of acrylamide polymerization [6].

PREVENTION OF SCHIFF BASE FORMATION

Mechanistic studies have proposed that N-glycosides and related compounds formed in the early phase of the Maillard reaction are key intermediates that lead to acrylamide [14,16]. The presence of monovalent and divalent cations in the reaction mixture influences the rate of decomposition of acrylamide precursors significantly due to the rate of glucose decomposition increased while most asparagine remains unreacted [8,9]. Addition of metal ions could give better high-temperature stability to asparagine matrix interactions and formation of a stable polymer network. This thereby renders the latter species unavailable for reaction with carbonyl precursors to produce acrylamide [11].

The first step in acrylamide production is the formation of a Schiff base between the carbonyl and α-amino group of asparagine by dehydration of the *N*-glycosyl compound [14,16]. The mass spectrometric analysis of pyrolysates demonstrates that in the corresponding extracted ion chromatograms the presence of Ca^{2+} prevents the formation of the Schiff base of asparagine (MW 294) and thus of acrylamide (MW 71) during heating [9].

CHANGES IN pH VALUE

Acrylamide formation and elimination are known to be pH sensitive with a maximum formation around a pH value from 7 to 9 [17]. Thus, lowering of pH has been suggested as a way to reduce acrylamide in foods. Asparagine is a critical participant in the production of acrylamide by the Strecker degradation mechanism of the Maillard reaction [14,18]. Schiff base formation is the initial reaction step in the Strecker degradation reaction. The first step is the nucleophilic addition of the free nonprotonated amine of asparagine to the carbonyl group. It was hypothesized that if the free nonprotonated amine were converted to a nonnucleophilic protonated amine, the reaction would be effectively blocked. The free nonprotonated amine is simply converted to a nonnucleophilic protonated amine by lowering the pH [19,20].

Metal cations have been known to cause a pH decrease when added to certain foods of food components. The mechanism of metal ion-induced pH lowering has been generally attributed to the competitive displacement of protons from ionizable functional groups, containing oxygen, nitrogen, or sulfur atoms that share electrons with hydrogen atoms, when the pH is high enough to permit mass action competition between protons and the metal cations, but the pH must not be so high that most protons have been removed prior to adding the metal ions [21]. The effect of the nonammonium-based raising agents is compared, for illustration, in Figure 2. This depicts different sodium, calcium, and potassium salts that could be explained entirely in terms of shifting the pH value in a model system [10]. Despite effective reduction by pH decreasing it should be taken into account that low pH is promoting formation of other contaminants such as 3-monochloropropane-1,2-diol (3-MCPD) [11] or HMF [7,22–24]. The divalent cations Ca^{2+} and Mg^{2+} were capable of inducing a significant pH reduction in the unheated model solutions. On the contrary, the pH increased with an increase in the CH_3COOK concentration. These pH changes would explain the acrylamide and HMF levels found in model solutions [7]. However, at the same pH value, the salts affected acrylamide and mainly HMF in a different manner, which would suggest that not only the pH but also the salt nature may influence the development of toxic molecules [7].

POLYMERIZATION OF ACRYLAMIDE

The amount of acrylamide generated during heating reaches an apparent maximum and then slowly decreases afterward [9]. Obviously, the content of acrylamide detected in the reaction mixture was the net result of the formation and degradation of this compound [25]. The decline of acrylamide is most likely due to polymerization [26]. The differential scanning calorimetry (DSC) record of pure acrylamide shows an endothermic peak with the onset at a temperature of 82 °C corresponding to the melting of acrylamide. At 156 °C, an exothermic peak starts, which is obviously connected with polymerization of acrylamide. In the DSC record of the mixture of acrylamide and NaCl, melting occurs at the same temperature as for pure acrylamide. However, the polymerization peak is shifted by 12 °C to lower temperatures. This thus indicates a strong acceleration of the acrylamide polymerization by NaCl. Moreover, DSC measurements indicated that the elimination of acrylamide was caused by its polymerization at elevated temperatures and stimulated by NaCl, even more by potassium ferrocyanide and potassium iodate, which are used as table salt additives. Polymerization reactions lead to the formation of a biologically inactive form of polyacrylamide. By comparing infrared spectra of acrylamide and the product formed, it was found that all inorganic compounds brought about polymerization of acrylamide molecules through new C–C linkages among vinyl groups, while carbonyl and amino groups of acrylamide molecules remained in the original state [27].

PROMOTION MECHANISM OF AMMONIUM SALTS

Baking agents are ingredients that mostly influence acrylamide formation in a sweet bakery. Ammonium hydrogen carbonate (NH_4HCO_3) is known to promote acrylamide formation during baking as confirmed by many researchers [13,28–31]. The effect of NH_4HCO_3 was investigated with respect to sugar fragmentation and formation of acrylamide from asparagine and sugar fragments in model systems under mild conditions. The presence of NH_4HCO_3 led to an increase in acrylamide and α-dicarbonyls from glucose and fructose, respectively. Ammonia from baking agent reacts with reducing sugars, whereby reactive sugar fragments, such as glyoxal, hydroxyethanal, and glyceraldehyde, are formed. These fragments in turn react with free asparagine, and deliver more acrylamide as compared to the reaction between asparagine and reducing sugars [15]. The enhancing effect of NH_4HCO_3 could be explained by (1) the action of NH_3 as a base in retroaldol reactions. This leads to sugar fragments, (2) facilitated retroaldol-type reactions of imines in their protonated forms and leads to sugar fragments, and (3) oxidation of the enaminols whereby glyoxal and other reactive sugar fragments are formed. These α-dicarbonyl and α-hydroxy carbonyl compounds may play a key role in acrylamide formation, especially under mild conditions [13].

APPLICATIONS OF INORGANIC SALTS ON ACRYLAMIDE MITIGATION IN FOODS

Acrylamide content in cereal products depends on their composition where asparagine content in flour and type of sugar are key determinants. The use of some types of flour such as oat, rye, or wholemeal with a higher content of asparagine or use of fructose in diabetic products showed to be considerably critical for acrylamide formation [4]. It is known that also baking agents strongly influence acrylamide formation in cereal foods [15]. To achieve the correct balance of gas release during baking and optimum texture, flavor, and color, combinations of NH_4HCO_3, $NaHCO_3$, and acidulant are often required [4]. Selection of inorganic salt addition in cereal products is limited by their influence on organoleptic properties, color, texture, and sensory preferences of product type, which should be evaluated individually [4].

GINGERBREAD

Gingerbread is a sweet bakery product in the form of a cake or a cookie, with various characteristics from crisp to a soft texture. Formulas for gingerbread dough vary depending on local customs. Generally, they comprise wheat or rye flour, sugar, honey, condiments, and chemical leavens [32]. Ammonium-based raising agents are widely used in biscuit manufacture, either individually or as a part of a mixed raising system providing browning, crisp, and porous crumb [11]. The amount of ammonium hydrogen carbonate (NH_4HCO_3) that is typically used in gingerbread has a very strong influence on acrylamide formation [15]. Therefore, the replacement of ammonium hydrogen carbonate by sodium hydrogen carbonate ($NaHCO_3$) led to a significant reduction of acrylamide content by around 60–70 % in the final product, which was observed by several researchers [15,29,33]. Using the mixture of various sodium salts such as combination of $NaHCO_3$ and $Na_2H_2P_2O_7$, or sodium acid pyrophosphate (SAPP) resulted in a substantial acrylamide of a >90% decrease from a concentration of 1200 to 70 μg/kg [34]. However, a proportion of ammonium hydrogen carbonate that could be replaced in gingerbread without undesirable organoleptic changes is limited and needs to be adjusted to each individual recipe. Total exclusion of ammonium raising agent altered the product to such an extent that caused unacceptability to consumers due to its unpleasant salty taste [30], fine porosity [29], and uncharacteristic pale color caused by suppressing of the browning intensity [13,29,33].

SHORT-DOUGH BISCUITS AND COOKIES

Biscuits and cookies represent a very variable category of cereal products with a moderate acrylamide content. Sugar and fat levels are relatively high in short-dough recipes. Baking temperatures and times are not excessive, and the amount of water that has to be removed is relatively low [1]. High asparagine flours such as oat, rye, or wholemeal ones as well as fructose content or some specific ingredients such as dry fruits, fruit fiber, or various nuts very often cause a significant increase of acrylamide content in this type of cereal products [4].

Addition of inorganic salts could moderate the acrylamide level in biscuits and cookies to some extent; however, the effect of inorganic salts depends on a particular formulation and processing conditions. For this reason, the effect of inorganic salts in biscuits and cookies is normally assessed specifically for each particular product. A high potential influence of various calcium inorganic salts on acrylamide elimination in biscuits and cookies was observed by many researchers [24,35–37]. Fortification of cookies with calcium lactate and calcium acetate produced darker colored cookies, while the addition of calcium citrate and calcium carbonate to the cookie recipe led to a lighter color [37]. Calcium carbonate added at a dosage of 1% to the dough was the most effective in reduction of acrylamide formation in a cereal model and cookies by 60% and 44%, respectively. This improved their overall acceptability [37]. Using calcium chloride, a 58% reduction of acrylamide in the cookies was achieved [24].

In sweet and savory baked and deep-fried leavened flour-based biscuits, the acrylamide content was reduced by 31% and 38%, respectively, when soda (Na_2CO_3) or baking soda ($NaHCO_3$) was used [38].

Individual inorganic salts in the mixture could influence each other in acrylamide formation. The synergic effect of salts, for example, $CaCl_2$ and $MgCl_2$ was observed with a 60% reduction of acrylamide in biscuits [7]. It was, however, pointed out that in the presence of potassium acetate (CH_3COOK), the addition of $CaCl_2$ and $MgCl_2$ was not effective on acrylamide mitigation due to possible interactions among the components [7]. The presence of two calcium derivatives, calcium lactate and $CaCl_2$, in commercially available products (Puracal Act 100 and Puracal Act 200) showed a significant acrylamide reduction in cookies up to 70% [22]. Puracal Act 100, which was characterized by a higher proportion of $CaCl_2$ and a lower calcium lactate content, was proposed in the study on addition up to 0.5%, which is suitable for maintaining of organoleptic properties of cookies. A higher addition of calcium increased acrylamide reduction up to

81.25%. However, a negative impact on color with increasing in lightness (L^*) and decreasing in redness (a^*) in a CIELab system using 1% addition of calcium salts and increasing of HMF were detected [22].

Sodium chloride (NaCl) is traditionally used in cereal products as it causes several important changes in rheological, technological, and sensory parameters. Rheological properties are influenced by NaCl since it is involved in the structure and formation of the gluten matrix; this increases stability, flexibility, and resistance of the dough and decreases stickiness and water absorption [39]. Decreasing sodium chloride concentration may induce less desirable properties such as a more liquid and stickier dough, lack of stability, and less resistance and extensibility. Regarding sensorial characteristics, NaCl is related to the perception of a salty taste and it also increases sweetness and masks its metallic and bitter taste [11]. In model systems, NaCl effectively reduced acrylamide formation [5,6]. However, the impact of NaCl on acrylamide formation in biscuits and cookies was ambiguous. The presence of 0.6% NaCl in biscuit formulation lowered acrylamide concentrations at 180 and 190 °C, but not at 200 °C. At all tested temperatures, NaCl led to a higher formation of HMF. The concentration of HMF after 15 min at 200 °C of baking was approximately four times higher in the recipe with NaCl in comparison to the control. Sucrose hydrolysis was a key step in acrylamide and HMF formations during biscuit baking [23]. Partial substitution of NaCl by commercial salt replacers is recommended for a 35–40% reduction of acrylamide in cookies. This fact could be explained by a possible synergic effect between salts when these are mixed together in the same dough [24].

NH_4HCO_3 application in biscuits is similar to gingerbread as stated above. The complete replacement of NH_4HCO_3 by $NaHCO_3$ reduced acrylamide content by >70% in biscuits prepared on an industrial scale. If NH_4HCO_3 was not fully replaced, the acrylamide content did not decrease. However, biscuits prepared without NH_4HCO_3 showed lesser leavening as compared to the standard product [29].

SOFT BREAD

Soft bread belongs to the most frequently consumed staple food. Acrylamide content in soft bread ranged between 30 and 75 μg/kg during 2007–2010 [40]. Despite the fact that acrylamide content in soft bread is low, it is an important acrylamide source in diet [41]. Acrylamide content in the crust of wheat-rye bread was found at concentrations up to 257 μg/kg of fresh weight, which resulted in an intake of acrylamide of 8.2 μg per day [31]. Acrylamide in bread is formed in the crust that is distinguished from the crumb in a colored dry and dense superficial structure [42,43].

Among various inorganic salts (chlorides, phosphates, and lactates of sodium, potassium, and calcium), the calcium chloride ($CaCl_2$) slightly suppressed acrylamide formation and concurrently improved sensorial and functional properties of bread loaves [31].

Calcium chloride ($CaCl_2$) is used in some countries such as the United Kingdom for standard calcium fortification required by the law for nutritional reasons [11] at concentrations of 0.3% in flour. This gives also a significant reduction in acrylamide level in the product [21,31]. From various inorganic salts that were tested as additives in a dry premix for homemade bread production in the concentration range from 2.5 to 3.0 g/kg, calcium chloride ($CaCl_2$) was the most effective in suppression of acrylamide formation in bread crust up to 10% with simultaneous improvement of sensorial and functional properties of the final product; this includes porosity, elasticity, and specific bread loaf volume [31], and therefore, $CaCl_2$ could be recommended for soft bread production [31]. Higher potential of calcium chloride in bread was observed using 0.04 M of $CaCl_2$ with simultaneous pH adjustment of dough to control, which gave up to 36% of acrylamide reduction [21]. Addition of table salt (NaCl) in the same conditions led to a 23% reduction in acrylamide content; meanwhile, $CaCO_3$ was not effective [21]. The presence of table salt (NaCl) in yeast-leavened bread formulation could play an ambiguous role, whereas low doses up to 2% lowered acrylamide by inhibition of the enzyme amylase activities; higher addition remarkably increased the contents due to an inhibition of yeast growth. Therefore, the optimal concentration of NaCl to minimize acrylamide levels in wheat bread and bread roll is 1–2 %, which is generally used in industrial production [33].

EXTRUDATES

Extrusion cooking is a low-cost, high-temperature, short time process, used worldwide for processing of a number of food products including snacks, ready-to-eat cereals, confectioneries, and crisp breads [19]. Acrylamide is easily generated in low moisture system. Extrusion processing of cereals and starch tuber flours is carried out under relatively lower moisture contents, and further, such extruded products have a very low moisture content of ≤5%. Inorganic salts showed to be effective in acrylamide mitigation strategy in extrusion technological processing of cereals [19,20,44]. Concentration of acrylamide in extruded snacks lowered acrylamide from 1500 ± 2 ppb to 526 ± 2 ppb (by 65%) using $CaCl_2$ at 50 μmol/g addition to flour blend of potato flour and semolina (70:30, w/w) without altering the sensory properties. Calcium lactate at the same concentration lowered 40% of acrylamide to a final concentration of 900 ± 10 μg/kg with a slightly lower sensory score [19].

Acrylamide content of wheat flakes toasted at 230 °C for 10 min showed that the presence of NaCl at a 2.5% addition resulted in a significantly lower acrylamide content in the product up to 30%. Regardless of the NaCl concentration, the acrylamide content increased during the first few minutes of toasting, after which acrylamide content remained stable or eventually decreased. It should be noted that acrylamide content did not follow color formation. The reason is that salts might have a different influence on Maillard reactions depending on the pathway taken. Acrylamide is a compound formed only via Maillard reactions, while color is formed via both Maillard and a caramelization reaction. Therefore, the different trends observed concerning the impact of NaCl on color and acrylamide might be due to different impacts of NaCl on each of these reactions. Indeed, NaCl might slow down the Maillard reaction pathway and lead to acrylamide formation while it could enhance the last few steps of the Maillard reactions to form melanoidins [44].

LIMITATIONS OF INORGANIC SALTS IN FOOD APPLICATIONS

Food formulation dictates flavor and color development and play an important role in controlling texture and mouthfeel in food products. The levels of Maillard reaction products also seem to be greatly affected by product formulation [4].

Replacement of the ammonium raising agent by sodium salts often results in a substantial decrease in acrylamide level. However, the problem with sensory impact of different raising agents emerged. The sensorial properties were altered to such an extent (tasty, softness, suppleness, porousness) that it caused unacceptability to consumers [34]. High doses of sodium bicarbonate will give biscuits an alkaline taste, a yellowish crumb, and surface coloration and an accompanying unpleasant taste known as soda bite [1].

The use of sodium bicarbonate as a leavening agent for bakery products is also limited by sodium intake as a potential risk factor for cardiovascular disease. Mitigation of acrylamide in biscuits and gingerbread was accompanied by a small increase in sodium intake. Around 1.3% of the population shifted from a sodium intake below to above 40 mg/kg bw/d [4].

Despite the promising efficiency of $CaCl_2$ in acrylamide reduction strategy in food, trials at a bench scale have given variable results; therefore, this approach has not yet been commercialized for biscuits [4]. The quality of products was strongly affected by $CaCl_2$, which hindered the rise and biscuits were unpalatable when tasted [7]. Differently to biscuits, fortification of flour by $CaCl_2$ improved final organoleptic properties, porosity and specific volume of soft bread [31].

One of points to be considered during the modification of cookie recipe is the increased formation of HMF when calcium derivatives are added [7,22,24,45]. Divalent metal ions added to food products may improve high-temperature stability to asparagine/matrix interactions, and decreases the availability for reaction with carbonyl precursors to generate acrylamide. Consequently, formation of the Schiff base, which is an essential intermediate of acrylamide formation, is prevented and the reaction path is changed toward to the dehydration of glucose. This leads to the formation of HMF and furfural [8].

SYNOPSIS OF ANALYTICAL TECHNIQUES

Numerous analytical methods have been developed to determine acrylamide in different heat-treated food samples. An effective procedure was established for the determination of acrylamide in cereal products by a liquid chromatography-tandem mass spectrometry (LC–MS–MS) with positive electrospray ionization [12,46]. Electrospray ionization is a sensitive technique that is used for the analysis and identification of small molecules. Manufacturers try to achieve and enhance sensitivity in electrospray ionization–MS by improving the desolvation and spatial focusing of ions [47]. Extraction conditions appear to be the most critical parameters in sample preparation for the detection of acrylamide in cereal products with inorganic salt content. Many factors can affect acrylamide extraction yield from food matrices such as sample particle size, defatting, extraction solvent, homogenization, application of mechanical forces, extraction temperature, and extraction time [48,49]. Water is the most efficient extraction solvent due to its high solubility and hydrophobicity of acrylamide. By using water also other unwanted compounds such as inorganic salts are extracted, which might interfere with the detection and degenerate the chromatographic system. Therefore, these interferences should be removed by a clean-up [48]. Using ethyl acetate to extract acrylamide from the aqueous phase enhanced the removal of interfering constituents such as salt and also sugars, starch, or amino acids [50]. The sample preparation is started by extracting the analyte from the ground sample with water or aqueous solution of acetic acid. Due to the hydrophilic nature of acrylamide, liquid–liquid extraction of acrylamide from water phase using ethyl acetate is needed to provide three times for a satisfactory extraction yield due to the hydrophilic nature of acrylamide. Ethyl acetate could be easily concentrated by rotary evaporation and quantitatively transferred back into the aqueous phase ready for the final injection [46].

KEY FACTS

- Inorganic salts influence acrylamide formation in the Maillard reaction. The addition of inorganic salts might change the optimal pH for acrylamide formation during baking of cereal products or influence the rate of decomposition of acrylamide precursors and Schiff base formation.
- The use of ammonium salts is critical due to the significant acceleration of acrylamide formation; therefore, their application as a baking agent should be avoided. For that reason, ammonium hydrogen carbonate (NH_4HCO_3) is typically replaced by sodium hydrogen carbonate ($NaHCO_3$), which represents a very effective way of avoiding high acrylamide content in gingerbread and biscuits.
- In cereal model systems, the application of divalent cations such as calcium salts was observed as the most effective way for acrylamide mitigation.

- There are limitations in the use of inorganic salts for acrylamide mitigation in cereal products, including health aspects, since the presence of inorganic salts may enhance the formation of other food contaminants such as HMF and 3-MCPD.
- Organoleptic properties, hardness, porosity, texture, color, and taste of cereal products are influenced by a raising agent type. Consumer preferences and line production standards also determine the selection of inorganic salts in recipes for each individual cereal product.

MINI-DICTIONARY

Leavening This is defined as a raising action that aerates dough or batter during mixing and baking so that the finished products are greater in volume and superior in texture with a porous structure. Leavening could be mechanical (incorporation of air by creaming or beating fat, sugar, and eggs), biological (conversion of simple sugars to carbon dioxide and alcohol by yeasts), or chemical (release of carbon dioxide).

Chemical leavening This includes the use of chemicals such as baking soda, ammonium bicarbonate, or baking powder; they are also called raising agents.

Baking soda This is also known as sodium bicarbonate and is used in recipes that contain an acidic ingredient and liberates carbon dioxide when heated or when mixed, either hot or cool.

Ammonium bicarbonate This dissociates during baking into carbon dioxide gas, ammonia gas, and water.

Baking powders These are combinations of baking soda (sodium bicarbonate) and dry acids or acid salts, usually with starch added for stability during storage.

Single-acting baking powders These release carbon dioxide at room temperature, and dough in which they are used must be baked immediately to avoid loss of most of the gas.

Double-acting baking powders These contain two acid salts, one reacts at room temperature to produce a rise as soon as the dough or batter is prepared, and the other reacts at a higher temperature, and causes a further rise during baking. Examples of high-temperature acid salts are aluminum salts, such as calcium aluminum phosphate.

Differential scanning calorimetry This is a thermal analysis technique that monitors heat effects associated with phase transitions and chemical reactions as a function of temperature. DSC allows the detection of transitions such as melts, glass transitions, phase changes, or polymerization of acrylamide. The peak temperature is associated with the temperature at which maximum reaction rate occurs.

SUMMARY POINTS

The addition of inorganic salts showed a great impact on the formation of acrylamide as a heat-induced food contaminant. The use of calcium and other divalent cations was determined as the most effective way of acrylamide elimination in model cereal systems and also in various cereal food products, mainly in soft bread. The use of ammonium salts as a raising agent typically led to a significant increase in the acrylamide level due to a change of the pH value to basic. The acrylamide content in many cereal products can be lowered up to 90% when the usage of ammonium hydrogen carbonate is avoided. It was observed that during the baking process, the formation of a Schiff base was prevented in the presence of inorganic salts, which represents a second mechanism of their action. The elimination of an already formed acrylamide is accelerated by a table salt, which is able to initiate polymerization reactions of this food contaminant. However, acrylamide mitigation strategies using inorganic salts are limited due to the change of the organoleptic properties of the cereal products and a simultaneous increase of other contaminants, mainly furan derivatives such as HMF or 3-MCPD.

ACKNOWLEDGMENT

This work was supported by the Research & Development Operational Programme funded by the European Regional Development Fund through implementation of the projects "Strategy of acrylamide elimination in food processing—ITMS 26240220091," "Establishment of HiTech Centre for Research of Formation, Elimination, and Assessment of Contaminants in Food—ITMS 26240120042," and by the Slovak Research and Development Agency under the contract No. LPP 0310-09.

REFERENCES

[1] Manley D. Additives as biscuit ingredients. In: Manley's technology of biscuits, crackers and cookies. Cambridge, UK: Woodhead Publishing; 2011. p. 223–34.
[2] Pop G. Researches regarding the chemical leavening agents' role in quality of bakery products. J Agroalim Process Technol 2007;8(1):105–12.
[3] Edwards WP. The science of bakery products. Cambridge, UK: RSC Publishing; 2007. 245 pp.
[4] CIAA acrylamide Toolbox. Brussels: Food Drink Europe 2014; 2013. Rev 13. http://ec.europa.eu/food/food/chemicalsafety/contaminants/toolbox_acrylamide_201401_en.pdf.
[5] Kolek E, Šimko P, Šimon P. Inhibition of acrylamide formation in asparagine/D-glucose model system by NaCl addition. Eur Food Res Technol 2006;24(2):283–4.
[6] Šimko P, Kolek E, Šimon P, Šimúth T, Marková L. Elimination of acrylamide by polymerization catalysed by inorganic food components. J Food Nutr Res 2009;48(1):8–13.
[7] Quarta B, Anese M. The effect of salts on acrylamide and 5-hydroxymethylfurfural formation in glucose–asparagine model solutions and biscuits. J Food Nutr Res 2010;49(2):69–77.
[8] Gökmen V, Şenyuva HZ. Effects of some cations on the formation of acrylamide and furfurals in glucose–asparagine model systems. Eur Food Res Technol 2007;225:815–20.
[9] Gökmen V, Şenyuva HZ. Acrylamide formation is prevented by divalent cations during the Maillard reaction. Food Chem 2007;103:196–203.
[10] Kukurová K, Ciesarová Z, Bednáriková A, Marková L. Effect of inorganic salts on acrylamide formation in cereal matrices. Czech J Food Sci 2009;27:S425–8.
[11] Sadd PA, Hamlet CG, Liang L. Effectiveness of methods for reducing acrylamide in bakery products. J Agric Food Chem 2008;56(15):6154–61.
[12] Ciesarová Z, Kukurová K, Bednáriková A, Morales JF. Effect of heat treatment and dough formulation on the formation of Maillard reaction products in fine bakery products—benefits and weak points. J Food Nutr Res 2009;48(1):20–30.
[13] Amrein TM, Andres L, Manzardo GG, Amado R. Investigation on the promoting effect of ammonium hydrogencarbonate on the formation of acrylamide in model systems. J Agric Food Chem 2006;54(26):10253–61.
[14] Stadler RH, Blank I, Varga N, Robert F, Hau J, Guy PA, et al. Acrylamide from Maillard reaction products. Nature 2002;419(6906):449–50.
[15] Amrein TM, Schönbächler B, Escher F, Amadò R. Acrylamide in gingerbread: critical factors for formation and possible ways for reduction. J Agric Food Chem 2004;52(13):4282–8.
[16] Zyzak DV, Sanders RA, Stojanovic M, Tallmadge DH, Eberhart BL, et al. Acrylamide formation mechanism in heated foods. J Agric Food Chem 2003;51:4782–7.

[17] De Vleeschouwer K, Van Der Plancken I, Van Loey A, Hendrickx ME. Impact of pH on the kinetics of acrylamide formation/elimination reactions in model systems. J Agric Food Chem 2006;54(20):7847–55.
[18] Mottram DS, Wedzicha BL, Dodson AT. Acrylamide is formed in the Maillard reaction. Nature 2002;419(6906):448–9.
[19] Mulla MZ, Bharadwaj VR, Annapure US, Singhal RS. Effect of formulation and processing parameters on acrylamide formation: a case study on extrusion of blends of potato flour and semolina. LWT-Food Sci Technol 2011;44(7):1643–8.
[20] Masatcioglu MT, Gokmen V, Ng PK, Koksel H. Effects of formulation, extrusion cooking conditions, and CO_2 injection on the formation of acrylamide in corn extrudates. J Sci Food Agric 2014;94(12):2562–8.
[21] Levine RA, Ryan SM. Determining the effect of calcium cations on acrylamide formation in cooked wheat products using a model system. J Agric Food Chem 2009;57(15):6823–9.
[22] Acar OC, Pollio M, Di Monaco R, Fogliano V, Gokmen V. Effect of calcium on acrylamide level and sensory properties of cookies. Food Bioprocess Technol 2012;5:519–26.
[23] Van Der Fels-Klerx HJ, Capuano E, Nguyen HT, Atac Mogol B, Kocadagli T, Goncuoglu Tas N, et al. Acrylamide and 5-hydroxymethylfurfural formation during baking of biscuits: NaCl and temperature–time profile effects and kinetics. Food Res Intar 2014;57:210–7.
[24] Mesías M, Holgado F, Márquez-Ruiz G, Morales FJ. Effect of sodium replacement in cookies on the formation of process contaminant and lipid oxidation. LWT-Food Sci Technol 2015;62(1):633–9.
[25] Elmore JS, Koutsidis G, Dodson AT, Mottram DS, Wedzicha BL. Measurement of acrylamide and its precursors in potato, wheat, and rye model systems. J Agric Food Chem 2005;53(4):1286–93.
[26] Stadler RH, Robert F, Riediker S, Varga N, Davidek T, Devaud S, et al. In-depth mechanistic study on the formation of acrylamide and other vinylogous compounds by the Maillard reaction. J Agric Food Chem 2004;52(17):5550–8.
[27] Kolek E, Šimko P, Šimon P, Gatial A. Confirmation of polymerisation effect of sodium chloride and its additives on acrylamide by infrared spectrometry. J Food Nutr Res 2007;46(1):39–44.
[28] Vass M, Amrein TM, Schonbachler B, Escher F, Amado R. Ways to reduce the acrylamide formation in cracker products. Czech J Food Sci 2004;22(special issue):19–21.
[29] Graf M, Amrein TM, Graf S, Szalay R, Escher F, Amado R. Reducing the acrylamide content of a semi-finished biscuit on industrial scale. LWT-Food Sci Technol 2006;39(7):724–8.
[30] Kukurová K, Marková L, Bednáriková A, Ciesarová Z. Tools for acrylamide mitigation in cereal products. Potravinárstvo 2010;4(2):317–21.
[31] Ciesarová Z, Kukurová K, Marková L. Success and limitations in acrylamide mitigation efforts Part 2: Impact of interventions in cereal food processing on exposure. Agro Food 2011;22(4):25–7.
[32] Cervenka L, Rezkova S, Kralovsky J. Moisture adsorption characteristics of gingerbread, a traditional bakery product in Pardubice, Czech Republic. J Food Eng 2008;84(4):601–7.
[33] Claus A, Mongili M, Weisz G, Schieber A, Carle R. Impact of formulation and technological factors on the acrylamide content of wheat bread and bread rolls. J Cereal Sci 2008;47:546–54.
[34] Ciesarová Z, Kukurová K, Bednáriková A, Marková L, Baxa S. Influence of food processing on acrylamide level in gingerbreads and cookies. Asp Appl Biol 2009;97(1):87–92.
[35] Gökmen V, Açar ÖC, Serpen A, Morales FJ. Effect of leavening agents and sugars on the formation of hydroxymethylfurfural in cookies during baking. Eur Food Res Technol 2008;226:1031–7.
[36] Acar O, Gokmen V. Investigation of acrylamide formation on bakery products using a crust-like model. Mol Nutr Food Res 2009;53:1521–5.
[37] Chang BK-L, Wang J-S, Sung W-C. Calcium salts reduce acrylamide formation and improve qualities of cookies. J Food Nutr Res 2014;2(11):857–66.
[38] Wang H, Feng F, Guo Y, Shuang S, Choi MMF. HPLC–UV quantitative analysis of acrylamide in baked and deep-fried Chinese foods. J Food Compos Anal 2013;31(1):7–11.
[39] Beck M, Jekle M, Becker T. Impact of sodium chloride on wheat flour dough for yeast-leavened products. I. Rheological attributes. J Sci Food Agric 2012;92(3):585–92.
[40] European Food Safety Authority (EFSA). Update on acrylamide levels in food from monitoring years 2007 to 2010. EFSA J 2012;10(10):2938–76.
[41] Ciesarová Z. Success and limitations in acrylamide mitigation efforts Part1: Relevance, occurrence and exposure. AgroFOOD 2011;22(3):30–2.

[42] Ahrne L, Andersson CG, Floberg P, Rosen J, Lingnert H. Effect of crust temperature and water content on acrylamide formation during baking of white bread: steam and falling temperature baking. LWT-Food Sci Technol 2007;40(10):1708–15.
[43] Marková L, Kukurová K, Ciesarová Z, Bednáriková A, Šimko P, Behan T, et al. Distribution of acrylamide in particular parts of bread. Chem Listy 2010;104:660–4.
[44] Moreau L, Lagrange J, Bindzus W, Hill S. Influence of sodium chloride on colour, residual volatiles and acrylamide formation in model systems and breakfast cereals. Inter J Food Sci Technol 2009;44(12): 2407–16.
[45] Fiore A, Troise AD, Mogol BA, Roullier V, Gourdon A, El Mafadi Jian S, et al. Controlling the Maillard reaction by reactant encapsulation: sodium chloride in cookies. J Agric Food Chem 2012;60:10808–14.
[46] Bednáriková A, Ciesarová Z. Validation of HPLC–ESI–MS–MS method for acrylamide determination in bakery products (Comparison of simple and improved electrospray ionization). Chem Listy 2012;106:252–6.
[47] Hoenicke K, Gatermann R, Harder W, Hartig L. Analysis of acrylamide in different foodstuffs using liquid chromatography–tandem mass spectrometry and gas chromatography–tandem mass spectrometry. Anal Chim Acta 2004;520:207–15.
[48] Petersson E, Rosén J, Turner C, Danielsson R, Hellenas K-E, Markides K. Critical factors and pitfalls affecting extraction of acrylamide from foods: an optimisation study. Anal Chim Acta 2006;557:287–95.
[49] Wenzl T, de la Calle B, Anklam E. Analytical methods for the determination of acrylamide in food products. Food Addit Contam 2003;20(10):885–902.
[50] Delatour T, P′erisset A, Goldmann T, Riediker S, Stadler RH. Improved sample preparation to determine acrylamide in difficult matrixes such as chocolate powder, cocoa, and coffee by liquid chromatography tandem mass spectroscopy. J Agric Food Chem 2004;52(15):4625–31.

CHAPTER 20

Inhibition of Acrylamide Formation by Vanadium Salt in French Fries and Potato Chips

Diganta Kalita, Sastry S. Jayanty
Department of Horticulture and Landscape Architecture, San Luis Valley Research, Colorado State University Center, CO, USA

INTRODUCTION

Since the discovery of potential acrylamide formation in carbohydrate-rich heated foodstuff [19,27,29], studies on acrylamide are receiving continuous attention worldwide. Acrylamide is known as a neurotoxin, genotoxin [1], and "probable carcinogen to humans (Group2A)" [8]. A number of studies have been reported toward the understanding the formation and mitigation of acrylamide [3,17,34]. It has been reported that acrylamide is formed via the Maillard reaction between asparagine and reducing sugars at high temperatures [25,27,28]. Acrylamide formation depends upon many factors such as content of reducing sugars, free asparagine, cooking process, including temperature, time, pH, and surface-to-volume ratio of food materials [3,26]. Citric acid and some organic acids, which are used for pretreatment to lower the pH of food, are reported to be effective acrylamide-reducing agents [9,11]. Amino acids or proteins, which can combine with acrylamide, are also found to be effective acrylamide reducing agents [18]. Recent studies indicated that monovalent, divalent, and trivalent metal cations efficiently reduce the acrylamide formation in an asparagine–glucose model reaction and in some food model systems, including potato and wheat [5,6,12].

Vanadium is a trace element usually present at low levels in plants and animals. Some foods such as mushrooms, shellfish, black pepper, parsley, dill weed, beer, wine, grain, and grain products have significant levels of vanadium. In mammals, vanadium is distributed in all tissues, and the highest concentration is found in nails, skin, plasma, and liver. Vanadium exists in many oxidation states (−I to +V); however, the most stable states are +IV and +V, which are known as vanadyl and vanadate. There are several biological and pharmacological applications of vanadium compounds in the treatments of various pathological conditions such as diabetes and hypertension, cancer, and inflammatory activities. It has been reported that vanadium and its compounds exhibit insulin-like activity. Consequently, different classes of vanadium compounds have been synthesized and tested for antidiabetic and anticancer properties. Vitamin B6- [20] and vitamin

Acrylamide in Food
http://dx.doi.org/10.1016/B978-0-12-802832-2.00020-6

E- [24] combined vanadium complexes have been reported to have better antidiabetic activity than vanadium alone. The chlorogenic acid vanadium complex has been shown to have antioxidant and anticancer activities [21]. These compounds reduce blood glucose by stimulating glucose uptake and metabolism, and glycogenesis. Vanadium-supplemented foods have been processed to improve antidiabetic and anticancer activity, for example, vanadium-supplemented fish oil has been proposed to have improved anticancer activity [15]. The oral administration of vanadium compounds normalizes blood glucose levels in streptozotocin (STZ)-induced diabetic rats, and similar effects have been observed in animal models that resemble type-2 diabetes condition. Vanadyl sulfate where vanadium is present in dicationic form is used as a nutritional supplement [22].

Various metal salts of sodium, potassium, and calcium have been shown to reduce the levels of acrylamide in cereals, French fries, and in an asparagine–glucose model system. However, there was no study on the inhibitory effect of vanadium in acrylamide until we investigated the effect of vanadium salt in the reduction of acrylamide in French fries and potato chips.

ASPARAGINE–GLUCOSE REACTION SYSTEMS AND EFFECT OF VO^{2+} ON ACRYLAMIDE FORMATION

Various research reports indicate that amino acids and reducing sugars are the major precursors of acrylamide formation. As a result, a mixture of amino acids (e.g., asparagine, cysteine, and glutamine) and reducing sugar (e.g., glucose, fructose, and galactose) has been tested in a model reaction for acrylamide formation. However, an aqueous solution of asparagine and glucose is widely used in model reactions to study the mechanism and kinetics of acrylamide formation. Moreover α-hydroxy carbonyls are much more reactive and form acrylamide with asparagine. It has been reported that an equimolar mixture of asparagine in water and/or buffer with different pHs are found to be one of the best models to study the kinetics of acrylamide formation. This reaction occurs via the Maillard reaction following a Schiff base intermediate rearrangement. However, the reaction condition depends on many other factors. Optimum acrylamide formation is seen at 1:1 asparagine and reducing sugars. Temperature-dependent studies revealed that acrylamide formation goes up until 120 °C, but after a certain temperature, it goes downward. Heating time also leads to the enhancement of acrylamide. We carried out a model reaction containing 12 mM of asparagine and glucose and heated the contents for 30 min. The amount of acrylamide formed was 1187.77 ± 29.89 ng/ml. Keeping the same ingredients, aliquots of vanadyl sulfate with a final concentration of 1, 3, and 6 mM $VOSO_4$ were added to it. The amounts of acrylamide formed are shown in Figure 1. A significant difference was found between the addition of different concentrations of $VOSO_4$ solution compared with the model system without the addition of $VOSO_4$ ($p < 0.05$). The highest reduction, that is, 97.69% occurred in the presence of 6 mM $VOSO_4$.

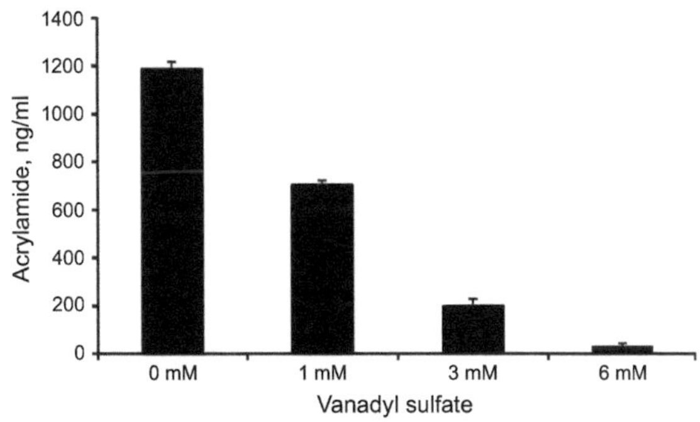

Figure 1 Acrylamide formation in the asparagine–glucose model reaction. The reaction mixture contains an equimolar concentration (12 mM) of asparagine and glucose with and without $VOSO_4$.

The number of model reactions in aqueous solutions has been reported to understand the possible mechanism of formation as well as mitigation of acrylamide [5,6,12,32,33]. The presence of monovalent, divalent, and trivalent cations, namely, Na^+, K^+, Ca^{2+}, Mg^{2+}, Zn^{2+}, Fe^{3+} has been studied by Gökmen et al. [6] to investigate the possible inhibitory effect on acrylamide formation by metal ions. The addition of cations significantly mitigated acrylamide formation besides enriching the finished food product. In another study, Koleke et al. [12] reported that the addition of sodium chloride in the asparagine–glucose model resulted in a significant reduction of the acrylamide formed. It has been postulated that the mitigation of acrylamide occurred due to the prevention of formation of the Schiff base from asparagine and sugar [5,6,13,14]. Lindsay and Jang [13] hypothesized that ionic associations of calcium cation and the charged groups of asparagine and related intermediate are likely to be involved during the course of the reaction. However, there was no direct evidence for the formation of such complex between metal cation and asparagine. To see if the reduction of acrylamide formed by vanadyl is due to the formation of an asparagine VO^{2+} complex, we carried out spectrophotometric studies for different combinations of the aforementioned model reaction precursors. Four different combinations were tested: (1) $VOSO_4$ solution; (2) $VOSO_4$ and asparagine; (3) $VOSO_4$, asparagine, and glucose; (4) $VOSO_4$ and glucose. The absorbances of these solutions were recorded after heating the solutions at 150 °C for 30 min (Figure 2). The spectrum of $VOSO_4$ showed an absorption maximum at 764 nm and a small shoulder at 625 nm. A significant observation was made with the reaction mixture containing $VOSO_4$ and asparagine, where it was found that the peak at 625 nm had shifted to 594 nm with a higher intensity. This clearly indicates the coordination of asparagine to the vanadyl ion. A similar shift in the absorption at 625 nm was also observed in the spectrum of the reaction mixture containing $VOSO_4$, asparagine, and glucose. However, the spectrum of the mixture containing $VOSO_4$ and glucose was found to be identical to the spectrum of $VOSO_4$, which suggests a complex that was not formed between glucose and $VOSO_4$. Thus, ultraviolet–visible (UV–Vis)

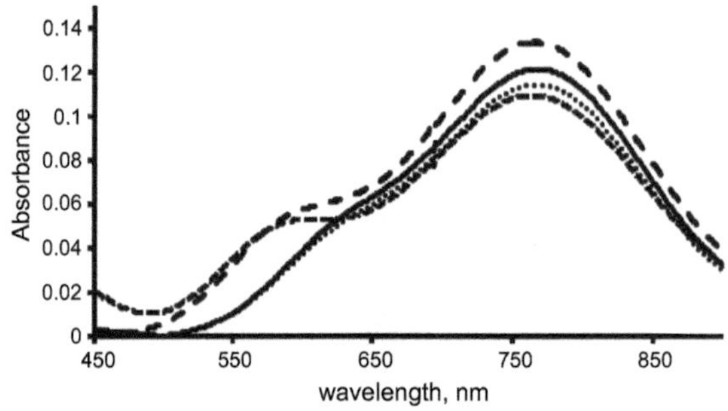

Figure 2 UV–Vis spectra of (1) $VOSO_4$ (—); (2) $VOSO_4$ and asparagine (- - -); (3) $VOSO_4$, asparagine, and glucose (- - -); (4) $VOSO_4$ and glucose (······).

spectrophotometric studies demonstrated the possibility of formation of such complexes with the addition of VO^{2+} to the reaction mixtures. Thus, the addition of $VOSO_4$ to a solution containing asparagine and glucose led to the formation of a vanadyl–asparagine complex, which inhibited Schiff base formation and ultimately hindered the course of the Maillard reaction.

ACRYLAMIDE FORMATION IN THE POTATO MODEL SYSTEM

A significant level of reducing sugars is accumulated in potato tubers throughout the growing periods; however, the minimum level is observed in matured tubers. Therefore, harvesting time and storage condition greatly influence the levels of these reducing sugars. Storing at temperatures <10 °C enhances the levels of sugars, which is known as cold-induced sweetening. Immature tubers accumulate higher reducing sugars than do matured ones. Other factors such as humidity also greatly influence the levels of sugars. Generally, during cold induced storage, sucrose is converted to glucose and fructose by the vacuolar enzyme acid invertase (VInv) [31]. Thus, genetic manipulation ofVInv can reduce cold-induced sweetening that occurs in potato tubers. It has been recommended that potato tubers for chip production should have a maximum concentration of sugars. It should have 0.30–0.35 mg/g fresh weight. Therefore, tubers intended for frying and chip processing are usually kept at 8–10 °C to minimize reducing sugar formation.

Potato tubers consist of many amino acids. Among them, the predominant amino acids are asparagine and glutamine. They both constitute 90% of the total free amino acid composition. The enzymes glutamine synthase and asparagine synthase are responsible for the accumulation of these amino acids. Therefore, triggering of these enzymes can affect the levels of these free amino acids in potato tubers. Aparagine levels greatly vary among potato cultivars. The concentration of asparagine was in the range of 1.8–9 mg/g across potato cultivars and advanced selections [10]. Other amino acids such as proline, alanine, valine, tyrosine, and histidine also vary among potato cultivars, but do not affect acrylamide formation during the cooking process.

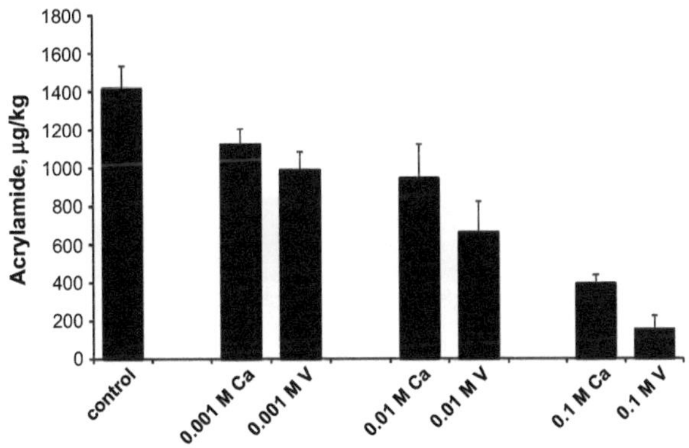

Figure 3 Amount of acrylamide in French fries after blanching with 0.001, 0.01, and 0.1 M $VOSO_4$ and $CaCl_2$.

Due to the presence of significant amounts of asparagine and sugars, there is a great potential for acrylamide formation in French fries and potato chips. We determined the content of asparagine and reducing sugars in the strips and slices that were used to make French fries and potato chips. The levels of asparagine were 16.1 ± 0.9 and 15.3 ± 1.7 mg/g in the strips and slices, respectively. The levels of reducing sugars were 14.6 ± 3.4 and 11.74 ± 3.0 mg/g in the strips and slices, respectively. To see the effect of $VOSO_4$ in the formation of acrylamide in these French fries and potato chips, we determined the acrylamide content of French fries with and without blanching with 0.001, 0.01, and 0.1 M $VOSO_4$. The amount of acrylamide was 987.01 ± 97.33, 661.18 ± 162.33, and 151.68 ± 72.21 µg/kg, respectively (Figure 3). The acrylamide formed in French fries without blanching with $VOSO_4$ was found to be 1417. 60 ± 116.1 µg/kg. The reductions in acrylamide were 30.37%, 46.64%, and 89.3% when strips were dipped in 0.001, 0.01 and 0.1 M $VOSO_4$ solution, respectively.

We also measured the acrylamide formation in chips (Figure 4). The amount of acrylamide formed in chips was found to be 3843.87, 2597.40, and 701.27 µg/kg after dipping in 0.001, 0.01, and 0.1 M $VOSO_4$ solutions, respectively. However, the acrylamide formed in control potato chips was found to be 9107.43 µg/kg. It has been observed that approximately three times higher amounts of acrylamide are formed in chips which was significant as compared to in French fries as observed by others [16]. This may be due to the high surface-to-volume area ratio in the case of potato chips compared to that of French Fries. Moreover, it has been observed that there were significant reductions in acrylamide formation in chips when the chips were dipped in $VOSO_4$ solution before frying, as seen in French fries. There were 57.7%, 71.48%, and 92.5% reductions in acrylamide formation when the chips were dipped in 0.001, 0.01, and 0.1 M $VOSO_4$ solution, respectively.

A comparative investigation of vanadyl sulfate and calcium chloride was made since both have dicationic metals and $CaCl_2$ was reported as an effective acrylamide formation inhibitor. In our experimental conditions, it has been found that Ca^{2+} and VO^{2+} induced

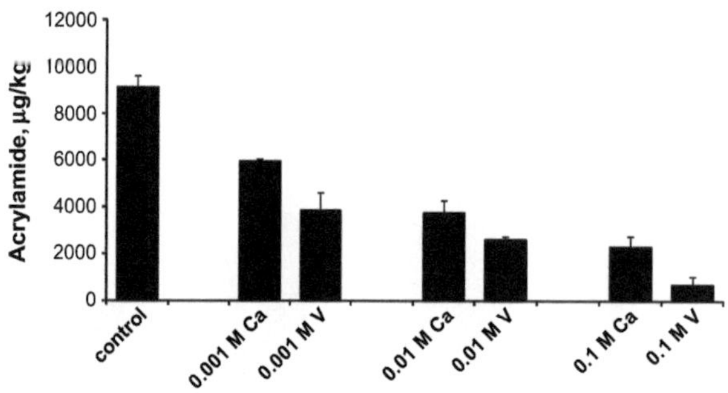

Figure 4 Amount of acrylamide in potato chips after blanching with 0.001, 0.01, and 0.1 M $VOSO_4$ and $CaCl_2$.

a maximum of 72.3% and 89.3% reduction of acrylamide when cut French strips samples were treated with 0.1 M $CaCl_2$ and $VOSO_4$, respectively. However, when the slices were treated separately with 0.1 M $CaCl_2$ and $VOSO_4$ solution, the reduction of acrylamide was found to be 79.8% and 92.5%, respectively. Thus, in our experimental conditions, $VOSO_4$ was found to be a stronger inhibitor than $CaCl_2$ in acrylamide reduction.

A number of raw material pretreatments were investigated for acrylamide inhibition. Although the mechanisms are not clear for all, some of the pretreatments inhibited acrylamide by (1) lowering of pH of foodstuff and/or (2) formation of an acrylamide adduct with additives. Lowering the pH of the foodstuff would block the nucleophilic addition of asparagine with a carbonyl compound, which could prevent the formation of the corresponding Schiff base, a key intermediate in the Maillard reaction [9,11,23]. Some amino acids reduce acrylamide formation by promoting competitive reactions and/or by covalently binding the acrylamide formed through Michael-type addition reactions. A number of monovalent, divalent, and trivalent cations such as Na^+, K^+, Ca^{2+}, Mg^{2+}, Zn^{2+}, and Al^{3+} have been reported to be effective acrylamide-reducing agents. The metal cation has long been known to cause pH reduction when added to certain foods or food components [7,30]. The mechanism for metal-ion-induced pH lowering has been generally attributed to the competitive displacement of protons from the ionizable functional groups containing oxygen, nitrogen, or sulfur atoms that share electrons with hydrogen atoms [2]. It is however not clear as to what is the main mechanism provoking a reduction in acrylamide formation. For instance, addition of Ca^{2+} ions to cut slices of potatoes is known to provoke a pH drop, since hydrogen chloride is produced when calcium crosslinks pectin.

It has been observed by others that treatments of foodstuffs with metal salts lower the pH. Blanching of potato strips and slices with $VOSO_4$ also led to a decrease in the pH of the samples (Figure 5). Moreover, it has been observed that higher levels of acrylamide forms at a pH of approximately 8 and that the amount of acrylamide formed is negatively correlated with pH. Reduction in acrylamide formation was also observed with a decrease in the pH of the potato samples. Treatment with $VOSO_4$ reduced acrylamide formation by 92%, in the

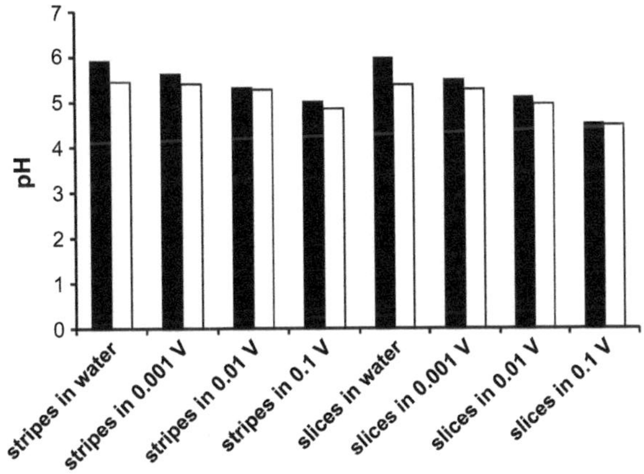

Figure 5 pH values of stripped and sliced potato samples after treating with $VOSO_4$ (black bar, before frying; white bar, after frying).

Table 1 Amount of vanadium in French fries and potato chips

Food product	Vanadium (mg) per serving
French fry 1	38.08
French fry 2	10.54
French fry 3	1.581
Potato chips 1	30.24
Potato chips 1	4.004
Potato chips 1	0.6328

Values are averages of three readings. French fries 1, 2, 3, and chips 1, 2, 3 are samples after blanching in 0.1, 0.01, and 0.001 M $VOSO_4$, respectively.

case of chips, and 89% in the case of French fries. Thus, the formation of a vanadium–asparagine complex in addition to a decrease in the pH of the potato samples after soaking with $VOSO_4$ makes $VOSO_4$ an effective agent in reducing acrylamide formation.

The use of vanadium in acrylamide inhibition is expected to have another advantage as food additives for French fries and potato chips since vanadyl sulfate is used as a nutritional supplement. Moreover, the reduction of acrylamide was concentration dependent on the blanching solution. Different levels of vanadium were absorbed by potato strips and slices because of their size and thickness. However, the level of vanadium was reduced after frying. Considering that the recommended service sizes of the French fries and potato chips are 85 and 28 g, respectively, the levels of vanadium in French fries and potato chips are shown per serving size (Table 1). The levels of vanadium found in French fries and potato chips (Table 1) are within safe limits of vanadyl sulfate doses. A therapeutic for management of type 2 diabetes is at least 50 mg of vanadyl sulfate twice daily. Moreover, in a study to investigate the efficacy of vanadyl sulfate in hyperglycemia, Goldfine et al. [4] found that doses of 75–300 mg taken daily for 6 weeks appeared to be safer for type 2 diabetic patients. Therefore, vanadium-enriched potato chips and French fries might be alternative sources of vanadium to meet required levels of vanadium.

CONCLUSION

The uses of vanadyl sulfate in an asparagine–glucose model reaction and potato foods showed that pretreatments with vanadyl sulfate before deep frying significantly reduced acrylamide formation in French fries and chips. A combined supplement of French fries and potato chips with vanadium might be an alternative source of dietary vanadium, which may reduce some major risks for type 2 diabetics, since potato, a glycemic form of carbohydrate, is hypothesized to increase insulin resistance and risk of type 2 diabetes.

SYNOPSIS OF ANALYTICAL TECHNIQUES

The asparagine–glucose model reaction was prepared from mixtures of solutions containing 12 mM of asparagine and 12 mM of glucose in a glass vial. An aliquot of vanadyl sulfate was added to obtain a final concentration of 1, 3, and 6 mM $VOSO_4$ and closed the vials carefully to each solution mixture. The reaction mixture was heated at 150 °C for 30 min. Each reaction mixture was immediately cooled down in an ice bath and then prepared for further acrylamide analysis.

French fries and potato chips were processed from strips and slices of the Rio Grande Russet cultivar (San Luis Valley Research Center, CO, USA). The strips and slices were soaked in a solution of vanadyl sulfate and calcium chloride at concentrations of 0, 0.001, 0.01, and 0.1 M for 60 min before frying at 180 °C for 2 min in 2 l of canola oil.

Acrylamide was extracted by adding water (20 ml) to ground French fries and chips (2 g) followed by spiking with 10 μl of ^{13}C acrylamide (1 mg/ml). The brominated derivative of acrylamide was prepared by adding brominating reagents (200 μl; Br_2 water, 5 ml + HBr, 0.8 ml + KBr, 15.2 g) to an aliquot of the aqueous extract (5 ml) at 4 °C. After completion of the bromination, excess bromine was titrated by adding sodium thiosulfate solution. The final products were extracted by adding ethyl acetate (20 ml), which was then reduced to a volume of 200 μl by passing nitrogen gas, and 1 μl was injected into a gas chromatograph mass spectrometer (GC-MS).

Acrylamide was quantified by using a Varian Saturn 2000 GC–MS with CP-Sil 24 CB column (dimension 30 m × 0.25 mm × 0.39 μm). The columns were held at 50 °C for 1 min, programed at 100 °C/min, and followed by 0.5 °C/min and again at 100 °C/min up to a final temperature of 235 °C. The carrier gas was helium with a flow rate of 1.0 ml/min. The temperature of the injector was 250 °C in the split-less mode. Data acquisition was performed by using the selected-ion monitoring mode with positive electron impact ionization. The qualitative analysis of acrylamide was conducted based on retention times and specific ions in 2,3-dibromopropenamide at 150 and 152 m/z. An isotopically marked [^{13}C]-acrylamide (153 and 155 for 2-bromo[$^{13}C_3$]propenamide was used as an internal standard. Quantitative analyses were performed using a five-point standard curve with an R^2 value 0.9963 through comparisons with the ratio of peak areas at 150 and 153 m/z.

Asparagine was determined by using a commercially available kit (K-ASNAM kit, Megazyme Wicklow, Ireland), by following the manufacturer's instructions.

Reducing sugars in the potato French fries and chips were determined by the dinitrosalicylic acid method by adding 120 μl dinitrosalicylic acid reagent (10 gl^{-1} dinitrosalicylic acid, 2 gl^{-1} crystalline phenol, a 10 gl^{-1} sodium hydroxide, and 0.5 gl^{-1} fresh sodium sulfite) to a tube containing 20 μl of the sample extract. After incubating this in a water bath at 100 °C for 20 min, 100 μl of the mixture was transferred to a 96-well flat bottom microplate containing 40 μl of 400 gl^{-1} of potassium sodium tartrate solution, and absorbances were measured at 570 nm. Reducing sugars were expressed in milligrams of glucose per gram of freeze-dried tuber.

At least three independent experiments were performed for each experimental condition. Results are expressed as mean ± SEM. Statistical differences were analyzed using Student's *t* test. A p value <0.05 was considered to be significant.

KEY FACTS

- Vanadium salts such as vanadate and vanadyl mimic the actions of insulin and lower blood glucose levels in various animal models. Vanadium compounds induce potent hypoglycemic effects in animals with type 1 and type 2 diabetes mellitus.
- Potato is the fourth major food after rice, wheat, and maize. Potato breeders have been developing new potato cultivars with distinctive flesh and skin colors, with high nutrition and phytochemicals. Purple and red potato cultivars are rich in polyphenols and have strong antioxidant activity.
- Along with starch, potato tubers are also rich in many health beneficial compounds such as protein, minerals (e.g., potassium, sodium, phosphorus, magnesium, iron, and zinc), vitamins (e.g., vitamins B1, B3, B6, folate, and riboflavin), and carotenoids.
- The process of cooking such as baking, boiling, and frying affect the nutrition levels of potato food products. The levels of minerals and vitamin C decreased when cooked by any of the cooking methods.
- Improper storage of potato tubers can decrease the quality of potato tubers. Storage at cold temperatures reduces sprout growth. However, cold storage enhances the levels of reducing sugars and thereby increases the levels of acrylamide when cooked.
- Diabetes mellitus is the most common endocrine disorder that affects 387 million people worldwide. It can lead to severe cardiovascular diseases and blindness.

MINI DICTIONARY

Antidiabetic compounds These compound lower blood glucose.
Asparagine This is an amino acid.
Carcinogen This is a cancer-causing compound.
Reducing sugars Glucose and fructose are reducing sugars containing aldehydic group.
$VOSO_4$ This is a vanadium salt.

SUMMARY POINTS

- Acrylamide forms by the reaction of asparagine and reducing sugars.
- Potato contains significant levels of these acrylamide precursors.
- French fries and potato chips produce significant levels of acrylamide.
- Vanadyl sulfate reduces the levels of acrylamide in French fries and potato chips.
- The combination of vanadium with asparagine and lowering of the pH of potato food species make $VOSO_4$ an effective acrylamide reducing agent.

REFERENCES

[1] Calleman CJ, Bergmark E, Stern LG, Costa LG. Environ Health Perspect 1993;99:221–23.
[2] Clydesdale FM. Minerals: their chemistry and fate in food. In: Smith K, editor. Trace minerals in foods. New York: Marcel Dekker; 1988. p. 57–94.
[3] Friedman M. J Agric Fodd Chem 2003;51:4504–26.
[4] Goldfine AB, Patti ME, Zuberi l, Goldstein BJ, LeBlanc R, Landaker E, et al. Metabolism 2000;49:400–10.
[5] Gökmen V, Senyuva HZ. Food Chem 2007;103:196–203.
[6] Gökmen V, Senyuva HZ. Eur Food Res Technol 2007;225:815–20.
[7] Hughes JC, Grant A, Faulks RM. J Sci Food Agric 1975;26:739–48.
[8] IARC. Acrylamide. In: IARC Monographs on the evaluation of carcinogenic risks to humans, vol. 60. Lyon (France): International Agency for Research on Cancer; 1994. p. 389–433.
[9] Jung MY, Choi DS, Ju JW. J Food Sci 2003;68:1287–90.
[10] Kalita D, Holm DG, Jayanty SS. Food Res Int 2013;54:753–9.
[11] Kita A, Bråthen E, Knutsen SH, Wicklund T. J Agric Food Chem 2004;52:7011–6.
[12] Kolek E, Šimko P, Simon P. Eur Food Res Technol 2006;224:2283–4.
[13] Lindsay RC, Jang SJ. Adv Exp Med Biol 2005;561:329–41.
[14] Lindsay RC, Jang SJ. Adv Exp Med Biol 2005;561:393–404.
[15] Manna S, Das S, Chatterjee M, Janarthan M, Chatterjee M. J Cell Biochem 2011;112:2327–39.
[16] Matthaus B, Hasse NU, Vossmann K. Eur J Lipid Sci Technol 2004;106:793–801.
[17] Medeiros Vinci R, Mestdagh F, DeMeulenaer B. Food Chem 2012;15:1138–54.
[18] Mestdagh F, Maertens J, Cucu T, Delporte K, Van Peteghem C, De Meulenaer B. Food Chem 2008;107:26–31.
[19] Mottram DS, Wedzicha BL, Dodson AT. Nature 2002;419:448–9.
[20] Mukherjee T, Pessoa JC, Kumar A, Sarkar AR. Inorg Chem 2011;50:4349–61.
[21] Naso LG, Valcarcel M, Roura-Ferre M, Kortazar D, Salado C, Lezama L, et al. J Inorg Biochem 2014;135:86–99.
[22] Nechay BR, Nanninga LB, Nechay PSE, Post RL, Grantham JJ, Macara IG, et al. Role of vanadium in biology. In: FASEB Fed Proc, vol. 45. 1986. p. 4349–61.
[23] Pedreschi F, Kaack K, Granby K, Troncoso E. J Food Eng 2007;79:1287–94.
[24] Refat MS, El-Shazly SA. Eur J Med Chem 2010;45:3070–9.
[25] Rommens CM, Yan H, Swords K, Richael C, Ye J. Plant Biotechnol J 2008;6:843–53.

[26] Rydberg P, Eriksson S, Tareke E, Karlsson P, Ehrenberg L, Törnqvist M. J Agric Food Chem 2003;51:7012–8.
[27] Stadler RH, Blank I, Varga N, Robert F, Hau J, Guy PA, et al. Nature 2002;419:449–50.
[28] Stadler RH, Scholz G. Nutr Rev 2004;62:449–67.
[29] Tareke E, Rydberg P, Karlsson P, Eriksson S, Törnqvist M. J Agric Food Chem 2002;50:4998–5006.
[30] Vadlamani KR, Seib PA. Cereal Chem 1999;76:355–60.
[31] Wiberely-Bradford AE, Busse JS, Jiang J, Bethke PC. BMC Res Notes 2014;7:801.
[32] Yuan Y, Shu C, Zhou B, Qi X, Xiang J. Food Res Int 2011;44:449–55.
[33] Zamora R, Delgado RM, Hidalgo FJ. J Agric Food Chem 2010;58:1708–13.
[34] Zhang Y, Ren Y, Zhang Y. Chem Rev 2009;109:4375–97.

CHAPTER 21

Impact of L-Asparaginase on Acrylamide Content in Fried Potato and Bakery Products

Zuzana Ciesarová
NPPC National Agricultural and Food Center, Food Research Institute, Bratislava, Slovak Republic

INTRODUCTION

Widespread acrylamide exposure from foods across consumers of different ages, locations, origins, and dietary habits challenges responsible bodies to find effective ways to diminish acrylamide burden to protect human health. The motivation to adjust personal dietary habits to reduce consumption of risky foods is often only a temporarily effective strategy, and a sustainable downward trend is hardly retained since consumer habits are quite resistant to changes. A more exact approach is to decrease acrylamide in foods to levels as low as are reasonably achievable. There are ways to either diminish the acrylamide content accumulated during heat treatment of raw food or to prevent acrylamide formation by interventions in food processing. Tools that have been applied in various levels of food production are summarized in a living document entitled the Acrylamide Toolbox released by the Food Drink Europe in 2013 [1].

If one compares different mitigation tools with various effectiveness, one finds that L-asparaginase enzyme treatment is a progressive way to prevent acrylamide formation. Since acrylamide is formed from two main precursors, L-asparagine and reducing sugars, L-asparaginase eliminates one of them. It hydrolyzes L-asparagine to L-aspartic acid, which does not enter into the reaction of acrylamide formation. An undeniable advantage of the use of L-asparaginase is its negligible impact on qualitative and sensorial properties of the final products and a relatively simple application in processing. However, it needs to be said that the appropriate adjustments of processing settings are inevitable.

The next chapter describes the following:

- Mechanism of L-asparaginase action
- Natural microbial and plant sources of L-asparaginase and a commercialization of its production
- Raw food materials with a high free L-asparagine level suitable for L-asparaginase treatment
- Advantages of L-asparaginase usage in food processing

Acrylamide in Food
http://dx.doi.org/10.1016/B978-0-12-802832-2.00021-8

- Verification of L-asparaginase application on a laboratory scale using model systems
- Trials of L-asparaginase treatment of potato-based and cereal-based products
- Commercialization of L-asparaginase production

MECHANISM OF L-ASPARAGINASE ACTION

Based on up-to-date knowledge, acrylamide is formed via the Maillard reaction generally described as a reaction of saccharides as providers of carbonyl compounds and the NH_2 group of amino acids. In the case of acrylamide, donors of carbonyl groups are represented by reducing saccharides, whilst amino groups originate from a particular amino acid, L-asparagine [2,3]. Because the removal of acrylamide from prepared foods is impractical, mitigation strategies have focused on limiting acrylamide formation. Elimination of both or at least one of those two precursors results in reduction of acrylamide formation and finally in the lowering of acrylamide yield. Carbonyls that enter the Maillard reaction can originate from various sources; thus, any elimination of them demands a comprehensive approach. Moreover, the concentration of sugars in most plants is very high, compared to the amount of free L-asparagine, which determines L-asparagine as a limiting factor in acrylamide formation. However, the source of the amino group is well identified. This allows the targeted focus on lowering of the only amino acid L-asparagine, which results in a significant acrylamide reduction [4].

Application of the enzyme L-asparaginase provides a possible alternative method for acrylamide mitigation, which should have a very limited effect on the general formation of Maillard products [5]. L-Asparaginase can selectively reduce the level of free L-asparagine by hydrolyzing it to L-aspartic acid and ammonia, which thereby specifically removes one of the essential acrylamide precursors (Figure 1). The majority of amino acids will not be affected, and the impact on the final food product quality should therefore be limited.

L-Asparaginase (L-asparagine amidohydrolase, EC 3.5.1.1, C.A.S. 9015-98-3, theoretical molecular weight 138.4) is a tetrameric protein composed of four identical subunits, each subunit contains 326 amino acid residues (Figure 2). The two threonine residues present at the active site are required for activity [6]. Optimal pH is determined at 8.6 [7]. The hydrolysis of L-asparagine catalyzed by L-asparaginase consists of two steps: In the first step, the nucleophilic threonine attacks the carbonyl of the amide substrate to generate an

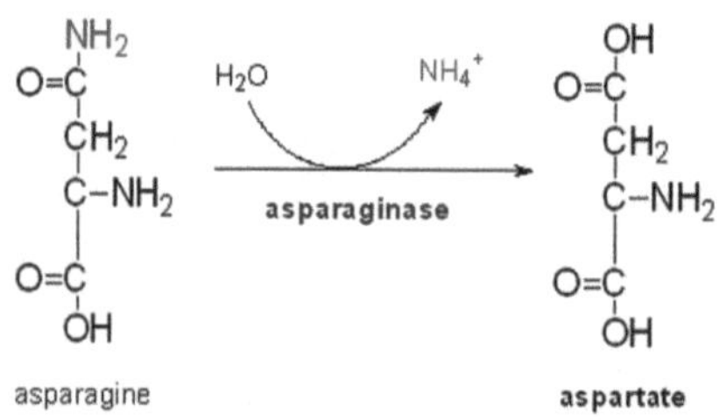

Figure 1 L-Asparaginase-assisted conversion of L-asparagine to L-aspartic acid.

acyl-enzyme intermediate. In this step, a molecule of ammonia is released. In the second step, water attacks the acyl–enzyme intermediate to produce L-aspartate [8]. The enzyme is also active with glutamine, and it cleaves its amino group off at a slower rate [9].

Bacterial type L-asparaginases are classified into subtypes I and II, which are defined by their intracellular or extracellular localization [8]. Type I (cytosolic) has a lower affinity to L-asparagine, while type II (periplasmic) has a high substrate affinity. EcAII exhibits a broad pH activity profile and a higher substrate affinity than does EcAI. Plant type L-asparaginases differ structurally and have a different evolutionary origin compared to bacterial L-asparaginases [10].

An L-asparaginase from *Escherichia coli* was developed as a pharmaceutical agent [11]. Asparaginase as a drug is considered to be a very effective therapy for those specific cases where blood cells become cancerous, such as in acute lymphoblastic leukemia. The enzyme cuts off the supply of asparagine in blood and pulls off its amine, to release aspartate and ammonia. The ammonia is subsequently reassimilated by the GS/glutamate synthase enzymes. It is normally involved in balancing the levels of amino acids for use in protein synthesis. However, if a large dose of this enzyme is introduced into the blood, it will circulate and continually break down all the asparagine that it finds, which will ultimately starve the cells that rely on the blood-borne supply. The abnormal cancer cells distinguish themselves from normal cells in their low level of L-asparagine [12]. Neoplastic cells are not able to synthesize their own L-asparagine due to the absence of L-asparagine synthetase [13]. For survival, tumor cells require greater amounts of L-asparagine, as they deplete all the L-asparagine from the circulatory blood, which leads to death of leukemic cells [11,14]. L-Asparaginase is an enzyme drug that is used to reduce the circulatory L-asparagine level. This thereby stops tumor cell growth, which is widely used in the treatment of acute lymphoblastic leukemia [14].

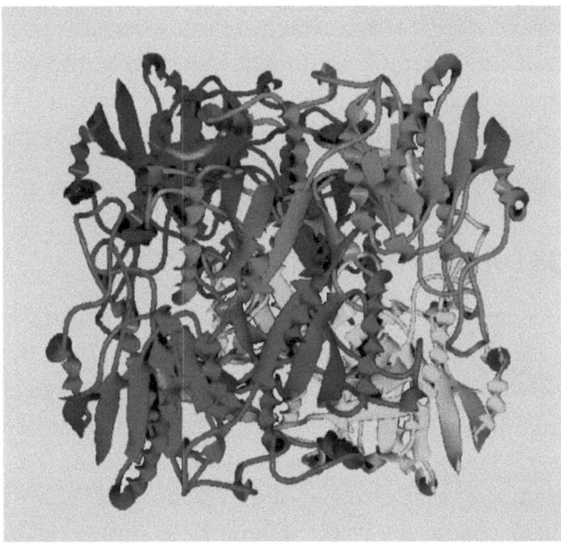

Figure 2 L-Asparaginase purified from bacterial cells composed of four identical subunits [11].

However, the use of a large protein as a drug poses several challenges. The human immune system is designed to destroy large, multisubunit proteins, so L-asparaginase is rapidly cleared from the blood in a day or so [11].

L-Asparaginase enzyme has been found in microorganisms [15] such as bacterial strains [9] of *E. coli*, *Erwinia chrysanthemi*, *Erwinia carotovora* [16], *Pseudomonas aeruginosa*, *Streptomyces griseus*, *Acinetobacter calcoaeticus*, *Proteus vulgaris* [17], and *Enterobacter aerogenes* [18]. The potential to produce L-asparaginase was observed also in eukaryotic microorganisms as yeasts *Saccharomyces cerevisiae* [19], and in fungi *Aspergillus tamari*, *Aspergillus nidulans* [20], *Aspergillus terreus* [15,21], *Penicillium* spp., and *Fusarium* spp. [22,23]. Further, L-asparaginase is also found in some plant materials as *Pisum sativum* [24], *Lupinus polyphyllus*, *Phaseolus multiflorus*, *Vicia faba*, *Zea mays*, and *Hordeum vulgare* [25]. Sodek et al. [26] reported that *Pisum sativum* has the highest L-asparaginase activity of the stated plant sources. Also, they found that L-asparaginase that is isolated from the plant materials mentioned, except for *Lupinus polyphyllus*, is dependent upon the presence of potassium ions for activity. Particularly high L-asparaginase activity was detected in legume seeds during the maturation process [27,28]. L-Asparaginase was detected also in animals, including guinea pigs [29], but it does not occur naturally in humans [8].

Extracellular production of L-asparaginase was successfully done and commercialized with strains *Aspergillus niger*, *Aspergillus oryzae* [30], *Trichoderma viride* sp. [14], and *E. coli* [31,32] by submerged fermentation. High-level recombinant expression of the extracellular *A. oryzae* L-asparaginase allows for production of sufficient amounts of enzyme to enable broad industrial implementation. The results presented demonstrate that it is possible to effectively reduce acrylamide formation in a range of food products by applying L-asparaginase during production [5]. The conditions tested show that the enzyme can be applied to wheat, rye, and potato products, in dough or to solid potato pieces. They also demonstrate that the *A. oryzae* L-asparaginase can work effectively from 10 to 55 °C, as used for crisp bread and potato experiments, respectively, and at pH values from 5.5 to 8.75, as applied in the treatment of potatoes at one extreme and semisweet biscuits at the other. High temperatures during the cooking process deactivate this enzyme by proteolysis.

OCCURRENCE OF L-ASPARAGINE IN PLANTS

L-Asparagine is a nonessential amino acid that naturally occurs in many raw plants in different extents [25]. Plants associated with a high L-asparagine level are potatoes, cereal grains (wheat, oat, rye, maize, rice, etc.), coffee beans, vegetables (asparagus, bell paprika, onion, broccoli, lupin, pumpkin), fruit (plum, buckthorn), other plants (chicory, green tea, etc.), and seeds (peanuts, soybean). The content of free L-asparagine depends on many factors, and it varies in species, varieties, seasonal changes, other agronomical factors such as fertilization, harvest time, and storage conditions. Plants accumulate L-asparagine during a period when there is a plentiful supply of reduced nitrogen and a shortage of soluble carbohydrates. L-Asparagine accumulation can be stimulated by external stresses that also increases the proportion of reduced nitrogen to carbon [25]. Typical ranges of free L-asparagine in raw and processed plants are as summarized in Table 1. Because the content of L-asparagine is generally in a good correlation with the final acrylamide content, the presence of this amino acid represents a clear potential for acrylamide formation [33].

Table 1 Occurrence of L-asparagine in food commodities

Commodity	Content of L-asparagine (g/kg)	References
Wheat grains	0.19–0.29	Ciesarova et al. [74]
Wheat bran	1.48	Fredriksson et al. [75]
Wheat germ	55.5–57.4	Fredriksson et al. [75]
Wheat flour	0.06–0.42	Ciesarova et al. [74], Claus et al. [76]
Whole wheat flour	0.18–0.42	Elmore et al. [36]
Spelt flour	0.06–0.23	Claus et al. [76], own data
Whole spelt flour	0.15–0.41	Own data
Oat grains	0.52–0.79	Ciesarova et al. [74]
Oat bran	0.71	Stadler [77]
Oat flour	0.50	Stadler [77]
Whole oat flour	0.79	Ciesarova et al. [57]
Rye grains	1.07	Fredriksson et al. [75]
Rye flour	0.32–0.66	Elmore et al. [36], Claus et al. [76]
Whole rye flour	0.57–0.86	Own data
Barley grain	0.21–0.41	Lea et al. [25]
Rice germ	0.23	Friedman [78]
Rice bran	0.28	Friedman [78]
Rice flour	0.07	Stadler [77]
Buckwheat flour	0.04	Markova et al. [79]
Potato tubers	0.02–17.0	Elmore et al. [36], DeWilde et al. [35], Chuda et al. [80], Becalski et al. [81], Muttucumaru et al. [33], Brunton et al. [82]
Asparagus	1.02–4.73	Lea et al. [25]
Pumpkin	0.32	Constantin et al. [83]
Eggplant	0.15	Constantin et al. [83]
Red bell peppers	0.13	Constantin et al. [83]
Buckthorn	2.48	Own data
Chicory	0.4–2.8	Loaec et al. [84]
Green coffee	0.28–0.96	Murkovic et al. [85]

To limit the formation of acrylamide, the food industry has adopted a strategy to better select raw materials. This has low contents of the two acrylamide precursors, L-asparagine and reducing sugars [1]. An important study on the strategies for mitigating acrylamide formation in potato-based and cereal-based products is now available [34–41]. In both food categories, preharvest parameters, such as cultivar selection and agronomy, have been studied to control the formation of precursors and subsequently to limit the formation of acrylamide.

L-ASPARAGINASE APPLICATIONS IN MODEL SYSTEMS AND IN FOOD MATRICES

The first documented application of L-asparaginase used for conversion of L-asparagine and concomitant acrylamide reduction in a potato matrix was published by Zyzak et al. [4]. Further, the effect of L-asparaginase was examined in gingerbread; this resulted in a decrease of 75% of free asparagine, 55% of acrylamide, and no negative effect on either taste or color [42]. After that, the effective use of asparaginase for acrylamide mitigation has been demonstrated in a number of studies done in simple systems of reagents in solution as well as in a number of food products [5,32,43,44]. However, a comparison of the results shows that only a little direct correlation is observed between the percentage reduction in L-asparagine levels in the ingredients and the percentage reduction in acrylamide formation in the various products. This is likely due to the impact of both processing conditions and other recipe components, which vary from product to product, in acrylamide formation. For this reason, simplified model systems containing sugars and L-asparagine or more complex systems resembling a specific food formulation were often applied for the investigation of the acrylamide formation in heated food products [45]. These models offered the advantages of reproducibility, simplicity, and feasibility to identify the key factors for acrylamide formation as well as to find ways for its elimination. In such models, the impact of L-asparaginase on the acrylamide content after heat treatment was investigated [44].

POTATO-BASED MATRICES FOR L-ASPARAGINASE TREATMENT

Concerning potato-based models, since the L-asparagine content in potatoes varies widely from 2.3 to 39.4 mg/g of dry weight, depending on factors such as variety, location, fertilization, storage, and processing [46], this fact was considered in the construction of a simulated potato model by Ciesarova et al. [44]. Using different concentrations of enzyme (2, 10, and 20 U/g), time, and temperature of the enzyme incubation (at 20, 37, 50, 60, and 70 °C for 5–60 min), the efficiency of L-asparaginase activity was compared. It was found that the incubation of the model matrix in the enzyme solution of 10 U/g for 30 min at 37 °C was most suitable for a 50% elimination of acrylamide. Higher acrylamide reduction (even 91%) was achieved with a 60-min-long exposure of the model matrix to the same enzyme solution [44].

In the treatment of French fries and sliced potato chips [30,47–49], key parameters were an enzyme dose and the method of application, because appropriate pretreatment, for example, blanching, was indicated for optimal performance. For French fries production, industrial implementation will require the modification of current production lines to include an enzyme bath. Soaking of the raw potato strips in a 10,000 U/L L-asparaginase solution at 40 °C for 20 min reduced acrylamide formation in French fries by 30% [30]. Increasing the soaking temperature of L-asparaginase solution to 50 and 60 °C, significantly (P > 0.05) diminished acrylamide reduction in French fries by only 20% and 22%, respectively. In contrast, when the potato strips are blanched at 75 °C for 10 min before the soaking treatment in L-asparaginase solution at 40 °C for 20 min, the acrylamide content of French fries surprisingly decreased by 62%. It is worth noting that when blanched strips are soaked in L-asparaginase solutions at temperatures >40 °C (50 and 60 °C), acrylamide formation in French fries was reduced significantly (P > 0.05) to 34% and 33%, respectively. This is due to the fact that the optimum temperature of the L-asparaginase activity measured at pH 7 is 60 °C (Novozymes A/S). However, at temperatures >60 °C, the enzyme activity decreases rapidly; therefore, a temperature range of 40–60 °C may be preferable. In general, the degradation of the L-asparaginase enzyme increases with an increase in temperature. Hence, the activity of the enzyme may last for a longer time if the temperature is set to a relatively low level. Of the two treatments considered in a study [30], 20 min of soaking or 1 min of dipping in the enzyme solution, the process with the shorter retention time was nearly as effective, should be easier to implement, and would be less likely to affect product quality. A 1-min enzyme treatment could be operated as a continuous process, if appropriate cooling of blanched potatoes is implemented, and an effective temperature control to ensure optimal enzyme performance and stability in the bath is established. The short retention time in the enzyme bath exploits the subsequent existing drying step used in industrial processing to provide time for the enzyme to act, and thus allow for enzyme treatment, whereas the total production time as such will not be severely affected.

The L-asparaginase soaking treatment had a significant effect in removing L-asparagine (P > 0.05). Soaking of blanched potato strips in L-asparaginase solution at 40, 50, and 60 °C significantly (P > 0.05) reduced their L-asparagine contents in 58%, 52%, and 41%, respectively. Soaking of blanched potato strips (75 °C, 10 min) in a 10,000 U/L L-asparaginase solution at 40 °C for 20 min is an effective way to significantly reduce acrylamide formation after frying by previously reducing the amount of L-asparagine. L-Asparaginase treatment duration was the same (10 min) at 50 and 60 °C and leads to less L-asparagine degradation at 60 °C. This could be due to the fact that even at 60 °C, which is the optimal temperature for L-asparaginase activity, the duration of the treatment was so long that the enzyme began to get inactivated. The best results were obtained when asparaginase was used at 40 °C for 20 min in this research [30]. One can conclude from this study that L-asparaginase pretreatment after some time/temperature/dosage optimization is realistic. It also depends on an authority's view or handling of acrylamide, which will affect the producers and their willingness to introduce process changes. Heat treatment like blanching during potato processing causes a change in the texture of potato strips (e.g., swelling of starch), which may improve the diffusion of L-asparagine toward the L-asparaginase solution surrounding the strips. Apparently, blanching produces microstructural changes in potato tissue, which allows L-asparagine and L-asparaginase to get in contact faster. Since heat treatment like blanching during potato processing causes a change in the texture of French fries (e.g., starch gelatinization) [50], this may improve the diffusion of L-asparagine toward the L-asparaginase solution surrounding the strips. Actually, in industrial practices, the texture of French fries is improved by various temperatures of the blanching water and blanching procedure, with the use of one, two, or three blanchers.

Cereal-Based Matrices for L-Asparaginase Treatment

In the simulated cereal model matrix constructed by Ciesarova et al. [51], three different concentrations of L-asparaginase (100, 500, and 1000 U/kg of dry matter) were tested, and resulted in 25, 90, and 96% decreases of acrylamide content, respectively. Nevertheless, the situation in genuine food products is more complicated since the presence of factors such as pH, water activity, moisture, and other facts that affect the behavior of enzyme, and the formation of acrylamide should be taken into consideration. These factors can stimulate or suppress the Maillard reaction process and consequently the final acrylamide content; therefore, the differences between the simulated cereal matrix and real samples can be expected.

L-Asparaginase effect on the acrylamide formation in cereal-based products was investigated in wheat cracker products [43], corn-based foods [52], short-dough biscuits [53,54], fried-dough pastry [51], gingerbread [42], and in bread [55–57]. All these authors reported a significant reduction in acrylamide levels upon L-asparaginase addition, although the specific conditions for the use of this enzyme should be finely tuned to make its use reliable in terms of costs at the industrial level.

In dough-based products, the key parameters influencing the enzyme effectiveness investigated were enzyme dose, reaction time, temperature, and water content. Moreover, the type of raising agent and its amount significantly influence the dough pH, which is important to take into account during the enzyme application regarding an individual formulation [51]. In this paper, the increase of the pH value from 5.82 in the control sample without raising reagent addition to 8.10 by sodium hydrogen carbonate caused up to a 65% decrease of L-asparaginase efficiency on acrylamide mitigation in cookies (for a sample baked at 205 °C for 15 min after a previous 15-min-long enzyme incubation). The impact of pH was not so evident when the incubation of the enzyme was prolonged (a 25–35 % decline in enzyme efficiency). Further, the enzyme activity is influenced by contact with the substrate. In fact, a limited mobility of the substrate and enzyme is responsible for incomplete hydrolysis of L-asparagine and only partial reduction of acrylamide formation. In this regard, food homogenization, as well as the water content of the reaction environment, could greatly affect L-asparaginase efficacy in reducing acrylamide formation. Considering this fact, an intermediate L-asparaginase concentration of 500 U/kg combined with the lowest time and temperature of incubation effectively reduced acrylamide formation in short-dough biscuits without affecting the color of the final products [53]. Similarly, L-asparaginase was successfully added in ginger honey biscuit production with a 90% reduction of acrylamide [58]. This was subsequently confirmed by a trial in a pilot-scale whole-grain biscuit production [59]. Industrial implementation in dough-based products will require some initial trials to establish the effective enzyme dosage for a specific product, but only minimal changes to existing production lines will in most cases be needed, because L-asparaginase can be added in the same manner as for other ingredients. Enzymes are already broadly used in the baking industry, and enzyme formulation, storage, and addition techniques are well established.

In a bread model system, the addition of L-asparaginase gave a large reduction, while no effect of glucose oxidase was found [55]. Capuano et al. [56] used a bread crisp model system made of flour, water, and yeast. The bread was toasted at different temperatures for different times. L-Asparaginase was added to 2000 U/kg of flour. The addition of L-asparaginase led to a reduction in acrylamide content, ranging from 70% to 88% at both 160 and 180 °C. It had no effect on browning development and antioxidant activity. In another case, in a newly developed wheat–oat bread, the increased level of acrylamide due to high L-asparagine amounts in oat flour was successfully diminished by the application of a commercial L-asparaginase solution on the surface of the bread loaf. This resulted in a 46% decrease of acrylamide in the bread crust [57]. Similarly, acrylamide in sweet bread was reduced when wheat-based dough was treated by L-asparaginase produced from *Cladosporium* sp. at different concentrations (50–300 U) [60]. There was no change in the rheological properties of wheat flour and physicosensory characteristics of bread with L-asparaginase treatment. At 300 U, there was a 97% and 73% reduction in acrylamide formation in the crust and crumb regions of bread, respectively.

The other approach to use L-asparaginase was applied in white wheat, wheat bran, and whole-grain wheat breads when the effect of pea (*Pisum sativum* L.)-originated L-asparaginase on acrylamide formation was also investigated [24]. Two-day germinated pea flour was used at 0%, 1%, 3%, and 5% levels for each bread type. It was found that acrylamide level can be reduced by 57% and 68% by the addition of 5% pea flour in wheat bran and whole-grain breads, respectively, without any negative impact on color and sensory properties.

Gingerbreads are known for their high acrylamide level, which results from accumulation of many factors that influence acrylamide formation: the presence of precursors counting L-asparagine from flour and single sugars fructose and glucose from honey, the use of ammonium hydrogen carbonate as a raising agent supporting acrylamide formation, and a high temperature up to 190°C generally used for gingerbread baking. For these reasons, acrylamide content in gingerbread ranged from 20 to 8000 μg/kg [42]. Besides other ways of acrylamide elimination in gingerbread, the addition of L-asparaginase during dough preparation to hydrolyze the amide group of L-asparagine was also tested. Application of L-asparaginase resulted in a decrease of 75% of free L-asparagine and 55% in acrylamide content. The taste and color were virtually identical to the normal product, which is a clear advantage of this approach.

For fried-dough pastry, Kukurova et al. [61] prepared samples according to a simplified recipe based on traditional Spanish rosquillas, with and without the addition of L-asparaginase. The samples were deep fried at 180 and 200°C for 4 and 8 min. These frying temperatures and frying times are usually applied to obtain a crispy product that is still moist inside without an excessive oil intake. As expected, acrylamide was formed during the frying of the dough used as control (range 50–215 μg/kg). Acrylamide content is strongly related to thermal input, and its final concentration is a result of concurrent formation and elimination (observed at temperatures >170°C). Increase of frying temperature by 20°C resulted in a 30–60% increase of acrylamide, and a doubling of the frying time led to almost a three times higher acrylamide content. However, a dramatic impact on acrylamide was recorded when L-asparaginase treatment was applied during the dough preparation. The acrylamide content in the final product was lower than the limit of quantitation (20 μg/kg) for all temperature–time combinations applied. The excellent results of the application of enzyme (originating from *A. oryzae*) at concentrations as low as 100 U/kg, resulting in a >90% acrylamide decrease in real samples, indicate the enormous potential for economically improving the manufacturing of cereal products. In other studies on the application of asparaginase in cereal products, a lower effect of asparaginase was achieved. Vass et al. [43] obtained a 70% acrylamide reduction in crackers, and Amrein et al. [42] reported only a 55% reduction in gingerbread using L-asparaginase. These findings were in part caused by an insufficient enzyme concentration (4 U/kg of dough in Amrein's study [42]). The method of enzyme application is also crucial. The addition of enzyme in the aqueous solution used for dough preparation is more effective than mixing with dough as reported by Amrein [42] due to better homogeneity in the system. L-Asparaginase is able to convert up to 96% of L-asparagine to L-aspartate when used at a concentration of 100 U/kg of flour and 97% with 500 U/kg of flour. This leads to final acrylamide elimination at both tested concentrations. In parallel, a 37% increment in glutamate content in the dough was observed when a higher concentration of enzyme was used. Although there is no direct explanation for this effect, it may be that the substrate specificity of L-asparaginase was more diversified, and it is reasonable to expect the presence of a glutaminase activity in asparaginase or an activity to other substrates [62,63]. In the mentioned study, also the impact of L-asparaginase treatment on browning of the products was evaluated. Thus, the use of L-asparaginase, in rosquillas, as a representative of deep-fried-dough products is a very effective method for acrylamide mitigation without any apparent negative impacts on the final product quality [61].

In the case of breakfast cereals, the producer of the commercial L-asparaginase announced that the simple addition of enzyme during ingredient mixing is most active during the cooking or preconditioning steps. This allows an acrylamide reduction of up to 50%. With this enzyme treatment, the healthy improvement of breakfast cereals is achieved without altering the tempting flavor or visual aspect of the products. This is a big bonus for food manufacturers looking for acrylamide-reducing solutions. Despite this proclamation, the Acrylamide Toolbox 2013 [1] indicates no significant asparagine reduction in a cooked toasted coarse grain cereal. Trials at laboratory and pilot scales confirmed that L-asparaginase was not effective for mitigation of acrylamide formation in breakfast cereal processes used. Breakfast cereal processes use a low moisture content, which makes enzyme penetration into the grain or food matrix difficult. Many breakfast cereal processes use coarse flours or chopped grains that are not readily penetrated by the enzyme.

Approval of L-Asparaginase Enzyme Usage by Responsible Institutions

Today, several companies are producing a commercial food-grade enzyme. As worldwide licences for patent holders, they may sublicense the rights to food manufacturing and processing companies to incorporate L-asparaginase in their food production processes to lower the amount of acrylamide [64].

The safety status of the products intended for usage in foods (Generally Recognized As Safe - GRAS) has been obtained from the US Food and Drug Administration (US FDA) for L-asparaginases produced from *A. niger*, *A. oryzae*, and *Bacillus subtilis* in 2006, 2007, and 2013, respectively. Joint Food and Agriculture Organization/World Health Organization Expert Committee on Food Additives reviewed L-asparaginase from *A. oryzae* at its 68th meeting in June 2007, and on the basis of available data and total dietary intake arising from its use concluded that asparaginase does not represent a hazard to human health [65]. Nevertheless, regulatory permission to apply L-asparaginase in foods requires clarification, nationally and internationally. Currently, many countries have approved it as a processing aid. A complete list of countries with no legal restrictions of L-asparaginase use is available in the Acrylamide Toolbox 2013 [1]. Despite the fact that L-asparaginase is commercially available just recently and it is sold today, the lack of motivation and any legislative pressure do not force producers to provide safer products using available tools of acrylamide mitigation. However, the results obtained, therefore, indicate that the L-asparaginase has the potential to be effective under industrial conditions. The cost of L-asparaginase treatment will depend on time/temperature/dosage optimization. Besides, the enzyme can be reused in other batches, which will bring down the enzyme dosage per kilogram of the final product, and thereby also costs. Finally, the content of L-asparaginase in the solution is low. This means that the total protein load of the water will not affect the overall quality of the process waste water.

SYNOPSIS OF ANALYTICAL TECHNIQUES

L-Asparaginase Assay

There are several methods for determination of L-asparaginase activity using a cationic glass electrode [66], a colorimetric technique [67], and spectrophotometric assay [68]. Essentially, the rate of hydrolysis of asparagine is determined by measuring released ammonia [69]. One unit releases 1 μmol of ammonia per minute at 37 °C and pH 8.6 under specified conditions.

Alternatively, L-asparaginase activity is determined by a coupled enzyme assay, which results in a colorimetric (570 nm)/fluorometric (λex = 535/λem = 587 nm) product, proportional to the aspartate generated. One unit of L-asparaginase is defined as the amount of enzyme that catalyzes the formation of 1 μmol of aspartate per minute at 25 °C. A novel assay for the determination of L-asparaginase activity in human plasma was recently described. This is based on the high-performance liquid chromatography quantitation of L-aspartic acid produced during enzyme incubation [70].

Determination of Acrylamide

Acrylamide in food is normally measured as "free water-soluble acrylamide." Detection methods are mainly mass spectrometry (MS) based, coupled with a chromatographic step using liquid chromatography (LC) or gas chromatography [71,72]. The most widely used method consists of aqueous extraction plus a cleaning step without derivatization of acrylamide and LC–MS/MS detection using an isotope-labeled internal standard [73]. Determination of acrylamide should be optimized with respect to specificity of different food matrices.

KEY FACTS

- L-Asparaginase has been proved as an effective tool for acrylamide mitigation in food.
- Safety of L-asparaginase as a processing aid is confirmed by GRAS status delivered by the US FDA and by an acceptation of particular national authorities.
- In French fries production, the application of L-asparaginase as a blanching solution resulted in a 62% acrylamide reduction.
- In biscuit production, the application of L-asparaginase into dough prior to baking led to a 90% acrylamide reduction.
- In fried-dough pastry products, the L-asparaginase application into dough caused a >90% acrylamide decrease.

- In bread production, L-asparaginase was successfully applied on the surface of bread loaves with 50% effectiveness on acrylamide diminishing.
- In bread toast, the effectiveness of L-asparaginase on acrylamide reduction was between 70% and 88%.
- In breakfast cereals, a 50% acrylamide reduction was achieved with L-asparaginase application.

MINI-DICTIONARY

L-Asparaginase This is L-asparagine amidohydrolase, EC 3.5.1.1, responsible for conversion of amino acid L-asparagine to L-aspartic acid.

L-Asparagine This is nonessential amino acid for humans, which means that it can be synthesized from central metabolic pathway intermediates and is not required in the diet. L-Asparagine was found in many animal and plant sources in different extents. L-Asparagine is required for development and functioning of the brain. It also plays an important role in the synthesis of ammonia. L-Asparagine is derived from aspartate via transamidation.

Aspartic acid This is a nonessential amino acid in mammals; it is produced from oxaloacetate by transamination. In plants and microorganisms, aspartate is the precursor to several amino acids, including four that are essential for humans:methionine, threonine, isoleucine, and lysine.

Acrylamide This is a suspected toxic substance that can be formed during high-temperature processing of foods such as French fries, potato crisps, biscuits, crackers, breakfast cereals, bread, and snacks. Acrylamide forms in foods as a result of the Maillard reaction, a chemical reaction between sugars (e.g., glucose, fructose, or lactose) and free asparagine, a naturally occurring amino acid. It typically requires a higher temperature during cooking (usually >120 °C).

GRAS This represents generally recognized as safe; status on safety of food additive delivered by the US FDA based on scientific testing.

SUMMARY POINTS

- This chapter described a progressive way of acrylamide mitigation by L-asparaginase enzyme application prior to the heat treatment of raw material in food processing.
- The natural content of free L-asparagine in different plants is summarized with special highlights on those known with a high asparagine level. Sources of L-asparaginase in microorganisms and plants suitable for the commercial production of the enzyme are bacteria, fungi, and plants. Commercially, L-asparaginase is produced by genetically modified microorganisms with an intense extracellular enzyme expression.
- The application of L-asparaginase prior to heat treatment of foods converts free asparagine to aspartic acid and diminishes the source of acrylamide; this thus prevents its formation in the Maillard reaction. L-Asparaginase action is associated with no significant effects on the qualitative properties of the final products such as color and texture changes. Despite unambiguous advantages of L-asparaginase treatment, the widespread usage of L-asparaginase in acrylamide prevention is limited by an increase of initial production costs and inevitable adjustment of processing conditions. This resulted in lower effectiveness of particular food production.

ACKNOWLEDGMENT

This work was supported by the Research & Development Operational Program funded by the European Regional Development Fund through implementations of the projects ITMS 26240220091 "Strategy of Acrylamide Elimination in Food Processing", and ITMS 26240120042 "The Center of Excellence for Contaminants and Microorganisms in Foods."

REFERENCES

[1] Acrylamide Toolbox. Food Drink Europe; 2013. p. 60.
[2] Mottram DS, Wedzicha BL, Dodson AT. Food chemistry: acrylamide is formed in the Maillard reaction. Nature 2002;419(6906):448–9.
[3] Stadler RH, Blank I, Varga N, Robert F, Hau J, Guy PA, et al. Food chemistry: acrylamide from Maillard reaction products. Nature 2002;419(6906):449–50.
[4] Zyzak DV, Sanders RA, Stojanovic M, Tallmadge DH, Eberhart BL, Ewald DK, et al. Acrylamide formation mechanism in heated foods. J Agric Food Chem 2003;51(16):4782–7.
[5] Hendriksen HV, Kornbrust BA, Ostergaard PR, Stringer MA. Evaluating the potential for enzymatic acrylamide mitigation in a range of food products using an asparaginase from *Aspergillus oryzae*. J Agric Food Chem 2009;57(10):4168–76.
[6] Carter P, Wells JA. Dissecting the catalytic triad of a serine protease. Nature 1988;332(6164):564–8.
[7] Zhang YQ, Tao ML, Shen WD, Zhou YZ, Ding Y, Ma Y, et al. Immobilization of L-asparaginase on the microparticles of the natural silk sericin protein and its characters. Biomaterials 2004;25(17):3751–9.
[8] Michalska K, Jaskolski M. Structural aspects of L-asparaginases, their friends and relations. Acta Biochim Pol 2006;53(4):627–40.
[9] Wade HE, Robinson HK, Philips BW. Asparaginase and glutaminase activities of bacteria. J Gen Microbiol 1971;69:299–312.
[10] Michalska K, Bujacz G, Jaskolski M. Crystal structure of plant asparaginase. J Mol Biol 2006;360(1):105–16.
[11] Goodsell DS. The molecular perspective: L-asparaginase. Oncologist 2005;10(3):238–9.
[12] Gallagher MP, Marshall RD, Wilson R. Asparaginase as a drug for treatment of acute lymphoblastic leukaemia. Essays Biochem 1989;24:1–40.
[13] Keating MJ, Holmes R, Lerner S, Ho DH. L-Asparaginase and PEG asparaginase- past, present, and future. Leuk Lymphoma 1993;10(S1):153–7.
[14] Lincoln L, More SS. Isolation and production of clinical and food grade L-asparaginase enzyme from fungi. J Pharmacogn Phytochem 2014;3(3):177–83.
[15] Arima K, Sakamoto T, Araki C, Tamura G. Production of extracellular L-asparaginases by microorganisms. Agric Biol Chem 1972;36:356–61.
[16] Shifrin S, Solis B, Chaiken I. L-Asparaginase from *Erwinia carotovora*. Physicochemical properties of the native and succinylated enzyme. J Biol Chem 1973;248:3464.
[17] Tosa T, Sano R, Yamamoto K, Nakamura M, Chibata I. L-Asparaginase from *Proteus vulgaris*. purification, crystallization, and enzymic properties. Biochemistry 1972;11:217.
[18] Mukherjee J, Joeris K, Reichel P, Scheper T. A simple methods for the isolation and purifications of L-asparaginase from *Enterobacter aerogenes*. Folia Microbiol 1999;44:15.
[19] Dunlop PC, Roon RJ. L-Asparaginase of *Saccharomyces cerevisiae*: an extracellular enzyme. J Bacteriol 1975;122(3):1017–24.
[20] Drainas D, Drainas C. A conductimetric method for assaying asparaginase activity in *Aspergillus nidulans*. Eur J Biochem 1985;151:591.
[21] De-Angeli C, Pocchiar F, Russi S, Tonolo A, Zurita VE, Ciaranf E, et al. Effect of L-asparaginase from *Aspergillus terreus* on ascites sarcoma in the rat. Nature 1970;225:549–50.
[22] Nakahama K, Imada A, Igarasi S, Tubaki K. Formation of L-asparaginase by *Fusarium* species. J General Microbiol 1973;75(2):269–73.
[23] Lee B, Yang HJ. Crystallographic studies on L-asparaginase from *Proteus vulgaris*: I. Prelim cryst data. J Biol Chem 1973;248(21):7620–1.

[24] Tuncel NB, YiLmaz N, Sener E. The effect of pea (*Pisum sativum* L.)-originated asparaginase on acrylamide formation in certain bread types. Int J Food Sci Technol 2010;45(12):2470–6.
[25] Lea PJ, Sodek L, Parry MAJ, Shewry PR, Halford NG. Asparagine in plants. Ann Appl Biol 2007;150(1):1–26.
[26] Sodek L, Lea PJ, Miflin BJ. Distribution and properties of a potassium-dependent asparaginase isolated from developing seeds of *Pisum sativum* and other plants. Plant Physiol 1980;65:22–6.
[27] Murray DR, Kennedy IR. Changes in the activity of enzymes of nitrogen metabolism in seedcoats and cotyledons during embryo development in pea seeds. Plant Physiol 1980;66:782–5.
[28] Lea PJ, Fowden L, Miflin BJ. The purification and properties of asparaginase from Lupinus species. Phytochemistry 1978;17:217–22.
[29] Broome J. Evidence that the L-asparaginase activity of guinea pig serum is responsible for its antilymphoma effects. Nature 1961;171:1114.
[30] Pedreschi F, Kaack K, Granby K. The effect of asparaginase on acrylamide formation in French fries. Food Chem 2008;109(2):386–92.
[31] Campbell H, Mashburn L, Boyse E, Old L. Two L-asparaginases from *E. coli* B: their separation, purification and antitumor activity. Biochemistry 1967;6:721.
[32] Amrein TM, Andres L, Escher F, Amadó R. Occurrence of acrylamide in selected foods and mitigation options. Food Addit Contam 2007;24(Suppl. 1):13–25.
[33] Muttucumaru N, Powers SJ, Elmore JS, Briddon A, Mottram DS, Halford NG. Evidence for the complex relationship between free amino acid and sugar concentrations and acrylamide-forming potential in potato. Ann Appl Biol 2014;164:286–300.
[34] Claus A, Carle R, Schieber A. Acrylamide in cereal products: a review. J Cereal Sci 2008;47(2):118–33.
[35] De Wilde T, De Meulenaer B, Mestdagh F, Govaert Y, Vandeburie S, Ooghe W, et al. Influence of storage practices on acrylamide formation during potato frying. J Agric Food Chem 2005;53(16):6550–7.
[36] Elmore JS, Koutsidis G, Dodson AT, Mottram DS, Wedzicha BL. Measurement of acrylamide and its precursors in potato, wheat, and rye model systems. J Agric Food Chem 2005;53:1286–93.
[37] Knutsen SH, Dimitrijevic S, Molteberg EL, Segtnan VH, Kaaber L, Wicklund T. The influence of variety, agronomical factors and storage on the potential for acrylamide formation in potatoes grown in Norway. LWT-Food Sci Technol 2009;42(2):550–6.
[38] Elmore JS, Briddon A, Dodson AT, Muttucumaru N, Halford NG, Mottram DS. Acrylamide in potato crisps prepared from 20 UK-grown varieties: effects of variety and tuber storage time. Food Chem 2015;182:1–8.
[39] Wicklund T, Østlie H, Lothe O, Knutsen SH, Bråthen E, Kita A. Acrylamide in potato crisp—the effect of raw material and processing. LWT-Food Sci Technol 2006;39(5):571–5.
[40] Weber EA, Koller WD, Graeff S, Hermann W, Merkt N, Claupein W. Acrylamide in vegetable food—agronomic reduction approaches for potatoes and cereals—a review. Pflanzenbauwissenschaften 2007;11(1):28–44.
[41] Williams JSE. Influence of variety and processing conditions on acrylamide levels in fried potato crisps. Food Chem 2005;90(4):875–81.
[42] Amrein TM, Schönbächler B, Escher F, Amadò R. Acrylamide in gingerbread: critical factors for formation and possible ways for reduction. J Agric Food Chem 2004;52(13):4282–8.
[43] Vass M, Amrein TM, Schönbächler B, Escher F, Amado R. Ways to reduce the acrylamide formation in cracker products. Czech J Food Sci 2004;22:19–21.
[44] Ciesarova Z, Kiss E, Boegl P. Impact of L-asparaginase on acrylamide content in potato products. J Food Nutr Res 2006;45(4):141–6.
[45] Ciesarova Z, Kiss E, Kolek E. Study of factors affecting acrylamide levels in model systems. Czech J Food Sci 2006;24(3):133–7.
[46] Martin FL, Ames JM. Formation of Strecker aldehydes and pyrazines in a fried potato model system. J Agric Food Chem 2001;49:3885–92.
[47] Pedreschi F, Kaack K, Granby K. Reduction of acrylamide formation in potato slices during frying. LWT-Food Sci Technol 2004;37(6):679–85.
[48] Pedreschi F, Kaack K, Granby K. Acrylamide content and color development in fried potato strips. Food Res Int 2006;39(1):40–6.

[49] Pedreschi F, Mariotti S, Granby K, Risum J. Acrylamide reduction in potato chips by using commercial asparaginase in combination with conventional blanching. LWT-Food Sci Technol 2011;44(6):1473–6.
[50] Lisińska G, Tajner-Czopek A, Kalum L. The effects of enzymes on fat content and texture of French fries. Food Chem 2007;102:1055–60.
[51] Ciesarova Z, Kukurova K, Bednarikova A, Morales FJ. Effect of heat treatment and dough formulation on the formation of Maillard reaction products in fine bakery products—benefits and weak points. J Food Nutr Res 2009;48(1):20–30.
[52] Teras LM, Zimmermen SP, Zyzak DV, Tak Lin PY, Stojanovic M, Sanders RA, et al. Method for the reduction of acrylamide in corn-based foods. Patent US20040265429, 2004.
[53] Anese M, Quarta B, Frias J. Modelling the effect of asparaginase in reducing acrylamide formation in biscuits. Food Chem 2011;126(2):435–40.
[54] Anese M, Quarta B, Peloux L, Calligaris S. Effect of formulation on the capacity of L-asparaginase to minimize acrylamide formation in short dough biscuits. Food Res Int 2011;44(9):2837–42.
[55] Bråthen E, Knutsen SH. Effect of temperature and time on the formation of acrylamide in starch-based and cereal model systems, flat breads and bread. Food Chem 2005;92(4):693–700.
[56] Capuano E, Ferrigno A, Acampa I, Serpen A, Açar OC, Gökmen V, et al. Effect of flour type on Maillard reaction and acrylamide formation during toasting of bread crisp model systems and mitigation strategies. Food Res Int 2009;42(9):1295–302.
[57] Ciesarova Z, Kukurova K, Mikusova L, Basil E, Polakovicova P, Duchonova L, et al. Nutritionally enhanced wheat-oat bread with reduced acrylamide level. Qual Assur Saf Crops Foods 2014;6(3):327–34.
[58] Ciesarova Z. Acrylamide issue: natural solution by enzyme pre-treatment of foods. Ernährung/Nutrition 2009;33(10):422–7.
[59] Ciesarova Z, Kukurova K, Markova L. Successes and limitations in acrylamide mitigation efforts. Part 2: Impact of interventions in cereal food processing on exposure. Agro Food Ind Hi-Tech 2011;22(4):25–7.
[60] Mohan Kumar NS, Shimray CA, Indrani D, Manonmani HK. Reduction of acrylamide formation in sweet bread with L-asparaginase treatment. Food Bioprocess Technol 2014;7(3):741–8.
[61] Kukurova K, Morales FJ, Bednarikova A, Ciesarova Z. Effect of L-asparaginase on acrylamide mitigation in a fried-dough pastry model. Mol Nutr Food Res 2009;53(12):1532–9.
[62] Allison JP, Mandy WJ, Kitto GB. The substrate specificity of L-asparaginase from Alcaligenes eutrophus. FEBS Lett 1971;14(2).
[63] Herrmann V, Rehm K-H, Schneider F. On the substrate specificity of L-asparaginase from E. coli. FEBS Lett 1974;39(2):214–7.
[64] Mariotti S, Pedreschi F, Carrasco JA, Granby K. Patented techniques for acrylamide mitigation in high-temperature processed foods. Recent Pat Food Nutr Agric 2011;3(3):158–71.
[65] Joint FAO/WHO Expert Committee on food additives: compendium of food additive specifications. 68th meeting, Geneva, 19–28 June 2007.
[66] Ferguson D, Boyd J, Phillips A. Continuous assays of L-asparaginase by coupling with glutamic dehydrogenase and by cationic glass electrode. Anal Biochem 1974;62:81.
[67] Cooney D, Moore A, Shagats R. A colorimetric technique for the demonstration of L-asparaginase in electropherograms. Clin Chim Acta 1972;40:249.
[68] Jayaram H, Cooney D, Jayaram S, Rosenblum L. A simple and rapid method for the estimation of L-asparaginase in chromatographic and electrophoretic effluents: comparison with other methods. Anal Biochem 1974;59:327.
[69] Mashburn L, Wriston J. Tumor inhibitory effect of L-asparaginase. Biochem Biophys Res Commun 1963;12:50.
[70] Nath CE, Dallapozza L, Eslick AE, Misra A, Carr D, Earl JW. An isocratic fluorescence HPLC assay for the monitoring of L-asparaginase activity and L-asparagine depletion in children receiving *E. coli* L-asparaginase for the treatment of acute lymphoblastic leukaemia. Biomed Chromatogr 2009;23(2):152–9.
[71] Becalski A, Lau BPY, Lewis D, Seaman SW, Sun WF. Determination of acrylamide in various food matrices: evaluation of LC and GC mass spectrometric methods. Adv Exp Med Biol 2005;561:271–84.
[72] Ciesarova Z, Balasova V, Kiss E, Kolek E, Simko P, Kovac M. Comparison of two method for acrylamide determination and dietary intake of acrylamide from potato crisps in Slovakia. Czech J Food Sci 2004;22:251–4.

[73] Bednarikova A, Ciesarova Z. Validation of HPLC—ESI—MS—MS method for acrylamide determination in bakery products (Comparison of simple and improved electrospray ionization). Chem Listy 2012;106:252–6.
[74] Ciesarova Z, Kukurova K, Bednarikova A, Hozlar P, Ruckschloss L. Amino acids profile of selected wholegrains important to acrylamide formation in cereal-based products. Chem Listy 2008;102:613–4.
[75] Fredriksson H, Tallving J, Rosén J, Åman P. Fermentation reduces free asparagine in dough and acrylamide content in bread. Cereal Chem 2004;81(5):650–3.
[76] Claus A, Schreiter P, Weber A, Graeff S, Herrmann W, Claupein W, et al. Influence of agronomic factors and extraction rate on the acrylamide contents in yeast-leavened breads. J Agric Food Chem 2006;54(23):8968–76.
[77] Stadler RH. The formation of acrylamide in cereal products and coffee. In: Skog K, Alexander J, editors. Acrylamide and other hazardous compounds in heat-treated foods. Abington, Cambridge, England: Woodhead Publishing Limited; 2006.
[78] Friedman M. Chemistry, biochemistry, and safety of acrylamide. A review. J Agric Food Chem 2003;51(16):4504–26.
[79] Markova L, Ciesarova Z, Kukurova K, Zielinski H, Przygodzka M, Bednarikova A, et al. Influence of various spices on acrylamide content in buckwheat ginger cakes. Chem Pap 2012;66(10):949–54.
[80] Chuda Y, Ono H, Yada H, Ohara-Takada A, Matsuura-Endo C, Mori M. Effects of physiological changes in potato tubers (*Solatium tuberosum* L.) after low temperature storage on the level of acrylamide formed in potato chips. Biosci Biotechnol Biochem 2003;67(5):1188–90.
[81] Becalski A, Lau BPY, Lewis D, Seaman SW, Hayward S, Sahagian M, et al. Acrylamide in French fries: influence of free amino acids and sugars. J Agric Food Chem 2004;52(12):3801–6.
[82] Brunton NP, Gormley R, Butler F, Cummins E, Danaher M, Minihan M, et al. A survey of acrylamide precursors in Irish ware potatoes and acrylamide levels in French fries. LWT-Food Sci Technol 2007;40(9):1601–9.
[83] Constantin OE, Kukurova K, Neagu C, Bednarikova A, Ciesarova Z, Rapeanu G. Modelling of acrylamide formation in thermally treated red bell peppers (*Capsicum annuum* L.). Eur Food Res Technol 2014;238(1):149–56.
[84] Loaëc G, Niquet-Léridon C, Henry N, Jacolot P, Volpoet G, Goudemand E, et al. Effects of variety, agronomic factors, and drying on the amount of free asparagine and crude protein in chicory. Correlation with the acrylamide formation during roasting. Food Res Int 2014;63:299–305.
[85] Murkovic M, Derler K. Analysis of amino acids and carbohydrates in green coffee. J Biochem Biophys Methods 2006;69(1–2):25–32.

CHAPTER 22

Alternative Technologies for the Mitigation of Acrylamide in Processed Foods

Burçe Ataç Mogol
Department of Food Engineering, Hacettepe University, Beytepe, Ankara, Turkey

INTRODUCTION

Since the discovery of acrylamide in foods, many efforts have been made to mitigate its formation. The government authorities, as well as organizations, have worked closely with industries to develop effective mitigation strategies. For example, the organization FoodDrinkEurope, which represents the European food and drink industry, developed a "toolbox" containing tools that can be used selectively by food producers in line with their particular needs to lower the acrylamide levels in their products [1].

To understand how to mitigate acrylamide formation, the factors affecting its formation should be described. These factors can be categorized as agronomical/recipe factors and processing factors. Many effective techniques were developed based on these factors, and some of them, that is, use of asparaginase enzyme and calcium chloride salt, are successfully established in industrial processes [2,3].

Thermal load is a well-known processing factor that directly affects acrylamide formation. Increasing the thermal load, either by process temperature or time, accelerates acrylamide formation [4]. Labuza and Baisier suggested that one of the measures to inhibit the Maillard reaction in cases where it is undesirable is the maintenance of the lowest possible temperatures [5]. To fit the purpose, the temperature profile of some industrial processes has been modified to decrease the thermal load. In contrast, new alternative techniques were introduced for mitigation of acrylamide formation mainly based on minimizing the thermal load. On the other hand, this chapter will summarize the alternative techniques, which were used as a mitigation strategy for acrylamide formation in different food products.

VACUUM PROCESSES

Baking, frying, and roasting are thermal processes taking place at high temperatures. High temperatures are known to induce chemical reactions in foods and lead to the formation of processing contaminants (e.g., acrylamide). In one sense, thermal processes

Acrylamide in Food
http://dx.doi.org/10.1016/B978-0-12-802832-2.00022-X

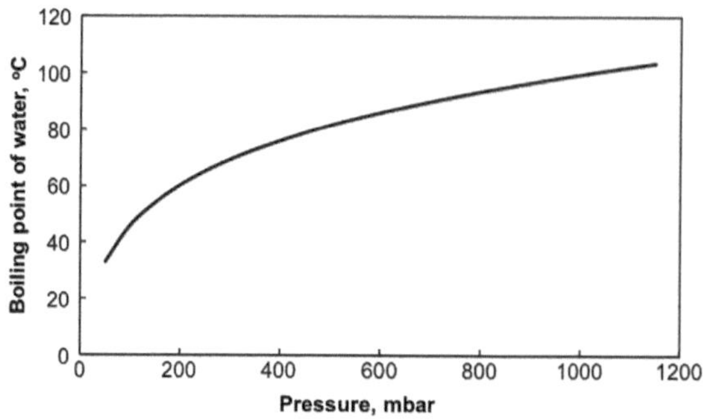

Figure 1 Change in the boiling point of water by pressure [8].

are also used for drying purposes. So, decreasing the temperature is not a viable approach to minimize processing effect, as sufficient drying would not be attained in the final product. Vacuum technology has been recently included in thermal processes to decrease the process temperature [6,7].

The theory behind the vacuum technology is based on the accelerated drying process at lower temperatures. Vacuum lowers the boiling point of water and it leads to evaporation at lower temperatures. By this way, drying can take place in the product without exposure to a high thermal load unlike in conventional atmospheric drying (Figure 1) [6]. Additionally, a vacuum has an increasing effect on internal vapor pressure of the material being dried. It leads to increased moisture migration, and thus to a higher drying rate [6]. Finally, vacuum application lowers the thermal load of the process by allowing the process to take place at a lower temperature.

Vacuum Baking and Combined Baking Technology

The conventional baking process is divided into two main stages: (1) development stage, (2) drying and coloring stage [9]. In the development stage, the dough temperature begins to increase and causes some physical changes in the dough. These include fat melting, gluten softening, and starch gelatinization. After a certain level of thermal load, both moisture and temperature conditions become favorable for browning reactions in the latter stage. These reactions are also responsible for the formation of thermal process contaminants (e.g., acrylamide). Hence, it is essential to minimize the process effect on the formation of acrylamide.

Vacuum technology has been introduced in the baking process of biscuits to decrease thermal load [7]. Biscuits were baked in either a conventional oven (180, 190, or 200 °C) or a vacuum oven (160, 170, 180 °C) operated at 500 mbar for different times. Another set of biscuits was baked in combination with these two baking processes.

Figure 2 shows the acrylamide formation in biscuits at different temperatures during conventional baking. Increasing baking temperature or time significantly increased the amounts of acrylamide formed in biscuits during conventional baking. For example, the acrylamide content of biscuits baked at 200 °C for 8 min was determined as 39 ng/g, while it increased to 211 ng/g when baked for 15 min. Similarly, the acrylamide content of biscuits baked at 180 °C for 13 min was found to be 21 ng/g and it increased to 83 and 191 ng/g at 190 and 200 °C, respectively. Other research studies have repeatedly indicated that the amount of acrylamide increased with temperature or duration of thermal treatment in different food matrices and model systems [4,10–12].

The study showed similar kinetic trends in biscuits during vacuum baking at 180 and 200 °C (Figure 3). However, no significant acrylamide formation (low limit of quantitation) was observed in biscuits during vacuum baking at 160 °C. Mogol et al. reported that the thermal load was limited to form acrylamide in biscuits under these conditions [7]. At 200 °C, the mean acrylamide concentration ranged from 11 to 51 ng/g in biscuits baked under a vacuum, while it ranged from 39 to 211 ng/g in biscuits baked in a conventional oven. The results indicated that the acrylamide concentrations attained during

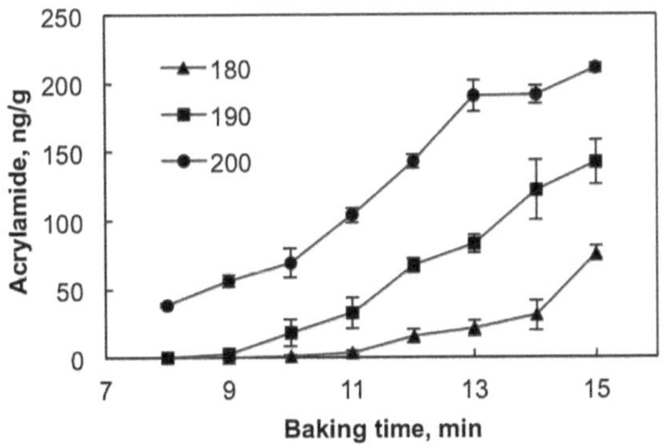

Figure 2 Change of acrylamide concentration in biscuits with time during conventional baking [7].

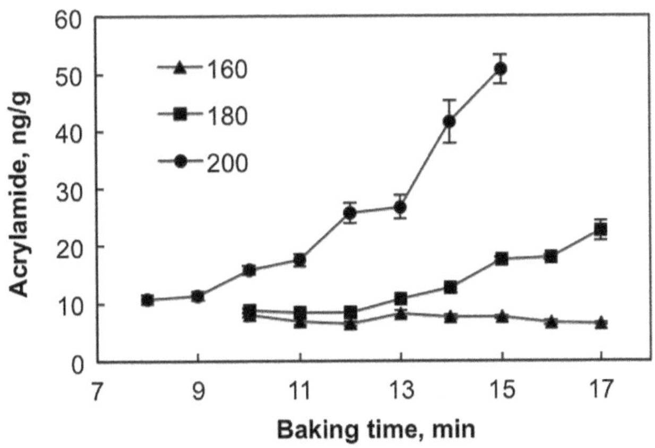

Figure 3 Change of acrylamide concentration in biscuits with time during vacuum baking [7].

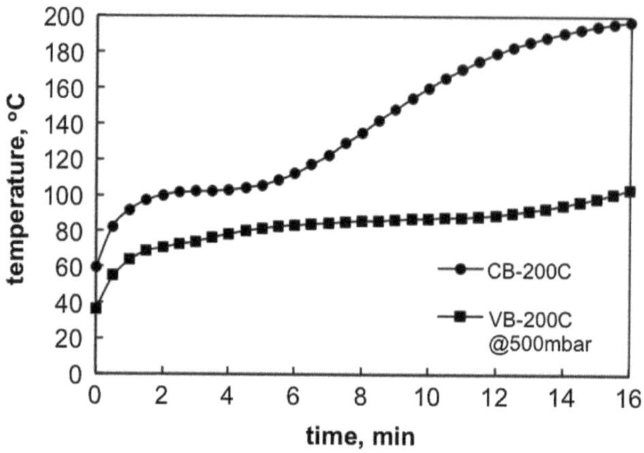

Figure 4 Time–temperature profiles of biscuits measured at the center [7].

vacuum baking were significantly lower than those attained during conventional baking at all temperatures studied ($p < 0.05$).

The time–temperature profiles at the center of biscuits baked in a conventional oven (CB) and vacuum (VB) oven are compared in Figure 4. Biscuits had a typical time–temperature profile during baking in such a way that the biscuit temperature rapidly rose to the boiling point of water within 2 min of baking; it then remained constant in the range 97–103 °C for 3 min until the moisture of the biscuits was evaporated largely. After a critically low moisture level was attained, the biscuit temperature began to increase again until it reached 200 °C at the end of the baking. The time–temperature profile of the biscuits was reported to be different during vacuum baking at 200 °C and at 500 mbar. The biscuit temperature rapidly rose to 71 °C in 2 min, and then slowly to 81 °C as the boiling point of water was reported as 81.6 °C at 500 mbar [13]. It continued to rise very slowly till it reached 105 °C at the end of the baking. In comparison to conventional baking, a noticeably low time–temperature profile of vacuum baked biscuits was the main reason for reduced acrylamide formation. So, applying a vacuum to the baking process allowed a lower temperature profile. This explained the low level of acrylamide formed in biscuits baked in a vacuum oven.

The study also revealed that a low pressure applied during vacuum baking accelerated the drying rate of biscuits. The drying rate of biscuits in conventional baking at 180 °C was found to be similar to that in vacuum baking at 160 °C. For example, moisture contents of biscuits baked at 180 °C for 8 and 11 min were 5.92% and 3.03% under conventional conditions, respectively, whereas moisture contents of biscuits baked at 160 °C for the same time were found as 5.79% and 2.89% under vacuum at 500 mbar.

Mogol et al. reported that biscuits baked by the vacuum process lack surface browning [7]. The reason was reported as the limited convection heat transfer. In conventional baking, three modes of heat transfer, namely conduction, convection, and radiation, are effective with certain contributions. Conduction is a heat transfer that takes place when the medium is stationary, like the heat transfer from a baking tray to the bottom of the biscuit dough, where convection occurs in a moving medium, as in heated air from inside the oven to the top of the biscuit dough [14]. Since oven air was partially removed in vacuum baking, convective heating was limited, but conduction and radiation took place inside the oven. It was stated that the convective heat transfer coefficient depends on the pressure of environment and it gradually decreases under vacuum [15].

Kocadağlı et al. showed that addition Maillard reaction products (MRPs) solved issues related to lack of color and flavor in biscuits baked with combined conventional baking and radiofrequency (RF) postdrying [16]. Mogol et al. applied the same approach to biscuits baked by the combined conventional and vacuum process [7]. When the dough was colored and flavored in the first instance, baking would be a matter of drying to obtain the final textural characteristics of the biscuits. A commercially available brown-colored powder, spray-dried (microgranulated) process flavor derived from a yeast extract, was used in different amounts (0.5% and 1.0%) to modify the recipe for biscuits prepared by means of a combined conventional and vacuum baking process. First, the dough was partially baked at 220 °C for a short time (2–4 min) in a conventional oven. Then, the partially baked biscuits were postbaked in a vacuum oven for accelerated drying at 180 °C and 500 mbar for 4–6 min until the desired final moisture content was attained. It was reported that in the combined process, the exposure of biscuits to high temperature long time conditions, which were essential for the formation of acrylamide, was prevented. The study demonstrated that there was no acrylamide formation (low limit of detection) in biscuits baked in the combined process. Control biscuits that were baked at 220 °C for 8 min in the conventional oven were found to have an acrylamide content of 40 ± 3 ng/g. The addition of 0.5% and 1.0% of the brown-colored powder significantly increased the acrylamide content of control biscuits to 84 ± 5 ng/g and 85 ± 3 ng/g, respectively ($p < 0.05$). This increase was attributed to the presence of reactive carbonyl species in the powder that could accelerate acrylamide formation in biscuits during conventional baking.

The sensory properties of food have to be considered when a new technology is involved in its process, as sensory properties of biscuits are of importance to consumer acceptability. Mogol et al. reported that biscuits containing brown-colored powder baked with combined technology were preferred to conventionally baked biscuits according to surface color, as they had a homogeneous color on the surface. No significant difference was observed between the taste of the biscuit containing a 1.0% brown-colored powder and the biscuit baked by conventional technology [7]. Additionally, no significant difference was observed between textures of the biscuits analyzed. Finally, overall sensorial analysis of biscuits showed that there was no significant difference between conventionally baked biscuits and biscuits baked with combined technology containing brown-colored powders ($p < 0.05$).

Mogol et al. also compared the color values of biscuits baked with combined technology with the conventionally baked ones, as color is an important attribute for biscuits [7]. Considering CIE $L^*a^*b^*$ space, L^* value typically decreases, where the a^* and b^* values increase for biscuits baked in the conventional process. According to the study, the L^* value significantly decreased, where the a^* and b^* values significantly increased when 1.0% of the brown powder was added to the recipe ($p<0.05$). So, the addition of the brown-colored powder to the recipe successfully simulated the color change of biscuits during baking.

As a conclusion, the study pointed out that combined conventional and vacuum baking processes could be considered as an alternative baking technology for biscuits as they effectively prevented the formation of acrylamide. The authors proposed the use of light colored biscuits for chocolate-coated products or adding brown-colored powders to biscuit formulations to solve lack of surface color. The combined process may be of importance for the production of baby biscuits in which the highest level of product safety is required in terms of thermal process contaminants.

Palazoğlu et al. investigated the effect of baking under partial vacuum on acrylamide content and quality attributes of cookies [17]. Cookies were prepared according to the AACC10-54 method and baked either completely under atmospheric pressure (101 kPa) or under reduced pressure after baking under atmospheric pressure (7.5 min under 101 kPa followed by 2.5 min under 60 kPa; 7.5 min under 101 kPa followed by 2.5 min under 40 kPa). According to the results, when the pressure was reduced to 60 kPa after 7.5 min of conventional baking, the acrylamide level of the resultant cookies was found to be 31.8 ng/g. This indicated a 45% lower acrylamide than in the control cookies (57.4 ng/g). Similarly, when the pressure was reduced to 40 kPa after 7.5 min of conventional baking, the acrylamide content of cookies was lowered by 53% (27.1 ng/g). Combined baking technology affects the surface color and partly textural properties of cookies. The use of vacuum after partial conventional baking resulted in significantly higher L^* values than in the control sample. This indicates a lighter surface color of cookies. The maximum force, a measure of hardness, was significantly higher in control cookies than in the cookies prepared by combined baking. Palazoğlu et al. stated that this might be attributed to the different time–temperature profiles experienced by the samples during different baking procedures. Further, no significant difference was observed between the samples in terms of deformation at maximum force, a measure of fracturability. Palazoğlu et al. have confirmed that combined conventional-vacuum technology holds great potential in limiting acrylamide formation in cookies through reduction in both baking time and temperature as the target moisture content is reached in a shorter period of time when a vacuum is applied [17].

Another approach in using vacuum for acrylamide mitigation is to remove the acrylamide formed physically from the product. Anese et al. investigated the physical removal of acrylamide from biscuits and potato chips, by considering different combinations of pressure, temperature, and time [18]. Ground commercial short dough biscuits and potato chips were subjected to vacuum treatment. Also, aliquots of the ground samples were hydrated before the vacuum treatment. Results showed that acrylamide removal was achieved only in samples previously hydrated at water activity values >0.83. The maximum acrylamide removal was reported to be between 5 and 15 min of vacuum treatment at 6.67 Pa and 60 °C. Vacuum application provided 43% and 18% physical removal of acrylamide from the biscuits and the potato chips, respectively. The study suggested implementation of this technique to tunnel ovens at the industrial level. The finished product coming from the tunnel ovens with a low water content is subjected to a hydration step (e.g., carried out by means of a spray of pressurized water) followed by a vacuum step, where both acrylamide and the excess of water are removed.

Vacuum Frying

Frying is one of the food processes carried out under atmospheric conditions and at temperatures between 165 and 190 °C. It includes simultaneous heat and mass transfer [19]. During this process, the moisture leaves the fried food as vapor and oil is absorbed at the same time. In the end, physical changes and chemical reactions occur. This brings the desired structural and textural characteristics to the final product. Frying of potato products, that is, potato chips and French fries, requires special attention as the Maillard reaction implies food safety concerns by forming acrylamide during frying [2,10,12]. An alternative technology, that is, vacuum frying, was introduced to fruit and vegetable products to minimize process-related effects on fried products [20,21]. Among these, potato chips and French fries attracted special attention as this technology decreases the formation of acrylamide by lowering the thermal load of the process.

Vacuum frying is a frying process operated below atmospheric pressure, preferably at <50 Torr (6.65 kPa) [19]. The boiling points of both the oil and the moisture in the food are lowered during vacuum frying due to this low pressure [22,23]. Eventually, the moisture in the fruit or vegetable is rapidly removed below the normal atmospheric boiling point of water and oil. A decreased boiling point provides some advantages including reduced oil content and preserved color and flavor in the product, and it keeps the good quality of oil for longer times [24]. For example, fruit and vegetable products fried under a vacuum showed better nutritional quality and color than the ones fried conventionally as the lack of air prevents oxidation [20,25].

Apart from these, the purpose of vacuum frying of potato products is to prevent acrylamide formation, as the highest levels of acrylamide have been found in French fries, potato chips, and other fried, deep-fat fried, or oven–cooked potato products [26]. Thermal load, a function of frying conditions, directly affects acrylamide formation, and thus should be controlled to limit the formation. Intense frying could be desired due to browning and flavor development, but it leads to higher acrylamide contents [27–29]. Decreasing thermal load results in increased frying times, which leads to higher fat absorption [30].

Granda et al. used different potato cultivars to fry potato chips both traditionally and under vacuum at different temperatures for different times [31]. The frying temperature of the vacuum process was kept lower than that of the traditional one. They revealed that as the frying temperature decreased from 180 to 165 °C, acrylamide content in potato chips reduced by 51% during traditional frying and by 63% as the temperature decreased from 140 to 125 °C during vacuum frying. They also found that potato slices fried under vacuum at 118 °C had a lower acrylamide content with the desirable yellow golden color and texture attributes compared with those fried traditionally. There was a 94% decrease in acrylamide content when potatoes were fried to the same final moisture content (1.5% + 0.3%) under vacuum compared to those fried under atmospheric conditions. The texture properties of potato chips fried under atmospheric pressure or vacuum were compared. They observed no significant differences in texture values as a function of varieties for both frying conditions. This implies that vacuum frying can produce potato chips with a low acrylamide concentration, with the same texture characteristics as those obtained under the traditional frying method. Finally, the results revealed that vacuum frying enabled frying at lower temperatures and acrylamide formation could be reduced up to 94% in potato chips.

Granda and Moreira investigated the formation kinetics of acrylamide in potato chips during vacuum frying [27]. The potato slices of the cultivar Atlantic was fried under vacuum at different temperatures, 118, 125, 140 °C, for different times ranging from 60 to 600 s. Potato chips were also fried traditionally at 150, 165, and 180 °C for 30–360 s. The results of the study confirmed that the acrylamide formation in potato chips fried traditionally or under vacuum increases either with time or with temperature. Figure 5 shows the acrylamide content of potato chips fried traditionally at 165 °C under atmospheric pressure and fried at 118 °C under vacuum (10 Torr).

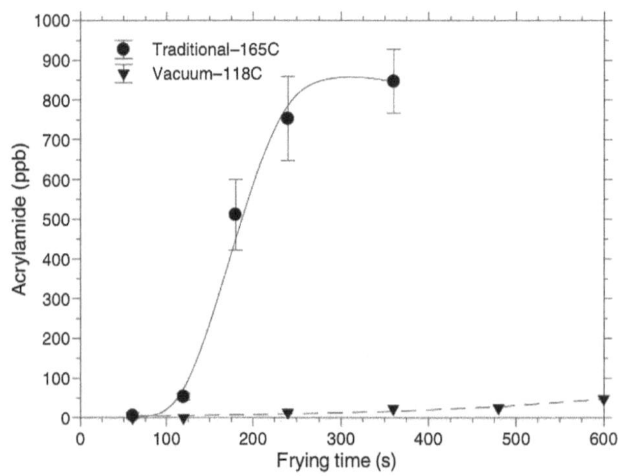

Figure 5 Effect of the frying method and frying time on acrylamide content in potato chips [27].

Vacuum frying led to a decrease in the acrylamide content of the potato chips. A 94% decrease (from 847 ± 80 ppb to 48 ± 1 ppb) was observed in acrylamide content when potatoes were fried under vacuum to the same final moisture content (1.5% ± 0.3% w.b.). This decrease was again associated with the lower temperature used during vacuum frying. Acrylamide formation, modeled kinetically, was found to be very different during traditional and vacuum frying. During traditional frying, acrylamide formation reached a maximum level and remained constant at longer frying times. This was expressed as acrylamide may not only be formed but may also be degrading while time progresses. However, the temperatures during vacuum frying were much lower; therefore, it took longer at lower temperatures for acrylamide to start degrading; thus, acrylamide increased exponentially for longer times. In contrast, the available amounts of reactants (asparagine and reducing sugars) in potato chips might be exhausted at a certain point, and acrylamide content remained constant [32]. Acrylamide accumulation under vacuum frying obeyed the first-order kinetics, while traditional frying fits well to the logistic kinetic model. As temperature increased, the acrylamide formation rate constant *k* increased as expected, because higher temperatures enhance the rate of acrylamide formation. The activation energies obtained from kinetic examination were very different for traditional (21,830.9 J/mol) and vacuum frying (61,059.2 J/mol) processes. This difference indicated that temperature had a stronger influence on the acrylamide content of potato chips fried under vacuum.

Vacuum Roasting

Potato products and bakery products are considered as the main source of acrylamide. However, coffee noticeably contributes to the dietary intake of acrylamide, especially in North European countries that consume higher amounts of coffee [33–35]. Acrylamide is formed in coffee beans due to the Maillard reaction during roasting that takes place at high temperatures, for example, in the range of 220–250 °C. Roasting causes acrylamide to form in coffee beans beyond the commonly accepted asparagine/sugar (or carbonyl) condensation [35]. Acrylamide is formed and destructed simultaneously during the roasting process. Many studies reported that the amount of acrylamide formed during roasting increases exponentially and reaches a maximum and then decreases rapidly [35–39]. Recently, Pastoriza et al. stated that acrylamide may react with coffee melanoidins during roasting, which leads to a net decrease in acrylamide content [40].

The degree and speed of roasting largely affects acrylamide formation due to formation kinetics [37]. Acrylamide Toolbox developed by the FoodDrinkEurope could not determine any effective strategy to mitigate acrylamide formation in coffee, as the importance of the sensory attributes of the product could not be overlooked. Recently, vacuum roasting was introduced to reduce the acrylamide content in coffee [41]. *Coffea arabica* beans were roasted for increasing lengths of time at atmospheric pressure (conventional roasting), under vacuum (i.e., 0.15 kPa; vacuum roasting) or combined conventional-vacuum roasting (at atmospheric pressure for 10 min followed by vacuum treatment 0.15 kPa). As expected, vacuum provided a higher rate of vaporization due to a low boiling point of water. A lower temperature profile was observed in vacuum roasting during the initial stages of the roasting process compared to that in conventional and combined roasting.

Roasting of coffee under different pressure conditions might be responsible for different rates of browning. Only slight differences in lightness were reported among the different samples [41]. However, hue angle decreases more in vacuum-processed coffee up to a certain thermal effect.

Typical acrylamide formation and reduction trends were observed among the roasting processes, with some differences (Figure 6). In particular, acrylamide concentration in the vacuum-roasted coffee sharply reached a maximum at a thermal effect value of 1.6 min and then decreased. The maximum acrylamide formation was observed in combined conventional-vacuum roasting at a range of thermal effect values of 1.5–4.2 min. However, after a certain thermal effect, no significant difference in acrylamide concentration was attained between different roasting techniques. Differences in acrylamide reduction were expressed as stripping effect of low-pressure conditions. A vacuum process caused a stripping effect toward water and water-soluble low-molecular weight compounds, including volatile compounds and acrylamide. Such a stripping effect would prevent acrylamide from being accumulated.

Anese et al. finally stated that medium-roasted coffee processed under low-pressure conditions contained 50% lower acrylamide levels, as compared to the conventionally roasted ones [41]. Although vacuum roasting decreases acrylamide formation in medium-roasted coffee, a change in the roasting process may cause a change in the sensory attributes of coffee. According to sensory analysis, medium-roasted coffee samples obtained by means

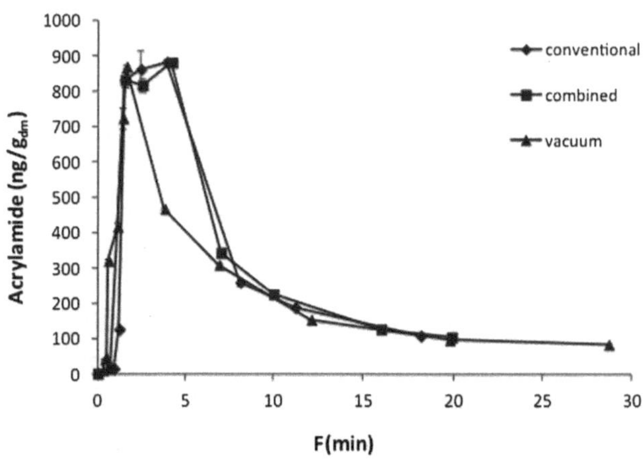

Figure 6 Acrylamide concentration of coffee beans subjected to conventional, combined conventional-vacuum and vacuum roasting as a function of thermal effect (F indicates thermal effect) [41].

of conventional and vacuum processes were not perceived as different. Conclusively, the study proposed the vacuum roasting process as an alternative roasting process to decrease acrylamide formation in medium-roasted coffee without affecting the sensory property [41].

RF PROCESSES

RF heating has been introduced as the final drying step to decrease thermal load applied to the product during thermal processes [16,42]. Doing so, the desired moisture content is attained in the final product.

The RF band covers a broad range of frequencies between 3 kHz and 1 MHz or between 1 and 300 MHz [43]. RF heating transfers electromagnetic energy into the product, and it causes frictional interaction between molecules generating heat within the product. The product as a dielectric medium is heated between two metal capacitor plates that are adjusted by a high-frequency alternating electric field either charged positively or negatively [43]. The polarity of the electric field changes rapidly and water molecules being polar try to continuously align themselves. Colliding neighboring molecules lead to kinetic energy and friction and cause heat within the product. So, heat causes water to evaporate, and the material that is heated by RF loses moisture and dries consequently. RF heating has been successfully applied in the food industry for drying, baking, and thawing. Especially, RF heating reduces baking times by up to 30% in bakery products without further color changes and provides lower residual moisture [44].

There are few research studies that focus on the effect of RF on acrylamide formation. Acrylamide is formed in the later stages of baking when a certain moisture content and temperature are achieved. Therefore, combined processes involving conventional and alternative techniques are designed. The RF dryer is very effective when used in combination with the conventional technique, such as baking and frying. It enables one to control the moisture content accurately and also has benefits in terms of productivity and quality. RF is known to generate heat inside the product, that is, cookies, and the moisture content of prebaked product decreases rapidly. Due to the lower dielectric constant, the temperature of the drying surface of cookies does not continue to increase on application of RF.

The effect of RF heating on acrylamide formation was investigated in leavened cakes and short dough biscuits [45]. Leavened cakes (thickness, 2 cm and diameter, 5 cm) and short dough biscuits (thickness, 0.3 cm and diameter 7 cm) were baked by conventional heating or different combinations of conventional and RF heating until the certain moisture content was achieved. RF-assisted baking was found to be a mitigation strategy for short dough biscuits with a water content of 3.0%. This alternative baking process provided 35% lower acrylamide in short dough biscuits than in those baked by conventional heating only. As acrylamide is formed in the later stages of baking, the use of the

RF technique in the last stages resulted in being a promising strategy to keep acrylamide levels low in bakery products. This technique was found to be more suitable for thin bakery products (e.g., biscuits).

Another study examined the effect of RF postdrying of partially baked cookies on the acrylamide content, texture, and color of the final product [42]. Cookies were partially baked for 8 or 9 min, and then postdried in a 27.12-MHz RF tunnel oven until the moisture content of control cookies was attained. The acrylamide content of biscuits baked in a conventional and combined conventional-RF oven is shown in Figure 7. In general, the use of RF heating as postdrying decreased the acrylamide formation in cookies. The acrylamide level was reduced by 52.3% (from 107.3 to 51.1 ng/g) in comparison to that in the control when 8-min conventional baking was combined with RF (45 s) postdrying. Similarly, acrylamide content of the biscuits baked in a conventional oven for 9 min and postdried in an RF oven (30 s) decreased by 30.5% [42]. The lower acrylamide formation is attributed to a lower temperature profile in RF heating as it removes moisture without overheating the surface of the food. This results in less heat damage and surface browning, and temperature increase is avoided at the product's dry surface [46].

The use of a new technology should be considered together with its effect on the product appearance. Limiting temperature increase by combined baking technology might decrease acrylamide formation, but it could affect the color and textural properties of the product. Comparing with the control sample, cookies prepared by the combination of conventional-RF baking had a lower L^* value, while hardness and fracturability of cookies were not affected. Sensory evaluation revealed that RF-treated samples had a less crumbly texture in comparison to that of the control, and a slightly uncooked flavor

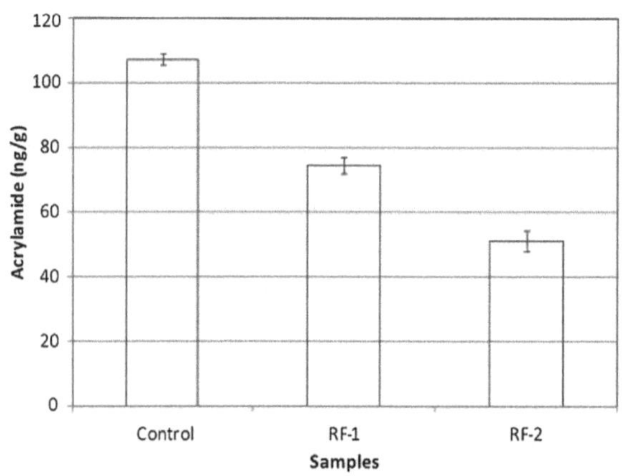

Figure 7 Acrylamide levels of cookies prepared by the conventional method and a combination of conventional-RF baking (control: 11 min at 205 °C, RF-1: 9 min at 205 °C + 30 s RF postdrying; RF-2: 8 min at 205 °C + 45 s RF postdrying) [42].

was noticeable in RF postdried samples that were conventionally baked for 8 min. However, cookies baked for 9 min in a conventional oven and postdried in RF were found to closely resemble control cookies. Palazoğlu et al. concluded that transferring of partially baked cookies from the conventional to the RF oven in the final stage of baking results in limited acrylamide formation, and this technique might be effectively used to design safer cookies [42].

It has been reported that using RF in combination with conventional baking results in less surface browning of cookies [42]. The reason was that introduction of RF as a postdrying step for biscuits decreases browning as browning takes place in the final stage of baking. In contrast, color is an important attribute for the acceptability of bakery products by the consumer. Therefore, the color of the cookies baked with combined technology needs to be standardized. For this purpose, another study investigated incorporating MRPs in dough [16]. MRPs gave a brown color to the dough in the first step, so it allows baking of the biscuits at lower temperatures. The MRPs were prepared by heating a binary mixture of arginine and glucose at 100 °C for 6 h (MRP-I) or by overbaking thin dough-layered disks (MRP-II). The biscuits were prepared with combined partial conventional baking and RF postbaking process. RF postbaking drying was carried out for 30 s (RF-1) and 45 s (RF-2) following baking of cookies at 205 °C for 9 and 8 min, respectively. The control was baked at 205 °C for 11 min. Kocadağlı et al. confirmed that RF treatment significantly decreased acrylamide formation. MRP-I used in lower temperature baking experiments resulted in an increase in acrylamide formation, while the addition of MRP-II had no significant effect on acrylamide formation ($p < 0.05$). The addition of MRP to the dough improved the visual acceptability of cookies baked by combined conventional baking (205 °C for 8 min) and RF postdrying process (45 s). Moreover, remarkable improvement in surface browning was observed in cookies by means of MRP-II. The L^* value of dough decreased to 52.7 from 70.48, while a^* value increased from 1.26 to 10.1 by the addition of MRP-II. Kocadağlı et al. stated that the MRP might be considered as useful recipe modifiers to have less acrylamide in cookies to be obtained by means of a combined conventional baking and RF postbaking drying process. Consequently, MRP-II delivers acceptable color and texture properties to cookies containing less acrylamide.

Fiore et al. investigated the acrylamide formation and the sensory properties of potato and sweet potato cubes cooked using RF or the conventional procedure [47]. Potato cubes were cooked in two different ways, in a conventional oven and in an RF combined oven. For the RF combined cooking, the oven temperature was set at 230 °C and potato cubes were cooked for 15 min with an energy transfer of 400 kJ. For the conventional oven (same oven without turning on the RF), the conditions were first cooking at 210 °C for 41 min and then lowering the temperature to 180 °C for 11 min for a total cooking time of 52 min. Sweet potato cubes were cooked in the conventional oven at 250 °C for 30 min or 250 °C for 15 min with a total energy of 374 kJ for the RF combined cooking. The authors stated that potatoes were cooked to reach the same consistency (i.e., to the complete starch gelatinization at the core of the potato cubes). The cooking time was found to be 30% shorter when RF was used in combination with conventional heating. Fiore et al. indicated that the acrylamide content of conventionally cooked potato cubes was double the amount of those cooked using an RF oven. In the case of sweet potato, the acrylamide concentration was 250% higher in the conventional cooked samples than in those cooked in a combined RF oven. They concluded that RF cooking could be proposed as an efficient mitigation strategy for acrylamide formation in potato samples.

RF has been proposed as a postdrying process combined with conventional frying [48]. Potato slices were partially fried at 180 °C for 95 s (RF-1) or 80 s (RF-2), and then postdried using RF for 5 min; the upper electrode voltage was operated at 3 kV. Control samples were fried at 180 °C for 110 s. Compared to the control sample, potato slices produced by combined technology have 26% and 32% lower acrylamide levels, respectively (Figure 8). No acrylamide was formed during RF drying. As shown in Figure 8, the combined technology not only decreased acrylamide content but also reduced oil content by 12% in potato slices fried for 80 s and then postdried by RF. However, the analyses showed that RF postdrying resulted in a lower hardness and less browning with no significant difference in terms of crispiness when compared to the control process. Sensory analyses showed that the panelists liked all the samples fried by conventional or combined process equally. Consequently, frying potato slices partially in hot oil followed by an RF drying treatment was found to be an applicable approach to limit acrylamide formation in potato chips due to the selective heating of the remaining moisture by RF drying.

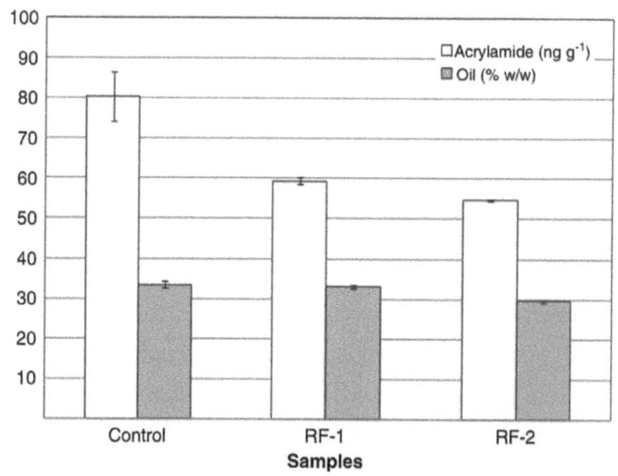

Figure 8 Acrylamide level and oil content of potato chip samples [48].

SYNOPSIS OF ANALYTICAL TECHNIQUES

Acrylamide analyses are performed with gas chromatography (GC)- or liquid chromatography (LC)-based methods. Sample preparation protocols differ in the method. LC-based methods require the extraction of acrylamide from the food matrix. In general, water is used as an extraction solvent. In some methods, water is slightly acidified to make ionization easy in mass spectrometry (MS). Besides single-step extraction, a multiple-stage strategy has been applied to acrylamide extraction for various foods, and it made acrylamide extraction more efficient by preventing underestimation of its concentration. The recovery is determined by adding isotopically labeled internal standards either [$^{13}C_3$]-acrylamide or [D_3]-acrylamide to the sample in the extraction step. The extract obtained is usually cleaned up by using solid-phase extraction (SPE) cartridges. Most clean-up procedures consist of a combination of several SPE cartridges. One approach is by the use of Oasis MCX (mixed-mode cation exchanger; Waters Milford, MA, USA) or Isolute Env+, 1 g (Biotage, Sweden) cartridge alone. Another approach is to combine Oasis HLB (Waters, Milford, MA, USA) and Bond Elut-Accucat (mixed-mode: C8, SAX and SCX) (Varian, Palo Alto, CA, USA). Prior to use, the SPE cartridges are conditioned by passing methanol and are equilibrated by passing water at a flow rate of approximately two drops per second. When the extract is loaded to the cartridge, the first seven to eight drops of the effluent are discarded to prevent any dilution of the sample by replacing water held in the sorbent void fraction with the sample effluent. After the clean-up step, acrylamide is separated from other coextractives. Cleaning up the extract significantly improves the chromatograms by avoiding the coextractives that interfere in MS detection.

For the chromatographic separation of acrylamide, most methods use reversed-phase chromatography, in which C18 columns are used. For example, Acquity UPLC HSS T3 (Waters, Milford, USA) is suitable for separation of polar and nonpolar compounds with a 100% aqueous mobile phase. Waters Spherisorb ODS2 (Waters, Milford, USA) and Inertsil ODS-3 column (GLC-Sciences, Tokyo, Japan) can also be used for acrylamide separation. The column temperature differs according to the method and varies from room temperature to 40 °C. Elution is mostly performed isocratically by using an aqueous mobile phase. Water, slightly acidified with formic acid, can be used alone or in combination with methanol. Addition of acid to the aqueous phase increases conductivity and provides a source of protons to facilitate the ionization process in MS.

MS, either single or tandem, is widely used to analyze acrylamide in foods. After column, the eluent is ionized in an ion source, mostly by electrospray ionization (ESI). ESI, a soft ionization technique, is suitable for acrylamide ionization. ESI produces ions by the use of an electrospray in which a high voltage is applied to a liquid to create an aerosol. LC–MS/MS provides a high selectivity in multiple reaction-monitoring mode (MRM). MRM monitors the transition from a precursor ion, produced in ion source, to a product ion. The precursor ion is separated in the first quadrupole, then the product ion is generated in a collision cell by collision with argon. The product ion enters the second quadrupole where it is selectively analyzed. The transition 72→55 is usually selected to quantify acrylamide because it shows a relatively high intensity. To confirm, other transitions, such as 72→44 and 72→27, are also used. For isotopically labeled acrylamide ($[^2H_3]$- and $[^{13}C_3]$-acrylamide), the transition of 75→58 is monitored. External calibration is used to quantify the acrylamide present in the food sample.

GC–MS is another technique for acrylamide analysis. The sample extract is usually put through a derivatization process, which is a useful step to produce a more volatile compound and to increase the selectivity of the determination. Another sample extraction method, that is, headspace solid-phase microextraction (SPME), can also be used to quantify acrylamide. In this method, acrylamide undergoes silylation with N,O-bis(trimethylsilyl)trifluoroacetamide to form volatile N,O-bis(trimethylsilyl)acrylamide (BTMSA). When formed, BTMSA can be extracted from the headspace over the silylation reaction using a poly(dimethylsiloxane) SPME fiber. In SPME analysis, the analytes are thermally desorbed from the fiber in the GC injector held at high temperatures and operating in either the splitless or split mode. Separation can be performed in a fused silica capillary column and helium is used as the carrier gas. Temperature needs to be programmed properly to resolve components completely. After separation, the ions are analyzed in a mass spectrometer operated in the single ion monitoring mode at m/z 128 for BTMSA and at m/z 131 for N,O-bis(trimethylsilyl)-$[^2H_3]$ acrylamide, isotopically labeled standard. The calibration is carried out by analyzing standard mixtures of acrylamide and $[^2H_3]$ acrylamide.

KEY FACTS

- Vacuum technology is used for the cost-efficient cooling of fruit and vegetables, as well as bakery products in a very short time. Application of a vacuum allows water to boil at lower temperatures. When the pressure is reduced to a level where water starts boiling at <5 °C, the boiling process takes away the heat from the product, which results in the product being cooled down to that temperature within 20–30 min.
- Vacuum changes heat transfer mechanisms that govern during baking. While the convective heat transfer coefficient decreases under vacuum, the transfer of heat energy by radiation is not affected by the vacuum environment.
- Due to the decreased thermal load during combined conventional-vacuum baking, the formation of another processing contaminant, 5-hydroxymethylfurfural, was prevented.

- Vacuum frying could be combined with any other drying techniques, like microwave to reduce the fat content of vacuum-fried products. A high positive correlation between the fat content and initial water content was reported. Due to consumer health concerns, the initial moisture content needs to be decreased by using combined systems to reduce oil absorption in fried products.
- In RF drying, energy absorption of the food material decreases exponentially as the moisture content decreased during drying. So, the material is expected to have a better homogeneous moisture content to dry it homogeneously.

MINI-DICTIONARY

Thermal load This is the amount of energy transferred to the product while it is processed thermally. It is a function of process temperature and time. Increasing either the temperature or duration of the process results in an increased thermal load.

Vacuum baking This is a baking process that takes place in an oven operated below the atmospheric pressure. The baking system unit consists of a vacuum oven and pump module. The vacuum pump provides a low pressure in the baking environment.

Convective heat transfer This refers to the heat transfer between a solid and a moving fluid.

Conductive heat transfer This refers to transfer of energy from the more energetic particles of a substance to the adjacent less energetic ones as a result of interactions between the particles. Conduction can take place in solids, liquids, or gases.

Radiation heat transfer This entails the energy emitted by matter in the form of electromagnetic waves (or photons) as a result of the changes in the electronic configurations of the atoms or molecules.

SUMMARY POINTS

- This chapter summarizes acrylamide mitigation strategies that are focused on alternative technologies.
- To decrease the acrylamide formation in processed foods, it is essential to decrease the thermal load.
- Decreasing of thermal load via lowering of temperature is not a viable approach when using conventional technologies.
- Alternative technologies, such as vacuum and RF, were successfully implemented to the conventional process to lower the thermal load.
- Application of a vacuum decreases the boiling point of water; thus, it accelerates moisture removal from the processed product. This enables one to process the product thermally at lower temperatures without affecting the final moisture and characteristics of the product.
- Combining vacuum or RF technology with the conventional method seemed to be a viable approach while keeping the quality of the final product.
- Lower thermal loads cause a lighter surface color of biscuits and cookies, which can be adjusted by adding a brown-colored powder or MRPs.

- RF decreased the thermal load, and consequently acrylamide formation.
- When using alternative techniques, the temperature of the product does not increase to the critical level that allows acrylamide formation as in conventional techniques.

REFERENCES

[1] FoodDrinkEurope, Acrylamide Toolbox, http://ec.europa.eu/food/food/chemicalsafety/contaminants/ciaa_acrylamide_toolbox09.pdf In 2011.

[2] Zyzak DV, Sanders RA, Stojanovic M, et al. Acrylamide formation mechanism in heated foods. J Agric Food Chem 2003;51(16):4782–7.

[3] Gökmen V, Şenyuva HZ. Effects of some cations on the formation of acrylamide and furfurals in glucose–asparagine model system. Eur Food Res Technol 2007;225(5–6):815–20.

[4] Mottram DS, Wedzicha BL, Dodson AT. Food chemistry: acrylamide is formed in the Maillard reaction. Nature 2002;419(6906):448–9.

[5] Labuza TP, Baisier WM. The kinetics of non-enzymatic browning. In: Schwartzberg HG, Hartel RV, editors. Physical chemistry of foods. New York: Marcel Decker Inc; 1992. p. 604–9.

[6] Nimmol C, Devahastin S. Vacuum infrared drying. In: Pan Z, Atungulu GG, editors. Infrared heating for food and agricultural processing. FL, USA: CRC Press; 2011. p. 141–67.

[7] Mogol BA, Gökmen V. Mitigation of acrylamide and hydroxymethylfurfural in biscuits using a combined partial conventional baking and vacuum post-baking process: preliminary study at the lab scale. Innov Food Sci Emerg Technol 2014;26(0):265–70.

[8] Straub J. NBS/NRC steam tables. Von L Haar, Gallagher JS, Kell GS. Hemisphere Publishing Corp. Washington–New York–London; 1984. 1. Aufl., XII, 320 s., Chem Ing Tech 1985;57(9):812.

[9] Manley D. Baking. In: Technology of biscuits, crackers and cookies. 3rd ed. Cambridge, England: Woodhead Publishing Limited; 2000.

[10] Becalski A, Lau BPY, Lewis D, et al. Acrylamide in Foods: occurrence, sources, and modeling. J Agric Food Chem 2003;51(3):802–8.

[11] Rydberg P, Eriksson S, Tareke E, et al. Investigations of factors that influence the acrylamide content of heated foodstuffs. J Agric Food Chem 2003;51(24):7012–8.

[12] Tareke E, Rydberg P, Karlsson P, et al. Analysis of acrylamide, a carcinogen formed in heated foodstuffs. J Agric Food Chem 2002;50(17):4998–5006.

[13] Wolfram|Alpha, http://www.wolframalpha.com/with query [water boiling point at 500 mbar], In 2014.

[14] Wong K-FV. Fundamentals of heat transfer. In: Intermediate heat transfer. CRC Press; 2003.

[15] Saidi M, Abardeh RH. Air pressure dependence of natural-convection heat transfer. London, U.K.: World Congress on Engineering; 2010.

[16] Kocadağlı T, Palazoğlu TK, Gökmen V. Mitigation of acrylamide formation in cookies by using Maillard reaction products as recipe modifier in a combined partial conventional baking and radio frequency post-baking process. Eur Food Res Technol 2012;235(4):711–7.

[17] Palazoğlu TK, Coşkun Y, Tuta S, et al. Effect of vacuum-combined baking of cookies on acrylamide content, texture and color. Eur Food Res Technol 2015;240(1):243–9.

[18] Anese M, Suman M, Nicoli MC. Acrylamide removal from heated foods. Food Chem 2010;119(2):791–4.

[19] Moreira RG. Vacuum frying versus conventional frying—an overview. Eur J Lipid Sci Technol 2014;116(6):723–34.

[20] Da Silva PF, Moreira RG. Vacuum frying of high-quality fruit and vegetable-based snacks. Food Sci Technol 2008;41(10):1758–67.

[21] Yagua CV, Moreira RG. Physical and thermal properties of potato chips during vacuum frying. J Food Eng 2011;104(2):272–83.

[22] Garayo J, Moreira RG. Vacuum frying of potato chips. J Food Eng 2002;55(2):181–91.

[23] Fan LP, Zhang M, Mujumdar S. Vacuum frying technology. In: Passos ML, Ribeiro CP, editors. Innovation in food engineering: new techniques and products. FL, USA: CRC Press; 2010. p. 411–32.

[24] Shyu S-L, Hau L-B, Hwang LS. Effect of vacuum frying on the oxidative stability of oils. J Amer Oil Chem Soc 1998;75(10):1393–8.
[25] Shyu S-L, Hwang LS. Effects of processing conditions on the quality of vacuum fried apple chips. Food Res Int 2001;34(2–3):133–42.
[26] EFSA. Update on acrylamide levels in food from monitoring years 2007 to 2010. 2012. pp. 2938–2976.
[27] Granda C, Moreira RG. Kinetics of acrylamide formation during traditional and vacuum frying of potato chips. J Food Process Eng 2005;28(5):478–93.
[28] Pedreschi F, León J, Mery D, et al. Color development and acrylamide content of pre-dried potato chips. J Food Eng 2007;79(3):786–93.
[29] Gökmen V, Şenyuva HZ. Study of colour and acrylamide formation in coffee, wheat flour and potato chips during heating. Food Chem 2006;99(2):238–43.
[30] Moreira RG, Castell-Perez ME, Barrufet MA. Deep-fat frying: fundamentals and applications. Gaithersburg, MD, USA: Aspen Publisher; 1999.
[31] Granda C, Moreira RG, Tichy SE. Reduction of acrylamide formation in potato chips by low-temperature vacuum frying. J Food Sci 2004;69(8):E405–11.
[32] Taubert D, Harlfinger S, Henkes L, et al. Influence of processing parameters on acrylamide formation during frying of potatoes. J Agric Food Chem 2004;52(9):2735–9.
[33] Dybing E, Farmer PB, Andersen M, et al. Human exposure and internal dose assessments of acrylamide in food. Food Chem Toxicol 2005;43(3):365–410.
[34] Granby K, Fagt S. Analysis of acrylamide in coffee and dietary exposure to acrylamide from coffee. Anal Chim Acta 2004;520(1–2):177–82.
[35] Guenther H, Anklam E, Wenzl T, et al. Acrylamide in coffee: review of progress in analysis, formation and level reduction. Food Addit Contam 2007;24(Suppl. 1):60–70.
[36] Bagdonaite K, Derler K, Murkovic M. Determination of acrylamide during roasting of coffee. J Agric Food Chem 2008;56(15):6081–6.
[37] Lantz I, Ternité R, Wilkens J, et al. Studies on acrylamide levels in roasting, storage and brewing of coffee. Mol Nutr Food Res 2006;50(11):1039–46.
[38] Taeymans D, Wood J, Ashby P, et al. A review of acrylamide: an industry perspective on research, analysis, formation, and control. Crit Rev Food Sci Nutr 2004;44(5):323–47.
[39] Senyuva HZ, Gökmen V. Study of acrylamide in coffee using an improved liquid chromatography mass spectrometry method: investigation of colour changes and acrylamide formation in coffee during roasting. Food Addit Contam 2005;22(3):214–20.
[40] Pastoriza S, Rufián-Henares JÁ, Morales FJ. Reactivity of acrylamide with coffee melanoidins in model systems. Food Sci Technol 2012;45(2):198–203.
[41] Anese M, Nicoli MC, Verardo G, et al. Effect of vacuum roasting on acrylamide formation and reduction in coffee beans. Food Chem 2014;145(0):168–72.
[42] Palazoğlu TK, Coskun Y, Kocadagli T, et al. Effect of radio frequency postdrying of partially baked cookies on acrylamide content, texture, and color of the final product. J Food Sci 2012;77(5):E113–7.
[43] Dev SRS, Raghavan VGS. Radio-frequency drying of food materials. In: Awuah GB, Ramaswamy HS, Tang J, editors. Radio-frequency heating in food processing, principles and applications. FL, USA: CRC Press; 2015. p. 215–30.
[44] Piyasena P, Dussault C, Koutchma T, et al. Radio frequency heating of foods: principles, applications and related properties—a review. Crit Rev Food Sci Nutr 2003;43(6):587–606.
[45] Anese M, Sovrano S, Bortolomeazzi R. Effect of radiofrequency heating on acrylamide formation in bakery products. Eur Food Res Technol 2008;226(5):1197–203.
[46] Fellows PJ. Food processing technology principles and practices. 2nd ed. New York: CRC Press; 2000.
[47] Fiore A, Di Monaco R, Cavella S, et al. Chemical profile and sensory properties of different foods cooked by a new radiofrequency oven. Food Chem 2013;139(1–4):515–20.
[48] Koklamaz E, Palazoglu TK, Kocadagli T, et al. Effect of combining conventional frying with radio-frequency post-drying on acrylamide level and quality attributes of potato chips. J Sci Food Agric 2014;94(10):2002–8.

SECTION D

Methods of Analysis

CHAPTER 23

Analysis of Acrylamide in Foods with Special Emphasis on Sample Preparation and Gas Chromatography–Mass Spectrometry Detection

Vural Gökmen
Department of Food Engineering, Hacettepe University, Beytepe, Ankara, Turkey

INTRODUCTION

In 2002, Swedish researchers first reported the formation of acrylamide in foods processed at elevated temperatures [1]. Presence of acrylamide in common heated foods has been considered an important food-related crisis by international authorities. It has highlighted the fact that the agrofood industry is poorly prepared to face such problems rattling the consumers' confidence in terms of quality and safety of thermally processed foods.

After the discovery of acrylamide, a lot of studies confirmed its presence in nearly all fried, baked, and roasted foods sold in Germany [2], the United Kingdom [3], Sweden [4], the Netherlands [5], Hong Kong [6], Japan [7], Austria [8], Australia [9], Turkey [10], Spain [11], Finland [12], Italy [13], Brazil [14], China [15], and Korea [16].

Detection of acrylamide levels in processed foods has been an intensive area of research. Acrylamide is a white crystalline solid having a molecular weight of 71 Da. It has a melting point of 84.5 °C and a high boiling point (136 °C at 3.3 kPa). As is evident from its structure ($CH_2{=}CH{-}CO{-}NH_2$) the limited conjugation involving π-electrons means that acrylamide lacks a strong chromophore for UV (ultraviolet) detection and does not fluoresce.

Rosén and Hellenäs [17] firstly reported the analysis of acrylamide in different heat-treated foods. They developed a mass spectrometry (MS) method for direct detection of acrylamide, which would unequivocally verify the presence of acrylamide in a range of heat-treated foods. Significant progress has been made to date on acrylamide research pertaining to analytical methods. Numerous methods have been developed so far to determine the acrylamide in foods. These methods are mainly based on MS as the determinative technique for better identification of acrylamide in processed foods, coupled with a chromatographic step either by liquid chromatography (LC) [1,3,8,18–22] or by gas chromatography (GC), the latter either after derivatization of the analyte [7,23,24], or in a few cases by analysis of the compound directly [25,26].

Acrylamide in Food
http://dx.doi.org/10.1016/B978-0-12-802832-2.00023-1

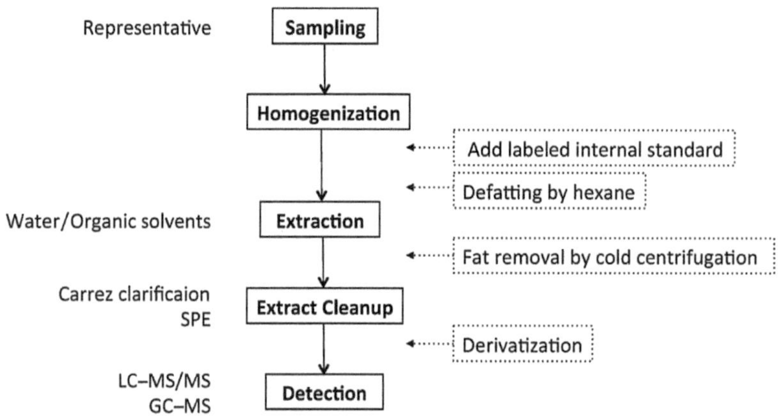

Figure 1 Major steps and variations in the determination of acrylamide in processed foods.

Due to the diversity of sample matrices, various measurement methods and sample preparation procedures have been applied so far. The scheme shown in Figure 1 indicates the main steps in the analysis of acrylamide with possible variations. Looking at this scheme, one can easily deduce that there are dozens of possibilities in the analysis of foods for acrylamide. Along with the nature of the foodstuffs to be tested, the physical and chemical properties of acrylamide have to be considered because they too influence the choice of analytical approaches.

SAMPLING AND HOMOGENIZATION

It is apparent that sample preparation had a great influence on the concentration of acrylamide measured in food samples. Therefore, the initial step during the analysis of acrylamide is to obtain a well-represented homogeneous sample. Acrylamide formation is a largely surface phenomenon where the conditions of high temperature and low moisture pertain during cooking or when thermally processing foods. The affected foods are mostly solids. Acrylamide is not distributed homogeneously throughout these foods, but is concentrated at the surface. There can be significant differences between individual food pieces such as potato crisps or French-fries. There can also be differences in acrylamide content within individual items since, for example, the edges and tips of fried potato products brown more than the rest of the item as does the crust of a bread loaf compared to its inner crumb. So to get a true result, the whole portion or serving of food as eaten should be homogenized thoroughly before a specimen is taken for extraction and analysis [27]. This is usually achieved by means of grinding the solid sample using a lab/kitchen grinder.

Disintegration of the samples to small particles prior to extraction makes additional homogenization, thus avoiding the need for vigorous mixing by ULTRA-TURRAX to increase the extraction yield [28]. A great disparity exists in the mechanical support of acrylamide extraction from food samples. These include an ultrasonic bath or shaker [29–32], horizontal shaker [12,33,34], rotating shaker [35,36], ULTRA-TURRAX homogenizer [37,38], vortex oscillator or vibrator [39,40], and thermostat water bath [41].

The internal standard is added to the specimen of food and a period of incorporation is allowed before commencing extraction. It is then assumed that the internal standard behaves in the same way as the native acrylamide. Addition of an internal standard significantly improves the accuracy and precision of measurements. Coelution of the internal standard with the analyte identifies the location of the analyte in the ion profiles. The presence of an internal standard response and the absence of an acrylamide response are used as evidence for the absence of acrylamide in the test portion [42]. If the internal standard response in a sample set degrades or disappears, the data are demonstrating an analytical problem that must be corrected before the analyses can proceed. In contrast, the data would have been construed as one of nondetects or of low results if the measurements were based on comparisons to external standard responses. Recovery of the internal standard is used to verify whether the analyte can be recovered from the matrix and an internal correction for instrument performance or sample preparation can be effected [42].

Widely used internal standards for the determination of acrylamide include both isotope-labeled internal standards such as deuterium-labeled ($^2H_3$-acrylamide or d_3-acrylamide) and carbon-labeled ($^{13}C_3$-acrylamide) and nonisotope-labeled internal standards such as methacrylamide, *N,N*-dimethylacrylamide, and acetamide [27,37, 43–45]. The isotope-labeled internal standards are ideal for use in MS-based analysis.

EXTRACTION

Different procedures were used for the extraction of acrylamide from food samples. Owing to its high polarity, acrylamide is freely soluble in water, lower alcohols such as methanol and ethanol, and in other polar organic solvents such as acetonitrile, ethyl acetate, and acetone. It is virtually insoluble in nonpolar organic solvents such as heptane and carbon tetrachloride. This solubility behavior dictates the extraction strategy used for analysis of foods and the low molecular weight and low volatility of acrylamide have consequences for the analytical measurement techniques used [27].

Water and polar solvents are commonly used to extract acrylamide from different food matrices [46]. Small portion of homogenized sample (0.5 or 1.0 g) is usually brought to extraction. There is no extraction method that is fully accepted for the extraction of acrylamide from various food samples. Extraction conditions appear to be the most critical parameter in sample preparation for the detection of acrylamide in a range of foods. To date, significant efforts have been directed to developing the best performing extraction procedures with a focus on chromatographic separation and MS detection conditions [47]. Many factors can affect acrylamide extraction yield from food matrices such as sample particle size, defatting, extraction solvent, solvent-to-sample ratio, ULTRA-TURRAX homogenization, the application of mechanical forces, extraction temperature, and extraction time [28,47–49].

Water at room temperature is a highly efficient solvent to extract acrylamide from foods [1,10,17,50,51]. This minimizes the dissolution of hydrophobic compounds in the food products, and also the coextraction of other unwanted compounds in the matrix, which might adversely affect chromatographic separation and interfere detection, if not removed by cleanup [28,36]. Aqueous extraction is usually conducted at room temperature with a solvent-to-sample ratio of 10:1 or 20:1 [1,17]. Sometimes elevated temperature is used for swelling of the matrix and to achieve better penetration of the extraction solvent into the food matrix [28]. However, extraction using hot water does not seem to give any higher recovery [52].

Organic solvents such as methanol, n-propanol, acetone, and acetonitrile can also be used for the extraction of acrylamide from foods. Biedermann and Grob [53] reported that n-propanol as an extraction solvent has several advantages compared to water. These advantages are as follows:

- it yields a clean solution mostly without centrifugation;
- it extracts lipids, which might hinder extraction with water;
- it can be evaporated easily; and
- it azeotropically removes some of water.

Şenyuva and Gökmen [51] reported that addition of water to homogenized potato sample results in thick slurry, rendering extraction difficult. Contrarily, methanol is very compatible with food matrices containing high amounts of starch and fat like fried potatoes. Since methanol does not extract starch, it yields a much clearer extract even without centrifugation. In addition, it can be easily evaporated under a gentle stream of nitrogen [46,54]. Koshnam et al. [46] proposed acetone to extract acrylamide from potato chips with minimum interferences. Takatsuki et al. [55] reported that a sufficient amount of water should be used to increase extraction yield when acetone is used. Delatour et al. [56] suggested liquid–liquid extraction of acrylamide with ethyl acetate to salt-out the aqueous mixture prior to evaporative concentration. Using ethyl acetate to extract acrylamide from the aqueous phase enhances also the removal of other coextractives such as sugars, starch, and amino acids [15].

Processed potato- and cereal-based food samples are usually composed of high amounts of colloids and fat, which should be separated after extraction with water [51]. Therefore, some laboratories have included a de-fatting step before or in combination with the extraction step [57]. This can be carried out by extraction with hexane, dichloromethane, petroleum ether, and cyclohexane [13,28,58]. Some researchers claimed that the removal of fats with n-hexane improves the swelling properties of the starchy matrix and improves acrylamide extraction yield during aqueous extraction. However, Petersson et al. [28] reported that the removal of fat by an organic solvent gives a negative effect to

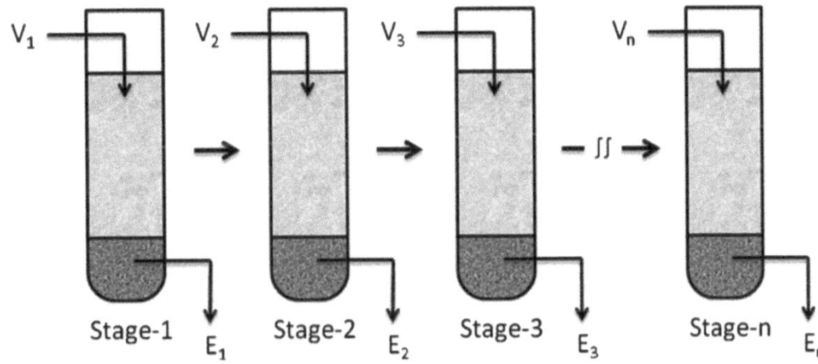

Figure 2 Scheme of multiple-stage extraction for acrylamide analysis in foods.

the extraction yield. Instead of defatting by organic solvent, Şenyuva and Gökmen [51] used cold centrifugation at 0 °C after aqueous extraction of fatty matrix to solidify fat as a layer on top that can be removed easily.

Single-stage extraction has been extensively applied to extract acrylamide from solid food matrices using slight differences in solvent type, solvent-to-sample ratio, extraction time, and temperature [1,3,8,10,20,27,42,59,60]. However, incomplete extraction can lead to erroneous results during the analysis of acrylamide when a single-stage extraction is applied. Therefore, Gökmen et al. [47] introduced a multiple-stage extraction strategy to improve the extraction yield of acrylamide for different food matrices (Figure 2).

Using 10 mM formic acid and methanol, various cereal- and potato-based foods were sequentially extracted determining acrylamide concentration at each extraction stage. As shown in Figure 3, extractability of acrylamide from the solid food matrix is an exponential function, regardless of the type of solvent and food.

This kind of extraction behavior may help us defining the conditions to obtain a well-represented extract during the analysis of acrylamide in foods. Considering that the extractability of acrylamide from a solid matrix is an exponential function, the acrylamide concentration in any extract in a stepwise operation may be related to the following equation:

$$\ln C = \ln C_0 - k_{ext} \cdot n$$

Here,

C: concentration of acrylamide extracted at the nth stage, ng/g;

C_0: original concentration of acrylamide, ng/g;

k_{ext}: the extraction yield constant "extractability," 1/extraction stage; and

n: the extraction stage number.

From an analytical point of view, 90% recovery of native acrylamide from a sample can be considered enough for the success of an extraction procedure. The exponential nature of its extractability makes it possible for the estimation of extraction steps required

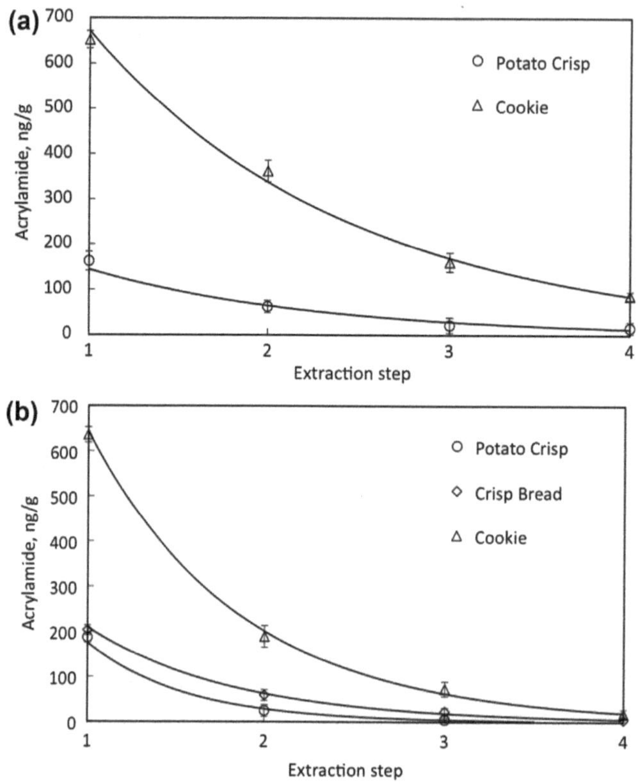

Figure 3 Results of multiple-stage extraction in different food matrices using (a) 10 mM formic acid and (b) methanol as extracting solvent.

in a multiple procedure. In this study, we defined E_{90} as the number of extraction steps to extract 90% of acrylamide from a food sample. Then, E_{90} is calculated by the following equation:

$$E_{90} = \frac{2.303}{k_{ext}}$$

Here, k_{ext} is determined from the experimental extraction data by fitting it to an exponential function. From a practical point of view, the number of extraction steps (N) required to extract at least 90% of acrylamide must be an integer function of E_{90}, and calculated as follows:

$$N = INT(E_{90}) + 1$$

Aqueous extraction using 10 mM formic acid was more effective than was methanolic extraction. It required lesser number of extraction steps than the methanol extraction for both potato- and cereal-based processed foods [47].

EXTRACT CLEANUP

Following approaches are often used in combination for cleanup:

- chemical deproteination with Carrez I and II reagents;
- solid phase extraction (SPE) with complementary stationary phases; and
- freeze–thaw precipitation and membrane filtration.

Chemical deproteination or clarification is needed for protein-rich sample matrices. Carrez reagents (potassium ferricyanide [I] and zinc sulfate [II]) are effective to precipitate proteins in the aqueous extract. Proteins can also be precipitated using organic solvents such as acetonitrile and methanol [33,54,59,61–64]. Mastovska and Lehotay [65] developed a fast and easy dispersive solid phase extraction (dSPE) procedure for the extraction of acrylamide from different food matrices. This method is based on addition of an internal standard to a homogenized sample together with hexane, water, acetonitrile, anhydrous magnesium sulfate, and sodium chloride. Water enhances the extraction of acrylamide, hexane serves for sample defatting, and the addition of salt significantly reduces the mutual miscibility, even resulting in phase separation of the acetonitrile from aqueous phase and forcing the majority of acrylamide into the acetonitrile layer. This extraction method has several advantages, including higher sample throughput and lower costs.

For a successful cleanup of extract, it is extremely important to know the proximate composition of the sample being analyzed, so the potential interferences. Fried potatoes, baked and roasted foods are common sources of dietary acrylamide. Carrez clarification may avoid the colloids, but potentially interfering smaller molecules may still be present in the extract. Therefore, extracts after clarification still need further cleanup by means of SPE that involves a liquid–solid partition, where the extracting phase is a solid sorbent, and it has been used extensively to remove and concentrate trace organic materials from liquid samples or solutions. A choice of sorbents is available using different mechanisms for extraction/retention of analytes. The use of the optimum SPE cartridge can have significant effects on recoveries [66–68]. Cleanup of crude extracts by means of SPE eliminates interfering compounds and increases precision and accuracy of instrumental measurements. A great diversity exists among the SPE cleanup procedures described in the literature. Most cleanup procedures consist of the combination of several SPEs to reduce ion suppression or the so-called matrix effects, and to achieve low limit of detection. One approach has been to combine Oasis HLB and Bond Elut-Accucat (mixed mode: C_8, SAX and SCX) cartridges. Ahn et al. [3] concluded that the passage of the food sample extract through a mixed-mode SPE cartridge containing apolar, strong cation exchanger and strong anion exchanger functional groups removed any entrained fats and largely eliminated the matrix effects. Becalski et al. [20] used a combination of three different cartridges—Oasis MAX (mixed-mode anion exchange), Oasis MCX (mixed-mode cation exchange), and ENVI-Carb (graphitized carbon). A similar combination of SPE cartridges consisting of Bond Elut C_{18}, Bond Elut Jr-PSA (anion exchange), and Bond Elut Accucat were chosen for cleanup of samples, which were measured by LC-MS with column switching [55]. Another group of laboratories have used Isolute Mixed Mode (C_{18}, SAX, and SCX) cartridges combined with filtration or centrifugation to avoid analytical column blockage. Zhang et al. [29] evaluated SPE cartridges including the nonpolar stationary phase and hydrophilic–lipophilic balanced copolymer. They found Oasis HLB useful in terms of good adsorbability and high recovery of acrylamide. Eerola et al. [12] used the combination of Oasis HLB and Accucat for cleanup of problematic roasted coffee extract. Şenyuva and Gökmen [51] identified certain amino acids as interfering coextractives in potato- and cereal-based foods extracts. It is very well known that amino acids can be either positively or negatively charged according to the pH of the environment because of their amphoteric property. Passing the acidified aqueous extract through an Oasis MCX cartridge could easily retain these interfering amino acids. In order to prevent matrix ion suppression effects, Rufián-Henares and Morales [60] assayed two types of cartridges, Isolute Multimode and mixed Oasis HLB and MCX, to purify acrylamide extracts from potato chips. Rosén et al. [50] reported stronger retention of acrylamide with the polymeric and graphitized carbon phases. The retention of polar compounds on graphitized carbon has been attributed to charge-induced interactions with the polarizable surface of graphite. In addition, its flat crystalline surface is different from the brush-like surface of silica-bonded phases. Bortolomeazzi et al. [69] used a single mixed SPE cartridge packed with C18, SCX, and SAX sorbents for the determination of acrylamide in roasted coffee. The SPE column involved all three types of interactions (hydrophobic, strong anionic, and cationic exchange) in order to eliminate the majority of the interfering compounds in coffee extract.

Solid-phase micro-extraction (SPME) technology is a recent advance in sample preparation for trace analysis. It is a solvent-free sample preparation technique that uses a fused silica fiber coated with an appropriate stationary phase attached to a modified microsyringe. SPME is essentially a two-step process, first the partitioning of analytes between the sample matrix and the fiber coating, and then desorption of the concentrated extract from the fiber into the analytical instrument, usually a GC, where the sample components are thermally desorbed. Lagalante and Felter [61] developed a headspace SPME procedure for the extraction of acrylamide from processed cereal products. In this method, acrylamide undergoes silylation reaction with *N,O*-bis(trimethylsilyl) trifluoroacetamide to form a volatile derivative, which is readily extracted from the headspace using a poly(dimethylsiloxane) SPME fiber. Lee et al. [70] used the direct immersion SPME method to determine acrylamide in aqueous matrices with a fiber coated with carbowax/divinylbenzene for extraction at pH 7 for 20 min.

GC–MS DETECTION

Acrylamide can be detected by using GC techniques with or without derivatization [29,49,57,63,64,71,72]. Most of the GC methods are based on derivatization to improve detection sensitivity and specificity [73,74]. Different detection techniques have been coupled with GC for the determination of acrylamide in foods such as MS, tandem mass spectrometry (MS/MS), high-resolution mass spectrometry (HRMS), ion-trap MS, high-resolution time-of-flight mass spectrometry (GC-TOF-MS), electron capture detection (ECD), micro electron capture detection (MECD), nitrogen phosphorus detection (NPD), and flame ionization detection (FID) [24,44,45,53,58,63,70,73,75]. Despite the fact that direct analysis of underivatized acrylamide by GC–MS is less laborious, it is rarely used [74,76]. Because, the direct analysis of acrylamide using GC–MS is troublesome due to the following reasons:

- Lack of characteristic ions in the mass spectrum of underivatized acrylamide and interference caused by the matrix, which increases background noise, and hence a low limit of detection is impossible to obtain [64,65].
- Potential generation of acrylamide in situ in the heated GC injector from coextracted precursors (asparagine and reducing sugars) when water is used for extraction [73,77].
- High solubility of acrylamide in water in comparison with organic solvents complicates sample preparation for GC [65,78].
- Owing to the higher polarity of underivatized acrylamide, it has poor retention on conventional GC columns [63,78].
- The compound 3-hydroxypropionitrile may be coeluted with acrylamide, causing overestimation of acrylamide [53,79].

Different analytical solutions have been developed to overcome the above-mentioned troubles associated with the direct analysis of acrylamide using GC–MS. These solutions include:

- The application of tandem GC–MS/MS in the SRM mode to eliminate interferences [64,70].
- Using n-propanol for extraction followed by solvent exchange with acetonitrile to avoid coextraction of acrylamide precursors that can generate additional analytes during injection [73,80].
- Use of capillary columns with polar stationary phases to improve gas chromatographic separation of acrylamide from interfering compounds [29,78].
- Use of a high-resolution time-of-flight mass analyzer (GC-TOF-MS) enabling partial or complete resolution of interfering matrix coextractives [73].
- Derivatization of acrylamide followed by extraction from the aqueous phase into ethyl acetate [74].

Bromination has been the most widely applied derivatization procedure for the determination of acrylamide using GC techniques [29,64]. However, other derivatization procedures have also been described [61,73,81,82]. It is a fact that acrylamide molecule undergoes bromination by adding bromine under acidic conditions forming 2,3-dibromopropionamide derivative [1,72,83].

The bromination of acrylamide has the following advantages [24,57,64,84,85]:

- When aqueous extraction is used, bromination produces less polar molecules than the native acrylamide, which is easily extracted by ethyl acetate or n-hexane, eliminating many water-soluble components.
- Bromination increases the molecular weight of acrylamide, which leads to improved sensitivity and specificity during GC–MS analysis.
- Bromination increases the volatility of acrylamide, which leads to improved separation during GC analysis.

Many laboratories have used a bromination agent consisting of saturated bromine solution, potassium bromide, and hydrobromic acid [1,24,84,86]. An alternative bromination procedure is to use the mixture of potassium bromide–potassium bromate in acidic medium, in which bromine is formed in situ by an oxidation–reduction reaction [19,63,85,87]. This procedure is considered more convenient and safer than the use of hazardous saturated aqueous bromine with hydrobromic acid. In addition, the reaction is performed in 30 min at cold storage temperature with very good reproducibility [19,24,57,63,78]. After derivatization, a thiosulfate solution is added to remove the excess bromine. Then, the derivative is extracted with an organic solvent such as ethyl acetate or the mixture of ethyl acetate and n-hexane. The organic phase is separated and dried by adding anhydrous sodium sulfate [24,71,88–91]. This organic phase can be injected directly into a GC instrument or subjected to further cleanup using fractionation on a Florisil adsorbent, gel permeation chromatography, or silica-based SPE cartridge [1,24,86].

The results indicated that the bromination product 2,3-dibromopropionamide is thermally unstable and may decompose in the GC injector or at the front end of the GC column during analysis into more stable to 2-bromopropenamide (2-BPA) [19,24,82,92]. The conversion of 2,3-dibromopropionamide to 2-BPA is partial and may adversely affect reproducibility and accuracy of the GC methods. Therefore, it is preferable to convert 2,3-dibromopropionamide to its stable form by adding 10% triethylamine to the final organic phase prior to GC analysis [19,24,49,63,75,92].

The GC–MS-based methods using different capillary columns for chromatographic separation have been applied to a wide range of heated food matrices with certain levels of analytical success. A quantitative GC–MS method has been developed for the determination of trace levels (<50 μg/kg) of acrylamide in cereal-based foods [24]. The acrylamide derivatization product 2-BPA was separated on a ZB-WAX capillary column. The LOD and LOQ were estimated at 2 and 5 μg/kg, respectively. A GC–MS method was developed for the determination of acrylamide in potato chips. Acrylamide was determined after derivatization to 2,3-DBPA [93]. The analytical separation was performed in the capillary column MDN-12. The MS detector was operated in EI mode. The LOD and LOQ were estimated to be 12.8 and 38.8 μg/kg, respectively.

The AA recovery from the spiked solutions varied between 98% and 111%. A GC–MS method was developed to determine acrylamide in roasted coffee products [91]. A layered SPE column was used for elimination of the main chromatographic interferences in the coffee extract. The analytical separation was performed in the capillary column DB 1301. The method was found applicable to a wide range of coffee products. A GC–MS method was optimized for the determination of acrylamide in green tea samples without the interference of bromination by tea catechins [90]. An ADB-17 MS capillary column was used for the separation of 2,3-DBPA derivative. The mass spectrometer was operated in EI ionization and SIM mode. A GC–MS method was used to analyze acrylamide contents of different malt samples after bromination [94]. The recoveries were found between 72% and 86% for different spiking levels, with an LOQ of 25 μg/kg. Casado and Montano [95] have used a GC–MS method to investigate the occurrence of acrylamide in black ripe olives. The determination of acrylamide was carried out using GC–MC after bromination. The chromatographic separation of 2-BPA was performed on a BPX70 capillary column. The LOD and LOQ of the method were estimated as 6 and 20 ng/mL, respectively. A high-resolution GC–MS method was developed for the determination of acrylamide in various food samples [55]. The capillary column Supelcowax-10 was used for separation of the 2-BPA derivative. The LOD was estimated as 5 ng/g. An ion trap GC–MS method has been employed to determine the acrylamide content in Chinese foods [75]. An HP–5MS column was used to separate the brominated derivative, 2-BPA. The analysis was performed using EI ionization and SIM. The recoveries were found between 102% and 110% when wheat flour dough was spiked with 10–50 μg/kg of acrylamide, with an LOD of 5 μg/kg. Soares and Fernandes [96] developed a GC–MS method for the determination of acrylamide in different food matrices. The analytical separation of brominated acrylamide was performed on an MDN-12 capillary column. The MS was used in the SIM mode. The LOD and LOQ values of the method were estimated as 5.2 and 15.7 μg/kg, respectively. The method combined with SPE cleanup prior to GC–MS analysis was also successfully used for the determination of acrylamide in coffee and coffee substitutes [71]. Daniali [89] has used the GC–MS method to determine acrylamide contents in the popular banana-based snacks in Malaysia. The 2-BPA derivative of acrylamide was separated on an Innowax capillary column. The LOD and LOQ values of the method were estimated as 5 and 15 μg/g, respectively. Bent et al. [87] have used a GC–MS method to determine acrylamide levels in commercial and homemade Caribbean foods. The 2-BPA derivative of acrylamide was separated on two capillary columns, DB-1701 and DB-VRX. The MS detector was operated in EI ionization with SIM mode. They reported that the LOD of the GC–MS method was dependent on the type of column. The LOD values were estimated as 20 and 4 μg/kg for the DB-1701 and the DB-VRX columns, respectively.

KEY FACTS

- Numerous methods have been developed so far to determine the acrylamide in foods. These methods are mainly based on MS as the determinative technique for better identification of acrylamide, coupled with a chromatographic step either by LC or by GC.
- Acrylamide formation is largely a surface phenomenon where the conditions of high temperature and low moisture pertain during heating. As a result, acrylamide is not distributed homogeneously within foods, but it is concentrated at the surface.
- An internal standard is added to the sample prior to extraction assuming that it behaves in the same way as the native acrylamide. This significantly improves accuracy and precision of the analytical method.
- Owing to its high polarity, acrylamide is commonly extracted with polar solvents like water and alcohols. Homogenization, defatting, solvent type, solvent-to-sample ratio, temperature, and time affect the extraction yield of acrylamide from food matrices. Although single-stage extraction procedures have been extensively used, multiple-stage extraction either with acidified water or with methanol has shown to improve significantly the yield of extraction.
- The first step of extract cleanup is clarification by adding Carrez I and II reagents. Further cleanup by means of SPE eliminates interfering compounds and increases precision and accuracy of instrumental measurements.
- Although acrylamide can be detected by means of GC techniques with or without derivatization, most of the GC methods are based on derivatization to achieve an increased detection sensitivity and selectivity. Bromination has been the most widely applied derivatization procedure.

MINI DICTIONARY

Multiple-stage extraction An extraction strategy based on sequential aqueous or alcoholic extraction of acrylamide from ground food sample. According to this strategy, two or three extraction steps are usually required depending on the solvent type used to achieve at least an extraction yield of 90%.

Internal standard In analytical chemistry, an internal standard is a chemical substance that is added in a constant amount to samples, the blank and calibration standards in a chemical analysis. This is done to correct for the loss of analyte during sample preparation.

Carrez clarification Carrez I (potassium ferricyanide) and Carrez II (zinc sulfate) reagents are commonly used to precipitate colloids, particularly proteins in aqueous food extracts.

Solid phase extraction (SPE) SPE is a sample preparation process by which compounds that are dissolved or suspended in a liquid mixture are separated from other compounds in the mixture according to their chemical properties. It is used to concentrate and purify samples for instrumental analysis.

Bromination Bromination is a derivatization procedure for the determination of acrylamide using GC techniques. Acrylamide molecule undergoes bromination by adding bromine under acidic conditions forming 2,3-dibromopropionamide derivative. The mixture of saturated bromine solution, potassium bromide, and hydrobromic acid is widely used as a brominating agent. A safer alternative for bromination is to use the mixture of potassium bromide–potassium bromate in acidic medium, in which bromine is formed in situ by an oxidation–reduction reaction.

SUMMARY POINTS

The analytical methods developed so far for the determination of acrylamide in foods are mainly based on MS as the determinative technique for better identification of acrylamide, coupled with a chromatographic step either by LC or by GC. During the last decade, GC–MS methods have been widely used to determine acrylamide in different foods such as fried potatoes, bakery products, roasted coffee, and traditional foods. Although direct determination of acrylamide using GC–MS seems quite simple, it has poor sensitivity and selectivity due to lack of characteristic ions for small and polar acrylamide molecule. Therefore, there is an absolute need for derivatization, mostly bromination prior to GC–MS analysis. The bromination procedure is time-consuming and may include handling of hazardous chemicals such as bromine. Therefore, care should be taken during the bromination reaction. To obtain accurate results in acrylamide analysis, extraction is considered as one of the critical steps in sample preparation. Using a multiple-stage extraction strategy is highly recommended to ensure high extraction yield of acrylamide.

REFERENCES

[1] Tareke E, Rydberg P, Karlsson S, Eriksson M, Törnqvist M. Analysis of acrylamide, a carcinogen formed in heated foodstuffs. J Agric Food Chem 2002;50:4998–5006.
[2] Gutsche B, Weisshaar R, Buhlert J. Acrylamide in food–screening results from food control in Baden-Wurttemberg. Dtsch Lebensm 2002;98(12):437–43.
[3] Ahn J, Castle L, Clarke D, Lloyd A, Philo M, Speck D. Verification of the findings of acrylamide in heated foods. Food Addit Contam 2002;19(12):1116–24.
[4] Svensson K, Abramsson L, Becker W, Glynn A, Hellenas KE, Lind Y, et al. Dietary intake of acrylamide in Sweden. Food Chem Toxicol 2003;41(11):1581–6.
[5] Konings EJM, Baars AJ, van Klaveren JD, Spanjer MC, Rensen PM, Hiemstra M, et al. Acrylamide exposure from foods of the Dutch population and an assessment of the consequent risks. Food Chem Toxicol 2003;41(11):1569–79.
[6] Leung KS, Lin A, Tsang CK, Yeung STK. Acrylamide in Asian foods in Hong Kong. Food Addit Contam 2003;20(12):1105–13.
[7] Ono H, Chuda Y, Ohnishi-Kameyama M, Yada H, Ishizaka M, Kobayashi H, et al. Analysis of acrylamide by LC-MS/MS and GC-MS in processed Japanese foods. Food Addit Contam 2003;20(3): 215–20.
[8] Murkovic M. Acrylamide in Austrian foods. J Biochem Biophysical Methods 2004;61(1–2):161–7.
[9] Croft M, Tong P, Fuentes D, Hambridge T. Australian survey of acrylamide in carbohydrate-based foods. Food Addit Contam 2004;21(8):721–36.
[10] Şenyuva HZ, Gökmen V. Study of acrylamide in coffee using an improved liquid chromatography mass spectrometry method: investigation of colour changes and acrylamide formation in coffee during roasting. Food Addit Contam 2005;22(3):214–20.
[11] Rufian-Henares JA, Arribas-Lorenzo G, Morales FJ. Acrylamide content of selected Spanish foods: survey of biscuits and bread derivatives. Food Addit Contam 2007;24(4):343–50.
[12] Eerola S, Hollebekkers K, Hallikainen A, Peltonen K. Acrylamide levels in Finnish foodstuffs analysed with liquid chromatography tandem mass spectrometry. Mol Nutr Food Res 2007;51(2):239–47.
[13] Tateo F, Bononi M, Andreoli G. Acrylamide levels in cooked rice, tomato sauces and some fast food on the Italian market. J Food Compos Analysis 2007;20(3–4):232–5.
[14] Arisseto AP, Toledo MC, Govaert Y, Van Loco J, Fraselle S, Weverbergh E, et al. Determination of acrylamide levels in selected foods in Brazil. Food Addit Contam 2007;24(3):236–41.

[15] Zhang Y, Ren YP, Zhao HM, Zhang Y. Determination of acrylamide in Chinese traditional carbohydrate-rich foods using gas chromatography with micro-electron capture detector and isotope dilution liquid chromatography combined with electrospray ionization tandem mass spectrometry. Anal Chim Acta 2007;584(2):322–32.
[16] Koh BK. Determination of acrylamide where science meets business content of food products in Korea. J Sci Food Agric 2007;86(15):2587–91.
[17] Rosén J, Hellenäs KE. Analysis of acrylamide in cooked foods by liquid chromatography tandem mass spectrometry. Analyst 2002;127:880–2.
[18] Höfler F, Maurer R, Cavalli S. Schnelle Analyse von Acrylamid in Lebensmitteln mit ASE und LC/MS. GIT Labor-Fachzeitschrift 2002;9:968–70.
[19] Nemoto S, Takatsuki S, Sasaki K, Maitani T. Determination of acrylamide in foods by GC/MS using ^{13}C-labeled acrylamide as an internal standard. J Food Hyg Soc Jpn 2002;43:371–6.
[20] Becalski A, Lau BPY, Lewis D, Seaman SW. Acrylamide in foods: occurrence, sources, and modelling. J Agric Food Chem 2003;51:802–8.
[21] Hoenicke K, Gatermann R, Harder W, Hartig L. Analysis of acrylamide in different foodstuffs using liquid chromatography–tandem mass spectrometry and gas chromatography–tandem mass spectrometry. Anal Chim Acta 2004;520:207–15.
[22] Gökmen V, Şenyuva HZ. A generic method for the determination of acrylamide in thermally processed foods. J Chromatogr A 2006;1120:194–8.
[23] Gertz C, Klostermann S. Analysis of acrylamide and mechanisms of its formation in deep-fried products. Eur J Lipid Sci Technol 2002;104:762–71.
[24] Pittet A, Périsset A, Oberson JM. Trace level determination of acrylamide in cereal-based foods by gas chromatography-mass spectrometry. J Chromatogr A 2004;1035:123–30.
[25] Biedermann M, Biedermann-Brem S, Noti A, Grob K, Egli P, Mandli H. Two GC–MS methods for the analysis of acrylamide in foods. Mittl aus Leb Hyg 2002;93:638–52.
[26] Tateo E, Bononi M. A GC/MS method for the routine determination of acrylamide in food. Italian J Food Sci 2003;15:149–51.
[27] Castle L. Determination of acrylamide in food: GC-MS, bromination method. In: Presentation at the workshop analytical methods for acrylamide determination in food, Belgium, Oud-Turnhout. 2003.
[28] Petersson EV, Rosen J, Turner C, Danielsson R, Hellenas K-E. Critical factors and pitfalls affecting the extraction of acrylamide from foods: an optimisation study. Anal Chim Acta 2006;557:287–95.
[29] Zhang Y, Jiao J, Ren Y, Wu X, Zhang Y. Determination of acrylamide in infant cereal-based foods by isotope dilution liquid chromatography coupled with electrospray ionization tandem mass spectrometry. Anal Chim Acta 2005;551:150–8.
[30] Bagdonaite K, Derler K, Murkovic M. Determination of acrylamide during roasting of coffee. J Agric Food Chem 2008;56:6081–6.
[31] Geng Z, Wang P, Liu A. Determination of acrylamide in starch-based foods by HPLC with pre-column ultraviolet derivatization. J Chromatogr Sci 2011;49:818–24.
[32] Mojska H, Gielecinska I, Stos K. Determination of acrylamide level in commercial baby foods and an assessment of infant dietary exposure. Food Chem Toxicol 2012;50:2722–8.
[33] Geng Z, Jiang R, Chen M. Determination of acrylamide in starch-based foods by ion-exclusion liquid chromatography. J Food Compos Analysis 2008;21:178–82.
[34] Liu J, Zhao G, Yuan Y, Chen F, Hu X. Quantitative analysis of acrylamide in tea by liquid chromatography coupled with electrospray ionization tandem mass spectrometry. Food Chem 2008;108:760–7.
[35] Tezcan F, Erim FB. On-line stacking techniques for the non-aqueous capillary electrophoretic determination of acrylamide in processed food. Anal Chim Acta 2008;617:196–9.
[36] Wang H, Lee AW, Shuang S, Choi MM. SPE/HPLC/UV studies on acrylamide in deep-fried flour-based indigenous Chinese foods. Microchem J 2008;89:90–7.
[37] Jezussek M, Schieberle P. A new LC/MS method for the quantitation of acrylamide based on a stable isotope dilution assay and derivatization with 2-mercaptobenzoic acid. Comparison with two GC/MS methods. J Agric Food Chem 2003;51:7866–71.
[38] Lasekan O, Abbas K. Analysis of volatile flavour compounds and acrylamide in roasted Malaysian Tropical Almond (*Terminalia catappa*) nuts using supercritical fluid extraction. Food Chem Toxicol 2010;48:2212–6.

[39] Karasek L, Wenzl T, Anklam E. Determination of acrylamide in roasted chestnuts and chestnut-based foods by isotope dilution HPLC–MS/MS. Food Chem 2009;114:1555–8.

[40] Shi Z, Zhang H, Zhao X. Ultrasonic-assisted precolumn derivatization-HPLC determination of acrylamide formed in Radix Asparagi during heating process. J Pharm Biomed Analysis 2009;49: 1045–7.

[41] Paleologos EK, Kontominas MG. Determination of acrylamide and methacrylamide by normal phase high performance liquid chromatography and UV detection. J Chromatogr A 2005;1077:128–35.

[42] Roach JA, Andrzejewski D, Gay ML, Nortrup D, Musser SM. Rugged LC–MS/MS survey analysis for acrylamide in foods. J Agric Food Chem 2003;51:7547–54.

[43] Tareke E, Rydberg P, Karlsson P, Eriksson S, Törnqvist M. Acrylamide: a cooking carcinogen? Chem Res Toxicol 2000;13:517–22.

[44] Kim SH, Hwang J-H, Lee K-G. Analysis of acrylamide using gas chromatography-nitrogen phosphorus detector (GC-NPD). Food Sci Biotechnol 2011;20:835–9.

[45] Sun S-Y, Fang Y, Xia Y-M. A facile detection of acrylamide in starchy food by using a solid extraction-GC strategy. Food Control 2012;26:220–2.

[46] Khoshnam F, Zargar B, Pourreza N, Parham H. Acetone extraction and HPLC determination of acrylamide in potato chips. J Iran Chem Society 2010;7:853–8.

[47] Gökmen V, Morales FJ, Atac B, Serpen A, Arribas-Lorenzo G. Multiple-stage extraction strategy for the determination of acrylamide in foods. J Food Compos Analysis 2009;22:142–7.

[48] Pedersen JR, Olsson JO. Soxhlet extraction of acrylamide from potato chips. Analyst 2003;128:332–4.

[49] Wenzl T, Lachenmeier DW, Gökmen V. Analysis of heat-induced contaminants (acrylamide, chloropropanols and furan) in carbohydrate-rich food. Anal Bioanal Chem 2007;389:119–37.

[50] Rosén J, Nyman A, Hellenäs K-E. Retention studies of acrylamide for the design of a robust liquid chromatography–tandem mass spectrometry method for food analysis. J Chromatogr A 2007;1172: 19–24.

[51] Şenyuva HZ, Gökmen V. Interference-free determination of acrylamide in potato and cereal-based foods by a laboratory validated liquid chromatography–mass spectrometry method. Food Chem 2006;97:539–45.

[52] Owen LM, Castle L, Kelly J, Wilson LA, Lloyd AS. Acrylamide analysis: assessment of results from six rounds of Food Analysis Performance Assessment Scheme (FAPAS) proficiency testing. J AOAC Int 2005;88:285–91.

[53] Biedermann M, Grob K. In GC–MS, Acrylamide from heated foods may be coeluted with 3-hydroxy propionitrile. Eur Food Res Technol 2008;227:945–8.

[54] Gökmen V, Şenyuva HZ, Acar J, Sarıoğlu K. Determination of acrylamide in potato chips and crisps by high-performance liquid chromatography. J Chromatogr A 2005;1088:193–9.

[55] Takatsuki S, Nemoto S, Sasaki K, Maitani T. Determination of acrylamide in processed foods by LC/MS using column switching. J Food Hyg Soc Jpn 2003;44:89–95.

[56] Delatour T, Périsset A, Goldmann T, Riedeker S, Stadler RH. Improved sample preparation to determine acrylamide in difficult matrixes such as chocolate powder, cocoa, and coffee by liquid chromatography tandem mass spectrometry. J Agric Food Chem 2004;52:4625–31.

[57] Wenzl T, Beatriz de la Calle M, Anklam E. Analytical methods for the determination of acrylamide in food products: a review. Food Addit Contam 2003;20(10):885–902.

[58] Gökmen V, Şenyuva HZ. Liquid chromatographic analysis of acrylamide in foods - matrix specific interferences encountered during chromatographic separation and mass spectrometric detection. Trends Chromatogr 2006;2:83–93.

[59] Riediker S, Stadler RH. Analysis of acrylamide in food by isotope-dilution liquid chromatography coupled with electrospray ionization tandem mass spectrometry. J Chromatogr A 2003;1020:121–30.

[60] Rufian-Henares JA, Morales FJ. Determination of acrylamide in potato chips by a reversed-phase LC–MS method based on a stable isotope dilution assay. Food Chem 2006;97:555–62.

[61] Lagalante AF, Felter MA. Silylation of acrylamide for analysis by solid-phase microextraction/gas chromatography/ion-trap mass spectrometry. J Agric Food Chem 2004;52:3744–8.

[62] Schieberle P, Köhler P, Granvogl M. New aspects on the formation and analysis of acrylamide. In: Friedman M, Mottram D, editors. Chemistry and safety of acrylamide in food. New York: Springer; 2005. p. 205–22.

[63] Zhang Y, Dong Y, Ren Y, Zhang Y. Rapid determination of acrylamide contaminant in conventional fried foods by gas chromatography with electron capture detector. J Chromatogr A 2006;1116:209–16.
[64] Oracz J, Nebesny E, Zyzelewicz D. New trends in quantification of acrylamide in food products. Talanta 2011;86:23–34.
[65] Mastovska K, Lehotay SJ. Rapid sample preparation method for LC-MS/MS or GC-MS analysis of acrylamide in various food matrices. J Agric Food Chem 2006;54:7001–8.
[66] Kole PL, Venkatesh G, Kotecha J, Sheshala R. Recent advances in sample preparation techniques for effective bioanalytical methods. Biomed Chromatogr 2011;25:199–217.
[67] Ridgway K, Lalljie SPD, Smith RM. Sample preparation techniques for the determination of trace residues and contaminants in foods. J Chromatogr A 2007;1153:36–53.
[68] Xu L, Zhang L, Qiao X, Xu Z, Song J. Determination of trace acrylamide in potato chip and bread crust based on SPE and HPLC. Chromatographia 2012;75:269–74.
[69] Bortolomeazzi R, Munari M, Anese M, Verardo G. Rapid mixed mode solid phase extraction method for the determination of acrylamide in roasted coffee by HPLC–MS/MS. Food Chem 2012;135:2687–93.
[70] Lee M-R, Chang L-Y, Dou J. Determination of acrylamide in food by solid-phase microextraction coupled to gas chromatography–positive chemical ionization tandem mass spectrometry. Anal Chim Acta 2007;582:19–23.
[71] Soares CMD, Alves RC, Casal S, Oliveira MBP, Fernandes JO. Development and validation of a matrix solid-phase dispersion method to determine acrylamide in coffee and coffee substitutes. J Food Sci 2010;75:T57–63.
[72] Lineback DR, Coughlin JR, Stadler RH. Acrylamide in foods: a review of the science and future considerations. Annu Rev Food Sci Technol 2012;3:15–35.
[73] Dunovská L, Cajka T, Hajslová J, Holadová K. Direct determination of acrylamide in food by gas chromatography–high-resolution time-of-flight mass spectrometry. Anal Chim Acta 2006;578:234–40.
[74] Wenzl T, Karasek L, Rosen J, Hellenaes K-E, Crews C, Castle L, et al. Collaborative trial validation study of two methods, one based on high performance liquid chromatography–tandem mass spectrometry and on gas chromatography–mass spectrometry for the determination of acrylamide in bakery and potato products. J Chromatogr 2006;1132:211–8.
[75] Cheng W-C, Hsiao S-W, Chou S-S, Hwang LS, Lu T-J, Yeh A-I. Determination of acrylamide in Chinese foods by GC-ion trap MS using 2-bromopropenamide and 2-bromopropenamide-$^{13}C_3$. J Food Drug Analysis 2006;14:207–14.
[76] Taeymans D, Wood J, Ashby P, Blank I, Studer A, Stadler RH, et al. Review of acrylamide: an industry perspective on research, analysis, formation, and control. Crit Rev Food Sci Nutr 2004;44:323–47.
[77] Weisshaar R. Acrylamide in heated potato products—analytics and formation routes. Eur J Lipid Sci Technol 2004;106:786–92.
[78] Castle L, Eriksson S. Analytical methods used to measure acrylamide concentrations in foods. J AOAC Int 2005;88:274–84.
[79] Zhang Y, Ren Y. New research developments on acrylamide: analytical chemistry, formation mechanism, and mitigation recipes. Chem Rev 2009;109:4375–97.
[80] Kepekci Tekkeli SE, Önal C, Önal AA. Review of current methods for the determination of acrylamide in food products. Food Anal Methods 2012;5:29–39.
[81] Totani N, Yawata M, Takada M, Moriya M. Acrylamide content of commercial frying oil. J Oleo Sci 2007;56:103–6.
[82] Yamazaki K, Isagawa S, Kibune N, Urushiyama T. A method for the determination of acrylamide in a broad variety of processed foods by GC–MS using xanthydrol derivatization. Food Addit Contam Part A 2012;29:705–15.
[83] Daughton CG. Quantitation of acrylamide (and Polyacrylamide): critical review of methods for trace determination/formulation analysis & future-research recommendations. Final Report No. CGD-02/88. Berkeley, CA: California Public Health Foundation; 1988.
[84] Zhu Y, Li G, Duan Y, Chen S, Zhang C, Li Y. Application of the standard addition method for the determination of acrylamide in heat-processed starchy foods by gas chromatography with electron capture detector. Food Chem 2008;109:899–908.
[85] Keramat J, LeBail A, Prost C, Soltanizadeh N. Acrylamide in foods: chemistry and analysis. A review. Food Bioprocess Technol 2011;4:340–63.

[86] Castle L. Determination of acrylamide monomer in mushrooms grown on polyacrylamide gel. J Agric Food Chem 1993;41:1261–3.
[87] Bent G-A, Maragh P, Dasgupta T. Acrylamide in Caribbean foods-Residual levels and their relation to reducing sugar and asparagine content. Food Chem 2012;133:451–7.
[88] Becalski A, Lau BPY, Lewis D, Seaman SW, Sun WF. Determination of acrylamide in various food matrices. In: Friedman M, Mottram D, editors. Chemistry and safety of acrylamide in food. New York: Springer; 2005. p. 271–84.
[89] Daniali G. Determination of acrylamide in banana-based snacks and effect of different maturity stages on formation of acrylamide in banana fritters [Ph.D. thesis]. Universiti Putra Malaysia: Serdang, Malaysia; 2010.
[90] Mizukami Y, Kohata K, Yamaguchi Y, Hayashi N, Sawai Y, Chuda Y, et al. Analysis of acrylamide in green tea by gas chromatography–mass spectrometry. J Agric Food Chem 2006;54:7370–7.
[91] Soares C, Cunha S, Fernandes J. Determination of acrylamide in coffee and coffee products by GC–MS using an improved SPE cleanup. Food Addit Contam 2006;23:1276–82.
[92] Andrawes F, Greenhouse S, Draney D. Chemistry of acrylamide bromination for trace analysis by gas chromatography and gas chromatography–mass spectrometry. J Chromatogr A 1987;399:269–75.
[93] Fernandes JO, Soares C. Application of matrix solid-phase dispersion in the determination of acrylamide in potato chips. J Chromatogr A 2007;1175:1–6.
[94] Mikulíková R, Sobotová K. Determination of acrylamide in malt with GC/MS. Acta Chim Slov 2007;54:98–101.
[95] Casado FJ, Montaño A. Influence of processing conditions on acrylamide content in black ripe olives. J Agric Food Chem 2008;56:2021–7.
[96] Soares CMD, Fernandes JO. MSPD method to determine acrylamide in food. Food Anal Methods 2009;2:197–203.

CHAPTER 24

Liquid Chromatographic Tandem Mass Spectrometry to Determine Acrylamide in Foods

Colin Crews
Fera Science Ltd., Sand Hutton, North Yorkshire, UK

INTRODUCTION

Acrylamide is formed in food during processing operations that involve heat, especially under dry heating which includes home cooking. When eaten, acrylamide has a variety of toxic effects but the major concern is the incidence of neoplastic effects that are an indication of carcinogenicity. The toxic agent is believed to be the acrylamide metabolite glycidamide. The presence of acrylamide in foods consumed in which it occurs at the highest levels and/or which are consumed in the greatest quantities presents a problem to human health. The CONTAM Panel of the European Food Safety Authority (EFSA) evaluated over 40,000 results for the acrylamide content of foods collected and analyzed since 2010 [1]. The highest levels were in the food category "Coffee and coffee substitutes" followed by the categories "Potato crisps and snacks" and "potato fried products." In view of these reasons the demand for analytical data is focused primarily on cooked potato products, roasted coffee, and baked cereals.

Acrylamide has been determined in these foods by a limited range of methods but for routine surveys and research studies the most popular has been liquid chromatography (LC) coupled with mass spectrometry (LC–MS) and in particular coupled with tandem mass spectrometry (LC–MS/MS). The application of LC–MS/MS to the acrylamide issue arose from initial work at the Swedish National Food Administration following the discovery of acrylamide in cooked foods in that country. The first important LC–MS-based method was provided by Rosén and Hellenäs [2]. In this method, acrylamide is extracted from food with water after addition of a deuterated acrylamide internal standard, and the extract is cleaned up by solid phase extraction (SPE) using a single multi-mode cartridge. The assay was applicable to the determination of acrylamide in potato and cereal products over a range of 30–10,000 μg/kg.

Validation of the method showed it to perform well, with better between-laboratory performance than gas chromatography–mass spectrometry (GC–MS) methods [3]. Many variations of this approach have followed, particularly in the adoption of different extraction

Acrylamide in Food
http://dx.doi.org/10.1016/B978-0-12-802832-2.00024-3

and cleanup procedures intended to deal with a variety of food matrices, and to improve the specificity and limit of detection through the removal of interferences caused by coeluting compounds. Advances in SPE and LC column manufacture have helped considerably to improve LC–MS methods. Most developments of the method have involved the use of two SPE cartridges of different phases to clean up an aqueous extract of acrylamide for final determination by LC–MS, with the mass spectrometry being carried out in a mode that fragments the ion obtained from acrylamide to permit confirmation by comparison of product ion ratios and to increase sensitivity by decreasing background noise. The LC–MS/MS analysis of acrylamide has been reviewed in some detail [4–12].

EXTRACTION METHODS

The performance of LC–MS and LC–MS/MS methods for acrylamide is strongly related to the extraction step used and is affected by the cleanup procedure but less so by the LC and MS parameters [3]. The extraction solvents used are limited in number but satisfactory for the task, but in subsequent cleaning up of the extract for LC-MS a considerable variety of procedures have been applied, and several problems encountered for the more difficult matrices [13–15].

Acrylamide is polar in nature and relatively soluble in water, enabling aqueous extraction which is desirable on account of the fact that the bulk of lipids and nonpolar compounds are not extracted simultaneously. An important disadvantage is the fact that the removal of water for the purpose of increasing the concentration of acrylamide in the extract is difficult. Such concentration can be made easier by incorporation of a water-miscible organic solvent; however, this can reduce the extraction efficiency and increase the quantity of coextracted material. Extraction with methanol alone has been used for coffee powders and for potato products with subsequent Carrez clarification and SPE cleanup [16]. Since methanol does not extract polysaccharides such as starch and pectin present in foods of plant origin, it yields a clearer extract than that of water even without centrifugation. Extraction with 10-mM aqueous formic acid was even more effective than water or methanol [17].

Organic solvent extraction other than with methanol is relatively rare in acrylamide analysis. Extraction with dichloromethane has been used to extract acrylamide from samples of potato chips, French fries, and coffee. After shaking the mixture was centrifuged, the supernatant was mixed with hexane, and the acrylamide back-extracted into water [18].

The extraction is affected by many factors including the particle size of the material, the solvent, the sample-to-solvent ratio, the extraction temperature and time, the mechanical method, and the use of a defatting step. The most suitable general methods are based on extraction by shaking with water at 25 °C for about 45 min [13]. Incomplete extraction is most likely to occur when the sample particle size is too large. Studies using two internal standards with one added to the sample and one after extraction and cleanup showed that ion-suppression was not affected by the extraction parameters but was, however, matrix dependent. There was little or no suppression with potatoes and their products but high suppression occurred with crispbread (20%–50%) and coffee (30%–40%). The degree of suppression was also related to the concentration of the extract.

Regardless of the solvent used, a first extraction is insufficient to extract quantitatively all of the acrylamide present in a food sample and a second or multiple extractions are much more effective in achieving recovery of all of the acrylamide present [17]. In general, the sample size should be reduced for matrices that are prone to give a high degree of interference such as some potato products and coffee, in order to minimize the background signal [19]; however, the swelling of the starch in potatoes and cereals means that a sample-to-solvent ratio of 10 or higher is required [17]. Regardless of the ratio of solvent-to-sample, the use of multiple extractions has a greater effect [17].

Extraction is carried out by combinations of standing in solvent, rapid shaking, ultrasonic treatment, and/or blending in a homogenizer at room temperature or slightly above. Automated techniques including pressurized liquid extraction (PLE, also known as accelerated solvent extraction, ASE) [20] have been used occasionally. PLE with solvents such as ethyl acetate works at elevated temperatures and pressure but does not perform well. Soxhlet extraction has also been tried and increases yield of acrylamide when carried out over a long period; however, this might be due to leaching and subsequent reaction of precursors [21].

QuEChERS (Quick, Easy, Cheap, Effective, Rugged, and Safe) extraction methods have so far not been much applied to acrylamide analysis. In this method, acrylamide is extracted by blending the sample with hexane, water, and acetonitrile. Salts (magnesium sulfate and sodium chloride) are added to partition the acrylamide into the acetonitrile prior to further SPE cleanup [22]. The method has been applied to a survey of Chinese foods of multiple classes [23].

Matrix solid phase dispersion (MSPD) has been used to extract acrylamide from potato chips [24]. In this procedure, the solid sample is disrupted by grinding with a solid support, packed into a tube and the acrylamide eluted with a solvent. In practice, the sample was mixed with C18 sorbent and blended with a glass pestle and mortar for 2–3 min. The mixture was then packed into an empty column with a polyethylene frit at the bottom and another frit was placed on top of the sample. The mixture was compressed with a syringe plunger and the mixture defatted with hexane in an SPE manifold. Acrylamide was extracted after drying using two aliquots of water. The sample was analyzed by GC–MS but the extraction procedure is clearly also suitable for LC–MS.

CLEANUP METHODS

Protein Precipitation and Defatting

Despite the poor solubility in water of many food components, minimization of the matrix effects (enhancement or more usually suppression of signal) that are often observed in the

LC–MS/MS determination of acrylamide in foods frequently requires some form of cleanup. Comparisons of cleanup procedures have been provided in the literature [25,26].

Foodstuffs in which acrylamide is most frequently found are those containing high levels of carbohydrate and fat (such as fried potatoes), and coffee. Many other foods can be analyzed without post-extraction cleanup but removal of fat and further cleaning by SPE may be advantageous.

For many samples, such as coffee and cocoa, a defatting step is normally used before water extraction or applied to the aqueous extract. In some samples fat can be removed by careful freezing [2], but blending with hexane or cyclohexane is more often used. It is important that a homogeneous slurry is formed when the sample is shaken with water with or without defatting solvent. However, despite solvent defatting being widely used, studies have shown that its contribution is actually of little value other than helping to preserve the condition of the LC column [27,28].

Protein and some other compounds can be precipitated by the incorporation of methanol or acetonitrile but is more usually achieved by adding Carrez reagents which are solutions of potassium hexacyanoferrate (Carrez I) and zinc sulfate (Carrez II). Precipitation of solid materials from the extract always requires subsequent centrifugation, three layers are produced, a top layer of fat, a middle aqueous layer containing acrylamide, and a lower precipitate. For many foods, a simple clarification with Carrez reagents followed by centrifugation at a cold temperature is very effective in providing a clean extract [23].

Large coextracted molecules can be removed by filtration using a molecular size cut-off filter of 3–5 kDa pore size [2,4], although this procedure has not been widely adopted.

Solid Phase Extraction

SPE involves passing a sample extract through a small cartridge containing usually a few 100 mg of an absorbent material that binds the analyte, allowing coextracted impurities to be washed off with solvent before the analyte is eluted with a stronger solvent.

It has become one of the most important aspects of the analytical procedure for acrylamide. In SPE, the binding mechanism can be based on an affinity related to polarity or on an ion exchange mechanism, or on mixtures of modes. The structure and high water solubility of acrylamide, however, means that it is only weakly bound to the conventional sorbents, and the technique is mostly used to retain unwanted matrix compounds while allowing acrylamide through the water it was applied with. SPE has the advantage of reducing matrix effects (notably ion suppression) in LC–MS/MS and in the case of acrylamide particularly for carbohydrate-based foods [15,25,27]. Many different SPE cartridges have been used in the analysis of acrylamide, the performance of 10 of which, based on silica, carbon, and polymer technologies including mixed modes, were compared in order to measure the acrylamide trapping ability for aqueous extracts [28]. It was found that silica-based phases all yielded poor retention regardless of the modification

(C18 etc.) but higher retention was achieved with the polymeric and graphitized carbon phases, with the Isolute ENV+ column giving the strongest retention—in that 20 ml of aqueous acrylamide solution per gram of SPE material could be passed through the cartridge without loss of more than 5% of the acrylamide.

The major types of SPE cartridge used are hydrophilic/lipophilic interaction (Oasis HLB), mixed-mode anion exchange (such as Oasis MAX), mixed-mode cation exchange (Oasis MCX), mixed-mode types with C18 (Isolute Multimode), mixed-mode anion/cation exchange (Bond Elut Accucat), strong cation exchange (SCX), polymeric strong cation exchange (Strata-X-C, Chromabond HR-XC), strong non-polar phases (ENV^+), and graphitized carbon (ENVI-Carb, Carbograph).

The HLB phase is a strongly hydrophilic polymer with balanced hydrophilic and lipophilic properties much used in methods for acrylamide. With HLB the elution volume of the water used to release acrylamide from the phase needed to be increased from the usual 1.5 ml to at least 3.0 ml to ensure good recovery, although for some foods there might be a risk of eluting undesirable polar materials [26]. The Oasis MAX cartridge contains a polymer-based mixed-mode phase with reversed-phase/strong anion-exchange properties, selective for acidic compounds. Bond Elut PSA is a mixed primary and secondary amino phase with high carbon content. The Strata-X-C, Oasis MCX, and ENV+ cartridges are strong cation-exchange sorbents based on a polymer with benzenesulfonic acid groups that retain basic compounds. These phases are intended for the retention of polar compounds from water, having multiple retention mechanisms such as hydrogen bonding, hydrophobic and cation exchange [25,29].

Multimode cartridges such as Bond Elut AccuCAT include C18, strong cation exchange (SO_3^-) and strong anion exchange ($-NR_3^+$) groups. Multimode cartridges contains octadecyl, quaternary amine, and sulfonic acid functionalities, enabling three different interactions (hydrophobic, strong anionic, and cationic exchange) to take place. The multimode cartridges appear to perform best with a relatively high proportion of C18 phase [30]. Silica-based phases are considered to be effective in providing clean extracts for LC–MS/MS [31] however they are not particularly efficient at dealing with difficult matrices that cause severe ion suppression, particularly extracts of coffee and cocoa. The use of mixed mode phases can be considered the most popular SPE approach [32–34]. Multimode cartridges have been prepared in-house by combining the components of individual columns into a single cartridge. For a water extract of roasted coffee, Bortolomeazzi et al. [30] used a single SPE column containing a mixture of C18, SCX, and SAX materials in a ratio determined by experiments aimed at eliminating the major interferences present in coffee including melanoidins, trigonelline, chlorogenic acids, and caffeine. The extract was passed through the cartridge and the acrylamide eluted in 1 ml of water.

Graphitized carbon phases are neutral resins with a layered carbon surface. When used under reversed-phase conditions they trap hydrophobic compounds but retain some hydrophilic activity.

Elution from most of the SPE cartridges is achieved by means of a relatively small volume of water. There are advantages in concentrating the eluate from the SPE cartridge when it can be assured that it is comparatively clean. However, evaporation of even small volumes of water is very time-consuming even when a moderate degree of heat (40 °C) or vacuum is applied. Care must be taken to avoid losses of acrylamide when evaporating methanol or other organic solvents as evaporation to dryness can result in a 30%–40% of loss in acrylamide [16]. This is especially true when the sample is warmed or the nitrogen gas flow rate is too high [17]. The evaporation time can be reduced by adding a miscible volatile solvent such as methanol to water either after SPE or before it, but the presence of methanol can reduce the efficiency of extraction of acrylamide from SPE cartridges. The highest proportion of methanol in a methanol–water mixture for the elution of acrylamide from the ENV+ column with good recovery was 60% [25]. With this, the most efficient solvent mixture, an elution volume of only 1 ml volume was required, which could be concentrated by evaporation more easily than 100% water. Evaporation of water even by azeotropic methods is still likely to cause losses of acrylamide, reducing absolute recoveries to less than 75% in some cereal products [35].

The recovery of acrylamide is dependent on the matrix and the SPE conditions with recovery for coffee, milk chocolate, and potato crisps typically being in the range 80%–100%, whereas for crispbread and raw potato (often using a different SPE method) it is sometimes lower, between 65% and70%.

Another approach to cleaning up the aqueous extract before SPE or direct analysis is to extract the analyte from water into a polar organic solvent, such as ethyl acetate which avoids the co-extraction of inorganic compounds, sugars and amino acids, and so on. The facile removal of the solvent can enable a considerable increase in sensitivity [37,38].

Figure 1 provides a simple schematic diagram of the typical extraction and cleanup process for LC–MS.

CHROMATOGRAPHY

LC–MS analysis in reversed-phase mode is conveniently compatible with the aqueous solvent that is best suited for extraction of acrylamide from foods, but on account of its small size and very polar nature acrylamide is poorly retained on most conventional LC reversed-phase material such as C18 [2,33] and the best retention is provided by the porous graphitic phase column when the organic component of the mobile phase is low or 100% water is used. The addition of some organic modifier is recommended to improve the signal stability in LC–MS/MS and this does not significantly affect the size of the response. Graphitized carbon stationary phases have been used by many analysts [2,3,19,25,35,39–43]. Polymeric resins also generally give good separation of polar compounds with mobile phases having a high water content.

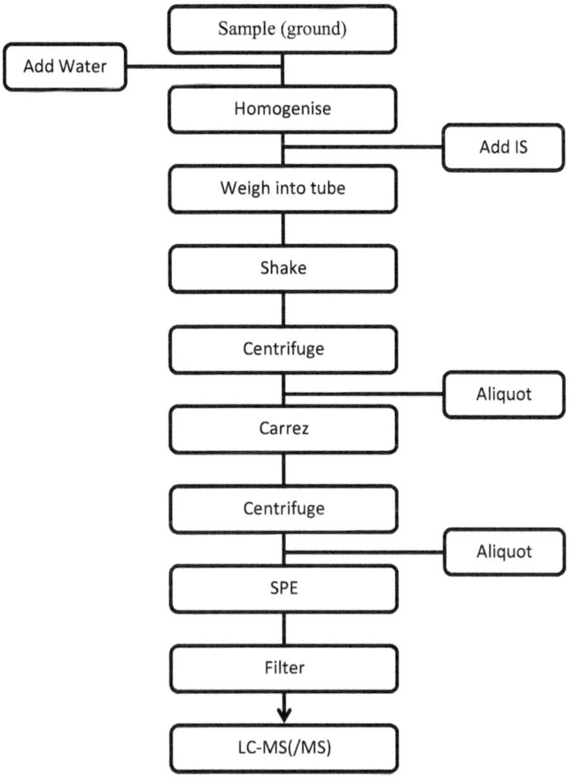

Figure 1 Schematic representation of the extraction and cleanup procedure.

LC columns benefit from protracted cleaning between batches of samples, even when these have been through series of SPE procedures. Graphitized carbon phases are easier to maintain through washing on account of their high stability. They are usually cleaned up by flushing with strong solutions (80%) in water of acetonitrile or methanol for a minimum of 30 min, and/or by reversing the direction of the solvent flow through the column after disconnecting it from the guard column and from the mass spectrometer, although a mixture of 50% tetrahydrofuran, 10% ammonia (ammonia solution 25%), and 40% water has been proposed as more effective [13].

Ultra performance liquid chromatography (UPLC) stationary phases have hybrid small-sized particles that can have unique selectivity. UPLC-based methods provide improvements in resolution and, therefore, in speed, and the compressed response increases the peak height, improving sensitivity where peak height is used to measure the response [44,45]. Silica-based UPLC columns with disfunctionally bonded C18 phases of 1.7-μm particle size are intended for the retention for polar and non-polar compounds with a reduced peak tailing and narrowed peak widths and can separate acrylamide sufficiently within a run time of only 3 min. The smaller volume of UPLC columns might enable a reduction in column cleaning procedures but this factor has still to be studied.

The most effective way of reducing interference in the MS detection to be applied at the chromatographic stage has been to connect four columns in series with switching of the elute flow to dump interfering compounds to waste [46]. The final switching was used to concentrate the acrylamide in a band at the head of the final column to regain a sharp peak shape.

Liquid Chromatography–Mass Spectrometry

Single ion trap mass spectrometers (mass detectors) are not much used for acrylamide determination as they have comparatively low sensitivity and precision. They have the advantage of low cost, rapid scanning rate, and the acquisition of full scan data. Single-stage mass spectrometry (LC–MS) has been applied to acrylamide determination but less frequently than LC–MS/MS although it is quite suitable for low complexity samples such as potatoes. Electrospray ionization (ESI) in positive mode is most often used in LC–MS but atmospheric pressure chemical ionization (APCI) is also very suitable. The ionization and in-source fragmentation of acrylamide in LC–MS produces a protonated molecular ion of at m/z 72 and ions at $[CH_2{=}CHCONH_2]^+$, m/z 55 (loss of amino), and m/z 27 (a subsequent loss of –CO).

LC–MS is used in selected ion monitoring mode (SIM) to record the signal from the protonated molecular ion of acrylamide at m/z 72 and the most intense fragment ion at m/z 55. Additional monitoring is included for the equivalent signals from isotopically labeled internal standards (m/z 73 and m/z 56 for acrylamide labeled with $^{13}C_3$ or $^2H_3$).

Liquid Chromatography–Tandem Mass Spectrometry

Acrylamide has a low mass and so the protonated molecular ion produced on ionization is prone to interference from ions of the same or similar mass resulting from the fragmentation of the very many coextracted compounds of higher mass. To improve confidence in the identification of the acrylamide peak tandem mass spectrometric procedures are most commonly used which produce fragments of the protonated molecular ion in constant ratios under the same conditions.

In LC–MS/MS the protonated molecular ion of acrylamide at m/z 72 $[CH_2{=}CHCONH_2]^+$ is fragmented by collision with a suitable gas such as argon to give product ions with m/z 55 $[CH_2{=}CHCNH]^+$ resulting from loss of NH_3, and at m/z 54 $[C_3H_4N]^+$. The transition of m/z 72 to 55 is the most abundant and is therefore invariably used to provide the signal for quantification. Supporting evidence of peak identity (confirmation transition), is obtained by monitoring the transition 72 > 54, 72 > 44 or 72 > 27 $[C_2H_3]^+$ [9].

The small size of the acrylamide molecule means that fragmentation is rather difficult and careful optimization of the collision gas pressure is recommended [47].

Mass spectrometers with triple quadruple analyzers (QqQ) have much more sensitivity, derived from the high selectivity reducing the background signal. In this instrument a selected ion, usually the protonated molecular ion is fragmented further and the resulting ions produced are measured.

ESI with triple quadrupole ion separation is now the most frequently used method for acrylamide determination but problems can arise when this is used with mobile phases of high water content. APCI mode was proposed for the determination of acrylamide in coffee [25,35,48] as it is less affected; less by the mobile phase. Ion trap analyzers have also been used to good effect [31,41,49].

Time of flight detection in LC–MS (LC–TOF–MS) does not have the sensitivity required for most acrylamide analyses, having an levels of detection (LOD) of about 30 ng/g in potato products, but is useful in identifying co-eluting interfering compounds.

With the newer technique of direct analysis in real-time (DART) analytes such as acrylamide can be measured without prior extraction or chromatography, under "ambient" conditions of room temperature and pressure. In this instrument, an ionizing of stream of metastable atoms or molecules of helium or nitrogen simultaneously desorbs and ionizes compounds from the surface of a sample at the mass spectrometer inlet. Used in combination with high-resolution mass spectrometry (HRMS) it provides data that can be interrogated later to reveal details of the composition. A DART–HRMS technique was applied to experimental biscuit samples, and statistical treatments used to accurately predict acrylamide concentrations in the biscuits. The results were in good agreement with a reference UHPLC–MS/MS method [50].

A combination of UPLC with tandem mass spectrometry (UPLC–MS/MS) is currently the best approach for the end determination of acrylamide as it has high reproducibility, selectivity, and sensitivity [51,52]. LODs and quantification (LOQ) for UHPLC–ESIMS/MS method applied to food such as potato chips are typically 1 and 3 μg/kg respectively.

Compounds that cause interferences at the end-determination by LC–MS/MS have been identified. Potatoes contain a considerable amount of valine, which is a significant source of interference as it produces the same precursor and product ions as acrylamide, and on account of its size avoids precipitation by Carrez reagent [17]. The role of valine in this interference was proposed by Şenyuva et al. [42] and confirmed by identification with ion trap and LC–TOF–MS measurements [25]. Its full scan mass spectrum has a base peak at m/z 118 with a minor ion at m/z 72 that was shown to be a product ion of m/z 118. Ion trap experiments confirmed a second valine product ion at m/z 55. These ions at m/z 72 and m/z 55 corresponded to the acrylamide precursor and product ions. Confirmation of the interfering signal as valine was achieved by high-resolution LC–TOF–MS detection which showed a base peak ion at m/z 118.0849 (valine = 118.0863) and a small peak corresponding to m/z 72.0821 (valine ion, acrylamide = 72.0444).

Coffee and cocoa have proven to be difficult matrices for the determination of acrylamide using LC–MS on account of deterioration of the chromatography after repeated injections of coffee extracts [53], matrix interferences, and severe ion suppression [34,36]. Some problems were associated with the saturation of the SPE materials and a few interferences persisted with reduced sample sizes. Several compounds that are coextracted with acrylamide from coffee cause problem in the LC–MS/MS determination through interference and matrix effects. Particular requirements for the analysis of acrylamide in coffee have been discussed [34,54–60]. Some difficulties with cocoa powder were linked to the defatting with organic solvent and the use of aqueous or organic solvent extraction [62].

Compounds in coffee that are known to be coextracted with acrylamide and retained on the LC column include trigonelline, chlorogenic acid, caffeine, and valine. Trigonelline does not coelute with acrylamide but its tailing elution pattern causes matrix effects that upset the acrylamide response. It has a quaternary positively charged nitrogen allowing its removal by shaking the coffee extract with cation exchange material.

Valine can be removed from most foods by passing the acidified aqueous extract through a cation-exchanger SPE cartridge [54]. In-house prepared mixed-mode phases containing various proportions of C18, SAX, and SCX materials have been used to retain trigonelline, chlorogenic acid, and caffeine in coffee extracts but commercially available Isolute Multimode SPE columns performed almost equally well. Using a small quantity of SPE material and a relatively small sample size in a high ratio of SPE phase to sample loading reduces the need for concentration and/or further cleanup. However, even with SPE cartridge cleanup using sequential cartridges such as HLB cartridge and then anion/cation exchangers the acrylamide ions monitored for coffee extracts are often impure [33].

Matrix effects in coffee analysis can be ameliorated by reducing the sample size but only if sensitivity permits. The decrease in response after repeated injections of coffee extracts affecting both acrylamide and isotopically labeled internal standards can be ameliorated by maintaining the LC at elevated temperature (35 °C) which helps to eliminate this problem. Many compounds with greater retention than acrylamide, notably chlorogenic acid and caffeine, accumulate at the font of the column to be eluted gradually with repeated cycles of the mobile phase gradient. This effect can be monitored by extending the run to watch for late eluting peaks, and avoided by careful setting of a delay time between injections. To further reduce these problems, the use of a guard column of the same phase is recommended and the analytical column should be washed with mixtures of water with methanol and/or acetonitrile, for a few hours if possible [33]. Elimination of acetic acid from the LC mobile phase gave a 3-fold increase in the response for acrylamide in coffee extracts.

Derivatization

A rather different LC–MS approach has been made by the derivatization of acrylamide with 2-mercaptobenzoic acid. This produces a thioether which gives a strong response

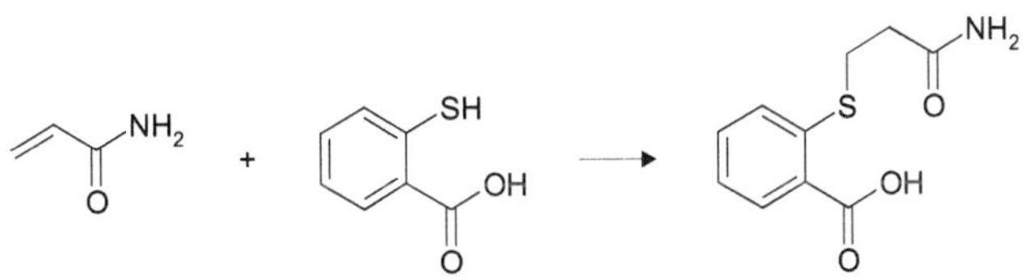

Figure 2 Derivatization of acrylamide with 2-mercaptobenzoic acid.

to LC–MS in positive electrospray mode using relatively inexpensive single-stage instruments [49]. The derivatization is illustrated in Figure 2.

This method was adapted to measure acrylamide in coffee by monitoring the responses for the protonated acrylamide derivative at *m/z* 226 and its sodium adduct at *m/z* 248 [62]. The LOD was 157 μg/kg and the LOQ 270 μg/kg. It was also used to provide an ionizable analyte for separation by capillary electrophoresis (CE) coupled to MS/MS for the analysis of acrylamide in food. The results were comparable with LC–MS/MS methods. Acrylamide can similarly be derivatized with cysteine, which gives a 20-fold increase in sensitivity under LC–MS/MS conditions [18].

QUANTIFICATION METHODS

Quantification of acrylamide in LC–MS is achieved by use of calibration standards with correction of the LC–MS response by use of an isotopically labeled acrylamide. Isotopically labeled internal standards are used to correct for quantitative losses but signal interference from the matrix does not necessarily affect different ions in the same manner and, for example, can interfere with the labeled ion signal and not the native acrylamide. Therefore, it is preferred to adapt cleanup procedures to remove coeluting matrix components. Both deuterium and ^{13}C labeled acrylamide are available. All of the carbon atoms on the acrylamide internal standard molecule should be fully labeled with ^{13}C or bonded with deuterium as in samples with high levels of acrylamide the natural abundance of ^{13}C acrylamide can contribute to the signal for the internal standard. Owing to this, optional ^{15}N labeling has not found favor.

A thorough study of extraction procedures [30] compared the responses for standard solutions and coffee extracts to calculate the absolute recovery of native and d3-acrylamide and also the suppression in an ion trap system. It was concluded that the fates of both acrylamide and d3-acrylamide are the same throughout the analytical procedure. The overall loss of response was 22%–32 % for both analytes with about 17% occurring in the SPE cleanup (using an in-house multi-mode cartridge). Losses due to matrix suppression were about 12%.

Calibration involves the use of a zero level point and five to seven acrylamide calibration standard solutions over the concentration corresponding to the expected levels in the sample extracts. These are typically 20- to 2000 mg/kg acrylamide in the sample and

2- to 200 ng/ml in the standard solution. For quality checks, the standards can be injected at the beginning and end of the analytical batch with one or more also injected regularly in the case of large sample batches.

Acrylamide is identified by a response at the same retention time of the standards and confirmed by comparison of the ratios of selected transition product ions to the major ion at *m/z* 72. The European Union has issued confirmation criteria for chemical residues [63] requiring the monitoring of two product ions, which the transitions described earlier can be used where additional evidence is required.

The reporting limit for analytical methods for acrylamide are set at the level at which the response can be confirmed as acrylamide by comparison of the quantitative and qualitative transitions with those of standards of similar concentration. The confirmation criteria set for chemical residues in animals and animal products [63] can provide good guidelines for application to acrylamide analysis.

QUALITY CONTROL AND METHOD PERFORMANCE

Certified reference materials (CRMs) and proficiency tests for testing the performance of analytical methods are extremely useful for assuring the performance of analytical methods and the laboratories applying them. CRMs for acrylamide have been developed for toasted bread [64], crispbread, and rusk [65]. A certified reference material for potato chips is also available and has been found to be stable for over six years [66,67].

The reproducibility of the method developed at the Swedish National Food Administration has been evaluated by collaborative trials for bakery and potato products [3] and has been accepted as an international standard LC–MS method [68].

Proficiency tests have been carried out regularly since 2002, usually with test materials containing acrylamide levels close to the median obtained in food surveys. The tests tend to reveal a similarity of performance between GC–MS methods and those based on LC–MS although they indicated considerable differences apparently linked to the extraction procedures used [61]. Many proficiency test participants report severe interference problems [69–72].

Current methods based on LC–MS/MS with standard LC columns typically have good recovery and within-laboratory repeatability with limits of detection of 3–20 μg/kg and limits of quantification of 10–50 μg/kg which is sufficient for the immediate requirements for food product monitoring and calculation of consumer exposure.

KEY FACTS

- Acrylamide is one of a number of food processing contaminants, but is comparatively unusual in being soluble in water. This is due to its high polarity derived from a combination of its small size and polar amide group. Solubility in water is advantageous in that it helps the separation of acrylamide from oils and fats present in many foodstuffs. Another advantage is that water is always a component of the liquid chromatography solvent (mobile phase). A disadvantage is that concentration of the extract by evaporation is lengthy and difficult.

- The chemical structure of acrylamide makes the food extracts that contain it poorly amenable to purification by SPE. Acrylamide is not well bound by most SPE phases, which are instead used to bind coextracted material. This is a reversal of the usual SPE procedure. The occurrence of acrylamide in many different food matrices has stimulated the testing of a wide range of SPE materials, often in combination.
- The performance of LC–MS methods in the determination of acrylamide is quite similar to that of the GC–MS methods. The major advantage of the latter is the lower cost of the instruments. LC–MS instrumentation has been developed and improved with great rapidity in recent decades. The performance of the instruments has led to lowering of limits of detection which in turn enable the setting of lower regulatory limits for food contaminants. The greater variation in working principles leads to a variety of methods being available. However, this has a downside in that a standardized approach to acrylamide analysis is made more difficult.

MINI DICTIONARY

Extraction This is the first treatment of a biological sample and it the important step in which a chemical contaminant is separated from the bulk of the sample. To achieve this, a representative and homogenous portion of the sample is mixed intimately with a liquid (solvent) that has been selected to have the properties of being a very good solvent for the contaminant, and a relatively poor solvent for other components of the sample. The bulk of most food samples comprises carbohydrate, protein, and fat and is often in large molecules that are insoluble in either water or organic solvent. The mixing is achieved by blending or shaking. The extract obtained usually needs further purification (cleanup) before instrumental analysis.

Solid phase extraction (SPE) This is a convenient form of cleanup in which the extract of a biological sample is passed through a solid medium which has an affinity for either the compound that is the target of the analysis or for the unwanted coextracted material. The solid medium is held in an inert plastic tube or cartridge and is typically based on very fine silica particles that have an active coating bonded to them chemically. The coating is activated by passing a solvent through the cartridge; then the coating is washed with the solvent used for the extraction. The sample is then passed through, during which passage either the analyte or the coextracted matrix is held on the cartridge phase. The required analyte or the unwanted material is then washed from the cartridge with a suitable solvent. If the analyte is held on the cartridge it can be recovered with a suitable solvent. The SPE cartridge phase can be one of several types, typically a non-polar wax or an ion exchange material.

Liquid chromatography/mass spectrometry (LC–MS) In liquid chromatography the extract containing the analyte and coextracted material is passed in a stream of solvent under pressure through a tube containing material (usually a silica-bound wax such as in SPE) through which the many components are separated in time, some passing through quickly and others more slowly. The solvent is usually a binary mixture of water and methanol or acetonitrile, the proportions of which are varied over time to optimize the separation of the component mixture. As they exit the column the compounds are ionized inside the mass spectrometer and detected as ions, or in LC–MS/MS they are fragmented into smaller pieces that are characteristic for each compound and can be used to identify and measure the target analyte.

Quantification It is the process by which the amount of a compound such as a contaminant that is in a sample is measured. This is usually achieved by comparison of the signal obtained from the sample by a measuring instrument with the signal obtained for an authentic standard of that contaminant over a range of concentrations that can be displayed as a calibration graph.

Validation It is the means by which a method of analysis can be shown to give the true result. This is usually demonstrated by measuring a series of samples including those with naturally zero, low, and high levels of analyte, and sample to which known amounts of analyte have been added. The results of multiple analyses of these samples are treated statistically to show factors such as accuracy, precision, and repeatability.

Limit of detection (LoD) It is the smallest quantity of an analyte that can be distinguished qualitatively from a blank value to a certain degree of confidence. It can be measured in various ways and is generally about three times the response obtained from the background signal.

Limit of quantification (LoQ) It is the smallest quantity of an analyte that can be measured with confirmation of its identity. It is generally approximately three times the LoD and 10 times the response obtained from the background signal. There are various means of assuring the identity, such as multiple signals derived from fragmentation in a mass spectrometer.

SUMMARY POINTS

Much effort has been expended in the field of acrylamide analysis by LC–MS methods and our understanding of the issues involved has increased rapidly; especially in the field of sample preparation and cleanup. Numerous collaborative trials and proficiency tests have helped to identify problems but there is still a greater variety in approach and of technique in acrylamide analysis than in the methods used for many other contaminants. Much of what has been achieved in recent years has been made possible by technical advances in sample preparation and analysis. Future generations of LC column technology and LC instrumentation will undoubtedly improve limits of detection and the quality and scale of data available, at the same time reducing the duration of the analysis. While these improvements are to be much anticipated and valued the biggest challenge is perhaps to build a consistent international approach to the analytical problem with frequently changing tools.

REFERENCES

[1] EFSA. Draft scientific opinion on acrylamide in food. EFSA panel on contaminants in the food chain (CONTAM). European Food Safety Authority (EFSA); 2013. http://www.efsa.europa.eu/en/consultations/call/140701.pdf.

[2] Rosén J, Hellenäs KE. Analysis of acrylamide in cooked foods by liquid chromatography tandem mass spectrometry. Analyst 2002;127(7):880–2.

[3] Wenzl T, Karasek L, Rosén J, et al. Collaborative trial validation study of two methods, one based on high performance liquid chromatography-tandem mass spectrometry and on gas chromatography-mass spectrometry for the determination of acrylamide in bakery and potato products. J Chromatogr A 2006;1132(1–2):211–8.

[4] Tareke E, Rydberg P, Karlsson P, et al. Analysis of acrylamide, a carcinogen formed in heated foodstuffs. J Agric Food Chem 2002;50(17):4998–5006.

[5] Wenzl T, de la Calle MB, Anklam E. Analytical methods for the determination of acrylamide in food products: a review. Food Addit Contam 2003;20(10):885–902.

[6] Wenzl T, Lachenmeier DW, Gökmen V. Analysis of heat-induced contaminants (acrylamide, chloropropanols and furan) in carbohydrate-rich food. Anal Bioanal Chem 2007;389(1):119–37.
[7] Taeymans D, Wood J, Ashby P, et al. A review of acrylamide: an industry perspective on research, analysis, formation, and control. Crit Rev Food Sci Nutr 2004;44:323–47.
[8] Castle L, Eriksson S. Analytical methods used to measure acrylamide concentrations in foods. J AOAC Int 2005;88(1):274–84.
[9] Zhang Y, Zhang GY. Occurrence and analytical methods of acrylamide in heat-treated foods - review and recent developments. J Chromatogr A 2005;1075(1–2):1–21.
[10] Oracz J, Nebesny E, Żyżelewicz D. New trends in quantification of acrylamide in food products. Talanta 2011;86:23–34.
[11] Arvanitoyannis IS, Dionisopoulou N. Acrylamide: formation, occurrence in food products, detection methods, and legislation. Crit Rev Food Sci Nutr 2014;54(6):708–33.
[12] Keramat J, LeBail A, Prost C, et al. Acrylamide in foods: chemistry and analysis. A review. Food Bioprocess Technol 2011;4(3):340–63.
[13] Petersson EV, Rosén J, Turner C, et al. Critical factors and pitfalls affecting the extraction of acrylamide from foods: an optimisation study. Anal Chim Acta 2006;557(1–2):287–95.
[14] Gökmen V, Morales FJ, Ataç B, et al. Multiple-stage extraction strategy for the determination of acrylamide in foods. J Food Compos Analysis 2009;22(2):142–7.
[15] Zhang Q, Ming-Zhao C, Zhi-Xin Z. Overview on sample preparation technique for the determination of acrylamide in foods. Food Sci Technol 2006;7:221–4.
[16] Gökmen V, Şenyuva HZ, Acar J, et al. Determination of acrylamide in potato chips and crisps by high-performance liquid chromatography. J Chromatogr A 2005;1088(1):193–9.
[17] Gökmen V, Şenyuva HZ. A generic method for the determination of acrylamide in thermally processed foods. J Chromatogr A 2006;1120:194–8.
[18] Lim H-H, Shin H-S. A new derivatization approach with D-cysteine for the sensitive and simple analysis of acrylamide in foods by liquid chromatography-tandem mass spectrometry. J Chromatogr A 2014;1361:117–24.
[19] Becalski A, Lau BP, Lewis D, Seaman SW. Acrylamide in foods: occurrence, sources, and modeling. J Agric Food Chem 2003;29;51(3):802–8.
[20] Cavalli S, Maurer R, Hofler F. Fast determination of acrylamide in food samples using accelerated solvent extraction (ASE®) followed by ion chromatography with UV or MS detection. LC GC N Am 2003:35–6.
[21] Pedersen JR, Olsson JO. Soxhlet extraction of acrylamide from potato chips. Analyst 2003;128(4):332–4.
[22] Mastovska K, Lehotay SJ. Rapid sample preparation method for LC-MS/MS or GC-MS analysis for acrylamide in various food matrices. J Agric Food Chem 2006;54:7001–8.
[23] Chen F, Yuan Y, Liu J, et al. Survey of acrylamide levels in Chinese foods. Food Addit Contam Part B-Surveillance 2008;1(2):85–92.
[24] Fernandes JO, Soares C. Application of matrix solid phase dispersion in the determination of acrylamide in potato chips. J Chromatogr A 2007;1175:1–6.
[25] Bermudo E, Moyano E, Puignou L, et al. Determination of acrylamide in foodstuffs by liquid chromatography ion-trap tandem mass-spectrometry using an improved clean-up procedure. Anal Chim Acta 2006;559(2):207–14.
[26] Cheng W-C, Kao Y-M, Shih DY-C, et al. Validation of an improved LC/MS/MS method for acrylamide analysis in foods. J Food Drug Analysis 2009;17(3):190–7.
[27] King R, Bonfiglio R, Ferandez-Metzler C, et al. Mechanistic investigation of ionization suppression in electrospray ionization. J Am Soc Mass Spectrom 2000;11(11):942–50.
[28] Rosén J, Nyman A, Hellenäs K-E. Retention studies of acrylamide for the design of a robust liquid chromatography-tandem mass spectrometry method for food analysis. J Chromatogr A 2007;1172(1):19–24.
[29] Konstantakopoulos IC, Andreopoulos AG. Polymer networks for solid phase extraction. J Macromol Sci Pure Appl Chem 1997;34(5):907–13.
[30] Bortolomeazzi R, Munari M, Anese M, et al. Rapid mixed mode solid phase extraction method for the determination of acrylamide in roasted coffee by HPLC-MS/MS. Food Chem 2012;135(4):2687–93.
[31] Ahn JS, Castle L, Clarke DB, et al. Verification of the findings of acrylamide in heated foods. Food Addit Contam 2002;19(12):1116–24.

[32] Riediker S, Stadler RH. Analysis of acrylamide in food by isotope-dilution liquid chromatography coupled with electrospray ionization tandem mass spectrometry. J Chromatogr A 2003;1020(1):121–30.

[33] Andrzejewski D, Roach JAG, Gay ML, et al. Analysis of coffee for the presence of acrylamide by LC-MS/MS. J Agric Food Chem 2004;52(7):1996–2002.

[34] Granby K, Fagt S. Analysis of acrylamide in coffee and dietary exposure to acrylamide from coffee. Anal Chim Acta 2004;520:177–82.

[35] Aguas PC, Fitzhenry MJ, Giannikopoulos G, et al. Analysis of acrylamide in coffee and cocoa by isotope dilution liquid chromatography-tandem mass spectrometry. Anal Bioanal Chem 2006;385(8):1526–31.

[36] Wenzl T, Szilagyi S, Rosén J, et al. Validation by collaborative trial of an isotope dilution liquid chromatographic tandem mass spectrometric method to determine the content of acrylamide in roasted coffee. Food Addit Contam Part A 2009;26(8):1146–52.

[37] Sanders RA, Zyzak DV, Stojanovic M, et al. An LC/MS acrylamide method and its use in investigating the role of asparagine. Presentation at the annual AOAC international meeting, Los Angeles, CA. September 22–26, 2002.

[38] Zyzak D, Sanders RA, Stojanovic M, et al. Acrylamide formation in heated foods. J Agric Food Chem 2003;51:4782–7.

[39] Yusà V, Quintás G, Pardo O, et al. Determination of acrylamide in foods by pressurized fluid extraction and liquid chromatography-tandem mass spectrometry used for a survey of Spanish cereal-based foods. Food Addit Contam 2006;23(3):237–44.

[40] Fohgelberg P, Rosén J, Hellenäs KE, et al. The acrylamide intake via some common baby food for children in Sweden during their first year of life–an improved method for analysis of acrylamide. Food Chem Toxicol 2005;43(6):951–9.

[41] Claus A, Weisz GM, Kammerer DR, et al. A method for the determination of acrylamide in bakery products using ion trap LC-ESI-MS/MS. Mol Nutr Food Res 2005;49(10):918–25.

[42] Şenyuva HZ, Gökmen V. Interference-free determination of acrylamide in potato and cereal-based foods by a laboratory validated liquid chromatography–mass spectrometry method. Food Chem 2006;97:539–45.

[43] Nielsen NJ, Granby K, Hedegaard RV, et al. A liquid chromatography-tandem mass spectrometry method for simultaneous analysis of acrylamide and the precursors, asparagine and reducing sugars in bread. Anal Chim Acta 2006;557(1–2):211–20.

[44] Zhang Y, Jiao J, Cai Z, et al. An improved method validation for rapid determination of acrylamide in foods by ultra-performance liquid chromatography combined with tandem mass spectrometry. J Chromatogr A 2007;1142(2):194–8.

[45] Churchwell MI, Twaddle NC, Meeker LR, et al. Improving LC-MS sensitivity through increases in chromatographic performance: comparisons of UPLC-ES/MS/MS to HPLC-ES/MS/MS. J Chromatogr B 2005;825(2):134–43.

[46] Takatsuki S, Nemoto S, Sasaki K, et al. Determination of acrylamide in processed foods by LC/MS using column switching. J Food Hyg Soc Jpn 2003;44(2):89–95.

[47] Douny C, Widart J, Maghuin-Rogister G, et al. Quantification of acrylamide in various Belgian potato products using solid phase extraction and liquid chromatography tandem mass spectrometry detection. Food Public Health 2012;2(5):137–41.

[48] Şenyuva HZ, Gökmen V. Study of acrylamide in coffee using an improved liquid chromatography mass spectrometry method: Investigation of colour changes and acrylamide formation in coffee during roasting. Food Addit Contam 2005;22(3):214–20.

[49] Jezussek M, Schieberle P. A new LC/MS-method for the quantitation of acrylamide based on a stable isotope dilution assay and derivatization with 2-mercaptobenzoic acid. Comparison with two GC/MS methods. J Agric Food Chem 2003;51:7866–71.

[50] Vaclavik L, Capuano E, Gökmen V, et al. Prediction of acrylamide formation in biscuits based on fingerprint data generated by ambient ionization mass spectrometry employing direct analysis in real time (DART) ion source. Food Chem 2015;173:290–7.

[51] Zhang Y, Fang H, Zhang Y. Study on formation of acrylamide in asparagine–sugar microwave heating systems using UPLC–MS/MS analytical method. Food Chem 2008;108:542–50.

[52] Zhang Y, Ren Y, Jiao J, et al. Ultra high-performance liquid chromatography-tandem mass spectrometry for the simultaneous analysis of asparagine, sugars, and acrylamide in Maillard reactions. Anal Chem 2011;83:3297–304.

[53] Roach JAG, Andrzejewski D, Gay ML, et al. Rugged LC-MS/MS survey analysis for acrylamide in food. J Agric Food Chem 2003;1(26):7547–54.
[54] Şenyuva HZ, Gökmen V. Survey of acrylamide in Turkish foods by an in-house validated LC-MS method. Food Addit Contam 2005;22(3):204–9.
[55] Delatour T, Perisset A, Goldmann T, Riediker S, Stadler RH. Improved sample preparation to determine acrylamide in difficult matrixes such as chocolate powder, cocoa, and coffee by liquid chromatography tandem mass spectroscopy. J Agric Food Chem 2004;52(15):4625–31.
[56] Granvogl M, Schieberle P. Quantification of 3-aminopropionamide in cocoa, coffee and cereal products. Eur Food Res Technol 2007;225(5–6):857–63.
[57] Guenther H, Anklam E, Wenzl T, et al. Acrylamide in coffee: review of progress in analysis, formation and level reduction. Food Addit Contam 2007;24:60–70.
[58] Pardo O, Yusà V, Coscollà C, Leon N, Pastor A. Determination of acrylamide in coffee and chocolate by pressurised fluid extraction and liquid chromatography-tandem mass spectrometry. Food Addit Contam 2007;24(7):663–72.
[59] Loaec G, Jacolot P, Helou C, et al. Acrylamide, 5-hydroxymethylfurfural and N-epsilon-carboxymethyl-lysine in coffee substitutes and instant coffees. Food Addit Contam Part A Chem Anal Control Expo Risk Assess 2014;31(4):593–604.
[60] Morales FJ, Mesias M. Analysis of acrylamide in coffee. In: Preedy V, editor. Coffee in health and disease prevention. Massachusetts: Academic Press; 2015. p.1013–21.
[61] Fauhl C, Klaffke H, Mathar W, et al. Acrylamide interlaboratory study, Federal Institute for risk assessment. 2003. Berlin, Germany. http://www.bfr.bund.de/cm/349/proficiency_testing_studie.pdf.
[62] Bagdonaite K, Derler K, Murkovic M. Determination of acrylamide during roasting of coffee. J Agric Food Chem 2008;56(15):6081–6.
[63] European Union: Commission Decision 2002/657/EC implementing Council Directive 96/23/EC concerning the performance of analytical methods and the interpretation of results.
[64] Dabrio M, Sejeroe-Olsen B, Musser S, et al. Production of a certified reference material for the acrylamide content in toasted bread. Food Chem 2008;110(2):504–11.
[65] Koch M, Bremser W, Koeppen R, et al. Development of two certified reference materials for acrylamide determination in foods. J Agric Food Chem 2009;57(18):8202–7.
[66] Kim B, Park S, Lee I, et al. Development of a certified reference material for the determination of acrylamide in potato chips. Anal Bioanal Chem 2010;398(2):1035–42.
[67] Kim B, Hyung S-W. Stability monitoring of acrylamide in potato chip certified reference material bulletin of the Korean. Chem Soc 2015;36:399–401.
[68] Nordic Committee on Food Analysis (NMKL). NMKL method 185:2007. Acrylamide: determination in potato and bakery products by liquid chromatography tandem mass spectrometry (LC-MS/MS). Oslo (Norway): Nordic Committee on Food Analysis.
[69] Clarke DB, Kelly J, Wilson LA. Assessment of performance of laboratories in determining acrylamide in crispbread. J AOAC Int 2002;85(6):1370–3.
[70] Owen LM, Castle L, Kelly J, et al. Acrylamide analysis: assessment of results from six rounds of food analysis performance assessment scheme (FAPAS) proficiency testing. J AOAC Int 2005;88(1):285–91.
[71] Klaffke H, Fauhl C, Mathar W, et al. Results from two interlaboratory comparison tests organized in Germany and at the EU level for analysis of acrylamide in food. J AOAC Int 2005;88(1):292–8.
[72] Wenzl T, Anklam E. Evaluation of results of an interlaboratory comparison test on determination of acrylamide in crispbread samples. J AOAC Int 2005;88(5):1413–8.

CHAPTER 25

Quantitation of Acrylamide in Foods by High-Resolution Mass Spectrometry

Antonio D. Troise[1,2], Vincenzo Fogliano[2]
[1]Department of Agriculture and Food Science, University of Naples "Federico II", Naples, Italy; [2]Food Quality and Design group, Wageningen University & Research Centre, Wageningen, The Netherlands

INTRODUCTION

High-resolution mass spectrometry (HRMS) has introduced several advantages in analytical chemistry, in particular for the detection of peptides and small molecules. The current developments in "foodomics" involve the analysis of very complex mixtures where the use of high resolution can be an innovative tool. In both "shotgun," targeted and untargeted based approach with or without chromatography separation, thousands of molecules with very large dynamic *m/z* range (such as the difference of the current ions associated with the most and least abundant molecules) have to be analyzed in several injections [1]. Moreover, several experimental parameters, such as needle temperature, capillary voltage, nebulizing gases, and vaporizer temperature, need to be tuned time to time. If the complex mixtures are analyzed by ion traps, tandem mass spectrometry (MS/MS), time of flight or other mass instruments having low resolving power, coeluting molecules with similar *m/z* ratios frequently overlap suppressing the target compounds [2]. Along with ion suppression, accurate mass analysis can be strongly influenced by the ion enhancement, where molecules can interact with each other leading to the formation of other compounds in source or just before the mass analyzer. In its ability to deliver low-ppm mass accuracy and extremely high resolution, with or without a time scale compatible with nano-LC separations, the HRMS analysis has emerged as a technique of choice for many applications, that not only involve proteomics, lipidomics, and metabolomics, but also the analysis of small molecules [3]. In addition to this, HRMS is one of the most valuable tools for the chemometrics thanks to its specificity and its ability to detect a broad range of molecules [4].

In the frame of "fast analysis, reliable results," the analytical strategies related to the monitoring of heat-induced toxicants, or Maillard reaction end products (MRPs), such as acrylamide, still represent one of the hottest topic of Food Chemistry with several implications in Food Toxicology and Food Engineering.

Since the first report highlighted the presence of acrylamide at high concentration in French fries, cereal-based foods, and coffee, the study of aspects linked to its mechanism of

Acrylamide in Food
http://dx.doi.org/10.1016/B978-0-12-802832-2.00025-5

formation, chemical kinetics, analytical detection, and quantification [5–7] has become one of the most emphasized area in food science, and in particular in Maillard chemistry. A huge variety of papers dealt with the chemical and analytical insights of acrylamide quantification and, as a consequence in the past decade, several methods have been reviewed [8–10]. Acrylamide analysis has three critical steps— sample preparation, including extraction, spiking with labeled-internal standard with or without derivatization, and cleanup of the samples [11]; chromatographic separation; and compounds detection. With regard to the first step several techniques have been developed providing satisfactory results—pressurized liquid extraction (PLE) cation-exchange cartridges, hydrophilic–lipophilic balanced cartridges, and C-18 cartridges [12,13].

The chromatographic separations should take into consideration, just like other small molecules with an amide group or an amino group such as glycine, several drawbacks; among them being high polarity as well as poor retention, matrix effect, and solvent interferences, such as acetonitrile, that can easily suppress the current ion associated with their signals [14–16]. Several efforts were undertaken to rule out these disadvantages. First, the derivatization procedures with mercaptobenzoic acid or through the use of the "bromination" method helped in increasing the selectivity in acrylamide determination, in particular with single-stage LC mass spectrometers [17,18]. Second, for the detection step the analytical techniques of choice were MS/MS acquisition performed by selected reaction monitoring (SRM) or multiple reaction monitoring (MRM), choosing the characteristic transitions *m/z* 72→55, 72→54, and 72→28 [19,20] Fourier transform infrared (FT-IR) analysis [21,22], or time of flight mass spectrometry (TOF/MS) [23], or the fluorescence method [24]. To overcome the matrix effect and the poor retention of acrylamide, and in addition to the above-mentioned analytical methods, many official institutions standardized their complete validated methods for the determination of acrylamide, most of which focused on LC–MS/MS analysis and isotope dilution with ^{13}C or ^{2}H (Swiss Federal Office of Public Health; US Food and Drug Administration) [19]. Despite the outstanding results obtained for the quantification of acrylamide in food matrices, there are still several strategies that can offer some advantages to the analytical aspects reviewed up to now. In particular, a promising approach is the use of DART and an Orbitrap analyzer, whose specificity and high mass accuracy can be further improved by the use of the internal recalibrating mass. The presence of contaminants in mass spectrometry is a negative outcome and in most cases the effects of the interferences have to be reduced or avoided. By using HRMS, the presence of contaminants with *m/z* near the one of the acrylamide can offer the unique chance to continuously recalibrate the instrument increasing the accuracy and as a consequence reducing the mass tolerance up to 2 ppm. This procedure was described by the group of Matthias Mann for complex peptide mixtures. They highlighted the benefits due to the introduction of an omnipresent contaminant from ambient air as an internal standard which they accumulate together with ions of interest before injecting this mixture into the Orbitrap. As described for acrylamide, absolute mass deviations of less than 1 ppm were achieved [2,3]. In this chapter, the underlying link between HRMS and acrylamide detection is deeply evaluated: on the one hand the performances of ultrahigh-pressure liquid chromatography coupled to Orbitrap high-resolution mass spectrometry (U-HPLC–HRMS) are unraveled and on the other novel approaches including DART–HRMS are reviewed in order to define new strategies for the accurate quantification of this compound. Moreover the present strategies are compared to the LC–MS/MS.

LC–HRMS SETUP

The potentialities of HRMS analysis for acrylamide quantification are related to the dynamic range over which accurate measurements of mass can be performed (extent of mass accuracy or mass tolerance setup). This aspect combined with the chromatographic separation of the small amide in a complex mixture determines the true utility of accurate mass capability for real-life applications by ruling out the problems associated with the solid phase extraction and, possibly, to the use of internal standard.

The first step deals with the optimization of chromatographic separation of acrylamide in combination with HRMS detection. It is well known that acrylamide is a very polar molecule with poor retention (k'<2.0) in conventional liquid chromatography reversed phase sorbents [13,25]. Several columns with different selectivity can give appreciable results according to chromatographic performances, including retention time and reproducibility. In general among the most common columns, including an HILIC phase, a pentafluorophenyl phase and a core shell C18 phase, a polar end-capped stationary phase allows a good reproducibility and avoids the interferences due to the solvent, to the matrix, and to the unretained compounds that eluted along with the solvent front. A typical chromatogram of acrylamide standard in LC–HRMS is presented in Figure 1. Along

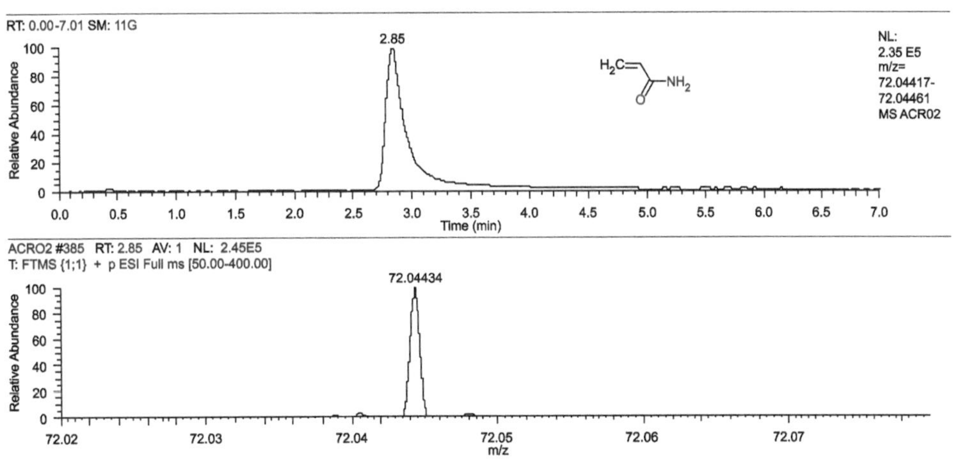

Figure 1 Typical extracted ions chromatogram and mass spectrum of acrylamide standard dissolved in water. Δppm = −0.7. Acrylamide theoretical mass: 72.04439. *Adapted from Troise et al. [27].*

with polar end-capped column, the use of a core shell C18 phase can ensure in some conditions and with defined food matrix, such as coffee beverages, a significant reproducibility for several subsequent injections even if the k' is always 1. From the chromatographic separation point of view, relevant results can be obtained with the HILIC phase in particular for the retention time, even if the use of acetonitrile interfered with the ionization of acrylamide. In this case a modified liquid interaction with methanol is mandatory.

By using a polar end-capped column, although the retention time is quite short (2.85 min), the separation from the first eluting impurities is ensured; the peak shape is generally maintained, and the deviation of retention time is <0.1 min among assays of calibration standards, real samples, and recovery samples.

The second critical step of acrylamide analysis is related to the setup of the HRMS condition. In order to ensure the highest abundance of the current ion associated with the m/z 72.04434 ($[M+H]^+$) a water solution of acrylamide (in the concentration range of 150–500 μg/mL) at a flow rate of 5 μL/min is pumped directly into the LC stream entering in the HRMS. The U-HPLC is coupled to an Exactive Orbitrap HRMS (Thermo Fisher Scientific, San Jose, CA, USA) equipped with a heated electrospray interface operating in the positive mode and scanning the ions in the m/z range of 50–400. The parameters used are listed in Table 1. Specifically, the resolving power is set to 50,000 full width at half maximum (FWHM, m/z 200) resulting in a scan time of 1 s. The automatic gain control is used to fill the C-trap and gain accuracy in mass measurements; maximum injection time is 50 ms. The interface parameters are as follows: the spray voltage 4.5 kV, the capillary voltage 42.5V, the skimmer voltage 14V, the capillary

Table 1 Mass spectrometry condition for LC–HRMS, each parameter is tuned according to the acrylamide exact mass (72.04439, $[M+H]^+$)

Parameters	Values
Mass scan range *(m/z)*	50–350
Resolving power (FWHM m/z 200)	50,000
Scan time (s)	1
All gain control (ions)	5×10^5
Maximum injection time (min)	50
Spray voltage (kV)	4.5
Capillary voltage (V)	42.5
Skimmer voltage (V)	14
Capillary temperature (°C)	300
Heater temperature (°C)	250
Sheath gas flow (AU)	40
Auxiliary gas flow (AU)	6

The results are obtained after direct infusion and LC stream infusion.
Adapted from Troise et al. [27].

temperature 300 °C, the heater temperature 250 °C, and a sheath gas flow 40 and auxiliary gas flow 6 (arbitrary units). All the above listed conditions are optimized to avoid the fragmentation of the molecule filling the C-trap (number of ions injected for each scan: 5×10^5 ions, ultimate mass accuracy mode) with the parental ions without the formation of daughter ions ensuring the highest abundance of *m/z* 72.04434. High resolution, high mass accuracy combined with low mass tolerance used require the optimal instrumental conditions to reduce the discrepancies between the theoretical and experimental mass; as a consequence the setup of the condition for mass accuracy optimization is one of the prominent points of acrylamide HRMS analysis. Before starting the acrylamide determination the instrument is externally calibrated by infusion with a solution that consists caffeine, Met–Arg–Phe–Ala (MRFA), Ultramark 1621, and acetic acid in a ratio of acetonitrile/methanol/water (2:1:1, v/v/v). Several reference masses (lock masses) can be evaluated for the mass analyzer recalibration: methanol ($[M_2+H]^+$, exact mass: 65.05971); methanol/water ($[A_2B_2+H]^+$, exact mass: 101.08084); sodium adduct of methanol/water ($[A_2B_2+Na]^+$, exact mass: 123.06278); [2,3,3-d_3]-acrylamide ($[M+H]^+$,exact mass: 75.06322); and diisooctyl phthalate ($[M+H]^+$, exact mass: 391.28429) [26]. In order to optimize the HRMS conditions and the mass accuracy, 200 µL of the above-mentioned calibration solution is spiked with 10 µg of [2,3,3-d_3]-acrylamide and the scanning window is set in the range of 70–1400 *m/z* only for the initial calibration procedure. In a first set of fine-tuning trials several lock masses are used to recalibrate the instrument and reduce the mass error [31]; and the mass to charge ratio of some contaminants or interference ions can be evaluated (Table 2), such as methanol ($[M_2+H]^+$, exact mass = 65.05971); methanol/water ($[A_2B_2+H]^+$, exact mass = 101.08084); and sodium adduct of methanol/water ($[A_2B_2+Na]^+$, exact mass = 123.06278). Then, the presence of [2,3,3-d_3]-acrylamide solution ($[M+H]^+$, exact mass = 75.06322) in the liquid chromatography flow with a final concentration >3 µg/mL can reduce the mass accuracy of acrylamide, even if the results are unsatisfactory for complex mixture. The solution

Table 2 Contaminants, interference ions, and amino acids used to recalibrate the instruments during the runs

m/z	Ion type	Formula for M	Compound ID	Origin
65.05971	$[M_2+H]^+$	CH_3OH	Methanol	Solvent
101.08084	$[A_2B_2+H]^+$	$[MeOH]_n[H_2O]_m$	Methanol/water	ESI solvents
123.06278	$[A_2B_2+Na]^+$	$[MeOH]_n[H_2O]_m$	Methanol/water	ESI solvents
75.06322	$[M+H^+]$	$C_3H_2D_3NO$	[2,3,3-d_3]-acrylamide	External infusion
76.03930	$[M+H^+]$	$C_2H_5NO_2$	Glycine	External infusion
90.05496	$[M+H^+]$	$C_3H_7NO_2$	Alanine	External infusion

[2,3,3-d_3]-acrylamide, glycine, and alanine can be used both during the runs both directly in the calibration mixture before each session.
Adapted from Keller et al. [26] and Troise et al. [27].

relies on the ability of the instrument to be daily calibrated according to the molecular ions of interest. By using this strategy it is possible to spike the calibration solution with the acrylamide internal standard and focus the sensitivity of the instrument on the low molecular weight molecules. The final results is the reduction of the mass accuracy up to 0.5 ppm where the presence of [2,3,3-d_3]-acrylamide in the calibration solution determined the reduction in the discrepancies between the observed mass ($[M+H]^+$, exact mass = 72.04434), and the theoretical mass of acrylamide ($[M+H]^+$, exact mass = 72.04439). The presence of deuterated standard in the calibration mixture is not mandatory. Generally the similarity toward the native molecule represents an advantage, but also other molecules with low m/z can be diluted in the calibration mixture. Results of the same order of magnitude were obtained with glycine, and isoleucine can set the instrument for the optimal detection of small molecules with $m/z < 80$ [27].

According to the instrumental conditions described above acrylamide extraction and sample preparation follow a simplified approach. On the one hand powdered samples, such as that of instant coffee, French fries, and cookies, after grinding in a knife mill, are dissolved in deionized water, Carrez reagent potassium salt, and Carrez reagent zinc salt to promote protein precipitation and the consequent polar compounds extraction. On the other hand, liquid samples, such as coffee, are diluted. The samples are vortexed for 10 min at 800 rpm at room temperature and after centrifugation (2500 × g, 4 °C, 10 min), 1 mL of the aqueous layer is accurately filtered and collected in a glass vial [27].

DART–HRMS AND THE INDIRECT ESTIMATION OF ACRYLAMIDE FORMATION

A simplified extraction procedure can be optimized also for the indirect estimation and prediction of the acrylamide formation in food matrices by the following two steps— the evaluation of the potential of high-throughput direct analysis in real-time high-resolution mass spectrometry (DART–HRMS) fingerprinting and the study of the spectra obtained in order to perform a chemometric analysis. This novel approach tries to circumvent the main problems related to the analysis of acrylamide (such as the need for extensive and laborious sample pretreatment) by collecting the rich mass spectral information of a methanol/water extract and clustering the data obtained in order to build a solid regression methodology for the prediction of the acrylamide formation.

The simplified DART–HRMS fingerprinting workflow typically involves the following steps: (1) extraction of the samples using a generic protocol to get in solution as many compounds as possible; (2) instrumental analysis of the samples to record rich mass spectra characterizing chemical composition of the test material; (3) processing of the mass spectral records optimizing the background subtraction and combining the technical replicates followed by selection of ions across the examined sample set to form data matrix; and (4) analysis of the data by chemometric tools [28].

The main advantages related to the introduction of ambient ionization techniques, such as DART and desorption electrospray ionization (DESI) [29], are related to ambient mass spectrometry that enables sample examination in an open laboratory environment by exposing the sample to a stream of desorbing/ionizing medium in proximity to the inlet of the mass spectrometer [28,30]. This experimental setup is characterized by minimal sample preparation, no gas or liquid chromatographic separation, and optimization of high-throughput measurements. Even if the absence of chromatographic separation can negatively influence the results, in DART, excited-state metastable atoms or molecules of gas (helium or nitrogen) are used as the medium for ionization. Sample introduction is typically performed with melting point capillaries that are used to transfer liquid or solid samples spread on their surface into the DART ionization region. In the paper by Vaclavik et al. this simplified procedure was employed for the analysis of two biscuit recipes baked at different temperatures. The DART analysis performed in both positive and negative ionization mode revealed a complex pattern of ions that contributed to the definition of the principal component analysis and to the construction of the regression model.

The analysis of mass spectra collected revealed numerous ions relative and/or absolute intensities of which increased or decreased with the prolonged heat-treatment during the baking of the biscuits according to the thermal loading of the recipe. These dynamics of changes in relative intensities and absolute responses observed in extracted ion chronograms can be described for the ions detected at m/z 90.0550 and 127.0387 in positive ionization mode. The resolving power of the mass spectrometer employed in this study allowed the estimation of their elemental compositions and the compounds were tentatively identified. Using the information on accurate masses, signals observed at m/z 90.0550 and 127.0387 were tentatively identified as $[M+H]^+$ ions of the aliphatic amino acid alanine and hydroxymethylfurfural (HMF), one of the most studied product of the Maillard reaction product. On the one hand the relative intensity of the ion assigned as alanine decreased with prolonged baking time, where the formation of dicarbonyls from the autoxidation of sugars and the presence of reducing sugars lead to the formation of deoxyfructosyl-alanine. On the other hand, the intensity of the ion assigned as HMF increased according to the reaction pathways described by Perez Locas and Yaylayan [31]. In addition to these two compounds, some additional positive and negative mode ions that exhibited dynamic changes in their intensities were also tentatively identified. These included amino acids, products, and intermediates of Maillard reaction, carbohydrates, and fatty acids. The overview of observed trends in their relative intensities during prolonged baking time is provided in Table 3 [32].

According to the data obtained the following step led to the construction of principal component analysis (PCA), where the unsupervised pattern recognition technique which allowed the reduction in the dimensionality of the multivariate dataset by retaining the

Table 3 Overview of tentatively identified ions used for the modeling of acrylamide concentration in DART–HRMS analysis

Ionization mode	*m/z*	Ion mode	Mass error (ppm)	Identification
DART +	90.0550	$[M+H]^+$	+0.5	Alanine
	109.0284	$[M-H_20+H]^+$	−0.1	HMF
	127.0387	$[M+H]^+$	−2.1	
	116.0705	$[M+H]^+$	−0.9	Proline
	118.0861	$[M+H]^+$	−1.3	Valine
	132.1016	$[M+H]^+$	−2.3	Leucine/isoleucine
	145.0492	$[M+H]^+$	−2.3	Acetylformoin
DART −	131.0347	$[M-H_20-H]^-$	+2.3	3-Deoxypentos-2-ulose
	149.0445	$[M-H]^-$	−3.4	Pentose
	161.0446	$[M-H]^-$	−2.5	3-Deoxy-D-erythro-hexosulose
	161.0446	$[M-H]^-$	−2.5	Hexose
	179.0552	$[M-H]^-$	−1.8	
	255.2327	$[M-H]^-$	+1.2	Palmitic acid
	272.2170	$[M-H]^-$	+1.0	Linolenic acid
	279.2327	$[M-H]^-$	+1.0	Linoleic acid
	281.2483	$[M-H]^-$	+0.9	Oleic acid
	305.0876	$[M-2H_20-H]^-$	+1.1	Sucrose
	323.0981	$[M-H_20-H]^-$	+0.8	
	341.1086	$[M-H]^-$	+0.6	

The formation/degradation of the compounds reported depends on the thermal treatment performed.
Adapted from Vaclavik et al. [32].

maximum structure in the data. Overlooking the aspects linked to the clustering of the dataset, where the different recipes and thermal treatments can be easily grouped the crucial step involves the construction of the regression model. Two different algorithms, namely partial least square regression (PLSR) and principal components regression (PCR) analyses, and three preprocessing techniques were tested in this study. A PLS model developed on the whole dataset showed that one of the biscuit samples baked at 200°C for 12min simultaneously exhibited an extremely high studentized residual and leverage exceeding the computed threshold values and was, therefore, excluded from further modeling as outlier. The regression models were then developed using a training set of 39 samples.

ACRYLAMIDE CONTENT IN FOOD: PREDICTION AND QUANTIFICATION

Food quality and acrylamide detection are strictly connected and each thermal process that involves carbohydrate, reducing sugars, and asparagine in a low water activity environment has to consider the acrylamide formation. Predictive models, such as

Table 4 Overview of pros and cons of the three techniques compared

Variables	LC/MS/MS	LC–HRMS	DART–HRMS
Technique	MRM	HR, mass error 0.5 ppm	Indirect prediction
Sensitivity	Up to 10 ppb	Up to 5 ppb	Markers dependent
Validation	Golden standard	Biscuits, French fries, coffee	Chemometric
Time analysis	Up to 20 min	Up to10 min	0.4
SPE	Yes	No	No
Chromatography	Yes	Yes	No
Internal standard	Yes	Yes/No	No
Recalibration	No	Yes, continuously	No

In the first column the golden standard LC–MS/MS is reported. The green (light gray in print versions) cells highlight the advantages of the techniques, while the red (gray in print versions) cells mark the drawbacks or the critical points.

kinetic modeling, have certainly contributed to the elucidation of the reaction mechanisms and they have illuminated a powerful tool able to monitor the formation of acrylamide in par-fried potatoes [33]. According to the predictive indirect outcomes, the use of DART has offered a new strategy for the measurement of acrylamide in cookies samples, but there are some limitations and the use of direct mass spectrometry quantification is the most common and reliable analytical technique of choice. The predicted values have to be compared to the measured values obtained by other techniques and the robustness of the method has to be evaluated time to time according to the matrix and to the thermal treatment. As a consequence, the golden standard for acrylamide detection is nowadays represented by MS/MS coupled to LC or gas chromatography in association with chemical derivatization. On the other hand, LC–HRMS is a smart tool for the acrylamide quantification and, as summarized in Table 4, the main advantages are related to the simplified extraction procedure, to the low mass error, and to the possibility to rule out the interferences of the matrix. By monitoring the current ion associated to the molecular ion of the acrylamide it is possible to limit the negative effects of the presence of coeluting compounds and as a consequence the use of solid-phase extraction and, in some cases, the use of the internal standard for the extraction procedure can be avoided.

As reported in Table 5, the LC–HRMS procedure here described is compared to MS/MS in three of the most commonly consumed acrylamide-containing food, namely French fries, coffee, and cookies. Even if the determination of acrylamide in these food matrices shows different requirements in terms of extraction and cleanup before chromatographic separation, both for DART and for LC-HRMS the extraction procedure is characterized by the same basic idea—the extremely simplified extraction procedure in order to include HRMS in routine analyses in food industries. By using LC–HRMS for French fries the concentration was extremely similar to the ones obtained through LC-MS/MS [20,33,34].

Table 5 Concentration of acrylamide in different foods by LC–HRMS with (ISTD) and without acrylamide internal standard (no ISTD), and their relationship toward LC–MS/MS analysis

Food	No ISTD	ISTD	LC–MS/MS	RSD No ISTD (%)	RSD ISTD (%)
Cookies 0 % NaCl	276 ± 3.5 ng/g	262 ± 4.0 ng/g	286 ± 15.3 ng/g	3.2	9.1
Cookies 0.65% NaCl	317 ± 9.2 ng/g	340 ± 5.6 ng/g	325 ± 9.3 ng/g	2.1	4.1
French fries (3 min)	270 ± 12.1 ng/g	254 ± 4.6 ng/g	280 ± 18.8 ng/g	3.7	10.2
French fries (7 min)	424 ± 10.3 ng/g	490 ± 10.0 ng/g	409 ± 11.1 ng/g	3.5	16.5
Brewed coffee	12 ± 0.1 ng/mL	15 ± 0.2 ng/mL	13 ± 0.2 ng/mL	8.7	13.3
Ground coffee	140 ± 3.6 ng/g	159 ± 8.1 ng/g	145 ± 3.2 ng/g	3.5	3.5

In the last column the relative standard deviation (RSD %) between the LC–HRMS and LC–MS/MS was reported.

The previous results were completely confirmed for the brewed coffee sample and ground coffee [35–38]. Acrylamide detection in coffee is characterized by several drawbacks such as the matrix effect and the roasting of coffee beans [38]. For this reason, for LC-MS/MS, the use of internal standard is a key point that can be easily overcome thanks to the HRMS. This is presented in Table 5 where not only the analysis performed through HRMS or MS/MS showed similar results, but also the results with and without the internal standard quantitation were of the same order of magnitude.

For cookies samples two different typologies of cookies prepared according to two recipes for acrylamide determination are compared with results of previous quoted papers and with those obtained by DART [32,34,39]. The values vary from 276 ± 3.5 ng/g (cookies with 0.65% sodium chloride) to 317 ± 9.2 ng/g (cookies without 0.65% sodium chloride). The use of the internal standard resulted in similar values: 262 ± 4.0 ng/g and 340 ± 5.6 ng/g for cookies with sodium chloride and without, respectively [39]. These data clearly showed the substantial agreement between the U-HPLC/HRMS method and the SPE LC/MS/MS method and interestingly similar results can be predicted by DART–HRMS.

COMMENTS AND TAKE HOME MESSAGE

There is the need to establish a procedure for monitoring the acrylamide formation in carbohydrate-rich products. The possibility to simplify the extraction procedure without limiting the reliability of the results can be an important tool for the improvement of food quality. A huge variety of papers has been published on quantitation procedure for acrylamide in foods. The methods here described by LC–HRMS and DART–HRMS proved to be very effective as it allowed to ride out the matrix effect and the time consumption due to solid phase extraction. The DART–HRMS technique enables high-throughput fingerprinting of biscuits in the control of some quality/safety parameters. The presented strategy requires considerably shorter time for both sample preparation (8 min per sample) and instrumental analysis (0.4 min per sample in both positive and negative ionization mode), as compared to conventional, target acrylamide analysis by chromatographic methods. The multivariate analysis of fingerprint data provided some useful information on the chemical differences between biscuits according to the recipes and to the thermal treatments. As described for acrylamide kinetic modeling in fried potato, the concentrations of acrylamide formed during the baking of the biscuit could be predicted with good accuracy based on combined use of multivariate data obtained by positive and negative mode DART–HRMS and chemometric modeling. However, the robustness of the model should be further investigated by studying variations in chemical compositions of raw materials such as flours used for the preparation of the biscuits or the concentration of the precursors such as reducing sugars and asparagine. Several aspects concur with the limitations of DART–HRMS technique—the necessity of an additional prediction model with the use of suitable sample sets, the study of the precursors influence, and the robustness in other foods require additional analysis.

The results obtained with LC–HRMS method are in most cases well in line with those present in the literature [34,40]. From the analytical point of view the use of [2,3,3-d_3]-acrylamide increased the robustness of the method, in particular the variability due to the ion suppression or to the matrix effect were eliminated for each analysis. Interestingly without the use of labeled internal standard the results were still satisfying and the differences between the two methods were less than 14% in solid foods, thus suggesting that in cases not requiring a high precision the use of HRMS method could theoretically also allow to avoid the use of internal standard. Moreover, the use of a modified calibration solution or the presence of some contaminants can lead to the optimal setup of the instrument and it improved enormously the mass accuracy for acrylamide detection resulting in robust and accurate quantification. Also the use of internal standard could be theoretically avoided with a negligible loss in method accuracy. From the analytical standpoint the performances were satisfactory—the relative standard deviation was less than 7%, the recovery in the three different matrices was always higher than 90%, and the limit of detection was comparable to the ones of MS/MS. These aspects can represent one of the most important advantages for the transferring of this method to routine analysis.

KEY FACTS

- The use of high-resolution mass spectrometry offers several advantages for the unambiguous detection of molecules in food samples. The "shotgun approach," targeted and untargeted analysis can rule out the interferences due to complex matrices and lead to the quantification of thermal induced toxicants, such as acrylamide.
- The presence of contaminants in mass spectrometry can be a powerful tool for the internal recalibration of high resolution mass spectrometer. Outstanding improvements in mass accuracy are obtained by selecting and monitoring a priori *or* a posteriori defined current ions associated with specific m/z; one of the most common example is the use of diisooctyl phthalate ($[M+H]^+$, exact mass: 391.28429) as lock mass.
- The presence of [2,3,3-d_3]-acrylamide in the calibration mixture promotes the anchoring of the instrument on low molecular weight masses. This strategy combined with the use of contaminants (or lock masses) as continuous recalibrating agents leads the mass accuracy up to 0.5 ppm. Along with proper chromatographic separation with polar end-capped columns, it is possible to introduce a single-stage hybrid high-resolution mass spectrometer as a new tool for the quantification of acrylamide in food. The results obtained are similar to the ones obtained by LC–MS/MS or GC–MS.
- DART–HRMS is an established technique for rapid mass spectral analysis of a large variety of samples. The possibility of analyzing the sample at atmospheric pressure, essentially in the open laboratory environment, allows a simplified extraction procedure that overcomes the matrix effects. Simultaneous positive and negative acquisition modes along with a multivariate regression technique obviate the indirect estimation of acrylamide in biscuits samples.
- The analysis of acrylamide in carbohydrate-rich foods is still one of the hottest topic of Food Chemistry. HRMS offers several advantages, in particular the reduction of time analysis, the reduction of the matrix effects, the improvements in mass accuracy, and robustness. By using DART–HRMS and LC–HRMS it is possible to avoid the solid phase extraction prior the analysis and in some conditions to limit the use of internal standard. These aspects can represent one of the most important advantages for the transferring of this method to routine analysis.

MINI DICTIONARY

Resolution It is the ability of a mass spectrometer to distinguish two peaks with slight difference in their m/z. For a single peak made up of singly charged ions at mass m in a mass spectrum, the resolution may be expressed as $m/\Delta m$ where Δm is the width of the peak at a height which is a specified fraction of the maximum peak height.

FWHM (Full width half maximum) It is the full width of a peak at half its maximum height where the peak width is the distance across a peak measured at a selected peak-height level, for FWHM at half of the height.

High-resolution mass spectrometry (HRMS) It is a mass spectrometry technique characterized by the ability to deliver low-ppm mass accuracy and extremely high resolution, up to 150,000 (FWHM, *m/z* 200) with a modern mass spectrometer.

Recalibrating agent (or lock mass) A known reference mass in the sample that is used to correct the mass spectral data in an accurate mass experiment. It performs a real time secondary mass calibration. The main purposes of using a recalibrating agent are to correct the masses of other peaks in a scan and to improve the mass accuracy. It deduces that for acrylamide mass accuracy, a lock mass with an *m/z* similar to the one of acrylamide works better than molecules with higher or lower *m/z*. The recalibrating agent can be internal, if it is analyzed during the same experiment and it is ubiquitous or contained in the samples or externally infused; the recalibrating agent can be external, if it is analyzed in a separate experiment.

Chemometrics The ability of extracting chemically relevant information from data produced in chemical experiments. This implies the use of mathematical models, and it is strictly linked to data representation and data interpretation.

Multivariate statistic It is a branch of statistics that involves the simultaneous analysis of more than one outcome (dependent) variable.

SUMMARY POINTS

The analysis of acrylamide in food is one of the most challenging procedures in Chemistry and several drawbacks concur to define this complex scenario—the high polarity of acrylamide, its poor retention in common stationary phases, its reactivity, and the interferences of the matrix. The use of high-resolution mass spectrometry defines new landscape in the accurate quantification of this small amide thanks to two main advantages— the ability in delivering low-ppm mass accuracy up to 0.5 ppm and the possibility to reach high resolution. In the frame "fast analysis, reliable results," HRMS combined with LC or with DART can theoretically avoid the use of both the solid phase extraction and deuterated acrylamide as internal standard. The use of HRMS can definitely promote the transferring of the acrylamide detection to routine analysis.

REFERENCES

[1] Olsen JV, de Godoy LMF, Li GQ, Macek B, Mortensen P, Pesch R, et al. Parts per million mass accuracy on an orbitrap mass spectrometer via lock mass injection into a C-trap. Mol Cell Proteomics 2005;4(12):2010–21.

[2] Mann M, Kelleher NL. Precision proteomics: the case for high resolution and high mass accuracy. Proc Natl Acad Sci 2008;105(47):18132–8.

[3] Scigelova M, Makarov A. Orbitrap mass analyzer–overview and applications in proteomics. Proteomics 2006;6(S2):16–21.

[4] Zhang T, Watson DG. A short review of the applications of liquid chromatography mass spectrometry based metabolomics techniques to the analysis of human urine. Analyst 2015;140(9):2907–15.

[5] Tareke E, Rydberg P, Karlsson P, Eriksson S, Tornqvist M. Acrylamide: a cooking carcinogen? Chem Res Toxicol 2000;13(6):517–22.

[6] Mottram DS, Wedzicha BL, Dodson AT. Acrylamide is formed in the Maillard reaction. Nature 2002;419(6906):448–9.

[7] Knol JJ, Van Loon WAM, Linssen JPH, Ruck AL, Van Boekel MAJS, Voragen AGJ. Toward a kinetic model for acrylamide formation in a glucose-asparagine reaction system. J Agr Food Chem 2005;53(15):6133–9.

[8] Wenzl T, de la Calle MB, Anklam E. Analytical methods for the determination of acrylamide in food products: a review. Food Addit Contam 2003;20(10):885–902.
[9] Zhang Y, Zhang GY, Zhang Y. Occurrence and analytical methods of acrylamide in heat-treated foods–review and recent developments. J Chromatogr A 2005;1075(1–2):1–21.
[10] Hu Q, Xu X, Fu Y, Li Y. Rapid methods for detecting acrylamide in thermally processed foods: a review. Food Control 2015;56:135–46.
[11] Jezussek M, Schieberle P. A new LC/MS-method for the quantitation of acrylamide based on a stable isotope dilution assay and derivatization with 2-mercaptobenzoic acid. Comparison with two GC/MS methods. J Agr Food Chem 2003;51(27):7866–71.
[12] Gokmen V, Senyuva HZ. A simplified approach for the kinetic characterization of acrylamide formation in fructose-asparagine model system. Food Addit Contam 2006;23(4):348–54.
[13] Gokmen V, Senyuva HZ. Study of colour and acrylamide formation in coffee, wheat flour and potato chips during heating. Food Chem 2006;99(2):238–43.
[14] Ono H, Chuda Y, Ohnishi-Kameyama M, Yada H, Ishizaka M, Kobayashi H, et al. Analysis of acrylamide by LC-MS/MS and GC-MS in processed Japanese foods. Food Addit Contam 2003;20(3):215–20.
[15] Cavalli S, Maurer R, Hofler F. Fast determination of acrylamide in food samples using accelerated solvent extraction followed by ion chromatography with UV or MS detection. LC GC Eur 2003:9–11.
[16] Rydberg P, Eriksson S, Tareke E, Karlsson P, Ehrenberg L, Tornqvist M. Investigations of factors that influence the acrylamide content of heated foodstuffs. J Agr Food Chem 2003;51(24):7012–8.
[17] Ahn JS, Castle L, Clarke DB, Lloyd AS, Philo MR, Speck DR. Verification of the findings of acrylamide in heated foods. Food Addit Contam 2002;19(12):1116–24.
[18] Tareke E, Rydberg P, Karlsson P, Eriksson S, Tornqvist M. Analysis of acrylamide, a carcinogen formed in heated foodstuffs. J Agr Food Chem 2002;50(17):4998–5006.
[19] Roach JAG, Andrzejewski D, Gay ML, Nortrup D, Musser SM. Rugged LC-MS/MS survey analysis for acrylamide in foods. J Agr Food Chem 2003;51(26):7547–54.
[20] Becalski A, Lau BPY, Lewis D, Seaman SW, Sun WF. Determination of acrylamide in various food matrices–evaluation of LC and GC mass spectrometric methods. Adv Exp Med Biol 2005;561:271–84.
[21] Yaylayan VA, Wnorowski A, Locas CP. Why asparagine needs carbohydrates to generate acrylamide. J Agr Food Chem 2003;51(6):1753–7.
[22] Yaylayan VA. Analysis of complex reaction mixtures: applications of Py-GG-MS and the microwave-assisted process. Am Lab 1999;31(9):30–1.
[23] Hamdan M, Bordini E, Galvani M, Righetti PG. Protein alkylation by acrylamide, its *N*-substituted derivatives and cross-linkers and its relevance to proteomics: a matrix assisted laser desorption/ionization-time of flight-mass spectrometry study. Electrophoresis 2001;22(9):1633–44.
[24] Schieberle P, Kohler P, Granvogl M. New aspects on the formation and analysis of acrylamide. Adv Exp Med Biol 2005;561:205–22.
[25] Gokmen V, Senyuva HZ, Acar J, Sarioglu K. Determination of acrylamide in potato chips and crisps by high-performance liquid chromatography. J Chromatogr A 2005;1088(1–2):193–9.
[26] Keller BO, Suj J, Young AB, Whittal RM. Interferences and contaminants encountered in modern mass spectrometry. Anal Chim Acta 2008;627(1):71–81.
[27] Troise AD, Fiore A, Fogliano V. Quantitation of acrylamide in foods by high-resolution mass spectrometry. J Agr Food Chem 2014;62(1):74–9.
[28] Gross JH. Direct analysis in real time-a critical review on DART-MS. Anal Bioanal Chem 2014;406(1):63–80.
[29] Takats Z, Wiseman JM, Gologan B, Cooks RG. Mass spectrometry sampling under ambient conditions with desorption electrospray ionization. Science 2004;306(5695):471–3.
[30] Cody RB, Laramee JA, Durst HD. Versatile new ion source for the analysis of materials in open air under ambient conditions. Anal Chem 2005;77(8):2297–302.
[31] Locas CP, Yaylayan VA. Isotope labeling studies on the formation of 5-(hydroxymethyl)-2-furaldehyde (HMF) from sucrose by pyrolysis-GC/MS. J Agr Food Chem 2008;56(15):6717–23.
[32] Vaclavik L, Capuano E, Gokmen V, Hajslova J. Prediction of acrylamide formation in biscuits based on fingerprint data generated by ambient ionization mass spectrometry employing direct analysis in real time (DART) ion source. Food Chem 2015;173:290–7.

[33] Parker JK, Balagiannis DP, Higley J, Smith G, Wedzicha BL, Mottram DS. Kinetic model for the formation of acrylamide during the Finish-Frying of Commercial french fries. J Agr Food Chem 2012;60(36):9321–31.
[34] Capuano E, Fogliano V. Acrylamide and 5-hydroxymethylfurfural (HMF): a review on metabolism, toxicity, occurrence in food and mitigation strategies. LWT-Food Sci Technol 2011;44(4):793–810.
[35] Andrzejewski D, Roach JAG, Gay ML, Musser SM. Analysis of coffee for the presence of acrylamide by LC-MS/MS. J Agr Food Chem 2004;52(7):1996–2002.
[36] Granvogl M, Schieberle P. Quantification of 3-aminopropionamide in cocoa, coffee and cereal products. Eur Food Res Technol 2007;225(5–6):857–63.
[37] Perez HL, Osterman-Golkar S. A sensitive gas chromatographic-tandem mass spectrometric method for detection of alkylating agents in water: application to acrylamide in drinking water, coffee and snuff. Analyst 2003;128(8):1033–6.
[38] Granby K, Fagt S. Analysis of acrylamide in coffee and dietary exposure to acrylamide from coffee. Anal Chim Acta 2004;520(1–2):177–82.
[39] Fiore A, Troise AD, Mogol BA, Roullier V, Gourdon A, Jian SE, et al. Controlling the Maillard reaction by Reactant Encapsulation: sodium chloride in cookies. J Agr Food Chem 2012;60(43):10808–14.
[40] Friedman M. Chemistry, biochemistry, and safety of acrylamide. A review. J Agr Food Chem 2003;51(16):4504–26.

CHAPTER 26

Detection of Acrylamide by Biosensors

Bhawna Batra, Chandra S. Pundir
Department of Biochemistry, M.D. University, Rohtak, Haryana, India

INTRODUCTION

Acrylamide is a well-known neurotoxin and potential carcinogen. It is a probable human carcinogen, which is formed during high-temperature cooking of many commonly consumed foods. It is derived from the well-known reaction between reducing sugars and proteins/amino acids (mainly asparagine). However, other routes are also possible. Various reports suggest that high temperatures (>120 °C) are needed for forming acrylamide in foods. High levels of this compound have been found in potato crisps, French fries, and several other common foods [1–3]. Other parameters such as oil type and amount, the glucose–asparagine ratio, and storage practices during potato frying influence acrylamide formation and its content in food. A very high amount of this toxic compound (4 mg/kg) has been found in potato chips. Among analytical methods used for determination of acrylamide levels, the expensive and time-consuming chromatographic techniques such as GC–MS [4], GC–MS–MS [5], HPLC–MS [5], and LC–MS–MS [6,7] predominate. These chromatographic methods require preparation of food samples, which involves extraction using water or methanol and the cleanup step typically consisting of a combination of several solid-phase extractions. GC–MS often needs an additionally laborious bromination step to form a more volatile acrylamide derivative and increase the selectivity of the determination. The growing demand for rapid and precise determination of acrylamide has stimulated to develop alternative methods for screening this neurotoxin cum carcinogen. A few techniques for detection of this toxic compound are available. The biosensors offer great promise for detection of biomolecules because of their simplicity, rapidity, and high sensitivity.

BASIC CONCEPT OF BIOSENSOR

A biosensor can be defined as a compact analytical device incorporating a biological or biologically derived sensing element either integrated within or intimately associated with a physicochemical transducer. A biosensor consists of a bio-element and a sensor-element. The bio-element may be an enzyme, antibody, living cells, tissue, or so on, and the sensing element may be electric current, electric potential, or so on. Different combinations of bio-elements and sensor-elements constitute several types of biosensors to

Acrylamide in Food
http://dx.doi.org/10.1016/B978-0-12-802832-2.00026-7

suit a vast pool of applications. The usual aim of a biosensor is to produce either discrete or continuous digital electronic signals, which are proportional to the concentration of single analyte or a related group of analytes. The key part of the biosensor is the transducer, which makes use of the physicochemical change accompanying the reaction. Several types of transducers have been used in the development of biosensors [8]. This chapter provides a comprehensive overview of recent advances in electrochemical acrylamide biosensors.

ACRYLAMIDE BIOSENSORS

Only two types of acrylamide biosensors namely amperometric and potentiometric have been described in the literature.

Classification of Acrylamide Biosensors

Amperometric Acrylamide Biosensors

Amperometric biosensors monitor current generated, when electrons are exchanged either directly or indirectly between a biological system and an electrode [9].

Principle: The idea of acrylamide biosensors relay on the reaction of hemoglobin (Hb) with acrylamide, which leads to the creation of a Hb–acrylamide adduct, which could alter the electroactivity of Hb. As a consequence, in parallel with an increase in Hb–acrylamide adduct concentration at the electrode surface, the current peak of cyclic voltammogram (CV) decreases. Thus, a decrease in current can be treated as the analytical signal, and could be a base for very selective and sensitive amperometric acrylamide. Therefore, hemoglobin serves as useful biomarker of human exposure to acrylamide.

Classification of Amperometric Acrylamide Biosensor Based on Types of Electrode

Carbon Electrode-Based Acrylamide Biosensors These can be classified further on the basis of whether nanomaterials were employed or not.

Acrylamide Biosensors Without Nanomaterials Stobiecka et al. [10] constructed a carbon-paste electrode modified with Hb, which contains four prosthetic groups of heme–Fe(III). Such an electrode displays a reversible reduction–oxidation process of Hb–Fe(III)/Hb–Fe(II). Acrylamide forms adduct with Hb as a result of the reaction with the alpha-NH_2 group of N-terminal valine of Hb. Interaction between Hb and acrylamide occurred through decreasing of the peak current of Hb–Fe(III) reduction. The electrodes presented a very low detection limit (1.2×10^{-10} M). The validation made in the matrix obtained by water extraction of potato chips showed that the electrodes were suitable for the direct determination of acrylamide in food samples.

Li et al. [11] proposed an acrylamide biosensor based on the interaction between acrylamide and DNA, which was confirmed by UV–visible spectroscopy. According to such an interaction, a label-free DNA biosensor for electrochemical determination of acrylamide was proposed. To fabricate the sensor, graphene oxide (GO) was first coated on glassy carbon electrode (GCE) surface. Then, DNA was immobilized on GO/GCE by electroadsorption. Due to large surface area of GO, DNA was effectively immobilized on the electrode surface. Moreover, the unique nanostructure and excellent electron transfer ability of GO promoted the direct electron transfer of DNA significantly. Thus, DNA showed two strong oxidation peaks on GO/GCE, which was the electrochemical signal for acrylamide sensing. The influences of adsorption potential and adsorption time on the immobilized DNA and pH for acrylamide sensing were systematically investigated. Under optimum conditions, the response of DNA/GO/GCE was linear to the concentration of acrylamide from 5.0×10^{-8} to 1.0×10^{-3} mol/L. Moreover, the sensor showed good reproducibility and high stability.

Acrylamide Biosensors Employing Nanomaterials Krajewaska et al. [12] presented the use of glassy carbon electrodes coated with single-walled carbon nanotubes (SWCNTs) and Hb for amperometric detection of acrylamide in water solutions. The presence of toxic acrylamide in a wide range of food products such as potato crisps, French fries, or bread was confirmed. The biosensor had a very low detection limit (1.0×10^{-9} M). The validation made in the matrix obtained by water extraction of potato crisps showed that the electrodes were suitable for direct determination of acrylamide in food samples.

Batra et al. [13] described the construction of a highly sensitive electrochemical biosensor for the detection of acrylamide, based on covalent immobilization of Hb onto carboxylated multiwalled carbon nanotube/copper nanoparticle/polyaniline (cMWCNT/CuNP/PANI) nanocomposite electrodeposited onto pencil graphite electrode. The enzyme electrode was characterized by CV, scanning electron microscopy (SEM), X-ray diffraction, transmission electron microscopy, Fourier transform infrared (FTIR) spectroscopy, and electrochemical impedance spectroscopy (EIS). The biosensor showed an optimal response at pH 5.5 (0.1 M sodium acetate buffer) and 35 °C when operated at 20 mV/s. It exhibited low detection limit (0.2 nM) with high sensitivity (72.5 μA/nM/cm^2), fast response time (<2 s), and wide linear range (5 nM–75 mM). Analytical recovery of added acrylamide was 95.40–97.56%. Within- and between-batch coefficients of variation were 2.35% and 4.50%, respectively. The enzyme electrode was used 120 times over a period of 100 days, when stored at 4 °C.

Gold Electrode-Based Amperometric Acrylamide Biosensor Batra et al. [14] described the construction of an improved amperometric acrylamide biosensor based on covalent immobilization of Hb onto nanocomposite of cMWCNT and iron oxide nanoparticles (Fe_3O_4NPs) electrodeposited onto Au electrode through chitosan (CHIT) film. The Hb/cMWCNT/Fe_3O_4NP/CHIT/Au electrode was characterized by SEM, FTIR spectroscopy, EIS, and differential pulse voltammetry at different stages of its construction. The biosensor was based on the interaction between acrylamide and Hb, which led to decrease in the electroactivity of Hb; that is, current generated during its reversible conversion [Fe(II)/Fe(III)]. The biosensor showed optimum response within 8 s at pH 5.0 and 30 °C. The linear/working range for acrylamide was 3–90 nM, with a detection limit of 0.02 nM and sensitivity of 36.9 μA/nM/cm^2. The biosensor was evaluated and employed for determination of acrylamide in potato crisps.

Potentiometric Acrylamide Biosensors

Principle: The potentiometric acrylamide biosensor is based on a direct biochemical interaction between acrylamide and intact bacterial cells. The biological recognition element consisted of whole cells of *Pseudomonas aeruginosa* containing intracellular amidase activity, which catalyzes the hydrolysis of acrylamide-producing ammonium ion (NH_4^+) detected by an ammonium ion selective electrode.

Silva et al. [15] developed a biosensor for toxic amides using whole cells of *P. aeruginosa* containing amidase activity, which catalyzes the hydrolysis of amides such as acrylamide-producing ammonia and the corresponding organic acid. Whole cells immobilized in several types of membrane in the presence of glutaraldehyde and an ammonium ion-selective electrode were used for biosensor development. This biosensor exhibited a linear response in the range of $0.1–4.0 \times 10^{-3}$ M of acrylamide, a detection limit of 4.48×10^{-5} M acrylamide, a response time of 55 s, a sensitivity of 58.99 mV/mM of acrylamide and a maximum $t_{1/2}$ of 54 days. The selectivity of this biosensor toward other amides was investigated, which revealed that it cross-reacted with acetamide and formamide, but no activity was detected with phenylacetamide, p-nitrophenylacetamide, and acetanilide. It was successfully used for quantification of acrylamide in real industrial effluents, and recovery experiments were carried out, which revealed an average substrate recovery of 93.3%. The biosensor was cheap, since whole cells of *P. aeruginosa* could be used as source of amidase activity.

The first electrochemical biosensor for acrylamide determination was based on a direct biochemical interaction between the analyte and intact bacterial cells, with intracellular enzymatic activity. The biological recognition element consisted of whole cells of *P. aeruginosa* containing intracellular amidase activity, which catalyzes the hydrolysis of acrylamide-producing ammonium ion (NH_4^+) and acrylic acid. The transduction process was accomplished by means of an ammonium ion selective electrode. Whole cells were first immobilized on single disks of polymeric membranes, such as polyethersulfone, nylon, and polycarbonate, which were, then, attached to the surface of the selective electrode. However, a significant loss of cells occurred, each time the biosensor was used, mainly at the beginning of the assay, when the membranes were attached to the ammonium electrode, and after the assay, when removed for storage purposes. This evidence determined a premature decrease in the biosensor's stability. Instead of using single membrane disks, a "sandwich" design with two membrane disks was considered. This way the cells remain contained between the membranes, never contacting the electrode's surface, preventing their premature loss. Consequently, the activity of the biosensor could be maintained for longer periods of time. The analytical performance of the biosensor was evaluated. The best results were obtained when polyethersulfone double membranes were used. A typical response of 120 mV (after 6 min reaction time), a Nernstian slope of 48 mV/decade, a limit of detection of 6.31×10^{-4} M and a half-life time of 27 days, were examples of some figures of merit observed for this biosensor [16].

A sensitive, simple, label-free cell-based electrochemical sensor was constructed to monitor the toxic effect of acrylamide on the pheochromocytoma cells. The surface of the electrode was modified with gold nanoparticles (AuNPs) and electrochemically reduced GO. The modified electrode was characterized by CV, EIS, and DPV. Reduced GO was proved to increase electron-transfer rate between the cell and the surface of electrode, while AuNPs retained cell bioactivity. The sensor exhibited good correlation to the logarithmic value of cell numbers ranging from 1.6×10^4 to 1.6×10^7 cells/mL, with RSD value of 1.68%. The value of DPV at cell adsorption concentration of 1.6×10^7 cells/mL decreased with the concentration of acrylamide in the range, 0.1–5 mM with the detection limit as 0.04 mM. SEM-based morphological and 3-(4,5-dimethylthiazol-2-yl)-2,5-diphenyltetrazolium bromide analysis confirmed the results of the electrochemical study. This sensor was proved to be a useful tool for probing the toxicity of cells, and assisted in the development of a labelling-free, simple, rapid, and immediate detection method [17].

A novel fluorescent sensing method based on acrylamide polymerization-induced distance increase between quantum dots (QDs) was proposed for detecting acrylamide in potato chips. The functional QDs were prepared by their binding with *N*-acryloxysuccinimide (NAS), which was characterized by FTIR spectra. The carbon–carbon double bonds of NAS modified QDs polymerized with assistance of photo initiator under UV irradiation, leading to QDs getting closer along with fluorescence intensity decreasing. Acrylamide in the sample participated in the polymerization and induced an increase of fluorescence intensity. This method possessed a linear range from 3.5×10^{-5} to 3.5 g/L ($r^2 = 0.94$) and a limit of detection of 3.5×10^{-5} g/L. Although the sensitivity and specificity could not be compared with standard LC–MS–MS analysis, this new method required much less time and cost, which was promising for online rapid detection of acrylamide in food processing [18].

Comparison of Various Acrylamide Biosensors

Table 1 summarizes the comparison of analytical characteristics of various acrylamide biosensors.

Table 1 A comparison of analytic characteristics of different acrylamide biosensors

Properties	Stobiecka et al. [10]	Krajewaska et al. [12]	Silva et al. [15]	Silva et al. [16]	Batra et al. [14]	Batra et al. [14]	Sun et al. [17]	Li et al. [11]
Type of electrode used	Carbon paste electrode	Glassy carbon electrode	Ammonium ion selective electrode	Ammonium ion selective electrode	Pencil graphite electrode	Gold electrode	Gold electrode	Glassy carbon electrode
Optimum pH	4.8	5.0			5.5	5.0		
Linearity	1.3×10^{-11} to 5.6×10^{-3} M	1.0×10^{-11} to 1.0×10^{-3} M	0.1 to 4.0×10^{-3} M		5×10^{-6} to 75×10^{-6} M	3×10^{-6} to 90×10^{-6} M	1×10^{-4} to 5×10^{-3} M	5.0×10^{-8} to 1×10^{-3} M
Detection limit	1.2×10^{-10} M	1.0×10^{-9} M	4.48×10^{-5} M	6.31×10^{-4} M	0.2×10^{-6} M	0.02×10^{-6} M	4×10^{-5} M	
Response time	–	–	55 s	3.5 min	2 s	8 s		
Sensitivity	–		58.99 mV/mM		72.5 μA/nM/cm^2	36.9 μA/nM/cm^2		
Storage stability	–		$t_{1/2}$-54 days	$t_{1/2}$-27 days	100 days	–		

CONCLUSION AND FUTURE OUTLOOK

The acrylamide biosensors worked optimally in following conditions: pH 7.5–7.75, concentration range 1–1950 μM, detection limit 2–111 μM, and storage stability 40–95 days. Recently, acrylamide bio-nanosensors were developed, but the futuristic goal of low-cost, high-throughput, multiplexed diagnostic laboratory on-a-chip devices is yet to be truly realized. A great deal of excitement has been generated by bio-nanosensors, due to their ability to detect a wide range of materials at incredibly small concentrations with high sensitivity. The growth in sensor technology to enter the commercial market has accelerated tremendously, due to nanotechnology.

KEY FACTS

- Acrylamide, a neurotoxin and potential carcinogen, is formed during high-temperature cooking of food.
- GC–MS, GC–MS–MS, and HPLC–MS are prominent methods but expensive, time-consuming, and cumbersome.
- Biosensing methods are simple, rapid, sensitive, and specific.
- Two types of acrylamide biosensors are amperometric and potentiometric biosensors.
- Acrylamide biosensors work optimally at pH 7.50–0.7.75 and acrylamide concentration range of 1–1950 μM.
- Detection limit and storage stability of acrylamide biosensor are in the range 2–111 μM and 40–95 days, respectively.

MINI DICTIONARY

Amperometric biosensor It measures signal in current (in amp.) and is linearly dependent to the amount of analyte.

Biosensor An analytical device that converts a biological response into a quantifiable and processable signal. It consists of a biological receptor, transducer, and processor.

Carbon nanotubes (CNTs) These are long thin cylinders of carbon. These are large macromolecules that are unique for their size, shape, and remarkable physical properties. CNTs with single cylindrical tubes are called SWCNTs and multiple-walled tubes are named MWCNTs. Functionalized/carboxylated MWCNTs are known as cMWCNT.

Chitosan A polysaccharide of glucosamine, obtained by partial deacetylation of chitin, is the major component of crustacean shell. It has free $-NH_2$ groups and is used widely for immobilization of enzymes, due to its excellent, biocompatibility, film-forming ability, physiological inertness, and high mechanical strength.

Copper nanoparticles (CuNPs) These are a kind of NPs that are formed by the reduction of cupric chloride in solution.

Iron oxide/magnetic nanoparticles (Fe_3O_4 NPs) These are a kind of NPs that have large surface area and good biocompatibility.

Nanomaterials The materials when reduced down to 1–100 nm in their dimensions show drastic changes with respect of their physical, chemical, magnetic, mechanical, and electrochemical properties. Different nanostructures include nanotubes, nanofibers, nanorods, nanoparticles (NPs), and thin films. NPs have attracted attention, due to their applications in construction of improved analytical devices.

Polyanline (PANI) It is a polymer of aniline, which is formed by cyclic voltammetry of analine.

Potentiometric biosensor It measures signal in potential (in volts) and has logarithmic relationship with the amount of product formed from analyte.

SUMMARY POINTS

- Acrylamide, neurotoxin and potential carcinogen, is formed during high-temperature (>120 °C) cooking of many commonly consumed foods. It is derived from reactions between reducing sugars and proteins/amino acids (mainly asparagine).
- Prominent chromatographic methods for determination of acrylamide levels such as GC–MS, GC–MS–MS, HPLC–MS, and LC–MS–MS are expensive, time-consuming, and require sample preparation.
- Biosensors offer great promise for the detection of acrylamide, because of their simplicity, rapidity, and high sensitivity.
- Two types of acrylamide biosensors are reported with or without nanomaterials: amperometric and potentiometric biosensors.
- The acrylamide biosensors worked optimally under following conditions: pH 7.5–7.75, concentration range, 1–1950 μM, detection limit 2–111 μM, and storage stability 40–95 days.
- The miniaturization of these bio-nanosensors is expected to provide smart sensing devices.

REFERENCES

[1] Tareke E, Rydberg P, Karlsson S, Eriksson M, Törnqvist M. Analysis of acrylamide, a carcinogen formed in heated foodstuffs. J Agric Food Chem 2002;50:4998–5006.
[2] Gökmen V, Palazoğlu TK, Şenyuva HZ. Relation between the acrylamide formation and time–temperature history of surface and core regions of French fries. J Food Eng 2006;77:972–6.
[3] Becalski A, Lau BP-Y, Lewis D, Seaman S. Acrylamide in foods: occurrence, sources and modeling. J Agric Food Chem 2003;51:802–8.
[4] Motaghi MM, Ardebili MS, Honarvar M, Mehrabani M, Baghizadeh A. Determination of acrylamide in selected types of Iranian breads by SPME technique. J Food Biosci Technol 2012;2:57–64.
[5] Ono H, Chuda Y, Ohnishi-Kameyama M, Yada H, Ishizaka M, Kobayashi H, et al. Analysis of acrylamide by LC-MS/MS and GC-MS in processed Japanese foods. Food Addit Contam 2003;20(3):215–20.
[6] Rosen J, Hellanas KE. Analysis of acrylamide in cooked foods by liquid chromatography tandem mass spectrometry. Analyst 2002;127:880–2.
[7] Cheng WC, Kao YM, Shih DYC, Chou SS, Yeh AI. Validation of an improved LC/MS/MS method for acrylamide analysis in foods. J Food Drug Anal 2009;17:190–7.
[8] Turner APF, et al. In: Biolayer modelling and optimization for SPARROW biosensor. Chapter 1. UMI. Ann Arbor (MI, USA): ProQuest Information & Learning Company; 1987. p. 1–18.
[9] Zang S, Wright G, Yang Y. Materials and techniques for electrochemical biosensor design and construction. Biosens Bioelectron 2000;15:273–82.

[10] Stobiecka A, Radecka H, Radecki J. Novel voltametric biosensor for determining acrylamide in food samples. Biosens Bioelectron 2007;22:2165–70.
[11] Li D, Xu Y, Zhang L, Tong H. A label free electrochemical biosensor for acrylamide based on immobilization on grapheme oxide–modified glassy carbon electrode. Int J Electrochem Sci 2014;9:7217–8.
[12] Krajewaska A, Rodecki J, Rodecka H. A voltametric biosensor based on glassy carbon electrode with SWCNT/Hb for detection of acrylamide in water extracts from potato crisps. Sensors 2008;8:5832–44.
[13] Batra B, Lata S, Sharma M, Pundir CS. An acrylamide biosensor based on immobilization of haemoglobin onto multiwalled carbon nanotubes/copper nanoparticles/polyaniline hybrid film. Anal Bichem 2012;433: 210–7.
[14] Batra B, Lata S, Pundir CS. Construction of an improved acrylamide biosensor based on hemoglobin immobilized onto cMWCNT/Fe_3O_4NPs/CHIT composite film. Bioprocess Biosys Engg 2013;36:1591–9.
[15] Silva N, Gil D, Karmali A, Matos M. Biosensor for acrylamide based on ion selective electrode using whole cells of *Pseudomonas aeruginosa* containing amidase activity. Biocat Biotransform 2008;27:143–51.
[16] Silva NAF, Matos MJ, Karmali A, Rocha MM. An electrochemical biosensor for acrylamide determination: merits and limitations. Port Electrochim Acta 2011;29:361–73.
[17] Sun X, Jin J, Jiang D, et al. Development of novel electrochemical sensor using pheochromocytoma cells and its assessment of acrylamide toxicity. Biosens Bioelectron 2013;44:122–6.
[18] Hu Q, Xu X, Zhang Y, Wang J, Fu Y, Li Y. Detection of acrylamide in potato chips using fluorescent sensing method based on acrylamide polymerization induced distance increase in quantum dots. Biosens Bioelectron 2014;15:54–64.

INDEX

Note: page numbers followed by "f", "t" and "b" denote figures, tables and boxes respectively.

B

C

F

G

I

J

K

L

M

Q

R

T

U

FSC
www.fsc.org
MIX
Papier | Fördert
gute Waldnutzung
FSC® C083411

Kolibri 360 GmbH
Ferdinand-Jühlke-Straße 7
99095 Erfurt, Deutschland
produktsicherheit@kolibri360.de